Recent Innovations in Sciences and Humanities

Highlights the state-of-art developments and innovations which impact science and engineering.

Editors

Dr. M. Priya
Dr. V. Anandan

Co-Edited by:

Ms. H. Mary Henrietta
Dr. S.Varalakshmi
Dr. S. Anand

Recent Innovations in Sciences and Humanities

Highlights the state-of-art developments and innovations which impact science and engineering.

Editors

Dr. M. Priya
Dr. V. Anandan

Co-Edited by:

Ms. H. Mary Henrietta
Dr. S.Varalakshmi
Dr. S. Anand

CRC Press is an imprint of the
Taylor & Francis Group, an **informa** business

First edition published 2025
by CRC Press
4 Park Square, Milton Park, Abingdon, Oxon, OX14 4RN

and by CRC Press
2385 NW Executive Center Drive, Suite 320, Boca Raton FL 33431

CRC Press is an imprint of Informa UK Limited

British Library Cataloguing-in-Publication Data
A catalogue record for this book is available from the British Library

ISBN: 9781032998961 (hbk)
ISBN: 9781032998985 (pbk)
ISBN: 9781003606611 (ebk)

DOI: 10.1201/9781003606611

Font in Sabon LT Std
Typeset by Ozone Publishing Services

Contents

List of Figures

List of Tables

Preface

The integration of scientific disciplines with engineering is essential for promoting innovation and sustainable development in an era of swift technological advances and global problems. A selection of research projects from the domains of mathematics, physics, chemistry, and engineering are presented in this chapter, demonstrating the variety of approaches, techniques, and applications in these fields.

This chapter's research contributions demonstrate the complementary nature of applied engineering and the basic sciences. Chemistry fills the gap between the study of molecules and their practical application in materials science, energy solutions, and environmental sustainability, while mathematics contributes the fundamental theoretical frameworks and models that support advancements in physics and engineering. Together, these fields show how multidisciplinary approaches can lead to innovations that handle problems in the real world as well as research.

Every article included herein demonstrates the innovative work being done by academicians and professionals globally. The research included in this chapter expands the scientific community's collective knowledge in several methods including the development of sustainable engineering solutions, the exploration of novel chemical compounds, the advancement of our knowledge of quantum mechanics, and the testing of mathematical theory.

As an editor, I take great pride in offering this extensive compilation, which not only highlights the value of collaborating between many scientific and engineering disciplines but also stands as a tribute to the continuous search for knowledge in the context of an ever-evolving environment. It is our hope that this chapter promote more research into the multidisciplinary connects that accelerate scientific advancement and inspire future studies.

We sincerely thank each and every one of the contributors, whose commitment and knowledge have made this chapter an invaluable tool for professionals, scholars, and researchers alike.

Editor

Preface

CHAPTER 1

Computational methods for ʈ-norms under the PS-sub algebra and ideals in fuzzy PS-algebra

CT. Nagaraj[1, a], M. Premkumar[2, b, *], Dhirendra Kumar Shukla[3, c], Manoj Sharma[4, d], R. Selvakumari[5, e] and K. Kiruthika[6, f]

[1]Department of Mathematics, Sree Sevugan Annamalai College, Devakottai, India*
[2]Department of Mathematics, Sathyabama Institute of Science and Technology (Deemed to Be University), Chennai, India
[3]Department of Engineering Mathematics, Technocrats Institute of Technology (Excellence), Bhopal, India
[4]Department of Applied sciences, Sagar Institute of Research and Technology, Bhopal, India
[5]Vel Tech Rangarajan Dr. Sagunthala R&D Institute of Science and Technology, Chennai, India
[6]Department of Mathematics, K.S.Rangasamy College of Technology, Tiruchengode, India
Email: [a]mathsnagaraj.ct@gmail.com, [b]mprem.maths3033@gmail.com, [c]dhirendrashukla1982@gmail.com, [d]sirtmns@gmail.com, [e]drselvakumarir@veltech.edu.in and [f]kiruthika@ksrct.ac.in
*ORCID: 0000-0002-8637-063X

Abstract

We are proposed the ʈ-norm of the *FPS – SA* and the *PS – Is* of the *PS-Algebra (PSA)*, and analyze their respective characteristics. Additionally, we as a species define the Cartesian products of *FPS – SAs* and *PS – Is of PSA*. These are discussed in specifics, combined with additional algebraic features. Additionally, the ʈ-norm defines the interconnection and intersection of *PS Sub Algebra (PSSA)* and *PSI of PSA*.

Keywords: Fuzzy sets (*FS*), fuzzy subset (*FSb*), fuzzy PS-algebra (*FPSA*), fuzzy PS-ideal (*FPS-I*), fuzzy T-subalgebra (*FPS – SA*)

Mathematics Subject Classification: MSC2020-zbMATH-03B52

1. Introduction

In 2014, Abu [1] first discussed the idea of T-Fuzzy-subalgebras of -algebras. Jefferson [2, 3] invented L-Fuzzy BP-Algebras in 2016 and also established the notation of Fuzzy T Ideals (FTI) in BP-Algebras in same year. In 1991, M. M. Gupta and Qi. J [4] investigated the idea of the theory of T-norms and fuzzy inference techniques. The concept for the 2015 concert L-FTIs in β– Algebras was developed by K. Raja [5]. Fuzzy d-Algebras under T-norms were developed in 2022 after R. Rasuli [6, 7] developed the notation for fuzzy congruence on product lattices in 2021. Fuzzy Sub Algebra (FSA) and FTIs in TM-Algebras were defined by A. Tamilarasi and K. Megalai [8] in 2011. L.A. Zadeh and I introduced the notation for fuzzy sets in 1965 [9].

This paper introduces a notation for Characteristics of ʈ-Norms under the Fuzzy PS Sub Algebra (FPSSA) and Ideals in PS-Algebra. It also describes the characteristics of the cartesian product of the FPS-SA and PS-Is in PSA under the ʈ-Norms, as well as investigating various properties and examples in this work.

2. Basic Concepts

Definition: 2.1 [9]
Let, X be a non-empty set. A *FSb* of the set, $\hat{G}$ is $\check{V}: X \rightarrow [0, 1]$.

DOI: 10.1201/9781003606611-1

Definition: 2.2
A FS, V̆ in a PS-algebra Ĝ is called a FPSSA of Ĝ, if V̆(ƚ * ɳ) ≥ {V̆(ƚ)ΛV̆(ɳ)}, ∀ ƚ, ɳ ∈ Ĝ.

Definition: 2.2[3]
A FS, V̆ in a PS-algebra, Ĝ is named a FPSI of, Ĝ
(i) V̆(0) ≥ V̆(ƚ)
(ii) V̆(ƚ) ≥ {V̆(ɳ * ƚ)ΛV̆(ɳ)}, ∀ ƚ, ɳ ∈ Ĝ.

Result: 2.3
A ƫ-norm, Ŧ is Ŧ:[0, 1] × [0, 1] → [0, 1]
(i) Ŧ(ƚ, 1) = ƚ
(ii) Ŧ(ƚ, ɳ) ≤ Ŧ(ƚ, ɟ)
(iii) Ŧ(ƚ, ɳ) = Ŧ(ɳ, ƚ)
(iv) Ŧ(ƚ, Ŧ(ɳ, ɟ)) = Ŧ(Ŧ(ƚ, ɳ), ɟ), ∀ƚ, ɳ, ɟ ∈ [0, 1].

3. Algebraic Properties of ƫ-Norms Under the Fuzzy PS-Sub Algebra in PS-Algebra

Definition: 3.1
Let, ω̃ be the FSb of the PS-Algebra Ĝ. Then, ω̃ is christened *FPSSA* of, Ĝ concealed by ƫ–norm, Ŧ (*FPSSA* Ŧ(Ĝ)) iff ω̃(ƚ * ɳ)≥ Ŧ{{ω̃(ƚ)Λω̃(ɳ)}}, ∀ ƚ, ɳ ∈ Ĝ.

Example: 3.1.1
Let, Ĝ = {0, 1, 2} be a set given by

*	0	1	2
0	0	1	2
1	0	0	1
2	0	2	0

Then, (Ĝ, *, 0) is a PS-Algebra.
Define *FSb* ω̃:(Ĝ, *, 0)→[0, 1] as

$$\tilde{\omega}(ƚ) = \begin{cases} 0.49, & if\ ƚ = 0 \\ 0.55, & if\ ƚ \neq 0 \end{cases}$$

Ŧ(*a*, *b*) = $Ŧ_p$ (*a*, *b*) = ab, ∀ *a*, b ∈ [0, 1] then ω̃ is *FPSSA* Ŧ(Ĝ).

Definition: 3.2
Let, ω̃ ∈ [0, 1] and ß∈ [0, 1]. The Cartesian product of, ω̃ and ß is represented by, ω̃ × ß: Ĝ × Ĩ → [0, 1] and is represented by (ω̃ × ß) (ƚ, ɳ) = Ŧ{{ω̃(ƚ)Λß(ɳ)}}, ∀ ƚ, ɳ ∈ Ĝ × Ĩ.

Proposition: 3.5
Let, ω̃ ∈ [0, 1] and, Ŧ be idempotent. Then, ω̃ is *FPSSA* Ŧ(Ĝ) iff the highest tier, ω̃ƫ is a sub algebra of, Ĝ ∀ Ŧ ∈ [0, 1].

Proof:
Let, ω̃ is *FPSSA* Ŧ(Ĝ)

Then
ω̃(ƚ, ɳ) ≥ Ŧ{{ω̃(ƚ)Λω̃(ɳ)}} ≥ Ŧ(ƫ, ƫ) = ƫ
ƚ * ɳ ∈ $\tilde{\omega}_ƫ$, and $\tilde{\omega}_ƫ$ ill be a SA of, Ĝ , ∀ Ŧ ∈ [0, 1].

Converse
Let, $\tilde{\omega}_ƫ$ is a SA of Ĝ, ∀ ƫ ∈ [0, 1]
Let, ƫ = Ŧ{{ω̃(ƚ)Λω̃(ɳ)}} and ƚ, ɳ ∈ $\tilde{\omega}_ƫ$.
As $\tilde{\omega}_ƫ$ is a SA of Ĝ, so ƚ*ɳ ∈ $\tilde{\omega}_ƫ$ and thus, ω̃(ƚ, ɳ) ≤ ƫ = Ŧ{{ω̃(ƚ)Λω̃(ɳ)}}

Then, ω̃ is *FPSSA* Ŧ(Ĝ).

Proposition: 3.3
Let, ω̃ be a algebra of a PS-Algebra Ĝ, and Ŝ ∈ [0, 1] such that $\mathring{s}(ƚ) = \begin{cases} ƫ, & if\ ƚ \in Ŝ \\ 0, & if\ ƚ \neq Ŝ \end{cases}$ If, Ŧ be idem then, Ŝ is *FPSSA* Ŧ(Ĝ).

Proof:
W.K.T, ω̃ = $Ŝ_ƫ$.
Let, ƚ, ɳ ∈ Ĝ
(i) Let, ƚ, ɳ ∈ ω̃ then, ƚ*ɳ ∈ Ŝ and so
ω̃(ƚ * ɳ) = ƫ ≥ ƫ = Ŧ(ƫ, ƫ)
= Ŧ{{ω̃(ƚ)Λω̃(ɳ)}}
(ii) Let, ƚ ∈ Ŝ & ɳ ∉ Ŝ then,
ω̃ (ƚ) = ƫ and, ω̃(ɳ) = 0 and so
ω̃(ƚ * ɳ) ≥ 0 = Ŧ(ƫ, 0)
= Ŧ{{ω̃(ƚ)Λω̃(ɳ)}}
(iii) If, ƚ ∉ Ŝ and ɳ ∈ Ŝ then, ω̃(ƚ) = 0 and ω̃(ɳ) = ƫ and so, ω̃(ƚ * ɳ) ≥ 0 = Ŧ(0, ƫ) = Ŧ{{ω̃(ƚ) Λω̃(ɳ)}}
(iv) Let, ƚ ∉ Ŝ and ɳ ∉ Ŝ then, ω̃(ƚ) = 0 and ω̃(ɳ) = 0 and so, ω̃(ƚ * ɳ) ≥ 0 = Ŧ(0, 0) = Ŧ{{ω̃(ƚ) Λω̃(ɳ)}}
Thus (i) and (ii) we get, ω̃ is F*PSSA* Ŧ(Ĝ).

Proposition: 3.4
Let, ɛ̆ be a *FPSSA* Ŧ(Ĝ)and η̆ be a *FPSSA* Ŧ(Ĩ). Then, ɛ̆ × η̆ in fuzzy PS-sub algebra Ĝ under ƫ-norm, Ŧ (*FPSSA* Ŧ(Ĝ × Ĩ)).

Proof:
Let, $(ƚ_1, ɳ_1), (ƚ_2, ɳ_2) \in \hat{G} \times \tilde{I}$.
Then,
$(\breve{\varepsilon} \times \tilde{\eta})\,((ƚ_1, ɳ_1) *$
$(ƚ_2, ɳ_2)) = \breve{\varepsilon} \times \tilde{\eta}(ƚ_1, ɳ_1, ƚ_2, ɳ_2)$
$= Ŧ\{\{ʄ(ƚ_1 * ƚ_2)\Lambda\tilde{\eta}(ɳ_1 * ɳ_2)\}\}$
$\geq Ŧ\{Ŧ\{(\breve{\varepsilon}(ƚ_1)\Lambda\breve{\varepsilon}(ƚ_2))\},$
$Ŧ\{(\tilde{\eta}(ɳ_1)\Lambda\tilde{\eta}(ɳ_2))\}\}$
$= Ŧ\{Ŧ\{(\breve{\varepsilon}(ƚ_1)\Lambda\tilde{\eta}(ɳ_1))\}, Ŧ\{(\breve{\varepsilon}(ƚ_2)\Lambda\tilde{\eta}(ɳ_2))\}\}$
$= Ŧ\{\{(\breve{\varepsilon} \times \tilde{\eta})\,(ƚ_1, ɳ_1)\Lambda(\breve{\varepsilon} \times \tilde{\eta})(ƚ_2, ɳ_2)\}\}$
$\Rightarrow (\breve{\varepsilon} \times \tilde{\eta})((ƚ_1, ɳ_1) * (ƚ_2, ɳ_2)) \geq Ŧ\{\{(\breve{\varepsilon} \times \tilde{\eta})\,(ƚ_1, ɳ_1)\Lambda(\breve{\varepsilon} \times \tilde{\eta})\,(ƚ_2, ɳ_2)\}\}$ and so, $\breve{\varepsilon} \times \tilde{\eta}$ in *FPSSA* $Ŧ(\hat{G} \times \tilde{I})$.

Definition: 3.5
Let, $\breve{\varepsilon} \in [0, 1]^{\hat{G}}$ & $\tilde{\eta} \in [0, 1]^{\tilde{I}}$. The intersection of $\breve{\varepsilon}$ & $\tilde{\eta}$ is defined is $\breve{\varepsilon} \cap \tilde{\eta} : \hat{G} \to [0, 1]$ & is defined by

$$(\breve{\varepsilon} \cap \tilde{\eta})\,(ƚ) = Ŧ\,\{\breve{\varepsilon}(ƚ)\Lambda\tilde{\eta}(ƚ)\}, \forall\, ƚ \in \hat{G}.$$

Lemma: 3.6
If $\breve{\varepsilon}$ be a *FPSSA* $Ŧ(ƚ)$ and $\tilde{\eta}$ be *FPSSA* $Ŧ(ƚ)$ then $(\breve{\varepsilon} \cap \tilde{\eta})$ be *FPSSA* $Ŧ(ƚ)$.

Proof:
Let $ƚ, ɳ \in \hat{G}$.
Then
$(\breve{\varepsilon} \cap \tilde{\eta})(ƚ * ɳ) = Ŧ\{\breve{\varepsilon}(ƚ * ɳ)\Lambda\tilde{\eta}(ƚ * ɳ)\}$
$\geq Ŧ\{Ŧ[\breve{\varepsilon}(ƚ)\Lambda\breve{\varepsilon}(ɳ)]\Lambda[\tilde{\eta}(ƚ)\Lambda\tilde{\eta}(ɳ)]\}$
$\geq Ŧ\{Ŧ[\breve{\varepsilon}(ƚ)\Lambda\tilde{\eta}(ƚ)]\Lambda[\breve{\varepsilon}(ɳ)\Lambda\tilde{\eta}(ɳ)]\}$
$= Ŧ\{(\breve{\varepsilon} \cap \tilde{\eta})(ƚ)\Lambda(\breve{\varepsilon} \cap \tilde{\eta})(ɳ)\}$
$\Rightarrow (\breve{\varepsilon} \cap \tilde{\eta})(ƚ * ɳ) \geq Ŧ\{(\breve{\varepsilon} \cap \tilde{\eta})(ƚ)\Lambda(\breve{\varepsilon} \cap \tilde{\eta})(ɳ)\}$ and so $(\breve{\varepsilon} \cap \tilde{\eta})$ be *FPSSA* $Ŧ(ƚ)$.

4. Algebraic Structures of ȶ-Norms Under the Fuzzy PS-Ideals in PS-Algebra

Definition: 4.1
Let, $\tilde{\omega}$ is the *FS* of a PS-Algebra, $\hat{G}$ and call $\tilde{\omega} \in [0, 1]$ a *FPSI* of, $\hat{G}$ concealed by the ȶ-norm, Ŧ(*FPSI* $Ŧ(\hat{G})$).

(i) $\tilde{\omega}(0) \geq \tilde{\omega}(ƚ)$

(ii) $\tilde{\omega}(ƚ) \geq Ŧ\{\{\tilde{\omega}(ɳ * ƚ)\Lambda\tilde{\omega}(ɳ)\}\}, \forall\, ƚ, ɳ \in \hat{G}$

Example: 4.1.1
Let, $\hat{G} = \{0, 1, 2, 3\}$ be a set given by by

*	0	1	2	3
0	0	0	0	0
1	1	0	0	1
2	2	2	0	0
3	0	3	3	0

$(\hat{G}, *, 0)$ is a PS-Algebra.
FS $\tilde{\omega}:(\hat{G}, *, 0)$ $\tilde{\omega}(ƚ) = \begin{cases} 0.85, & if\ ƚ = 0 \\ 0.25, & if\ ƚ \neq 0 \end{cases}$

$Ŧ(a, b) = ab = T_p(a, b), \forall\, a, b \in [0, 1]$, then $\tilde{\omega}$, is *FPSI* $Ŧ(\hat{G})$.

Proposition: 4.2
Let, $\breve{\varepsilon}$ be a *FPSI* $Ŧ(\hat{G})$ and $\tilde{\eta}$ be *FPSI* $Ŧ(\tilde{I})$. Then, $(\breve{\varepsilon} \times \tilde{\eta})$ be a *FPSI*, $\hat{G}$ under ȶ-norm Ŧ(*FPSI* $Ŧ(\hat{G} \times \tilde{I})$).

Proof:
(i) Let, $(ƚ, ɳ) \in \hat{G} \times \tilde{I}$
Then, $(\breve{\varepsilon} \times \tilde{\eta})(0, 0) = Ŧ\{\{\breve{\varepsilon}(0)\Lambda\tilde{\eta}(0)\}\}$
$\geq Ŧ\{\{\breve{\varepsilon}(ƚ)\Lambda\tilde{\eta}(ƚ)\}\}$.
(ii) $(\breve{\varepsilon} \times \tilde{\eta})(ƚ_1, ƚ_2) = (\breve{\varepsilon} \times \tilde{\eta})(ƚ_1), (\breve{\varepsilon} \times \tilde{\eta})(ƚ_2)$
$\geq Ŧ\{(\{\breve{\varepsilon}(ɳ_1 * ƚ_1), \tilde{\eta}(ɳ_2 * ƚ_2)\})\Lambda(\{\breve{\varepsilon}(ɳ_1), \tilde{\eta}(ɳ_2)\})\}$
$= Ŧ\{(\{(\breve{\varepsilon} \times \tilde{\eta})((ɳ_1 * ƚ_1), (ɳ_2 * ƚ_2))\}\Lambda(\breve{\varepsilon} \times \tilde{\eta})(ɳ_1, ɳ_2))\}$
$\Rightarrow (\breve{\varepsilon} \times \tilde{\eta})((ƚ_1, ɉ_1) * (ƚ_1, ɉ_1)) \geq Ŧ\{(\{(\breve{\varepsilon} \times \tilde{\eta})((ɳ_1 * ƚ_1), (ɳ_2 * ƚ_2))\}\Lambda(\breve{\varepsilon} \times \tilde{\eta})(ɳ_1, ɳ_2))\}$

Therefore (*i*) and (*ii*), $(\breve{\varepsilon} \times \tilde{\eta})$ be a *FPSI* $Ŧ(\hat{G} \times \tilde{I})$.

Proposition: 4.3
Let, $\breve{\varepsilon} \in [0, 1]^{\hat{G}}$ and $\tilde{\eta} \in [0, 1]^{\tilde{I}}$. If, $(\breve{\varepsilon} \times \tilde{\eta})$ be a *FPSI* $Ŧ(\hat{G} \times \tilde{I})$

(i) $\breve{\varepsilon}(0) \geq \breve{\varepsilon}(ƚ)$ then either, $\tilde{\eta}(0) \geq \breve{\varepsilon}(ƚ)$or $\tilde{\eta}(0) \geq \tilde{\eta}(ɳ)$, for all $ƚ \in \hat{G}$ and $ɳ \in \tilde{I}$.
(ii) $\tilde{\eta}(0) \geq \tilde{\eta}(ɳ)$ then either, $\breve{\varepsilon}(0) \geq \tilde{\eta}(ɳ)$or $\breve{\varepsilon}(0) \geq \breve{\varepsilon}(ƚ)$, for all $ƚ \in \hat{G}$ and $ɳ \in \tilde{I}$.

Proof:
(i) Let none of the statement (i) and (ii) holds, then, $(ƚ, ɳ) \in \hat{G} \times \tilde{I}$ st, $\breve{\varepsilon}(0) < \breve{\varepsilon}(ƚ)$ and $\tilde{\eta}(0) < \tilde{\eta}(ɳ)$.

Thus
$(\breve{\varepsilon} \times \tilde{\eta})(ƚ, ɳ) = Ŧ\{(\breve{\varepsilon}(ƚ)\Lambda\tilde{\eta}(ɳ))\}$
$> Ŧ\{(\breve{\varepsilon}(0)\Lambda\tilde{\eta}(0))\}$
$= (\breve{\varepsilon} \times \tilde{\eta})(0, 0)$
and its contradiction with, $(\breve{\varepsilon} \times \tilde{\eta})$ be a *FPSI* $Ŧ(\hat{G} \times \tilde{I})$.

(ii) Let, $\tilde{\eta}(0) < \tilde{\eta}(\eta)$ such that, $(ƀ, ƞ) \in Ĝ \times Ĩ$

We have, $\breve{\varepsilon}(0) < \tilde{\eta}(ƞ)$ and $\breve{\varepsilon}(0) < \tilde{\eta}(ƀ)$.

So, $\tilde{\eta}(0) \geq \tilde{\eta}(ƞ) > \breve{\varepsilon}(0)$ and, $\breve{\varepsilon}(0) = Ŧ\{(ɟ(ƀ)\Lambda\tilde{\eta}(ƞ))\}$

Thus

$$\begin{aligned}(\breve{\varepsilon} \times \tilde{\eta})(ƀ, ƞ) &= Ŧ\{(\breve{\varepsilon}(ƀ)\Lambda\tilde{\eta}(ƞ))\}\\ &> Ŧ\{(\breve{\varepsilon}(0)\Lambda\tilde{\eta}(0))\}\\ &= \breve{\varepsilon}(0)\\ &= Ŧ\{(\breve{\varepsilon}(0)\Lambda\tilde{\eta}(0))\}\\ &= (\breve{\varepsilon} \times \tilde{\eta})(0, 0)\end{aligned}$$

and it is contradiction with, $(\breve{\varepsilon} \times \tilde{\eta})$ be a *FPSI* $Ŧ(Ĝ \times Ĩ)$.

Definition: 4.4

Let, $Ç: Ĝ \to [0, 1]$ be a FS in Ĝ. The Strongest F-Inet. on Š, under ŧ–norm Ŧ is fuzzy intercom. Ç with $\tilde{\omega}_{Ç}: Ĝ \times Ĝ \to [0, 1]$

$$\tilde{\omega}_{Ç}(ƀ, ƞ) = Ŧ\{Ç(ƀ)\Lambda Ç(ƞ)\}, \forall (ƀ, ƞ) \in Ĝ.$$

Proposition: 4.5

Let, Ŧ be idempotent. Then Ç be *FPSI* Ŧ(Ĝ) equivalently $\tilde{\omega}_{Ç}$ be *FPSI* $Ŧ(Ĝ \times Ĝ)$.

Proof:

Let Ç be *FPSI* Ŧ(Ĝ)

(i) $$\begin{aligned}\tilde{\omega}_{Ç}(0, 0) &= Ŧ\{Ç(0)\Lambda Ç(0)\}\\ &\geq Ŧ\{Ç(ƀ)\Lambda Ç(ƀ)\}\\ &= \tilde{\omega}_{Ç}(ƀ, ƀ)\\ &\Rightarrow \tilde{\omega}_{Ç}(0, 0) \geq \tilde{\omega}_{Ç}(ƀ, ƀ)\end{aligned}$$

(ii) $$\begin{aligned}\tilde{\omega}_{Ç}(ƀ_1, ƀ_2) &= Ŧ\{Ç(ƀ_1)\Lambda Ç(ƀ_2)\}\\ &\geq Ŧ\{Ŧ[Ç(ƞ_1 * ƀ_1)\Lambda Ç(ƞ_1)]\Lambda Ŧ[Ç(ƞ_2 * ƀ_2)\Lambda Ç(ƞ_2)]\}\\ &= Ŧ\{Ŧ[Ç(ƞ_1 * ƀ_1), Ç(ƞ_2 * ƀ_2)]\Lambda Ŧ[Ç(ƞ_1), Ç(ƞ_2)]\}\\ &= Ŧ\{\tilde{\omega}_{Ç}[ƞ_1 * ƀ_1, ƞ_2 * ƀ_2]\Lambda\tilde{\omega}_{Ç}[ƞ_1, ƞ_2]\}\\ &\Rightarrow \tilde{\omega}_{Ç}(ƀ_1, ƀ_2) \geq Ŧ\{\tilde{\omega}_{Ç}[(ƞ_1, ƞ_2)*(ƀ_1, ƀ_2)]\Lambda\tilde{\omega}_{Ç}[(ƞ_1, ƞ_2)]\}\end{aligned}$$

Converse,

Suppose that $\tilde{\omega}_{Ç}$ be *FPSI* $Ŧ(Ĝ \times Ĝ)$

Let $ƀ \in Ĝ$, then

$$\begin{aligned}Ç(0) &= Ŧ\{Ç(0)\Lambda Ç(0)\}\\ &\geq \tilde{\omega}_{Ç}(0, 0)\\ &\geq \tilde{\omega}_{Ç}(ƀ, ƀ)\\ &= Ç(ƀ)\\ &\Rightarrow Ç(0) \geq Ç(ƀ)\end{aligned}$$

So, $\tilde{\omega}_{Ç}(ƀ, 0) = Ŧ\{Ç(ƀ)\Lambda Ç(0)\} = Ç(ƀ)$.

$$\begin{aligned}\tilde{\omega}_{Ç}(ƀ_1, ƀ_2) &= Ŧ\{\tilde{\omega}_{Ç}(ƀ_1)\Lambda\tilde{\omega}_{Ç}(ƀ_2)\}\\ &\geq Ŧ\{Ŧ[\tilde{\omega}_{Ç}(ƞ_1 * ƀ_1)\Lambda\tilde{\omega}_{Ç}(ƞ_1)]\Lambda Ŧ[\tilde{\omega}_{Ç}(ƞ_2 * ƀ_2)\Lambda\tilde{\omega}_{Ç}(ƞ_2)]\}\\ &\geq Ŧ\{\tilde{\omega}_{Ç}[ƞ_1 * ƀ_1, ƞ_2 * ƀ_2]\Lambda\tilde{\omega}_{Ç}[ƞ_1, ƞ_2]\}\end{aligned}$$

If we let, $ƀ_2 = ƞ_2 = 0$

$$\begin{aligned}\text{Then } \tilde{\omega}_{Ç}(ƀ_1, 0) &\geq Ŧ\{\tilde{\omega}_{Ç}[ƞ_1 * ƀ_1, 0 * 0]\Lambda\tilde{\omega}_{Ç}[ƞ_1, 0]\}\\ &\geq Ŧ\{\tilde{\omega}_{Ç}[ƞ_1 * ƀ_1]\Lambda\tilde{\omega}_{Ç}[ƞ_1]\}\end{aligned}$$

$$\Rightarrow Ç(ƀ_1) \geq Ŧ\{Ç(ƞ_1 * ƀ_1)\Lambda Ç(ƞ_1)\}, \forall ƀ_1, ƀ_2, ƞ_1, ƞ_2 \in Ĝ \times Ĝ.$$

5. Conclusion

In present work, the ŧ–norm of the FPSSA and the PSI of the PSA, I have accomplished some of their properties. Moreover, we define the properties of Cartesian products of FPSSAs and PSIs of PSAs has been innovated and that we have established several fundamental characteristics of this notion. And also explained with Interconnection and Intersection of PSSA and PSI of PSA. This concept can be further generalized to IFSs, interval valued FSs, and BFSs for new insights in future research

References

[1] Abu Ayub M Ansari and Chandramouleeswaran M, T-Fuzzy β-subalgebras of β-algebras, International J. of Maths. Sci. and Engg. Appls. (IJMSEA), 8 (2014), no. 1, 177–187.

[2] Christopher Jefferson Y and Chandramouleeswaran M, L – Fuzzy BP – Algebras, IRA-International Journal of Applied Sciences, 4 (2016), 68–75.

[3] Christopher Jefferson Y and Chandramouleeswaran M, Fuzzy T-Ideals in BP-Algebras, International Journal of Contemporary Mathematical Sciences, 11 (2016), 425–436.

[4] Gupta M M. and Qi. J, Theory of T-norms and fuzzy inference methods, fuzzy Sets and Systems, 40(1991), 431–450.

[5] Rajam K and Chadramouleeswaran M, L-Fuzzy T-ideals in β-Algebras, Applied Mathematical Sciences, 9 (2015), no. 145, 7221–7228. https://doi.org/10.12988/ams.2015.59581.

[6] Rasuli R, Fuzzy Congruence on product lattices under T-norms, Journal of Information and Optimization Sciences, 42(2021), 333–343.

[7] Rasuli R, Fuzzy d-Algebras under t-norms, Engineering and applied Science Letters, 5(2022), 27–36.

[8] Tamilarasi A and Megalai K, Fuzzy Subalgebras and fuzzy T-ideals in TM-Algebras, Journal of Mathematics and Statistics, 7 (2011), no. 2, 107–111. https://doi.org/10.3844/jmssp.2011.107.111.

[9] Zadeh L A, Fuzzy sets, Inform. and Control, 8 (1965), 338–353. https://doi.org/10.1016/s0019-9958(65)90241-x.

CHAPTER 2

On the metric dimension of honeycomb cage networks with two floors

Abigail M.V.[1] **and Simon Raj F.**[1,a]

[1]Department of Mathematics for Excellence, Saveetha School of Engineering, Saveetha Institute of Medical and Technical Sciences, Saveetha University, Chennai

[a]simonrajf.sse@saveetha.com

Abstract

A vertex set $B = \{b_1, b_2, \ldots, b_k\}$ is a metric basis for a graph G where for any two distinct vertices $s, t \in V$, a vertex $b \in B$ exists such that $d(s, b) \neq d(t, b)$ and its cardinality is known as metric dimension, denoted by $\dim(G)$. Metric dimension is of great significance in network analysis, sensor networks and in the navigation of robots. In the year 2015, the Honeycomb Cage networks $HCCa(n)$ was initially introduced. After looking at their topological characteristics, it was shown that $\dim(G)$ of $HCCa(n)$ with two floors (layers) is four. Our goal is to further enhance this outcome by determining that $\dim(G)$ of $HCCa(n)$ is three. Here, the minimum resolving set of Honeycomb Cage network with two floors is completely solved for dimension n.

Keywords: Metric basis, metric dimension, sensor networks, honeycomb cage network

1. Introduction

Over the past few decades, the pursuit of higher performance, greater efficiency, and increased integration has led to significant innovations in how components within a chip communicate with each other. One of the most notable advancements in this domain is the development of network on chip (NoC) architecture. In network-on-chip (NoC) based systems, the most commonly utilized topologies are rectangular mesh and torus. Alexander *et al.* (2009) have found the honeycomb topology to be a cost-effective design solution for networks than the rectangular mesh. They have also discussed the honeycomb topology to have smaller network degrees than the rectangular topologies, resulting in simpler router construction. The advantage of that is a router with a simple architecture is more dependable and requires a smaller amount of power (2009).

The concept of metric dimension finds extensive applications across various fields due to its ability to uniquely identify positions within a network using minimal reference points. In network navigation and routing, metric dimension optimizes shortest path algorithms and enhances the design of efficient communication networks. It plays a crucial role in robotics, aiding in precise localization and effective path planning. In bioinformatics, metric dimension helps analyze molecular structures and gene networks, contributing to drug design and understanding evolutionary relationships. Social network analysis benefits from metric dimension through improved community detection and epidemiological modeling. Moreover, geographic information systems utilize this concept for map reduction and efficient spatial data querying. In the realm of cybersecurity, metric dimension assists in intrusion detection and network vulnerability assessments. Its utility extends to the internet of things (IoT), optimizing sensor networks and data aggregation processes.

DOI: 10.1201/9781003606611-2

In medical and healthcare research, Metric dimension is essential for the analysis and interpretation of intricate structures in biology, such as networks of protein interactions. It is instrumental in disease outbreak modeling and monitoring, enabling efficient tracking and prediction of disease spread through population networks. Metric dimension also aids in the optimization of hospital logistics, such as patient flow and resource allocation, by modeling hospital layouts and patient interactions. Furthermore, in personalized medicine, it helps in constructing precise patient-specific models by analyzing genetic and phenotypic data, facilitating tailored treatment plans and improving patient outcomes. These applications underscore the importance of metric dimension in advancing medical research and enhancing healthcare delivery.

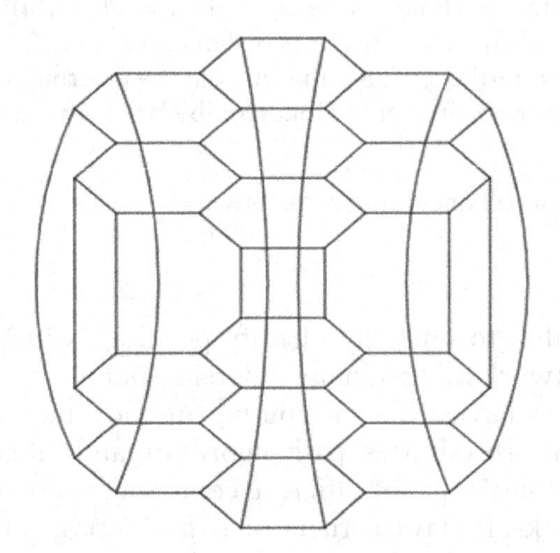

Figure 2.1: 3D view of $HCCa(2)$ with two floors

Paul *et al.* (2008) discovered the metric dimension problem of honeycomb network. Simon and George (2015) introduced the Beside Honeycomb Cage networks include (See Figure 2.1), which is formed by joining every boundary vertex of degree 2 of one Honeycomb network to another Honeycomb network of the same dimension. One interesting property that this network is Hamiltonian and has minimum six Hamiltonian cycles which is not found in the Honeycomb network. Further, Honeycomb Cage network can also be drawn in a different way (Bricks transformation) (See Figures 2.2&2.3) which is far efficient than grid networks.

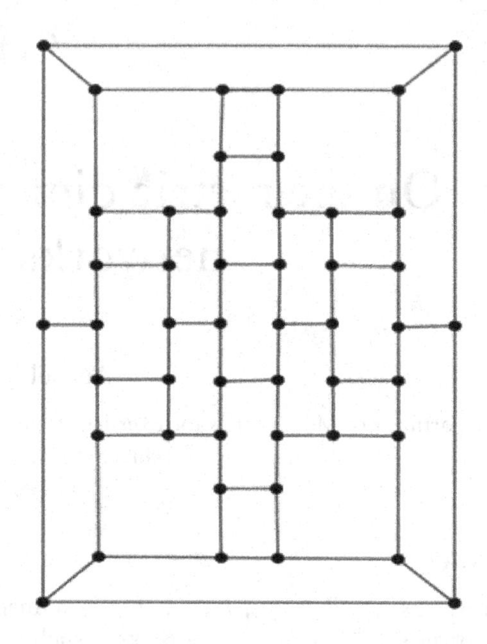

Figure 2.2: Brick Transformations of $HCCa(2)$ with two floors

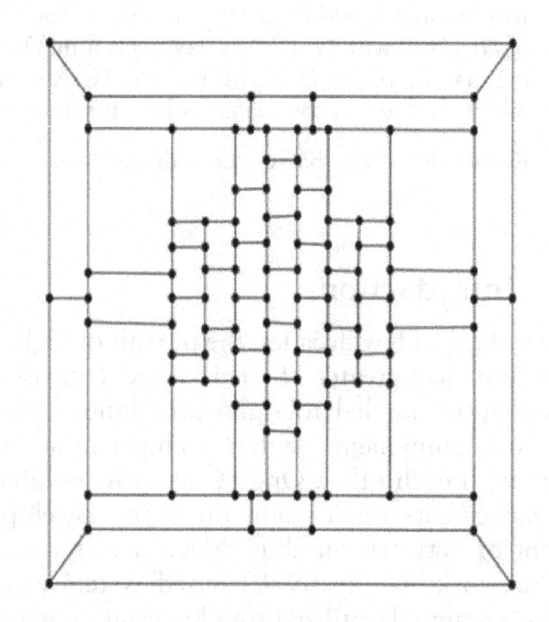

Figure 2.3: Brick Transformations of $HCCa(3)$ with two floors

2. Literature review

Melter and Harary (1976) were the first to examine the metric dimension problem. Tomescu and Melter (1984) investigated the locating number for grid graphs and their results were extended by Khuller *et al.* (1996). The minimum resolving set of unicyclic graphs were studied by Chartrand *et al.* (2000) and they have also given a new proof for the dimension of trees. Paul and Indra (2011) have determined the metric dimension of Silicate networks. Paul *et al.* (2008) have given the solution for the metric dimension of $HX(n)$, $HC(n)$ and have introduced an open problem with two Hex-derived networks (*HDN1* and *HDN2*). Xu and Fan (2014) have partially solved the open problem for *HDN* of type 1 and it has been completely solved by us in the year (2024). Simon *et al.* (2023,2024) have determined all possible metric basis of Honeycomb network and Enhanced Honeycomb network for dimensions two to six. The metric dimension problem for Torus networks (2006), Petersen graphs (2003), Silicate Stars (2015) have also been studied.

3. Results

The minimum resolving set of Honeycomb Cage network with 2 floors is found after determining the minimum resolving set of Hexagonal Cage network with 2 floors. Hexagonal network is formed by joining the boundary vertices of one Hexagonal network to another Hexagonal network of the same dimension (2015). The concept of neighborhood (2008) is used to prove the following results.

Lemma 1:

The n-dimensional Hexagonal Cage network $HXCa(n)$ with 2 floors has $dim(G) > 2$.

Proof. Let us prove this lemma by applying the concept of vertex neighborhood in $HX(n)$. By definition, $N_r(t) = \{s \in v : d(s, t) = r\}$ is an $r-$neighborhood of t. Consider V to be the vertex set of the Beside $HXCa(n)$ with 2 floors, include (See Figure 2.4). Stojmenovic introduced a coordinate system for honeycomb networks. Nocetti *et al.* implemented this technique to provide coordinates for hexagonal networks. In this system, the X, Y, and Z axes run parallel to the three edge directions, with each pair forming a 120-degree angle.

The above approach assigns coordinates $(x, y, z, 0)$ to any vertex of the first floor $HX_1Ca(n)$.

Designate P_X as a section of the X line made up of points that have a fixed x coordinate. That is, $P_X = \{(x_0, y, z, 0)/\ y_1 \le y \le y_2, z_1 \le z \le z_2\}$, $P_Y = \{(x, y_0, z, 0)/\ x_1 \le x \le x_2, z_1 \le z \le z_2\}$ and $P_Z = \{(x, y, z_0, 0)/\ x_1 \le x \le x_2, y_1 \le y \le y_2\}$. Additionally, designate $P'(X)$ a section of the X' line in the second floor of Hexagonal Cage network $HX_2Ca(n)$, which consists of points $(x', y', z', 1) = (x, y, z, 1)$ with a fixed x coordinate which is $P'_X = \{(x_0, y, z, 1)/\ y_1 \le y \le y_2, z_1 \le z \le z_2\}$, $P'_Y = \{(x, y_0, z, 1)/\ x_1 \le x \le x_2, z_1 \le z \le z_2\}$ and $P'_Z = \{(x, y, z_0, 1)/\ x_1 \le x \le x_2, y_1 \le y \le y_2\}$.

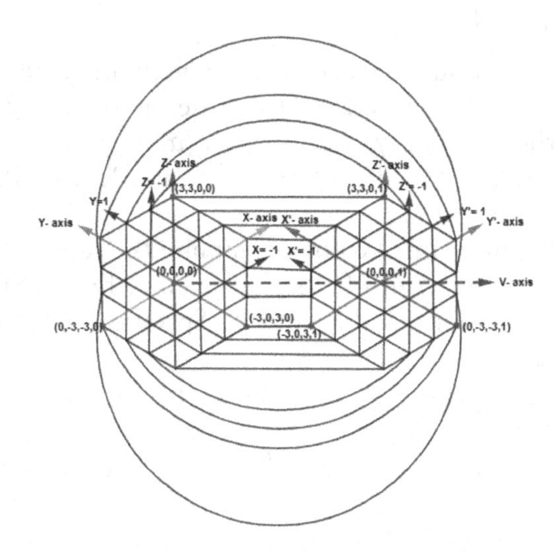

Figure 2.4: Co-ordinate vertices in $HXCa(4)$ with two floors

Theorem 2:

For any $HX(n)$, we have $N_r(\alpha) = P_Y \circ P_Z$, $N_r(\beta) = P_X \circ P_Z$, $N_r(\gamma) = P_X \circ P_Y$. From the previous theorem 2, the subsequent theorem 3 is obvious.

Theorem 3:

For any second floor $HX_2Ca(n)$, $N_r(\alpha') = P'_Y \circ P'_Z$, $N_r(\beta') = P'_X \circ P'_Z$, $N_r(\gamma') = P'_X \circ P'_Y$.

It is known that $\{\alpha, \beta, \gamma\}$ is a minimum resolving set for $HX(n)$. We show that *dim* (G) of Hexagonal Cage network with 2 floors is three and $\{\alpha, \gamma, \sigma,\}$ is a minimum resolving set for Hexagonal Cage network network with 2 floors for n dimensions.

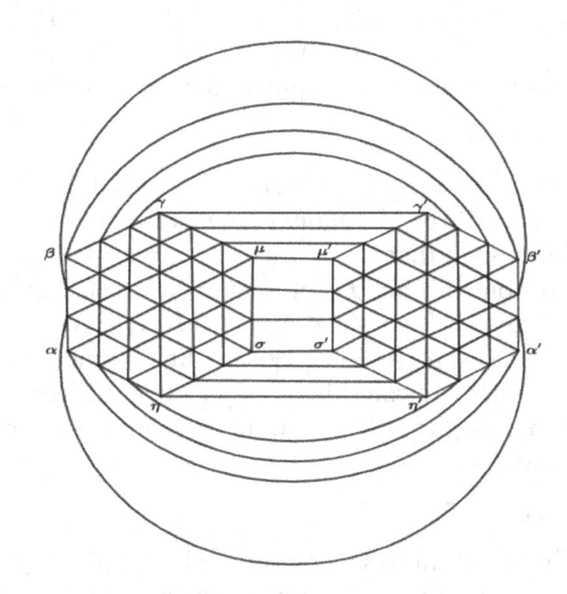

Figure 2.5: Hexagonal Cage network $HXCa(4)$ with 2 floors

Lemma 4:

In any Beside Hexagonal Cage network with 2 floors (See Figure 2.5), let us define the neighborhood of α, γ and σ as $N_r(\alpha) = P_Y \circ P_Z \cup N_{r-1}(\alpha')$, $N_r(\gamma) = P_X \circ P_Y \cup N_{r-1}(\gamma')$ and $N_r(\sigma) = P_X \circ P_Z \cup N_{r-1}(\sigma')$.

It can be inferred by examining the topology of the $HXCa(n)$ of first and second floor that $N_r(\alpha)$ comprises of P_Y and P_Z. Specifically, for $1 \leq r \leq n - 1$,

$P_Y = \{(r - a, -(n - 1) + r, -(n - 1) + a, 0), 0 \leq a \leq r\}$ and

$P_Z = \{(-b, -(n - 1) + r - b, -(n - 1) + r, 0), 1 \leq b \leq n - 1\}$

For $n \leq r \leq 2n - 2$,

$P_Y = \{((n - 1) -a, -(n - 1 - r), -(2n - 2 - r) + a, 0), 0 \leq a \leq r\}\}$ and

$P_Z = \{(-b, -(n - 1 - r) -b, -(n - 1 - r), 0), 1 \leq b \leq n - 1\}$

The neighborhood of α, $N_r(\alpha)$ of the Hexagonal Cage network in the first floor is $P_Y \circ P_Z$ and additionally in the second floor, α has neighbors at a distance of r which is denoted by $N_{r-1}(\alpha')$. Hence for any r, $N_r(\alpha) = P_Y \circ P_Z \cup N_{r-1}(\alpha')$. Similarly, it is also possible to prove $N_r(\gamma) = P_X \circ P_Y \cup N_{r-1}(\gamma')$ and $N_r(\sigma) = P_X \circ P_Z \cup N_{r-1}(\sigma')$.

Lemma 5

The intersection of $N_{r1}(\alpha)$ and $N_{r2}(\gamma)$ is either singleton or empty or a Y-line segment for any r_1 and r_2.

Proof. It is clear from the previous lemma, $N_{r1}(\alpha) = P_Y \circ P_Z \cup N_{r1-1}(\alpha')$ and $N_{r2}(\gamma) = P_X \circ P_Y \cup N_{r2-1}(\gamma')$. Thus, $N_{r1}(\alpha) \cap N_{r2}(\gamma)$ is either singleton or empty or a Y-line segment for any r_1 and r_2.

Corollary 6

Consider $s = (x_1, y_1, z_1, 0), t = (x_2, y_2, z_2, 0)$, as the vertices of the Hexagonal Cage network with 2 floors such that $x_1 \neq x_2$, $y_1 \neq y_2$, $z_1 \neq z_2$. Then at most one of s and t is present in $N_{r1}(\alpha) \cap N_{r2}(\gamma)$.

Proof. If both $s, t \in N_{r1}(\alpha) \cap N_{r2}(\gamma)$, then by Lemma 5, $N_{r1}(\alpha) \cap N_{r2}(\gamma)$ is a line segment of P_Y, which indicates that $y_1 = y_2$, a contradiction. Hence $N_{r1}(\alpha) \cap N_{r2}(\gamma)$ contains at most one of s and t.

Theorem 7

The minimum resolving set of Hexagonal Cage network $HXCa(n)$ with 2 floors is 3 for n dimensions.

Proof.

Case 1:

Consider the vertices $s = (x_1, y_1, z_1, 0)$, $t = (x_2, y_2, z_2, 0)$, in the first floor $HX_1Ca(n)$ of the Hexagonal Cage network $HXCa(n)$ with 2 floors. Then we see that,

When $x_1 = x_2$, $s, t \in P_X$ which implies $d(s, \alpha) \neq d(t, \alpha)$.

When $y_1 = y_2$, $s, t \in P_Y$ which implies $d(s, \sigma) \neq d(t, \sigma)$.

When $z_1 = z_2$, $s, t \in P_Z$ which implies $d(s, \gamma) \neq d(t, \gamma)$.

Consider the case where $x_1 \neq x_2$, $y_1 \neq y_2$, $z_1 \neq z_2$, Suppose $d(s, \alpha) = d(t, \alpha)$, then $s, t \in N_{r1}(\alpha)$ for some r_1. Let us claim that $d(s, \gamma) \neq d(t, \gamma)$. On the contrary assume that $d(s, \gamma) = d(t, \gamma)$ which implies $s, t \in N_{r2}(\gamma)$ for some r_2. Now, we have $s, t \in N_{r1}(\alpha) \cap N_{r2}(\gamma)$ which is contrary to Corollary 6. Hence, $d(s, \gamma) \neq d(t, \gamma)$.

Case 2:

Consider $s = (x_1, y_1, z_1, 0)$ to be a vertex in the first floor $HX_1Ca(n)$ of $HXCa(n)$ and $t = (x_2, y_2, z_2, 1)$ to be a vertex in the second floor $HX_2Ca(n)$ of $HXCa(n)$.

Then, consider $x_1 = x_2$, $s \in P_X$, $t \in P'_X$. Suppose that $d(s, \alpha) = d(t, \alpha)$, then $s, t \in N_{r1}(\alpha)$ for some r_1. Let us claim that $d(s, \gamma) \neq d(t, \gamma)$. On the contrary assume that $d(s, \gamma) = d(t, \gamma)$ which implies $s, t \in N_{r2}(\gamma)$ for some r_2. Now, we have $s, t \in N_{r1}(\alpha) \cap N_{r2}(\gamma)$, a contradiction to Corollary 6. Hence, $d(s, \gamma) \neq d(t, \gamma)$

Similarly, if $y_1 = y_2$, $s \in P_Y$, $t \in P'_Y$. Suppose that $d(s, \sigma) = d(t, \sigma)$, then $s, t \in N_{r3}(\sigma)$ for some r_3. Let us claim that $d(s, \alpha) \neq d(t, \alpha)$. On the contrary assume that $d(s, \alpha) = d(t, \alpha)$ which implies $s, t \in N_{r1}(\alpha)$ for some r_1. Now, we have $s, t \in N_{r1}(\alpha) \cap N_{r3}(\sigma)$ which is contrary to Corollary 6. Hence, $d(s, \alpha) \neq d(t, \alpha)$

Also, if $z_1 = z_2$, $s \in P_Z$, $t \in P'_Z$. If $d(s, \gamma) = d(t, \gamma)$, then $s, t \in N_{r2}(\gamma)$ for some r_2. Let us claim that $d(s, \sigma) \neq d(t, \sigma)$. On the contrary assume that $d(s, \sigma) = d(t, \sigma)$ which implies $s, t \in N_{r3}(\sigma)$ for some r_3. Now, we have $s, t \in N_{r2}(\gamma) \cap N_{r3}(\sigma)$ contradicting Corollary 6. Hence, $d(s, \sigma) \neq d(t, \sigma)$

Finally, let us consider the case where $x_1 \neq x_2$, $y_1 \neq y_2$, $z_1 \neq z_2$. Suppose $d(s, \alpha) = d(t, \alpha)$, then $s, t \in N_{r1}(\alpha)$ for some r_1. Let us claim that $d(s, \gamma) \neq d(t, \gamma)$. On the contrary assume that $d(s, \gamma) = d(t, \gamma)$ which implies $s, t \in N_{r2}(\gamma)$ for some r_2. Now, we have $s, t \in N_{r1}(\alpha) \cap N_{r2}(\gamma)$, a contradiction to Corollary 6. Hence, $d(s, \gamma) \neq d(t, \gamma)$.

Case 3:

Consider $s = (x_1, y_1, z_1, 1)$ and $t = (x_2, y_2, z_2, 1)$ to be any two vertices in the second floor $HX_2Ca(n)$ of $HXCa(n)$. Then we see that,

When $x_1 = x_2$, $s, t \in P'_X$ implying that $d(s, \alpha) \neq d(t, \alpha)$.

When $y_1 = y_2$, $s, t \in P'_Y$ implying that $d(s, \sigma) \neq d(t, \sigma)$.

When $z_1 = z_2$, $s, t \in P'_Z$ implying that $d(s, \gamma) \neq d(t, \gamma)$.

Consider the case where $x_1 \neq x_2, y_1 \neq y_2, z_1 \neq z_2$. Suppose $d(s, \alpha) = d(t, \alpha)$, then $s, t \in N_{r1}(\alpha)$ for some r_1. Let us claim that $d(s, \gamma) \neq d(t, \gamma)$. On the contrary assume that $d(s, \gamma) = d(t, \gamma)$ which implies $s, t \in N_{r2}(\gamma)$ for some r_2. Now, we have $s, t \in N_{r1}(\alpha) \cap N_{r2}(\gamma)$ which is contrary to Corollary 6. Hence, $d(s, \gamma) \neq d(t, \gamma)$. Thus, the metric dimension of $HXCa(n)$ with 2 floors is 3 for n dimensions.

4. Minimum resolving set of Honeycomb Cage network with 2 floors

Here, the minimum resolving set of Honeycomb Cage network with 2 floors is established as three. The strip that lies between two successive X-lines in $HXCa(n)$ is known as an X-channel, and it is represented by C_X. The definitions of C_Y and C_Z are similar to C_X (See Figure 2.6). The behaviour of an X-channel of $HCCa(n)$ is identical to that of an X-line of $HXCa(n)$. The proofs of Lemma 8 and Lemma 9 are equivalent to those of Lemma 4 and Lemma 5.

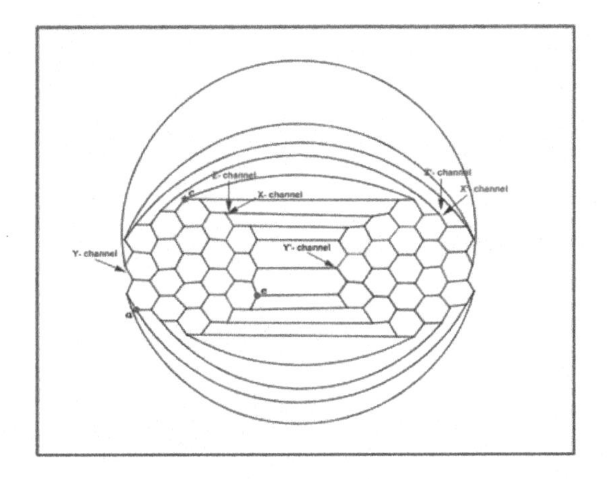

Figure 2.6: Honeycomb Cage network *HCCa*(3) with 2 floors

Lemma 8

Consider a, c and e to be the vertices in the Honeycomb Cage network $HCCa(n)$ with 2 floors. Then $N_r(a) \subset C_Y \circ C_Z \cup N_{r-1}(a')$, $N_r(c) \subset C_X \circ C_Y \cup N_{r-1}(c')$ and $N_r(e) \subset C_X \circ C_Z \cup N_{r-1}(e')$.

Lemma 9

$N_{r1}(a) \cap N_{r2}(c)$ is either empty or singleton or a Y- channel's line segment for any r_1 and r_2.

Minimum resolving set of $HC(n)$ has been proved in (2008). From theorem 7, it can be seen that the minimum resolving set of $HCCa(n)$ with 2 floors is $\{a, c, e\}$.

Theorem 10

Let G be an n-dimensional Honeycomb Cage network with 2 floors. Then $dim(G)$ is three.

5. Conclusion

We have enhanced the outcome of the result proposed in the previous research paper and have proved the minimum resolving set of Hexagonal Cage network with 2 floors and Honeycomb Cage network with 2 floors to be three. Henceforth, we denote it as $HX(n, 2)$ and $HC(n, 2)$ to further investigate the minimum resolving set of Hexagonal Cage network of dimension n with f floors $HX(n, f)$ and Honeycomb Cage network of dimension n with f floors $HC(n, f)$.

References

[1] Abigail, M. V., and Raj, F. S. (2024). An algorithmic approach for determining metric dimension of Hex-derived network of type 1 (HDN1). International Conference on Mathematics, Computers and Engineering Sciences, Malaysia, 25.

[2] Alexander, W. Y., Thomas, C. X., and Pasi, L. H. T. (2009). Exploration of honeycomb topologies for Network on Chip. Sixth IFIP International Conference on Network and Parallel Computing, 73–79.

[3] Xu, D., and Fan, J. (2014). On the metric dimension of HDN. J. Discrete Algorithms, 26, 1–6.

[4] Chartrand, G., Eroh, L., Johnson, M. A., and Oellermann, O. R. (2000). Resolvability in graphs and the metric dimension of a graph. Discrete Applied Mathematics, 105(1–3), 99–113.

[5] Harary, F., and Melter, R. A. (1976). On the metric dimension of a graph. Ars Comb, 2, 191–195.

[6] Khuller, S., Raghavachari, B., and Rosenfeld, A. (1996). Landmarks in graphs. Discrete Applied Mathematics, 70, 217–229.

[7] Libin Bright, J. M., and Raj, F. S. (2024). An almost deterministic algorithm for finding all possible metric bases for Hamming graphs $H_{3,4}$, $H_{4,3}$, $H_{2,6}$ and $H_{6,2}$, Proceedings of the International Conference on Mathematics, Computers & Engineering Sciences, Malaysia, 24.

[8] Manuel, P., Rajan, B., Rajasingh, I., and Monica M., C. (2008). On minimum metric dimension of honeycomb networks. J. Discrete Algorithms, 6, 20–27.

[9] Manuel, P., and Rajasingh, I. (2011). Minimum metric dimension of silicate networks. Ars Combin, 98, 501-510.

[10] Melter, R. A., and Tomescu, I. (1984). Metric bases in digital geometry. Computer Vision, Graphics and Image Processing, 25, 113–121.

[11] Rajan, B., Rajasingh, I., Cynthia, J. A., and Manuel, P. (2003). On minimum metric dimension. Proceedings of the Indonesia-Japan Conference on Combinatorial Geometry and Graph Theory, September 13–16, Bandung, Indonesia.

[12] Manuel, P., Rajan, B., Rajasingh, I., and Monica M., C. (2006). Landmarks in torus networks. Journal of Discrete Mathematical Sciences & Cryptography, 9, 263–271.

[13] Raj, F. S., and George, A. (2015). Topological properties of hexagonal cage networks. Global Journal of Pure and Applied Mathematics, 11(3), 1413–1424.

[14] Raj, F. S., and George, A. (2015). Topological properties of honeycomb cage networks. Fifth International Conference on Advances in Information Technology and Mobile Communication – AIM 2015.

[15] Raj, F. S., and George, A. (2015). On the metric dimension of silicate stars. ARPN Journal of Engineering and Applied Sciences, 10(5), 2187–2192.

[16] Raj, F. S., Abigail, M. V., Bright, J. M. L., and George, A. (2023). A deterministic algorithm for finding all metric bases of honeycomb network and enhanced honeycomb network of dimension two with O(2^n) time complexity. International Conference on Computational Engineering (ICCE) Proceedings, 308.

[17] Raj, F. S., and Abigail, M. V. (2024). All possible positions of minimum number of satellites for honeycomb mesh and enhanced honeycomb networks of dimension two to six using BIGS algorithm. International Conference on Mathematics, Computers and Engineering Sciences, Malaysia, 23.

CHAPTER 3

Strategic Variations in DeepStack for No-Limit Heads-Up Texas Hold'em: An In-Depth AIVAT Analysis

Nivedhitha Balaji[1,a], Umamaheswari S.[2,b,*], Sai Kumar M.[3,c]

[1]UG Student, Department of Mathematics and Physics, School of Advanced Sciences, Vellore Institute of Technology, Chennai, Tamil Nadu, India

[2*]Assistant Professor (Sr), Department of Mathematics, School of Advanced Sciences, Vellore Institute of Technology, Chennai, Tamil Nadu, India

[3]Research Scholar, Department of Mathematics, School of Advanced Sciences, Vellore Institute of Technology, Chennai, Tamil Nadu, India

Email: [a]nivedhithabalaji04@gmail.com, [b]umamaheswari.suk@vit.ac.in, [c]saikumar.m2022@vitstudent.ac.in

Abstract

Games are mathematically classified into perfect, imperfect, and incomplete information games based on the extent of knowledge available regarding the payoffs of all players for every possible action plan. Perfect information games, characterized by the complete visibility of all elements of the game, can achieve optimal equilibrium through holistic modelling using advanced AI solutions. In contrast, optimizing solutions for non-trivial incomplete and imperfect information games with extensive game states remains challenging. Significant progress has been made in abstracting and modelling imperfect information games, such as poker. Given their non-deterministic and zero-sum nature, these games provide a robust testing ground for AI models. Several models, including LOKI, POKI, PSOPTI, and VEXBOT, have been instrumental in advancing the field by enhancing knowledge bases, opponent modelling, and game tree search methodologies. Although two-player heads-up limit Texas Hold'em (HUNL) poker is considered weakly solved, the highly complex state space of bets in heads-up no-limit Hold'em poker (especially with non-integral bet sizes) continues to pose substantial challenges. This paper aims to elucidate the betting strategy variations employed by DeepStack in no-limit heads-up Texas Hold'em poker through a comprehensive AIVAT (Action-based Value and Importance Testing) analysis, thereby demonstrating its efficacy in opponent modelling.

Keywords: AIVAT analysis, DeepStack, imperfect games, limits heads-up Texas Hold'em, opponent modelling, reinforcement learning

AMS subject classifications. 91-00; 91B26

Article type: Original Research Article

1. Introduction

Poker is a non-deterministic, imperfect information game. The variation known as HUNL is challenging to model due to the extensive scaling of game states and action histories. The game is played with a deck of 52 cards and involves the distribution of two-hole cards (dealt face down - pre-flop) and subsequently five community cards (dealt face up) over three rounds (flop, turn, river). There are four possible actions at each turn: betting, calling, raising, and checking. Players bet, raise, or call if the pot odds seem appropriate for a win. They check if they wish to pass on the betting opportunity in

DOI: 10.1201/9781003606611-3

that round and fold if they seek to drop out of the match (in which case their contribution to the pot would be lost).

The decision between various betting strategies can be made statistically sound by examining the pot odds and table odds of each round. An effective strategy that exploits the opponent's weakness is bluffing, which can be efficient if properly executed. Human players seek to exploit the opponent's stance, expressions, and prior actions to make a move. Machine models that later replicated successful strategies in minor variants of poker have an efficient system of opponent modeling and game tree analysis.

The first projects that aimed to model Poker were LOKI, POKI, VEXBOT, etc. were able to attain a breakthrough in many aspects essential to winning poker including agent modelling, risk management, counter-deception and prediction. Davidson [1] underscored the accomplishments of AI in perfect-information games and the difficulties in imperfect-information arenas such as poker. His work with the POKI program focused on opponent modeling using neural networks and decision trees, attaining human-level performance and tackling the inherent uncertainty and deception in poker. Building on this, Papp [2] utilized poker as a testbed for decision-making under uncertainty, developing a program that managed all aspects of the game and employed adaptive opponent modeling. This introduced new techniques for evaluating poker hands and advanced agent modeling and risk management in scenarios with imperfect knowledge. Schauenberg [3] examined opponent modeling and counter-strategy formulation in poker, implementing a program that utilized expectimax search, evaluated performance in Leduc Hold'em and Texas Hold'em, and progressed strategies in domains with elements of chance and imperfect information.

Zinkevich et al. [4] introduced counterfactual regret for extensive games, demonstrating that minimizing it also minimized overall regret. His poker-specific algorithm efficiently computed approximate equilibria for large abstractions, surpassing previous methods and winning the 2006 AAAI Computer Poker Competition, significantly enhancing poker AI performance. Gilpin et al. [5] created Tartanian, a game theory-based player for HUNL poker, using a discretized betting model and automated abstraction. Victor [6] tackled AI challenges in poker by modeling opponents using clustering instead of Bayesian learning in Kuhn poker. Utilizing the EM algorithm, it swiftly learned opponents' strategies, and empirical experiments demonstrated that clustering outperformed Bayesian methods in limited-time games. Johanson [7] proposed an efficient algorithm for exact size computation of Annual Computer Poker Competition (ACPC) no-limit poker games, avoiding exhaustive tree traversal and covering variants since 2007, highlighting the need for a testbed domain to spur research in state-space abstraction for these expansive domains.

Bard [8] examined agent behavior optimization in multiagent environments, contrasting explicit and implicit modeling, introducing an efficient implicit modeling framework, a decision-theoretic clustering algorithm, and analyzing agent representation granularity. Empirical validation in poker domains showed the effectiveness of implicit modeling for real-time adaptation in complex, human-scale interactions. Bard et al. [9] developed an efficient greedy algorithm for decision-theoretic clustering, enhancing agent performance in extensive-form games by surpassing traditional k-means clustering in capturing agent behaviors, offering computational feasibility and improved strategic decision-making in complex environments. Burch et al. [10] developed AIVAT, a novel technique for value estimation in imperfect information games that combined heuristic value functions, game structure knowledge, and player strategy insights. AIVAT achieved unbiasedness and significant variance reduction, demonstrating up to three times lower standard deviation compared to existing methods, allowing for practical experiments with significantly less data, crucial in scenarios like human-bot interactions. Bowling et al. [11] declared HUNL poker weakly solved, marking the first nontrivial imperfect information game to be solved competitively. This achievement stemmed from algorithmic advances in game-theoretic reasoning, with implications for security systems and medical decision support, highlighting the importance of tackling uncertainty in decision-making.

Lisy and Bowling [12] introduced the Local Best Response method for quickly approximating exploitability in large poker strategies. The study demonstrated that existing poker bots, even top performers from ACPC 2016, exhibited high exploitability when using card abstraction, losing over 3 big blinds per hand on average against their worst-case opponents, highlighting the need for a significant paradigm shift to develop bots that approach equilibrium in full no-limit Texas Hold'em. Moravčík et al. [13] demonstrated DeepStack's success in defeating professional poker players by leveraging a paradigm shift in solving large, sequential imperfect-information games. This approach, focused on real-time decision-making and trained value functions, closed the gap between perfect- and imperfect-information games, extending implications to various real-world problems involving information asymmetry. Morrill [14] developed Regression Counterfactual Regret Minimization (CFR) (Regression Counterfactual Regret Minimization (RCFR)), an innovative algorithm leveraging regressors to estimate regret, replacing static abstractions in game-theoretic solutions. This approach scaled to complex, dynamic tasks, enhancing solution accuracy and efficiency, particularly in negotiation and auction scenarios, outperforming traditional methods by reducing exploitability in human-scale games.

Ganzfried and Yusuf [15] presented a novel machine learning formulation that computed human-understandable game-theoretic strategies using private information Cumulative Distribution Function (CDF) values and strategy probability vectors. This approach, applicable beyond poker to any multiplayer imperfect-information games, introduced effective distance metrics for strategies, enhancing applications like opponent exploitation and bot detection. Bhetuwal et al. [16] highlighted the limitations of the Zero Risk strategy and linear programs in long-term risk minimization and winning insights. The research advanced by analyzing betting patterns with continuous stack functions, suggesting a potential optimal minimal risk strategy through iterative strategy evaluation.

2. Methodology

We consider a two-player poker game (Table 3.1) between a human and DeepStack, an AI model of no limit heads up Texas Hold'em.

For the sake of the simplicity of abstraction techniques, we have considered games that extend to the flop stage (games proceeding to the river and turn stages have been omitted). We have extracted the AIVAT analysis of the games between International Federation of Poker (IFP) Pros and DeepStack. This allows us to isolate 6 rounds, the former 3 of which are against a loose aggressive play (Game 1) and the later 3 between a tight passive play (Game 2), in which DeepStack demonstrates a multitude of strategies that differ from those offered by other AI tools such as Cepheus.

The pre flop strategies employed by these IFP Pro Players could be compared against the pre flop strategies recommended by Sklansky and Malmuth [17] in order to categorize their gameplay. The variations in calling bluffs or raising the pot (action frequencies) that is demonstrated by DeepStack is observed to classify the strategy as tight or loose and as passive or aggressive. The information thus obtained is compared to the strategy engaged by DeepStack and the differences in modelling the opponent is obtained. AIVAT is a low variance analysis which relies on imitation learning from expert data, uses refined adversarial training to analyze

Table 3.1: A brief comparison about Poker models

Author(s)	Model	AIVAT	Pre-Flop (Isolated examination)	3-Betting	Action Frequency Analysis
Davidson [1]	POKI	No	No	Yes- aggressive	Yes
Papp [2]	LOKI	No	Yes	Rare-Optimal	Yes
Burch et al. [10]	All Poker Models	Yes	No	Analysis of play	Yes
Bowling et al. [11]	Cepheus	No	No	Yes- often	Yes
Moravčík et al. [13]	DeepStack	Yes	No	Yes- Adaptive	Yes
This paper*	DeepStack	Yes	Yes	Analysis of play	Yes

and modify strategies through the course of the game. Thus, it can evolve an exploitative strategy in a smaller depth of the game space and play more effectively against opponents.

Observations

Game 1: Bachmann Jeurgen Vs DeepStack
The AIVAT Analysis as put forth in a play-off against DeepStack is as follows:

The following examples demonstrate the three different betting strategies demonstrated by DeepStack (fold, raise, call):

Fold

Opponent Hand	QcKc
DeepStack Hand	4d2h
Pre-Flop moves	DeepStack folds the hand
Flop moves	NA

It is seen that the pre-flop hand of 4d2h is unsuited and of lower rank. Therefore, the round could be passed with a bluff or a fold and since the probability of a winning hand is meagre to none, DeepStack folds the hand.

Raise

Opponent Hand	Ah9c
DeepStack Hand	7d3d
Pre-Flop moves	DeepStack raises by 300, opponent raises by 750 followed by DeepStack calling
Flop moves	Opponent raises the pot but DeepStack folds

It is seen that a hand of 7d3d is suited. Although calling the hand might be a better option, DeepStack proceeds to raise the pot and then call. Thus, a mildly aggressive approach to the round is observed.

Call

Opponent Hand	4d5s
DeepStack Hand	7h10h
Pre-Flop moves	DeepStack calls followed by a raise of 300 from the opponent. DeepStack then folds
Flop moves	NA

With a hand of 7h10h, which is suited, it is better to call (similar to the previous case), as is implemented by DeepStack.

It is seen that out of 1088 games with 5 rounds and possible 3-betting and raising, the player folds 19.9%, calls 40.8% and raises 39.3% of the time. DeepStack, on the other hand is seen to fold 17.02%, calls 39.35%, raises 43.63% of the time. The system is seen to play an optimal game as the call frequencies exceed the raise and fold frequencies. The learning trend of the system is apparent from the similarity of trend in the higher iterations. DeepStack is seen to win an average of 53 per 100 games when it takes on the role of the big blind.

Game 2: Philip Laak Vs DeepStack

The AIVAT Analysis demonstrating the three different betting strategies (fold, raise, call) demonstrated by DeepStack is given below:

Fold

Opponent Hand	Js6d
DeepStack Hand	2d4c
Pre-Flop moves	DeepStack folds
Flop moves	NA

It is seen that the pre-flop hand of 2d4c is unsuited and of lower rank. Therefore, the round could be passed with a bluff or a fold and since the probability of a winning hand is meagre to none, DeepStack folds the hand.

Raise

Opponent Hand	4d5d
DeepStack Hand	4h3h
Pre-Flop moves	DeepStack raises by 300, followed by the opponent calling
Flop moves	Opponent calls whereas DeepStack raises following which the opponent folds

It is seen that a hand of 4h3h is suited. The opponent follows an optimal strategy where they call. This is followed by DeepStack raising the pot as the pot odds are high. The opponent is then forced to fold, unable to gauge the hand strength of DeepStack.

Call

Opponent Hand	7h3d
DeepStack Hand	2dQs
Pre-Flop moves	DeepStack and the opponent call the blind bet
Flop moves	The opponent raises the pot following which DeepStack forfeits (folds)

With a hand of 2dQs, which is unsuited and far apart, it is better to fold. However, DeepStack calls and the opponent raises, hoping for a straight, but DeepStack folds.

It is inferred that out of 3003 games with 5 rounds and possible 3-betting and raising, the player folds 28%, calls 29% and raises 44% of the time. DeepStack, on the other hand is seen to fold 20.82%, call 38.66%, raise 40.51% of the time. Thus, it could be perceived that the player is aggressive and DeepStack also counters with an aggressive play where the fold and call frequency are similar but the raise frequency is high. It could also be seen that the instances of raising and bluffing increases with the number of iterations. Therefore, the system is seen to learn from the previous action frequencies.

Inferences and Discussion

The AIVAT analysis of the DeepStack games reveals distinct strategic adjustments based on opponent behavior. Against a loose-passive adversary, DeepStack adopts a conservative approach, engaging aggressively only with strong hands and minimizing bluffing and slow play both pre- and post-flop. Conversely, when facing a tight-passive opponent, the frequency of bluffs and aggressive bets on the flop increases. In countering a tight-aggressive competitor, DeepStack reduces the frequency of calling and raising based on the predicted strength of the opponent's hand. Against a loose-aggressive player, frequent 3-betting and folding are observed. These insights are further enriched when considering how DeepStack responds to players with personalized gameplay that borders on aggressive strategies. The algorithm's response to different hands of similar potential provides valuable insights into its opponent modeling capabilities.

The AIVAT analysis also elucidates the impact of luck components inherent in the distribution of table and hole cards on player strategy, shedding light on the role of luck in decision-making. The DeepStack algorithm, with its advanced methods for modeling imperfect information games, not only demonstrates its effectiveness against various player types but also offers a robust framework for examining variations in both imperfect and incomplete information scenarios. The win rate of DeepStack against these aggressive plays highlights its efficacy in adopting suitable strategies, with potential applications extending beyond poker to fields such as auction theory and related disciplines.

The paper presents a rigorous analysis of generalized strategies encompassing calling, raising, and folding, utilizing data derived from the AIVAT analysis to quantify action frequencies across these betting methods. This empirical analysis underscores the adaptability of DeepStack's gameplay, as evidenced by nearly 3,000 games against distinct adversaries. Unlike other models that either rigidly map the entire game space or fail to employ regret minimization for leveraging historical data, DeepStack exhibits a more dynamic, human-like response, enhancing the unpredictability of its strategies to opponents.

The complexity of modeling such a vast game space brings forth several significant challenges. These include adapting to varied betting strategies, employing context-dependent tactics while managing non-integral bet sizes, optimizing decisions across multiple betting rounds while remaining exploitative of the opponent's strategy, and integrating psychological and emotional factors into gameplay. Additionally, the resilience of DeepStack to counter-exploitation by sophisticated adversaries remains an ongoing challenge, highlighting the intricate balance between adaptability and strategic robustness in imperfect information games.

Data Availability Statement

The data that support the findings of this study are openly available in DeepStack at https://www.deepstack.ai/s/DeepStack_vs_IFP_pros.zip

References

[1] Davidson, J. A. (2002). Opponent modeling in poker: Learning and acting in a hostile and uncertain environment.

[2] Papp, D. R. (1998). Dealing with imperfect information in poker.

[3] Schauenberg, T. C. (2006). Opponent modelling and search in poker.

[4] Zinkevich, M., Johanson, M., Bowling, M., & Piccione, C. (2007). Regret minimization in games with incomplete information. *Advances in neural information processing systems, 20*.

[5] Gilpin, A., Sandholm, T., & Sørensen, T. B. (2008, May). A heads-up no-limit Texas Hold'em poker player: Discretized betting models and automatically generated equilibrium-finding programs. In *Proceedings of the 7th international joint conference on Autonomous agents and multiagent systems-Volume 2* (pp. 911–918).

[6] VICTOR, A. S. (2008). Learning in Simplified Poker by clustering opponents. *University of Manchester*.

[7] Johanson, M. (2013). Measuring the size of large no-limit poker games. *arXiv preprint arXiv:1302.7008*. https://doi.org/10.48550/arXiv.1302.7008

[8] Bard, N. D. (2016). Online agent modelling in human-scale problems. https://doi.org/10.7939/R32N4ZV1T

[9] Bard, N., Nicholas, D., Szepesvári, C., & Bowling, M. (2015, April). Decision-theoretic clustering of strategies. In *Workshops at the twenty-ninth AAAI conference on artificial intelligence*.

[10] Burch, N., Schmid, M., Moravcik, M., Morill, D., & Bowling, M. (2018, April). Aivat: A new variance reduction technique for agent evaluation in imperfect information games. In *Proceedings of the AAAI Conference on Artificial Intelligence* (Vol. 32, No. 1). https://doi.org/10.1609/aaai.v32i1.11481

[11] Bowling, M., Burch, N., Johanson, M., & Tammelin, O. (2017). Heads-up limit hold'em poker is solved. *Communications of the ACM, 60*(11), 81–88. https://doi.org/10.1145/3131284

[12] Lisy, V., & Bowling, M. (2017, March). Eqilibrium approximation quality of current no-limit poker bots. In *Workshops at the Thirty-First AAAI Conference on Artificial Intelligence*.

[13] Moravčík, M., Schmid, M., Burch, N., Lisý, V., Morrill, D., Bard, N.,... & Bowling, M. (2017). DeepStack: Expert-level artificial intelligence in heads-up no-limit poker. *Science, 356*(6337), 508–513. https://doi.org/10.1126/science.aam6960

[14] Morrill, D. R. (2016). Using regret estimation to solve games compactly. https://doi.org/10.7939/R3NZ80Z2Z

[15] Ganzfried, S., & Yusuf, F. (2017). Computing human-understandable strategies: deducing fundamental rules of poker strategy. *Games, 8*(4), 49. https://doi.org/10.3390/g8040049

[16] Bhetuwal, U., Chavez, A., & Dobes, I. (2018). Minimal Risk Betting Analysis in Poker Tournaments.

[17] Skalinsky, D., & Malmuth, M. (1994). Hold'em poker for advanced players. *Two Plus Two Publishing LLC*.

CHAPTER 4

Optimizing frequency allocation in mobile networks using local antimagic labeling

A graph theoretic approach

Shankar Rathinavel

Assistant Professor, Department of Mathematics, School of Arts and Science, Vinayaka Mission's Chennai Campus, Vinayaka Mission's Research Foundation (Deemed to be University), Paiyanoor-603 104, India
Email: rathinavelshankar@gmail.com

Abstract

This paper investigates local antimagic labeling in graphs, emphasizing its use to optimize frequency allocation in mobile networks. The approach provides towers with distinct labels depending on their connections, reducing interference and increasing communication dependability. A systematic, algorithmic strategy is proposed for resolving connection challenges while ensuring adequate frequency consumption. The method improves overall network performance by making mobile networks more reliable and lowering data transmission mistakes.

Keywords: Local antimagic labeling, local-chromatic number, complete graph, mobile networking

1. Introduction

Let $G=(V,E)$ be a graph structure which is finite and undirected having no pseudograph. In G, the order denoted by $|V|=p$ and size denoted by $|E|=q$. To access graph logical terms, please see the work of Chartrand, Lesniak [3].

The thought of antimagic labeling of graph G was proposed in [5]. Consider a graph $G=(V,E)$ which satisfies isometry $f:E\to\{1,2,\dots,q\}$. The weight of each nodex x in the set $V(G)$ is determined by the total of the weights of incident edges to x, denoted as $E(x)$. f is referred to as an antimagic labeling for any two unique nodes x and y in $V(G)$ such that $w(x)\neq w(y)$. Such graphs are known as antimagic graphs.
Further conjectures were proffered in [5].

Conjecture 1.1 [5] All graphs that are connected and trees are antimagic excluding K_2.

Several researchers demonstrated that these two hypotheses were somewhat correct. However, they remain unsolved. Chapter 6 of [4] has a full examination of these hypotheses.

The new parameter local-chromatic number of graphs was presented in [1][2]. Let f be an isometry for edge sets $:E\to\{1, 2, 3,\dots, q\}$, a local antimagic labeling defined if for any $xy\in E$, the weight of x, denoted as $w(x)$, is not equal to the weight of y, denoted as $w(y)$, where $w(x)$ is defined as the sum of $f(m)$ for every $m\in E(x)$. A graph G is considered to be local antimagic if it has a local antimagic labeling. The local antimagic chromatic number $\chi_{la}(G)$ is the least possible colors needed to color G generated by a local antimagic labelling of G. In [1], the authors demonstrated χ_{la} of cycles C_n with $n=3$, a connected graph G has C_3, then $\chi_{la}(G)$ is at least 3. In addition, they demonstrated that $\chi_{la}(G)$, where G is a complete

DOI: 10.1201/9781003606611-4

graph with p nodes is c-colors and wheel graph W_n. They conjectured that no disconnected

graph with at the minimum of 3 nodes permit a local-vertex antimagic labeling. This postulation were halfway breakout was done in [2]. This faced many researchers intend to perform efforts to resolve this postulation. Finally, Haslegrave proved it by way of strongest maching probability tools [6].

Recently, researchers have been more focused on local vertex antimagic labeling and they published in [7], [8], [9], [10], [11], [12], [13], [14], [16].

There were massive utilization of mathematical labeling in universe nowadays. For an enhanced scheduling process to emigrant assignment is drawn through the graph local antimagic total labeling. This scheduling technique ensures that none of the members get same assigned time in same sections effectively. The sections with familiar emigrants will arrive at different allocated times, corresponded to the distint colors of the nodes in set forth graph noted that the colors are unique but is limited to the least [15].

Nowadays, local antimagic labeling is one of the most exciting research areas in graph labeling. Local antimagic labeling is because it connects the most exciting research area graph labeling and graph coloring. The chromatic number otherwise denoted as proper coloration in graph G, thus inducing the local antimagic node labeling. In general, labeling has many applications in many fields like security surveillance, encryption, and decryption in coding theory, network analysis. This manuscript reveals a real-life utilization of local antimagic labeling on graphs.

2. Mobile Networking

Mobile networking dates back to the late 1970s, when Nippon Telegraph and Telephone (NTT) developed the first commercial cellular network in Tokyo, Japan, ushering in the 1G generation. This analogue-based wireless communication technology quickly swept across Japan, transforming telecommunications. By the early 1990s, there had been a global movement from analogue to digital systems, resulting in the introduction of the second generation and the deployment of the first commercial digital cellular network. This change dramatically improved voice quality, decreased interference, and paved the way for data transfers in mobile networks. Cellular base stations, necessary for connecting mobile devices to the network, grew with these technical breakthroughs, offering dependable coverage and allowing smooth communication services for consumers worldwide.

The move from analogue to digital networks in the late 1980s and early 1990s marked the beginning of the Second Generation (2G) era. The first commercial digital cellular network was introduced in 1991, offering increased call quality, enhanced security measures, and data transmission capabilities. The 2G period saw the rise of technologies such as Mobile Communications globally and Code Division Multiple Access, which settled the structure more sophisticated to mobile services.

The 3G era began in the early 2000s and marked breakthroughs in data rates, multimedia capabilities, and mobile internet access. Technologies such as Universal Mobile Telecommunications System and CDMA2000 were released, allowing video calling, mobile surfing, and improved sound quality.

The switch to 4G networks began in 2009, providing higher data speeds, lower latency, and more network efficiency. Long-Term Evolution and access to Worldwide Interoperability for Microwave were critical technologies in the 4G era. They allowed high-speed mobile internet and supported applications like HD video streaming and online gaming.

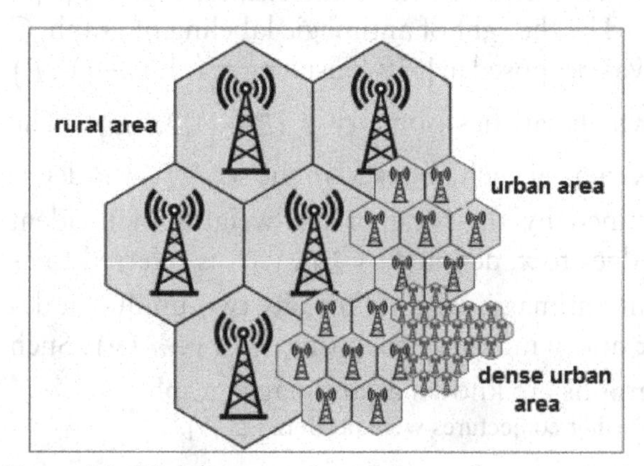

Figure 4.1: Mobile tower networking system

The current period centres on 5G technology, which promises ultrafast speeds, massive connections for IoT devices, and compatibility for new applications such as augmented reality (AR) and self-driving cars. 5G networks use mmWave (millimetre wave) technologies, beamforming, and network slicing to improve performance and connection.

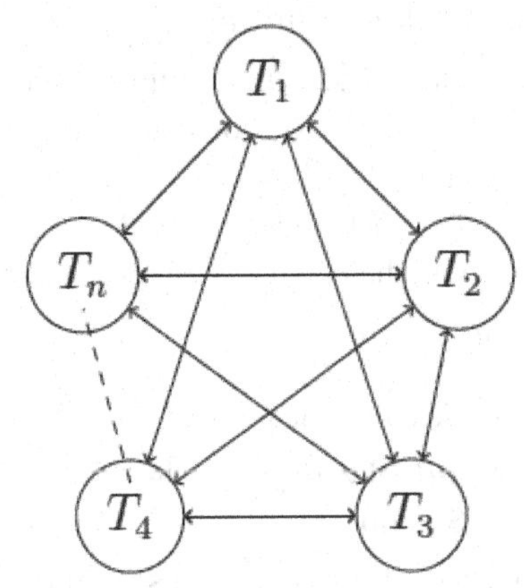

Figure 4.2: Graph structure of mobile tower networking system

Consider this scenario: a graph represents mobile base towers as nodes, while the network that connects them is denoted as edges. Each edge is clearly labeled, and each nodes (tower) allocates a weight, determining its color. Thus, depending on their weights, we get a suitable coloring of the nodes. Now, we define the local chromatic number of above graph G.

Graph G is the least count of colors be in need to suitably color the nodes so that no two neighboring nodes (connected towers) have equal color and no nodes shares a color with any of its neighbors.

Mobile networks may fail to connect for many reasons, including technical, logistical, or regulatory. Here are some of the primary reasons:

1. Technical issues: Signal interference can occur in densely crowded locations or areas with a high concentration of mobile devices, resulting in dropped connections or poor network performance. During high usage times or in places with insufficient network infrastructure, network congestion may prohibit devices from joining or keeping a steady connection. Faulty equipment or software problems in mobile devices, base stations, or network infrastructure can cause connectivity issues.

2. Geographic Challenges: Geographical characteristics such as mountains, valleys, and dense urban settings can obscure signals and make it harder for mobile devices to communicate with neighboring towers. In remote or rural places, a lack of infrastructure can lead to limited or no network coverage, making it challenging to connect devices.

3. Regulatory Constraints: Regulations on spectrum allocation affect network coverage and connections. Limited spectrum availability may result in crowded frequencies and connectivity concerns. International roaming agreements and legislative variances between nations can impact seamless connectivity when users travel across borders.

4. Service Provider Factors: Service provider maintenance, upgrades, or unexpected outages may disrupt users' connectivity. Service providers may encounter capacity constraints, particularly during peak demand periods, which can result in consumer connectivity issues.

5. User-related issues: Incompatibility between mobile devices and network technologies (e.g., 2G, 3G, 4G, and 5G) might cause connectivity problems. Incorrect network settings, deactivated network services, and SIM card difficulties can all hinder devices from connecting to mobile networks.

6. Security and Privacy Concerns: Strong network security mechanisms, such as firewalls or authentication systems, can occasionally mistakenly block genuine connections, resulting in connectivity concerns. Specific applications or services may have network connectivity issues due to user privacy settings or data access limitations.

3. Observations

- In the context of mobile tower networking, this idea might be read as ensuring that nearby towers run on distinct frequencies to minimize interference and that towers nearby use frequencies that their neighbors do not use.
- To calculate the local-chromatic number for this network structure, we need particular information on the weights of nodes and connections between towers. Parameters such as the number of towers, connection patterns, and interference limits would determine the local-chromatic number.

- Addressing these connection issues frequently requires infrastructure improvements, technological updates, network optimization, legislative changes, and user education and support.
- A complete graph, also known as a completely connected graph, has a unique edge connecting every pair of different nodes, and each node is pairwise adjacent.
- The local antimagic chromatic number of K_n with n nodes. In K_n, all nodes are adjacent. Therefore the local antimagic chromatic number of complete graph $\chi_{la}(K_n)=n$.

In terms of the local antimagic chromatic number, K_n has optimum connectivity, which means that every node is linked to all other nodes, resulting in a local-chromatic number is equal to node-counts in the network.

1. When one or more towers in a mobile network do not link to other towers, the graph structure changes, resulting in isolated or pendent nodes. These isolated nodes indicate towers not part of the linked network and have no direct communication ties with other towers. In contrast, pendent nodes show towers that are connected to any of the one tower.
2. An isolated vertex has a degree of 0, indicating that it does not connect to any other vertex. The isolated vertex creates a disconnected subgraph inside the main graph representing the mobile network. The pendant vertex has a degree of one, indicating that it is next to just one vertex in the graph.
3. Here is the algorithm to solve the issue of isolated towers in a mobile network graph using the local antimagic chromatic number approach:
   I. **Input:** Mobile network graph $G(V,E)$, where V is node-set (towers) and E is edge-set (connections).
   II. **Output:** A labeling scheme for nodes such that each tower has a unique label and satisfies the conditions of the local antimagic chromatic number.
   III. Identify isolated nodes and non-isolated nodes.
   IV. Assign labels to non-isolated nodes:
      – Initialize the edge labels from $\{1,2,3,\ldots,q\}$.
   V. Check conditions:
      - Initialize an empty set sumSet (weight) to store sums of labels and neighbor labels.
      - For each node x, y in V.
      - Get the neighbors edge labels of node v.
   VI. Solution:
      - Calculate the sum $w(x)=\sum_{e\in E(u)} f(e)$
      - $w(x)\neq w(y)$; otherwise, return "No Solution".

This algorithm first identifies isolated nodes and then assigns labels to non-isolated nodes, ensuring no adjacent nodes have the same label. It then assigns labels to isolated nodes and checks the sum conditions required for the local antimagic chromatic number. If the sum conditions are satisfied, output the labelling scheme as the solution; otherwise, it indicates no solution exists.

Problem 3.1. *Construct the mobile network project data for the activities* T_1, T_2, T_3, T_4, T_5, T_6, T_7, T_8, T_9 *and* T_{10} *the connectivity of the network towers are* $T_1 < T_3, T_6$; $T_2 < T_5, T_7, T_9$; $T_3 < T_5, T_6$; $T_4 < T_5, T_7, T_8, T_{10}$; $T_5 < T_6, T_7, T_9$; $T_6 < T_9$; $T_7 < T_9$; $T_8 < T_{10}$.

1. *If a tower connect to at least another two tower, how does this affect the graph structure of mobile networking?*
2. *Solve this issue, while using the method of local antimagic chromatic number?*

Solution: We can build the network diagram using these (network connectivity) links. However, because the exact structure and style of the diagram may differ, I'll offer a textual description of the network:

Step 1:
Network towers are T_1, T_2, T_3, T_4, T_5, T_6, T_7, T_8, T_9 and T_{10}. The edges indicate the connections between activities.

Table 4.1: Mobile tower network connectivity

Network Towers (nodes)		Connectivity Networks (Edges)
T_1	<	T_3, T_6
T_2	<	T_5, T_7, T_9
T_3	<	T_5, T_6
T_4	<	T_5, T_7, T_8, T_{10}
T_5	<	T_6, T_7, T_9
T_6	<	T_9
T_7	<	T_9
T_8	<	T_{10}

Step 2:

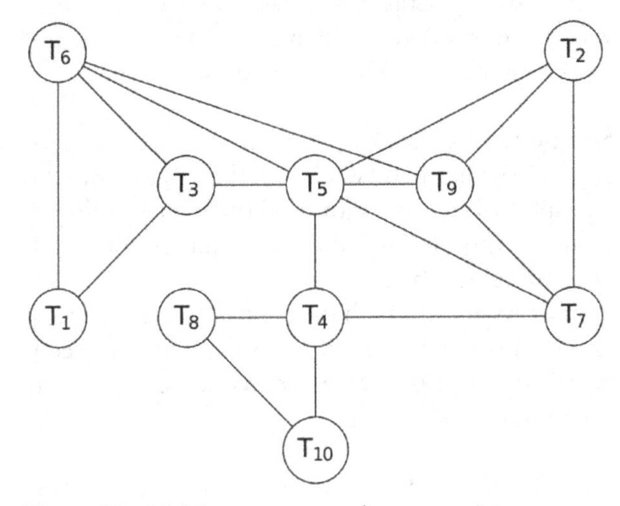

Figure 4.3: Mobile tower network construction

For example, T_1 is connected to T_3 and T_6, T_2 is connected to T_5, T_7, T_9, etc. Now, let's solve the issue using the local antimagic chromatic number.

Step 3: Check the conditions:
Now, we need to check if the labeling scheme satisfies the conditions of the local chromatic number:

- No two adjacent towers should have the same weight.
- For every tower and its neighbors, the weight must be unique.

Step 4: Final Result:
Using the local antimagic chromatic number strategy, we may address the problem by assigning suitable labels to towers depending on their dependencies and ensuring unique label sums for surrounding towers. This method aids in improving frequency allocation in mobile networks to reduce interference and provide optimal communication.

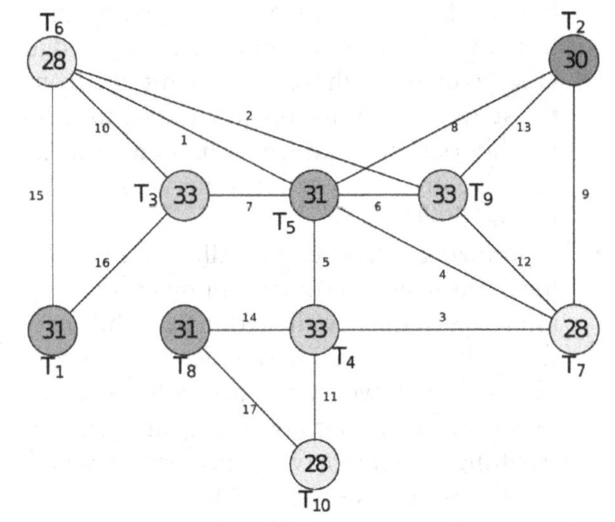

Figure 4.4: LACN for mobile tower network construction

As a result, the local chromatic number of the provided mobile network is four (4) since we chose four colors to ensure that no two neighboring towers share the same color and that no tower shares a color with any of its neighbors.

Indeed, assigning suitable colors (local-chromatic number is 4) to towers based on their connection and providing distinct label sums for surrounding towers is critical for four (4) different optimizing frequency allocation in mobile networks.

Table 4.2: Best optimizing mobile network frequency

Optimizing Frequency	Frequency	Network Towers
1(28)	Yellow	T_6, T_7, T_{10}
2(30)	Blue	T_2
3(31)	Pink	T_1, T_5, T_8
4(33)	Green	T_3, T_4, T_9

3.2. Benefits of this Method

This technique is consistent with the local antimagic chromatic number, which seeks to reduce interference and improve tower communication.

Here's an overview of how the local antimagic chromatic number solution addresses the problem:

- **Distinct Weights:** The primary notion behind the local antimagic chromatic number is to ensure the sum of a tower's edge label and its neighbor's labels differs from the sums of labels for other towers and their neighbors. This uniqueness in sums aids in effective frequency allocation while minimizing interference.
- **Optimizing Frequency Allocation:** The local antimagic chromatic number strategy improves frequency allocation in mobile networks by avoiding interference using suitable labeling. Towers with various frequencies may coexist without producing interference, resulting in better network performance and communication dependability.
- **Efficient Communication:** By optimizing frequency allocation and reducing interference, towers can communicate more efficiently and effectively. This is critical for ensuring stable connections, reducing lost calls and data transfer problems, and improving overall network performance.

4. Satellite Communication

Communication satellite that relays television or radio programming to the viewers directly to the Earth. From Figure 4.3, we know that the communication satellite act as the transmitter from the program channel and the receiver at the viewer's different location on Earth. The transmitter channels are the uplink earth stations and the receivers are the downlink earth stations. The satellite dish or antenna receives the signals at the viewer's location. Whenever there is destruction in the signal due to climate change, we could not get the channels clearly on the television due to the disturbance in the frequency. Nowadays, a small dish with a diameter of less than a meter with frequencies 12–18 GHz perfectly relays programs from the communication satellite. Thus the satellite television network can be used to find the local antimagic chromatic number of graphs.

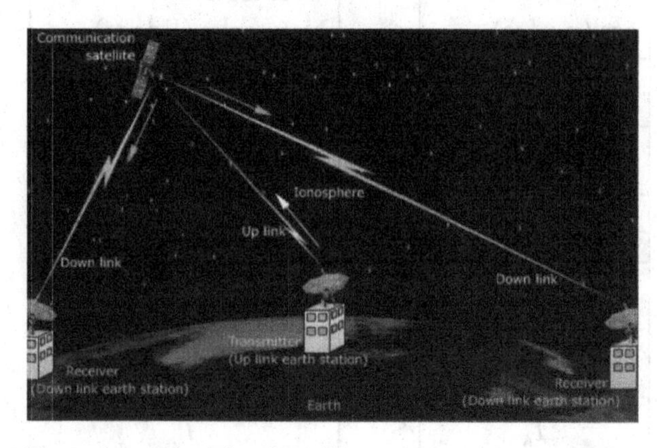

Figure 4.5: Principle of satellite communication

4.1 Problem

Company X runs a television network with n-unique channels. Programs are sent to a satellite and then broadcast to m-different subscribers. The company must assign programs to these channels in such a way that it reduces any potential interference or conflicts between channels for each subscriber.

Solution: To solve the problem of assigning programs to channels while minimizing conflicts for subscribers, we can approach the problem using graph theory and the concept of local chromatic number.

Step 1: Assign each channel program as vertices [1,5] and edges represents a conflict between two vertices. If programs are assigned to two or more channels.

Step 2: We find the local chromatic number for the optimal assignments of program by the channels (colors).

Step 3: Considering the different viewing aspects and patterns of subscribers [6,14], we need to provide the assignments to channels.

The whole setup of the satellite communication is converted to a graph structure. The transmitter channels and receivers at earth stations and the communication satellite are vertices and links of the frequencies are the edges of the graph structure in Figure 4.6.

Step 4: To find LACN:
The graph structure of satellite communication is transformed to a star graph. The local chromatic number of the star graph is $\chi_{la}(S_n) = n+1$. Thus, satellite communication is applying the local chromatic number of star graph. Hence the LACN of satellite communication of the graph G is $\chi_{la}(G) = n+m+1$, where n-unique channels and m-different subscribers.

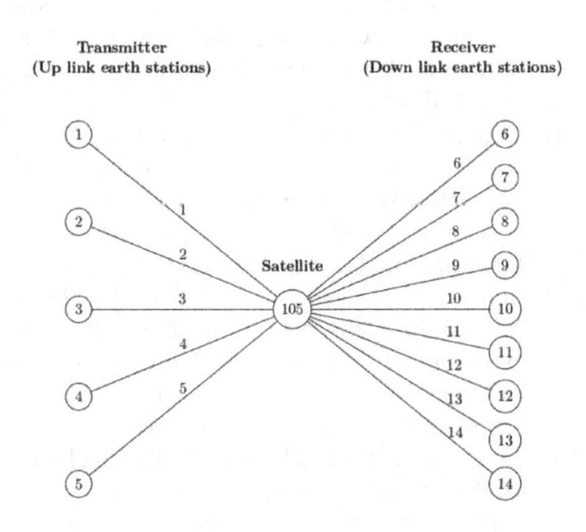

Figure 4.6: LACN for satellite communication

Step 4: Final Result:
The final result is a channel-to-program assignment that reduces the local chromatic number, providing each subscriber with the best possible viewing experience by minimizing interference or conflicts between the channels they access.

4.2 Benefits of this Method

Here are three main benefits of using the local chromatic number method:

1. **Minimized Interference**: By focusing on the local chromatic number, this method reduces the likelihood of conflicts between channels for each subscriber, leading to a clearer and more reliable viewing experience.
2. **Efficient Channel Utilization:** The approach ensures that channels are assigned in a way that maximizes the use of available resources, minimizing unnecessary overlap and making the most efficient use of the network's channel capacity.
3. **Enhanced Subscriber Satisfaction**: With fewer conflicts and smoother program delivery, subscribers are more likely to enjoy uninterrupted service, leading to higher satisfaction and potentially reducing churn rates.

5. Conclusion

The local antimagic chromatic number strategy provides a systematic and efficient solution to frequency allocation problems in mobile networks. This strategy reduces interference, optimizes frequency consumption, and improves communication dependability by allocating tower labels based on their interconnections and ensuring unique sums of labels for adjoining towers. The local antimagic chromatic number strategy helps mobile networks run more effectively by carefully identifying and considering connections, decreasing lost calls and data transfer mistakes, and enhancing overall network performance.

6. Future Scope

Furthermore, while the local-chromatic number sheds light on graph coloring schemes, its applications extend beyond mobile networks to scheduling algorithms, resource allocation, and distributed computing.

7. Acknowledgment

The author thanks our Mathematics Department colleagues at Vinayaka Mission's Research Foundation for their tremendous assistance and insights. I thank our family for their unwavering support and understanding throughout this study.

References

[1] Arumugam, S., Premalatha, K., Martin Baca and Andrea Semanicova-Fecnovcıkova. (2017). Local Antimagic Vertex Coloring of a Graph, Graphs and Combinatorics, 33:275–285.

[2] Bensmail, J., Senhaji, M., and Szabo Lyngsie, K. (2017). On a combination of the 1-2-3 conjecture and the antimagic labeling conjecture, Discrete Math. Theor.Comput. Science, 19(1):22.

[3] Chartrand, G., and Lesniak, L. (2005). Graphs and Digraphs, Chapman and Hall, CRC, 4th edition.

[4] Gallian., J. A. (2020). A dynamic survey of graph labeling, Electron. J. Combinatorics.

[5] Hartsfield., N, Ringel., G. (1994). Pearls in graph theory, Academic Press, INC., Boston.

[6] Haslegrave., J. (2018). Proof of a local antimagic conjecture, Discrete Math. Theorotical Computer Science, 20(1), #18.

[7] Lau., G. C, Ng., H. K. and Shiu., W. C. (2020). Affirmative Solutions on Local Antimagic Chromatic Number, Graphs and Combinatorics 36, 1337–1354.

[8] Marsidi and Agustin., I. H. (2019). The local antimagic on disjoint union of some family graphs, Journal Matematika MANTIK, 5(2), 69–75.

[9] Nalliah., M. and Shankar., R and Tao-Ming Wang. (2022). Local antimagic vertex coloring for generalized friendship graphs, Journal of Discrete Mathematical Sciences and Cryptography, 1–16.

[10] Setiawan and Sugeng., K. A. (2022). Local antimagic vertex coloring of corona product graphs $P_n \circ P_k$, in International Conference on Mathematics, Geometry, Statistics, and Computation (IC-MaGeStiC 2021), Atlantis Press, 65–70.

[11] Shaebani., S. (2020). On Local antimagic Chromatic Number of Graphs, Journal of Algebraic Systems, 7(2), 245–256.

[12] Shankar., R. Nalliah., M. (2022). Local Vertex Antimagic Chromatic Number of Some Wheel Related Graphs, Proyecciones (Antofagasta, On line), 41(1), 319–334.

[13] Shankar., R. Nalliah., M. (2002). Local antimagic chromatic number for the corona product of wheel and null graphs, Vestn. Udmurtsk. Univ. Mat. Mekh. Komp. Nauki, 32(3), 463–485.

[14] Silitonga., M. N. B. and Sugeng., K. A. (2022). Local antimagic vertex coloring of gear graph, International Conference on Mathematics, Geometry, Statistics, and Computation (IC-MaGeStiC 2021), Atlantis Press, 71–75.

[15] Utami., W. Wijaya., K. and Slamin. (2020). Application of the local antimagic total labeling of graphs to optimise scheduling system for an expatriate assignment, Journal of Physics: Conference Series, 1538(1), 13–20.

[16] Yang., X. Bian. Yu., H. and Liu., D. (2021). The local antimagic chromatic numbers of some join graphs, Mathematical and Computational Applications, 26(4), 1–13.

CHAPTER 5

Independent domination number of rooted product of graphs

B. Vasuki[1], S. Geethamalini[2], S. Sangeetha[3], D. Anandhababu[4,a*]

[1]SRM Valliammai Engineering College, Chennai, India

[2]Department of Mathematics, Sathyabama Institute of Science and Technology, Chennai, India

[3]Department of Mathematics, SRM Institute of Science and Technology, Chennai, India

[4]Department of Mathematics, Vel Tech Multi Tech Dr. Rangarajan Dr. Sakunthala Engineering College, Chennai, India

[a]ananddevendiran123@gmail.com*

Abstract

A graph with a designated root vertex is referred to as a rooted graph to distinguish it from other vertices. The rooted product of any two trees always results in a tree. Various graph products exist, such as the Cartesian product, tensor product, and lexicographic product. This article specifically investigates the rooted product of certain tree-related graphs such as star graphs, double star graphs, and binary trees. Additionally, it determines the independent domination number for these graphs.

Keywords: Star graphs, double star graphs, binary trees, independent domination number

1. Introduction

Let G = (V, E) be a basic, undirected graph. "Then V is the vertex set, and E is the edge set. Let S be a subset of V. If S is considered to be dominant set then each vertex in V-S is adjacent to vertex set S. The minimal cardinality of a dominating set in G is known as the dominance number. It is denoted by $\gamma(G)$. If $\alpha(G)$ is called independent set no 2 vertices are adjacent in the set". A set that is both independent and dominant is known as an independently dominant set. The minimum cardinality independent dominating set is called independent domination number. A rooted graph is one in which one vertex is designated as the root vertex, separating it from the other vertices. Any two trees produce a tree as their rooted product. There are numerous graph products, namely cartesian product, tensor product, lexicographic product and so on. This chapter explores the rooted product of some tree related graphs like star graph, double star graph and binary tree. Also, the independent domination number is determined for the above mentioned graphs.

2. Main Results

Definition 1. "Let G be a graph with n vertices, and H be another graph with v as the root vertex. The rooted product of G and H is a graph with one copy of G and n copies of H that connects the vertex ui of G to the vertex v in the ith copy of H for each $1 \leq i < n$."

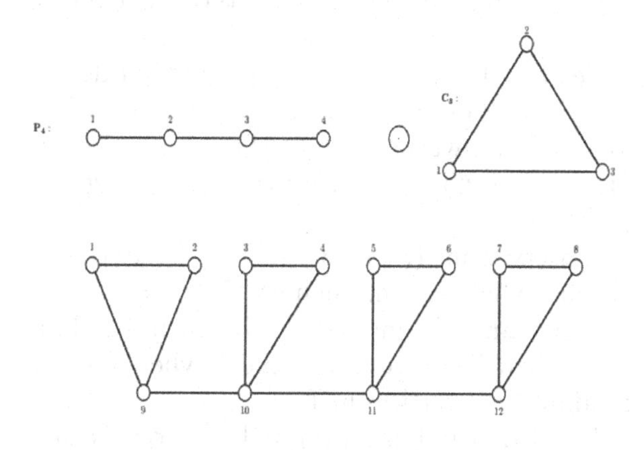

Figure 5.1: Rooted product of C_3 and P_4

DOI: 10.1201/9781003606611-5

Definition 2. [8] "The double star $S_{r,q}$ is the graph obtained by linking two star graphs $K_{1,r}$ and $K_{1,q}$ by an edge to the central vertices."

Theorem 1. [8] "For every graph T, the following are equal. (a) T is a tree (i.e., it is connected and round). (b) T is connected, and m = n minus 1. (c) T is acyclic, with m = n – 1. (d) For every pair of vertices v and w in V (T), there exists a special v - w path."

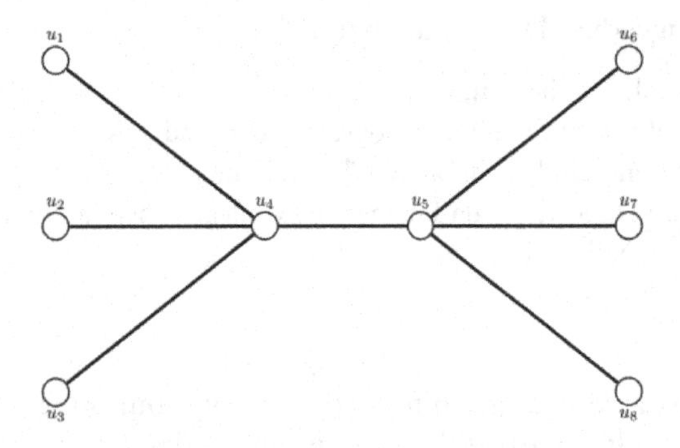

Figure 5.2: Double star graph

Theorem 2. [8] "Every tree of order n ≥ 2 has at least two end vertices."

For the tree T , the vertex v is the most central with respect to the eccentricity measure, namely, rad(T) = ecc(v) = 4 ≤ ecc(u) for every u ∈ V (T).

Theorem 3. [8] "For any tree T with n ≥ 2, there exists a vertex v ∈ V , such that γ(T – V) = γ(T).

Proof. Clearly the result is true if T = K_2. Therefore, assume that T has at least one vertex v with deg(v) ≥ 2 that is adjacent to at least one end vertex and at most one internal vertex. If v is adjacent to two (or more) end vertices v_1 and v_2, then v is in every γ -set for T and γ(T – u_1) = γ(T). If not, then v is adjacent to one endvertex u and deg(v) = 2.

Let T' = T – v – u. For any graph G, if deg(u) = 1, then γ(G – u) ≤ γ(G). Hence, γ(T') ≤ γ(T – u) ≤ γ(T). However, γ(T) ≤ γ(T) – 1. If γ(T) – 1, then γ(T) = γ(T – v). Otherwise, γ(T') = γ(T) = γ(T – u).

Observation 1. Let S_n and S_m be two star graphs, where n ≤ m. Then $i(S_n \odot S_m) = n - 1$.

Theorem 4. If any tree T , i(T) = n – Δ(T) if and only if T is a wounded spider, where n is the total number vertices in T".

Proof that if T is an injured spider, it is easy to demonstrate that i(T) + Δ(T) = n. Assume i(T)+Δ(T) = n, and v ∈ i(T) is a minimum degree vertex in T. If T N[v] = φ, then T is the star K1 with t ≤ 0 (a wounded spider). So assume T N[v] contains at least one vertex. If α(t) is the largest independent set of <T – [v] >, then α(T) ∪ {v} is an independent dominating set for T. Therefore, n = i(T) + Δ ≤ |α(T)| + 1 + Δ(T) ≤ n, indicating that V –N(v) is an independent set. Furthermore, N(v) is an independent set since any edge in N(v) forms a cycle. T is connected to each vertex and other vertices in V – N[v], implying that it is a closed neighbor of at least one vertex in N(v). A cycle forms when a vertex in V – N[v] is close to two or more vertices in N(v). As a result, each vertex in V – N[v] is next to one vertex in N(v). To ensure that Δ(T) + 1 vertices dominate T, there should be not less than one vertex in N(v) that isn't contiguous to any additional vertex in V – N[v], and each vertex in N(v) has either 0 or 1 neighbours in V – N[v]. As a result, T is an injured spider.

Proposition 1. "If Tl is a perfect binary tree, then the independent domination number is,

$$i(T_l) = \sum_{m=2}^{\lceil \frac{l-1}{3} \rceil} 2^{l-m}$$

Proof. Let (l – 1) be the maximum degree vertex stage, it is the maximum number of sequence of vertices adjacent to private neighbor and consider any two stages i and j which have common neighbours where i ≠ 1, j ≠ 1.

Now, consider $\{u_1, u_2, ...u_n\}$ as the set of vertices of perfect binary tree and $\{u_1, u_4, u_{7,...} u_{l-m}\}$ is the independent dominating set of perfect binary tree. This independent dominating set dominates its private neighbours. $i(T_l) = 2^{(l-1)} + 2^{(l-4)} + 2^{(l-7)} + 2^{(l-10)} + ... + 2^{(l-m)}$ where l is the number of stages and m is any positive integer. This is the minimum independent domination number of perfect binary tree."

Then we have $i(T_l) = \sum_{m=2}^{\lceil \frac{l-1}{3} \rceil} 2^{l-m}$

Corollary 1. [8] For any graph G,
γ(G) ≤ i(G) ≤ β_0(G) ≤ Γ(G).

Theorem 5. [8] "If G is a graph containing no induced subgraph isomorphic to either $K_{1,3}$ or the A – L graph, then ir(G) = γ(G) = i(G)."

Theorem 6. "Let Tl be a perfect binary tree with n vertices. If $(T_l \odot T_l)$ is a rooted product of two graphs with n vertices, then the independent domination number is,"

$$i(T_l \odot T_l) = \begin{cases} \left\lceil \frac{n}{3} \right\rceil + \sum_{m=2}^{\frac{l-1}{3}} 2^{l-m}, & \text{if } l = 0 \pmod 3 \\ \left\lceil \frac{2n}{3} \right\rceil + n\sum_{m=2}^{\frac{l-1}{3}} 2^{l-m}, & \text{if } l = 1 \pmod 3 \\ n\sum_{m=2}^{\frac{l-1}{3}} 2^{l-m}, & \text{if } l = 3 \pmod 3 \end{cases}$$

PROOF: The goal of this proof is to identify a separate dominant set with the lowest cardinality. Assume that the set vertices of a perfect binary tree are {u11, u12……u1n}, {u21, u22…… u2n},… {un1, un2……uln}. Attach a perfect binary tree with l stages at each vertex. Let Tl be an ideal binary tree with the following definitions for its vertices: For l phases, the vertex set of Tl is specified. In stage 0, there is only one vertex, u0. There are two vertices in Stage 1: u11 and u12. There are four vertices in Stage 2: u21, u22, u23, and u24. In a similar manner, the stage has two vertices, ul1, ul2,…, ull. There are 2j vertices at each stage j, and there are n vertices in Tl, where |n| = 20 + 21 +…+ 2l.

"Attaching an ideal binary tree of order n to each vertex of a perfect binary tree Tl yields the rooted product Tl ʧ Tl. The jth stage vertices of a perfect binary tree Tl are more numerous than the vertices of (j + 1)th and (j − 1)th stage where j ≠ l and 0."

Case (i): The minimal dominating set is given when l ≡ 0 (mod 3), where all the vertices of the alternate stages of the perfect binary tree are attached to the original graph Tl starting from l − 1, and the vertices of the original graph Tl starting from stage 0. Since these rooted vertices are in charge of themselves,

$$i(T_l \odot T_l) = (2^{(l-1)} + 2^{(l-4)} + 2^{(l-7)} + 2^{(l-10)} + \dots + 2^{(l-m)})$$
$$+ (2^{(l-1)} + 2^{(l-4)} + 2^{(l-7)} + 2^{(l-10)} + \dots + 2^{(l-m)}) + \dots + (2^{(l-1)} + 2^{(l-4)} + 2^{(l-7)} + 2^{(l-10)} + \dots + 2^{(l-m)}) \text{ (n-times)}$$
$$+ u_{l-1} + u_{l-4} + \dots + u_{l-1}$$
$$i(T_l \odot T_l) = n(2^{(l-1)} + 2^{(l-4)} + 2^{(l-7)} + 2^{(l-10)} + \dots + 2^{(l-m)}) + \left\lceil \frac{n}{3} \right\rceil = \left\lceil \frac{n}{3} \right\rceil + n\sum_{m=1}^{\frac{l-1}{3}} 2^{l-m}$$

Case (ii): "The independent dominant set is given by all the vertices of the alternate stages of the perfect binary tree attached to the original path Tl, starting from stage l -1, if q ≡ 2(mod 3). The origin dominates all of the child vertices".

$$i(T_l \odot T_l) = (2^{(l-1)} + 2^{(l-4)} + 2^{(l-7)} + 2^{(l-10)} + \dots + 2^{(l-m)})$$
$$+ (2^{(l-1)} + 2^{(l-4)} + 2^{(l-7)} + 2^{(l-10)} + \dots + 2^{(l-m)}) + \dots + (2^{(l-1)} + 2^{(l-4)} + 2^{(l-7)} + 2^{(l-10)} + \dots + 2^{(l-m)}) \text{ (n-times)}$$
$$= n(2^{(l-1)} + 2^{(l-4)} + 2^{(l-7)} + 2^{(l-10)} + \dots + 2^{(l-m)})$$
$$i(T_l \odot T_l) = n\sum_{m=1}^{\frac{l-1}{3}} 2^{l-m}$$

(i.e) "The rooted vertices also included in dominating set"

Observation 2. If $S_{q,r}$ is the double star graph with $q \le r$, then $i(S_{q,r}) = q$ (Figure 5.3).

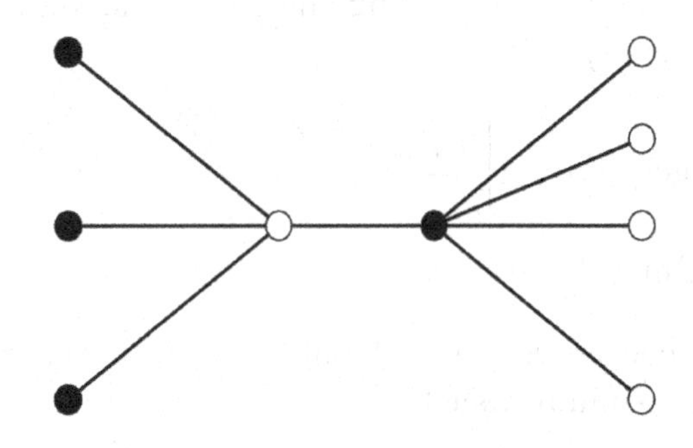

Figure 5.3: Independent domination set of double star graph

Remark 1. Let $S_{n,n}$ be a double star graph, then $i(S_{n,n} \odot S_{n,n}) = \frac{n^2}{2}$.

Theorem 7. If G is Double rooted binary tree of maximum stage q, then

$$i(G) = \begin{cases} 2\left[\frac{1 + 2(2^{q-(3i+1)} - 1)}{3}\right], & \text{for } q = 2m,\ m \in N \\ 2\left[\frac{1 + 4(2^{q-(3i+1)} - 1)}{3}\right], & \text{for } q = 2m - 1,\ m \in N \end{cases}$$

Proof. Consider the q stages namely $\{Q_0, Q_1, Q_2, Q_{q-1}, Q_q\}$ of G at each level, where Q_i $(0 \le i \le q)$ has 2^q vertices. We consider the following cases.

Case (i): For q=2m. The vertices in Q_{q-1} will be dominated by Q_q and Q_{q-2}. Hence $Q_{q-1} \in i(G)$ and Q_{q-4} dominates Q_{q-3} and Q_{q-5}, $Q_{q-4} \in i(G)$.

Continuing the process all the vertices are dominated expect root vertex. So the independent domination number of double rooted binary tree is $i(G) = 2(2^{q-1} + 2^{q-4} + \ldots\ldots 2^1) + 1$. By rearranging and finding geometric sum of the series, we get $2[2^1 + 2^4 + 2^7 + \ldots\ldots + 2^{q-1}]$.

Hence i(G) =

$$2\left[\frac{1+2(2^{q-(3i+1)}-1}{2^{3-1}}\right]=1+2\left[\frac{1+2(2^{q-(3i+1)}-1}{3}\right]$$

for $q = 2m$ for $m \in N$.

Case (ii): For q=2m-1. The vertices in Q_{q-1} will be dominated by Q_q and Q_{q-2} then Q_q , $Q_{q-2} \in i(G)$, which is a maximal independent set and continuing this same way, the independent domination number of the double rooted binary tree is $i(G) = 2[1 + (2^{q-1})2^{q-4} + \ldots\ldots + 2^2)]$. By rearranging and finding the geometric sum of the series,

we get $i(G) = 2\left[\frac{1+2(2^{q-(3i+1)}-1}{2^{3-1}}\right]$ for

$q = 2m-1$, for $m \in N$.

Theorem 8. If G is Double rooted k-ary tree of maximum stage l , then

$$i(G) = \begin{cases} 2\left[\frac{1+2(k^{l-(3i+1)}-1}{3}\right], & \textit{for } l \textit{ is even} \\ 2\left[\frac{1+4(k^{l-(3i+1)}-1}{3}\right], & \textit{for } l \textit{ is odd} \end{cases}$$

Proof. Consider the l stages namely $\{L_0, L_1, L_2, L_{l-1}, L_l\}$ of G at each level. Where L_i $(0 \le i \le l)$ has k^l vertices. We consider the following cases.

Case (i): For l is even the vertices in L_{l-1} will be dominated in L_l and L_{l-2} Stages. Hence $L_{l-1} \in i(G)$, L_{l-4} dominates L_{l-3} , L_{l-5} and then $L_{l-4} \in i(G)$ continuing the process all the vertices are dominated except the root vertex. So the independent domination number of double rooted binary tree is $i(G) = 2(k^{l-1} + k^{l-4} + \ldots\ldots k^1) + 1$. By rearranging and finding geometric sum of the series. $2[k^1 + k^4 + k^7 + \ldots\ldots + k^{l-1}]$.

Hence i(G) =

$$2\left[\frac{1+2(k^{l-(3i+1)}-1}{k-1}\right]=1+2\left[\frac{1+2(k^{l-(3i+1)}-1}{k-1}\right].$$

Case (ii): For l stages the odd vertices in L_{l-1} stage will be dominated by L_{l-1} and L_{l-2} , where L_l, $L_{l-2} \in i(G)$, which is a maximal independent set continuing this same way, the independent domination number of double rooted binary tree is $i(G) = 2[1 + (k^{l-1})k^{l-4} + \ldots\ldots + k^2)]$. By rearranging and finding the geometric sum of the series we get

$$i(G) = 2\left[\frac{1+2(k^{l-(3i+1)}-1}{k-1}\right].$$

Theorem 9. If G is Double rooted binary tree of maximum stage q, then

$$\alpha(G) = \begin{cases} 2\left[\frac{1+2(2^{q-(2i+1)}-1}{2}\right], & \textit{for } q = 2m,\ m \in N \\ 2\left[\frac{1+4(2^{q-(2i+1)}-1}{2}\right], & \textit{for } q = 2m-1,\ m \in N \end{cases}$$

Proof. Consider the q stages namely $\{Q_0, Q_1, Q_2, Q_{q-1}, Q_q\}$ of G at each level. The stage Q_i $(0 \le i \le q)$ has 2^q vertices. We consider the following cases.

Case (i): For q=2m, the vertices in Q_{q-1} is adjacent to Q_q and Q_{q-2}. Hence $Q_q \in \alpha(G)$ and Q_{q-4} is adjacent to Q_{q-2} and Q_{q-4}, then $Q_{q-3} \in \alpha(G)$ continuing the process all the vertices are adjacent to except the root vertex. So the independent number of double rooted binary tree is $\alpha(G) = 2(2^q + 2^{q-2} + \ldots\ldots 2^1) + 1$. By rearranging and finding the geometric sum of the series. $2[2^2 + 2^4 + 2^6 + \ldots\ldots + 2^q]$.

Hence $\alpha(G) = 2\left[\frac{1+2(2^{q-(2i+1)}-1}{2}\right]$, $= 1 +$

$2\left[\frac{1+2(2^{q-(2i+1)}-1}{2}\right]$ for $q=2m$ for $m \in N$.

Case (ii): For q=2m-1, the vertices in Q_{q-1} which is adjacent to Q_q and Q_{q-2} then Q_q , $Q_{q-2} \in$

α (G), which is a maximal independent set and continuing this same way, the independent number of double rooted binary tree is $\alpha(G) = 2[1 + (2^q)2^{q-2} + \ldots\ldots + 2^1)]$. By rearranging and finding the geometric sum of the series we have $\alpha(G) = 2\left[\frac{1+2(2^{q-(2i+1)}-1}{2}\right]$ for q= 2m-1, for m ∈ N.

Theorem 10. If G is Double rooted k-ary tree of maximum stage l , then

$$i(G) = \begin{cases} 2\left[\frac{1+2(k^{l-(2i+1)}-1}{2}\right], & \textit{for } l \textit{ is even} \\ 2\left[\frac{1+4(k^{l-(2i+1)}-1}{2}\right], & \textit{for } l \textit{ is odd} \end{cases}$$

Proof. Consider the l stages namely $\{L_0, L_1, L_2, L_{l-1}, L_l\}$ of G at each level. The stage L_i $(0 \leq i \leq l)$ has k^l vertices. We consider the following cases.

Case (i): For l is even, the vertices in L_l will be adjacent to L_{l-1}. Hence $L_l \in \alpha$ (G) and L_{l-4} is adjacent to L_{l-3} and L_{l-5} then $L_{l-3} \in \alpha$ (G) by continuing the process,all vertices are adjacent to except the root vertex. So the independent number of double rooted k-ary tree is $\alpha(G) = 2(k^l + k^{l-2} + \ldots\ldots k^1) + 1$. By rearranging and finding the geometric sum of the series. $2[k^2 + k^4 + k^6 + \ldots\ldots + k^l]$.

Hence $\alpha(G) = 2\left[\frac{1+2(k^{l-(2i+1)}-1}{k-1}\right] = 1 + 2\left[\frac{1+2(k^{l-(2i+1)}-1}{k-1}\right]$.

Case (ii): For l is odd, the vertices in L_l will be dominated by L_{l-1} then $L_l \in \alpha$ (G), which is a maximal independent set continuing this same way, the independent number of double rooted k-ary tree is $\alpha(G) = 2[1 + (k^l + k^{l-3} + \ldots +k)]$. By rearranging and finding the geometric sum of the series we have

$$\alpha(G) = 2\left[\frac{1+2(k^{l-(2i+1)}-1}{k-1}\right].$$

3. Conclusion

In this article, the independent domination number of the rooted product of graphs were determined. The independent domination number of a rooted product of the perfect binary tree was also examined.

References

[1] Katseff H., Incomplete hyper cubes, IEEE Transactions on Computers 37 (1988) 604–608

[2] Boals A.J., Gupta A.K., Sherwani N.A. , Incomplete hyper cubes: algorithms and embeddings, The Journal of Supercomputing 8(1994)263–294

[3] Harper L.H., Global Methods for Combinatorial Isoperimetric Problems, Cambridge University Press

[4] Kuziak Magdalena, Lema´nska·Ismael G. Yero3 Bull. Malays. Math. Sci. Soc. (2016) 39:199–217. DOI 10.1007/s40840-015-0182-5

[5] Xu M., Topological Structure and Analysis of Interconnection Networks , Kluwer Academic Publishers, 2001

[6] Allan R.B. and LaskarR. On domination and independent domination numbers of a graph. Discrete Math., 23:73–76, 1978

[7] Berge C. Theory of Graphs and its Applications. Methuen, London, 1962

[8] Haynes T.W., Hedetniemi S.T. and Slater P.J. Fundamentals of Domination in Graphs. Marcel Dekker Inc., New York, 1998

CHAPTER 6

Impact of viscosity variation with MHD couple stress fluid between porous rough infinitely long rectangular plates

S. Sangeetha[1,a], S. Geethamalini[2,b], Pragya Pandey[1,c] and D. Anandhababu[4,d]

[1]Department of Mathematics SRM Institute of Science and Technology, Ramapuram
[2]Department of Mathematics, Sathyabama Institute of Science and Technology, Chennai
[3]Department of Mathematics, Vel Tech Rangarajan Dr. Sagunthala R&D Institute of Science and Technology, Chennai
[a]sangeethasekar@yahoo.com, [b]geetha_malini@hotmail.com, pragyapondy@gmail.com,
[d]ananddevendiran123@gmail.com

Abstract

The impact of viscosity variation in the presence of a magnetic field between porous rough infinitely long rectangular plates with couple stress fluid is analyzed. Expression describing the characteristics of the squeeze film are derived. Numerical simulations indicate that roughness on the infinitely long rectangular plates leads to increased pressure accumulation but also adversely affects squeeze film performance due to the effect of negatively skewed roughness pattern in load carrying capacity with film thickness. Moreover, variations in Hartmann number, couple stress parameter and viscosity parameter enhance the effectiveness of squeeze film lubrication compared to the non-viscous classical scenario.

Keywords: Couple stress fluid, surface roughness, porous, infinitely long rectangular plate, viscosity variation

1. Introduction

A number of studies have applied the magneto hydrodynamic (MHD) effect using couple stress fluid to analyze various hydrodynamic lubrication problems (Sangeetha et al., 2022). The study of (Rajashekar and Biradar Kashinath 2012) demonstrates that the influence of MHD between a sphere and a porous plane surface was examined. The findings suggest that in comparison to a solid instance the porosity parameter influences load-carrying capacity and squeeze-film time. (Bujurke and Naduvinamani et al., 2011) explored the existence of roughness in the magnetized field between rectangular plates. A roughness pattern is taken into consideration. According to the findings, time and load carrying efficiency are improved by the negatively skewed roughness. Additionally, in comparison to the non-conductive lubricant the bearing efficiency is enhanced by the magnetic effect represented by the Hartmann number.

The influence of roughness and MHD on plane slider bearing have been analyzed by (Syeda Tasneem Fathima et al., 2016). (Fathima et al., 2014) derives the mathematical solution for rough porous elliptical plates with MHD couple stress fluid. The pressure equation for porous journal bearing is solved using a finite difference technique by (Naduvinamani and Patil 2009). In the literature mentioned above, viscosity was considered to be constant, despite the fact that viscosity changes with temperature and pressure. (Byeon Haewon et al., 2023) analyze Christensen's stochastic theory for rough surfaces to determine the impact of surface roughness, MHD, and viscosity variations on couple stress squeeze film properties of curved circular and flat plates. Henceforth, no research has been accomplished to explore the impact of magnetic fields using couple stress fluid across

DOI: 10.1201/9781003606611-6

infinitely long porous rough rectangular plates with viscosity variation.

2. Mathematical Formulation

The top plate, which has a rough porous surface by nature approaches the bottom plate at a constant velocity. The couple stress fluid is determined as the lubricant in the porous field. (Sparrow et al., 1972) states that the magnetic field B_o is applied at the porous face in the y-direction. In the porous medium's flow is said to follow Darcy's law according to the theory of hydro magnetic lubrication. Because the top plate is squeezing motion towards the bottom plate, the velocity $-\frac{\partial h}{\partial t}$ is negatively oriented.

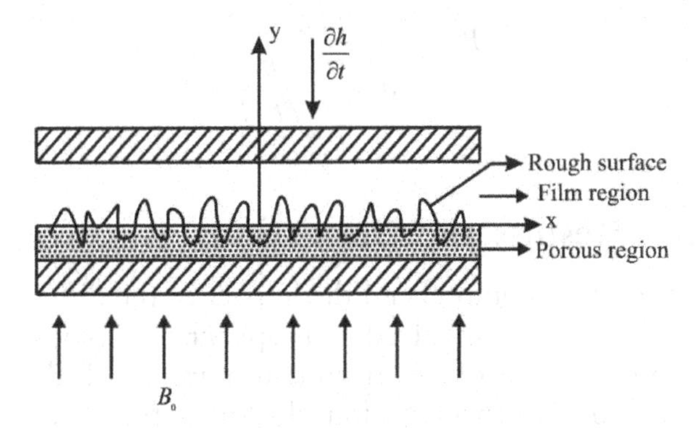

Figure 6.1: Physical representation of the problem

$$\mu\frac{\partial^2 u}{\partial y^2}-\eta\frac{\partial^4 u}{\partial y^4}-\sigma B_o^2 u=\frac{\partial p}{\partial x} \tag{1}$$

$$\frac{\partial p}{\partial y}=0 \tag{2}$$

$$\frac{\partial u}{\partial x}+\frac{\partial v}{\partial y}=0 \tag{3}$$

Darcy's law has been modified to determine the flow of lubricant in porous regions using (Fathima et al., 2014)

$$u^*=\frac{-k}{\mu\left(1-\beta+\dfrac{kM^{*2}}{M'h^{*2}}\right)}\frac{\partial p^*}{\partial x^*} \tag{4}$$

$$v^*=\frac{-k}{\mu\left(1-\beta+\dfrac{kM^{*2}}{M'h^{*2}}\right)}\frac{\partial p^*}{\partial y^*} \tag{5}$$

$$\frac{\partial u^*}{\partial x}+\frac{\partial v^*}{\partial y}=0 \tag{6}$$

For the velocity components, the corresponding boundary conditions are

(i) Across the bottom plate surface $y=0$ (7)

$$u=0,\ \frac{\partial^2 u}{\partial z^2}=0$$

(ii) Across the top plate surface $y=h$ (8)

$$u=0,\ \frac{\partial^2 u}{\partial z^2}=0,\ v=-\frac{\partial h}{\partial t}$$

Solving the equation (1) with respect to the condition (7) and (8) velocity in the x direction is acquired as

$$u=-\frac{h_o^2}{\mu_o e^{\beta p}M^2}\frac{\partial p}{\partial x}\left\{\frac{1}{\left(\zeta_1^2-\zeta_2^2\right)}\left(\frac{\zeta_2^2\cosh\zeta_1\dfrac{(2y-h)}{2l}}{\cosh\left(\dfrac{\zeta_1 h}{2l}\right)}-\frac{\zeta_1^2\cosh\zeta_2\dfrac{(2y-h)}{2l}}{\cosh\left(\dfrac{\zeta_2 h}{2l}\right)}\right)+1\right\} \tag{9}$$

$$\zeta_1=\left\{\frac{1+\left(1-4l^2M^2\right)^{\frac{1}{2}}}{2}\right\}^{\frac{1}{2}};\zeta_2=\left\{\frac{1-\left(1-4l^2M^2\right)^{\frac{1}{2}}}{2}\right\}^{\frac{1}{2}}$$

The continuity equation is obtained by substituting u and integrating. In order to determine the characteristics of the porous squeeze film the Reynolds Equation for the infinitely long rectangular plates has been obtained in a non-dimensional form:

$$\frac{\partial}{\partial x}\left\{\frac{h_o^2}{M^2}\left(\tau(h,\mathrm{L},M)+\frac{k\delta}{D_1^2}\right)e^{-\beta p}\frac{\partial p}{\partial x}\right\}=\frac{dh}{dt} \tag{10}$$

$$\tau(h,\mathrm{L},M)=\frac{2L}{(\zeta_1^2-\zeta_2^2)}\left(\frac{\zeta_2^2}{\zeta_1^2}(\tanh(\zeta_1/2L))-\frac{\zeta_1^2}{\zeta_2^2}(\tanh(\zeta_2 h/2L))\right)+h$$

For the longitudinal one-dimensional roughness, the film thickness assumes the form using (Naduvinamani et al., 2007) $H=h(t)+h_s(x,\xi)$

$$\tau(h_s)=\begin{cases}\dfrac{35}{32n^7}\left(n^2-h_s^2\right)^3 & -n\le h_s\le n\\ 0 & elsewhere\end{cases} \tag{11}$$

$$\alpha = E(h_s) \tag{12}$$

$$\sigma = E\left[(h_s - \alpha)^2\right] \tag{13}$$

$$\varepsilon = E\left[(h_s - \alpha)^3\right] \tag{14}$$

$$E(.) = \int_{-\infty}^{\infty} (.)\,\tau(h_s)\,dh_s \tag{15}$$

Taking the stochastic average of equation (10) with respect to $\tau(h_s)$ we obtain

$$\frac{\partial}{\partial x}\left\{\frac{h_o^2}{M^2}\left(\tau(h,L,M) + \frac{k\delta}{D_1^2}\right)e^{-\beta E(p)}\frac{\partial E(p)}{\partial x}\right\} = \frac{dh}{dt} \tag{16}$$

Integrating the nondimensional modified lubrication equation (16) with respect to x and using the pressure boundary conditions $P = 0$ & $x^* = \pm 1$

$$p = \frac{-1}{G} In\left\{\frac{G}{2\left[\tau(H,L,M,\alpha,\varepsilon,\sigma) + \frac{\psi}{D_1^2}\right]}(1 - x^2) + 1\right\} \tag{17}$$

The load carrying capacity equation in dimensionless form is acquired as

$$E(w) = \int_{-a}^{a} E(p)dx$$

$$W = \left\{\frac{-E(w)h_o^3}{\mu\frac{dh}{dt}}\right\}$$

$$= \int_{-1}^{1} \frac{-1}{G} In\left\{\frac{G}{2\left[\tau(H,L,M,\alpha,\varepsilon,\sigma) + \frac{\psi}{D_1^2}\right]}(1 - x^2) + 1\right\}dx \tag{18}$$

where

$$\tau(H,L,M,\alpha,\varepsilon,\sigma) = \frac{1}{M^2}\left[\frac{L}{(\zeta_1^2 - \zeta_2^2)}\left(\begin{array}{c}\frac{\zeta_2^2}{\zeta_1}\left(\tanh\left(\frac{\zeta_1 H}{L}\right) + g_1\right)\\ -\frac{\zeta_1^2}{\zeta_2}\left(\tanh\frac{\zeta_1 H}{L} + g_2\right)\end{array}\right) + H\right]$$

$$g_1 = \frac{1}{24L^3}\left[1 - \tanh^2\left(\frac{\zeta_1 H}{2L}\right)\right]\left[12L^2\zeta_1\alpha - \zeta_1^3\left(\varepsilon + 3\alpha\sigma^2 + \alpha^3\right)\right]$$

$$g_2 = \frac{1}{24L^3}\left[1 - \tanh^2\left(\frac{\zeta_2 H}{2L}\right)\right]\left[12L^2 E\zeta_2\alpha - \zeta_2^3\left(\varepsilon + 3\alpha\sigma^2 + \alpha^3\right)\right]$$

$$D_1^2 = \left(1 - \beta + \frac{\psi M^2}{\delta M'}\right),\ \psi = \frac{k\delta}{h_0^3},\ \delta^* = \frac{\delta}{h_o},$$

$$H = \frac{h}{h_o}\ L = \frac{l}{h_o},\ x = \frac{x^*}{a},\ G = \frac{\beta\mu_o a^2\left(-\frac{\partial h}{\partial t}\right)}{h_0^3},$$

$$P = -\frac{ph_0^3}{\mu_o a^2\left(\frac{-\partial h}{\partial t}\right)}$$

3. Results and Discussions

In order to comprehend the impact of the different parameters involved in this paper, a graphical representation of the Hartmann number M, the mean of the stochastic film thickness, the standard deviation of the film thickness σ, the permeability parameter ψ, the couple stress parameter L, and a measure of the symmetry of the stochastic random variable ε are provided. The following ranges of the values for these parameters are used in the numerical computations of the results. $\alpha^* = -0.1,\ -0.05,\ 0.05,\ 0.1$; $L = 0.1,\ 0.2,\ 0.3,\ 0.4$; $M = 1 - 5$; $G = 0.01, 0.02, 0.03, 0.04$.

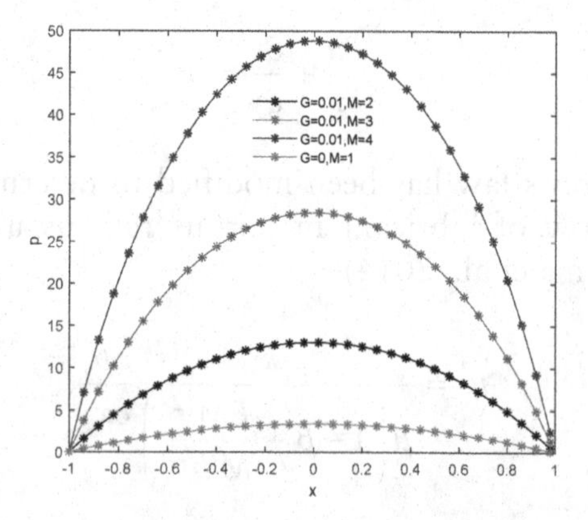

Figure 6.2: Variation with distance x for different values of Hartmann number M on pressure with α = 0.05, β = 0.3, δ^* = 0.01, ψ = 0.01, ε = 0.01, M′ = 0.15, σ = 0.2, L = 0.4, H = 0.4

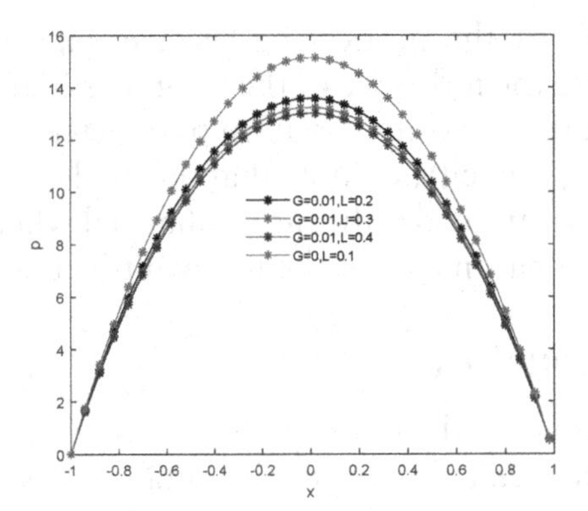

Figure 6.3: Variation with distance x for different values of couple stress parameter L on pressure with $\alpha = 0.05$, $\beta = 0.3$, $\delta^{*} = 0.01$, $\psi = 0.01$, $\varepsilon = 0.01$, $M' = 0.15$, $\sigma = 0.2$, $M = 0.2$, $H = 0.4$

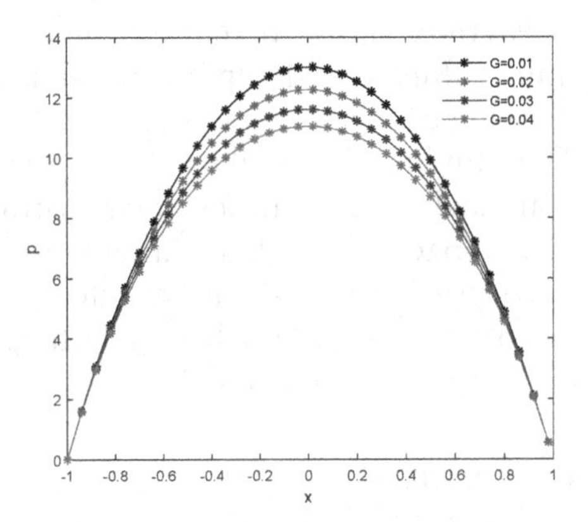

Figure 6.4: Variation with distance x for different values of viscosity variation parameter G on pressure with $\alpha = 0.05$, $\beta = 0.3$, $\delta^{*} = 0.01$, $\psi = 0.01$, $\varepsilon = 0.01$, $M' = 0.15$, $\sigma = 0.2$, $M = 2$, $H = 0.4$, $L = 0.4$

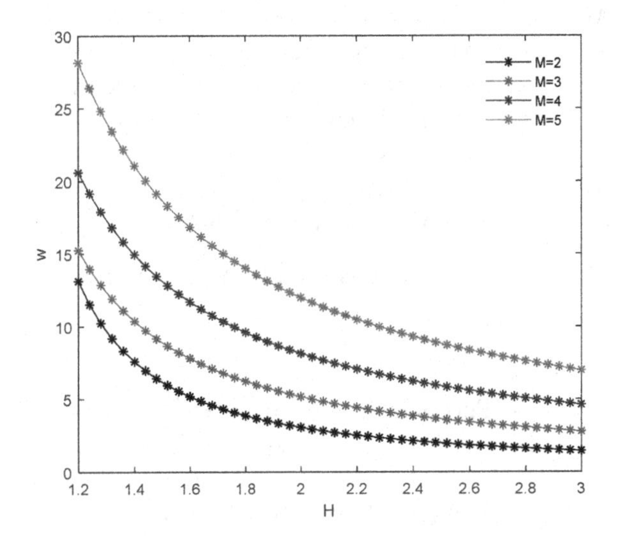

Figure 6.5: Variation with distance x for different values of Hartmann number M on on load carrying capacity w with $\alpha = 0.2$, $\beta = 0.3$, $\delta^{*} = 0.01$, $\psi = 0.01$, $\varepsilon = 0.01$, $M' = 0.15$, $\sigma = 0.2$, $G = 0.01$, $L = 0.1$

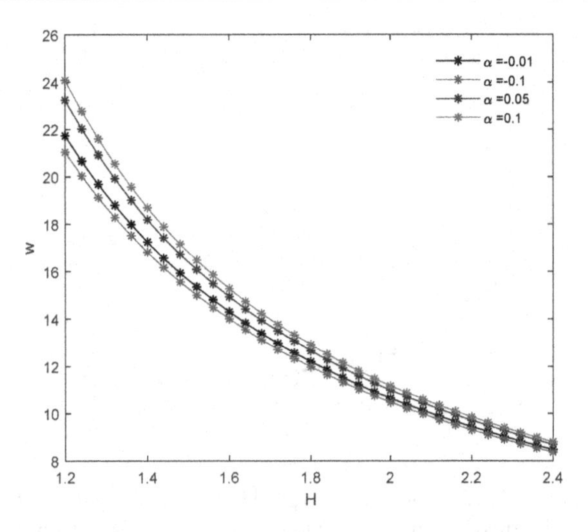

Figure 6.6: Variation with height H for different values of stochastic film thickness α on load carrying capacity w with $M = 4$, $\beta = 0.3$, $\delta^{*} = 0.01$, $\psi = 0.01$, $\varepsilon = 0.01$, $M' = 0.15$, $\sigma = 0.2$, $G = 0.01$, $L = 0.1$

Figure 6.2 describes variation on pressure p with distance in x direction for different values of Hartmann number M. As x increases from -1 to 1 the pressure begins to increase as it goes from $x = -1$ to a limit at the midpoint $x = 0$ and then decreases when it goes to $x = 1$. The pressure p is maximum for Hartmann number *M varying values with viscosity variation parameter.* When $G = 0$ the equation reduces to non -viscous case. An increase in the viscosity variation parameter gradually improves the squeeze film when compared to the non-viscos situation. Figure 6.3 shows how pressure varies with distance in x direction for varying couple stress parameter values. When the couple stress function takes on distinct values, pressure rises. For enhancing viscosity variation parameter values, Figure 6.4 illustrates how pressure varies with distance in x direction. It is demonstrated that as the viscosity variation parameter increases, pressure rises as well.

Figure 6.5 demonstrates how the load carrying capacity W varies with height H for increasing Hartmann number values. The lubricant gets more magnetized as Hartmann number values increases and this interacts with the applied magnetic field. Additionally, in the film region, a magnetic field applied perpendicular to the flow path reduces the fluid's velocity while keeping a substantial amount of fluid which generates the pressure distribution. When the mean of the stochastic film thickness varies, Figure 6.6

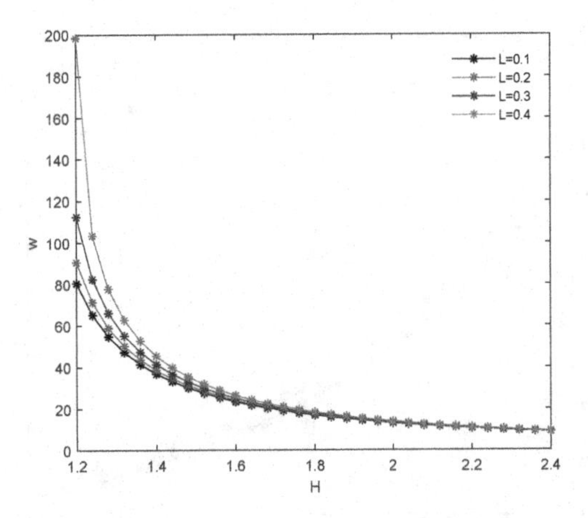

Figure 6.7: Variation with height H for different values of couple stress parameter L on load carrying capacity w with $\alpha = 0.2$, $\beta = 0.3$, $\delta^{*} = 0.01$, $\psi = 0.01$, $\varepsilon = 0.01$, $M' = 0.15$, $\sigma = 0.2$, $G = 0.01$, $M = 2$

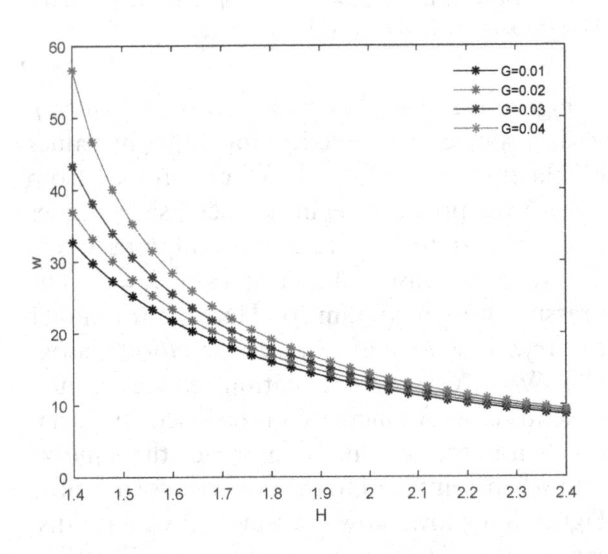

Figure 6.8: Variation with height H for different values of viscosity variation parameter G on load carrying capacity w with $\alpha = 0.2$, $\beta = 0.3$, $\delta^{*} = 0.01$, $\psi = 0.01$, $\varepsilon = 0.01$, $M' = 0.15$, $\sigma = 0.2$, $L = 0.1$, $M = 2$

illustrates the way load carrying capability varies with height. Evidently, the impact of negatively skewed roughness and couple stress parameter leads to an increase in load carrying capacity.

The load carrying capacity changes with height for various couple stress parameter values is displayed in Figure 6.7. Squeeze film behaviors are gradually improved by the couple stress parameter. This is due to the fact that there is less chance of direct contact among the contacting surfaces when the couple stress parameter increases with viscosity and increases the strength of the lubricating surface. Figure 6.8 illustrate the influence of the roughness parameter and the viscosity variation parameter in load carrying capacity. we note that the load-carrying capacity can be significantly enhanced when the viscosity parameter of the lubricant increases.

4. Conclusion

Following are the findings that are derived from numerical calculations of the obtained results.

- For infinitely long rectangular plates, the expression for pressure and load carrying capacity are found using the influence of viscosity variation with MHD and roughness at the permeable interface. This study shows that the couple stress fluid influences the pressure and load carrying capacity by increasing it
- The findings are illustrated through graphical presentations demonstrating that impact of roughness along viscosity variation substantially affect the performance of infinitely long rectangular plates under magnetic field.

Nomenclature

B_0 – Magnetic field

δ – Porous layer thickness

M – Hartmann number = $\left(B_0 h_0 \left(\frac{\sigma}{\mu}\right)^{1/2}\right)$

M' – Porosity

E – Expectancy Operator

α – Mean of the stochastic film thickness

σ – Standard Deviation of the film thickness

σ^2 – Variance

h – Film thickness = $(h + h_s)$

H – Non-dimensional film thickness = $\left(\frac{h}{h_0}\right)$

β – Percolation parameter= $\frac{\eta}{\mu k}$

l – Couple stress parameter = $\left(\frac{\eta}{\mu}\right)^{1/2}$

L – Non-dimensional couple stress parameter

x, y – Cartesian co-ordinates

u, v – Velocity components in film region
u^*, v^* – Modified Darcy velocity components in x, y directions respectively
η – Material constant characterizing couple stress
ψ – Permeability parameter = $\frac{k\delta}{h_0^3}$
μ – Viscosity co-efficient
p – Non-dimensional pressure
W – Non-dimensional load carrying capacity

References

[1] Bujurke, N.M., Naduvinamani, N.B., and Basti, D.P., (2011), "Effect of surface roughness on magneto hydrodynamic squeeze film characteristics between finite rectangular plates" Tribology International, pp. 916–921.

[2] Byeon, Haewon, Y. L. Latha, B. N. Hanumagowda, Vediyappan Goninan, A. Salma, Sherzod Abdullaev, Jagadish V. Tawade, Fuad A. Awwad, and Emad AA Ismail. "Magnetohydrodynamics and viscosity variation in couple stress squeeze film lubrication between rough flat and curved circular plates." Scientific Reports 13, no. 1 (2023): 22960.

[3] Fathima, S.T., Naduvinamani, N.B, and Kumar, J.S., Naganagowda, H.B., (2014), "Derivation of modified Mhd-stochastic Reynolds Equation with conducting couple stress fluid on the squeeze film lubrication of porous rough elliptical plates," International Journal of Mathematical Archive., pp. 135–145.

[4] Naduvinamani, N.B., Basti, D.P., (2009), "Numerical Solution of finite modified Reynolds Equation for couple stress squeeze film lubrication of porous journal bearings," Computers and Structures, pp. 1287–1295.

[5] Naduvinamani, N.B., Siddangouda, A., (2007), "Effect of surface roughness on the hydrodynamic lubrication of porous step-slider bearings with couple stress fluids," Tribology International, pp. 780–793.

[6] Rajashekar, M., and Kashinath, B., (2012), "Effect of Surface Roughness on MHD Couple Stress Squeeze Film Characteristics between a Sphere and a Porous Plane Surface" Advances in Tribology, pp. 1–10.

[7] Sangeetha S, L.Sivakami and S. Lakshmi Priya., Effect of MHD with Micropolar Fluid Between Conical Rough Bearings, International Journal of Heat and Technology., 2022, 40(5), pp. 1210–1216.

[8] Sangeetha S, Govindarajan A. Effect of Magneto Hydrodynamic using Micropolar Fluid between Rough Annular Plates, AIP Conference Proceedings, 2022.

[9] Syeda Thasneem Fathima., Sree kala, C.K., and Hanumagowda, B.N., (2016), "Effect of Surface Roughness on Plane Slider Bearing Lubricated with Couple Stress Fluid in the Presence of Transverse Magnetic Field," International Journal of Engineering Science and Computing., pp. 2120–2126.

[10] Sparrow, E.M., Beavers, G.S., and Hwang, I.T., (1972), "Effect of velocity slip on porous-walled squeeze films. Journal of lubrication technology" pp. 260–264.

CHAPTER 7

Independent and connected domination numbers of Halin graph constructed from increasing perfect k-ary tree

K. Marimuthu[1,a], G. Suresh[1,b], T. Vivekanandan[2], P. Murugaiyan[3] and D. Anandhababu[3,c,*]

[1]Department of Mathematics, Vel Tech High Tech Dr. Rangarajan Dr. Sakunthala Engineering College, Avadi, Chennai, Tamil Nadu, India

[2]Department of Mathematics, Assistant Professor, VSB Engineering college, Karudayambalayam, Karur, Tamil Nadu, India

[3]Department of Mathematics, Vel Tech Multi Tech Dr. Rangarajan Dr. Sakunthala Engineering College, Avadi, Chennai, Tamil Nadu, India

[a]marimuthu.k@velhightech.com, [b]sureshg@velhightech.com, [c]ananddevendiran123@gmail.com

[*]ananddevendiran123@gmail.com

Abstract

In graph theory, a Halin graph, also known as a Hamiltonian graph, can be constructed to exhibit an exponential number of Hamiltonian cycles, a topic that has been extensively studied. Many combinatorial problems become tractable when applied to this specific class of graphs. This article explored a comparative study between the independent domination number and the connected number of Halin graphs.

Keywords: Halin graph, independent domination, connected number

1. Introduction

A Halin graph is also known as a Hamiltonian graph. It is possible to construct a "Halin graph with an exponential number of Hamiltonian cycles which have been studied extensively; many combinatorial problems are easy to solve when restricted to this class of graphs". In this chapter, a comparative study is made between "the independent domination number and independent number of Halin graphs. The independent domination number is a maximal independent set and also a minimal dominating set".

G. Indulal and R. Balakrishnan, studied about Distance spectrum of Indu–Bala product of graphs. A characterization of well covered cubic graphs was discussed by S.R. Campbell, et.al., E.J. Cockayne and C.M. Mynhardt explained Independence and domination in 3 connected cubic graphs. The independent domination number of random regular graphs was studied by W. Duckworth and N.C. Wormald. M. Farber explored independent domination in chordal graphs. N.I. Glebov and A.V. Kostochka explained the independent domination number of graphs with given minimum degree. W. Goddard and M.A. Henning explored Nordhaus-Gaddum bounds for independent domination. Independent domination in hypercubes was discussed by F. Harary and M. Livingston. Also, F. Harary and M. Livingston explained Characterization of trees with equal domination and independent domination numbers.

The A rooted perfect k-array tree is represented by T_q. It is at level 1, and its descendant k_2 vertices are at level 2. There are $k_3$verticesat level 3 "and so on. Let P be the center path of the Halin graph is the vertex of P. Then contains a rooted binary tree. Vertex includes a rooted binary perfect tree, and so on". So covers the perfect rooting k-array tree.

DOI: 10.1201/9781003606611-7

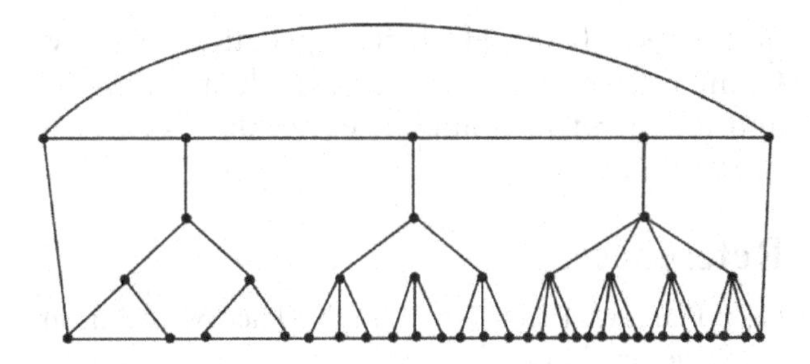

Figure 7.1: Halin graph with increasing arry tree

2. Results

Theorem 1: Haynes et al., 1998

"If G is an isolated free graph with n vertices, then" i(G) ≤ n + 2 - γ(G) - $\left\lceil \frac{n}{\gamma(G)} \right\rceil$.

Theorem 2: Haynes et al., 1998

For any tree T with n ≥ 2 vertices,

i(T) ≤ (n + l(T))/3

and this bound is obtained if and only if T ∈ $\mathcal{F}$.

The property that is equivalent to the property that <S> has no P_2, that is, $\langle S \rangle$ is P_2-free. Forbidden subgraph conditions placed on $\langle S \rangle$ are open avenues for study. For example, Haynes and Henning have extended the property of being P_2-free to being P_3-free. Obviously, this idea can be extended to any subgraph H.

Theorem 3:

The independent domination number is

$$i(H) = \begin{cases} \frac{p}{3} + \sum_{k=2}^{p} \sum_{r=1}^{\frac{q-1}{3}} K^{q-r} & if\ q \equiv 0(mod\,3) \\ P + \sum_{K=2}^{P} \sum_{r=1}^{\frac{q-1}{3}} K^{q-r} & if\ q \equiv 1(mod\,3) \\ 1 + \sum_{k=2}^{p} \sum_{r=1}^{\frac{q-1}{3}} K^{q-r} & if\ q \equiv 2(mod\,3) \end{cases}$$

Proof: Suppose u is a vertex within the central path of H. Assume that the vertices of the tree are in stages q, q1, and q2. T_q Layer l1 dominates these vertices. We can prove that the dominant vertex divides the level of T_q into three groups in the same way. Now we need to find out if these vertices exist. As the result of, the vertex of the intermediate path is dominated by the link lemma, and for the above reasons, we get,

$$|S| = (2^{l-1} + 2^{l-4} + \ldots + 2) + (3^{l-1} + \ldots + 3) + \ldots$$
$$(k^{l-1} + \ldots + k) + \left\lceil \frac{p}{3} \right\rceil$$

$$\therefore \sum_{k=2}^{p-1} \sum_{m=1}^{\left\lceil \frac{l-1}{3} \right\rceil} K^{l-m} + \left\lceil \frac{p}{3} \right\rceil$$

When $l \equiv 1(\text{mod}\,3)$, all internal vertices of the center path are dominated, and the start vertex and end vertex must be combined into a dominance set.

$$p + \sum_{k=2}^{p-1} \sum_{m=1}^{\left\lceil \frac{l-1}{3} \right\rceil} K^{l-m}$$

Theorem 4: For any connected graph G, $\left\lceil \frac{diam(G)+1}{3} \right\rceil$ ≤γ(G).

Proof: Suppose that S is the set γ of the connected graph G. Consider the diameter of the length (G) of any path (G). At most, two edges of the induced subgraph are contained in this path of diameter, each one. Also, since γ, the path of diameter contains at most γ (G)- 1 edge S that connect neighboring vertices. Therefore, diam (G) ≤ γ (G) + 2γ (G)- 1 = 3γ (G)- 1 and the required results are as follows.

Theorem 5: Haynes et.al., 1998

If G theγ (G) ≤ 3, we end this section with the upper limit of the plan with a diameter of 3

Observation 6: If the graph G has no isolated vertices, then

γ(G) ≤ α_0(G),

γ(G) ≤ β_0(G),

γ(G) ≤ β_1(G),

γ(G) ≤ α_1(G).

Here α_0(G) is independent number, β_0(G) is covering number, β_1(G) is matching number, α_1(G) is connectivity number.

Theorem 7: Indulal, G et al., 2016

If H is a Halin graph, then $\left\lceil \frac{n}{1+\alpha_0} \right\rceil$ ≤ i(G) ≤ n - α_0.

Proposition 8: Campbell et al., 1993

Let P_n and C_n be a path graph and cycle graph, respectively, with $n \geq 3$ vertices. Then, $\gamma_c(P_n) = \gamma_c(C_n) = n - 2$.

Theorem 9: "If G is a connected graph and n ≥ 3, then γc (G) = n-εT (G) ≤ n-2.

Proof: Let T be the spanning tree of G, where T (G) is the endpoint, and let L be the set of endpoints. Then T-L is a connected dominating set with n-εT vertices, that is, γc (G) ≤ n-εT (G)". In contrast, "let S be a set. Since is connected, has a spanning tree Ts. The spanning tree T of G is formed by adding the remaining n γc (G) vertices of V-S to Ts and adding the edges of G, so that each vertex in V-S coincides with a vertex in S Adjacent. Now T has at least n γc (G) extremes. Therefore, εT (G) ≥ n γ_c (G) ≥ n-εT (G)".

Kleitman and West studied connected graphs with multi-leaf spanning trees. Their results combined with Theorem 1.

Theorem 10: Every connected graph G andδ (G) ≥ k has at least one spanning tree, which has at least n 3[n / (k + 1)] + 2 leaves.

Corollary 11: F or any connected graph G andδ (G) ≥ k, γc (G) ≤ 3[n / (k + 1)] 2.

Corollary 12: For any connected graph G andδ (G) ≥ 3, γc (G) ≤ 3n / 4- 2.

Corollary 13: For any graph G connected with δ (G) ≥ 4, γc (G) ≤ 3n / 5- 2.

Theorem 14:

If H is a Halin graph which contains a K-ary tree, then γ_c(H) =P$\left[\frac{K^{q-1}-1}{K-1}\right] - q - 2$.

Proof:

Let $V_1, V_2, \ldots, V_{p+2}$ is the set of vertices in central path tree. Then each vertices having the K-ary tree as a branch with l-stages and connect with outer cycle. The pendent vertices of K-ary tree no need to as dominating set that mean l^{th} stage vertices is not including the connected dominating set. Using the proposition (1) then we have remaining vertices is the connected dominating vertices.

3. Conclusion

In this chapter, some special families of the Halin graph is constructed using a perfect K-ary tree. Determine the number of independent dominance from Harlem's special genealogy. Comparative study of independent numbers and independent domination number is carried out in this article.

References.

[1] Harary, F. (1972) Graph Theory, Addison Wesley, Massachusetts.

[2] Haynes, T.W., Hedetniemi, S.T. (1998) and Slater, P.J., "Domination in Graphs: Advanced Topics". Marcel Dekker Inc., New York.

[3] Haynes, T.W., Hedetniemi, S.T. and Slater, P.J. (1998), "Fundamentals of Domination in Graphs". Marcel Dekker Inc., New York.

[4] Indulal, G. and Balakrishnan, R. Distance spectrum of Indu–Bala product of graphs, AKCE Int. J. Graphs Comb. 13 (2016) 230–234.

[5] Campbell, S.R., Ellingham, M.N., and Royle, G.F., A characterization of well covered cubic graphs. J. Combin. Math. Combin. Comput., 13:193–212, 1993.

[6] Cockayne, E.J., and Mynhardt, C.M., Independence and domination in 3 connected cubic graphs. J. Combin. Math. Combin. Comput., 10:173–182, 1991.

[7] Duckworth, W., and Wormald, N.C., On the independent domination number of random regular graphs. Combin. Probab. Comput., 15:513–522, 2006.

[8] Farber, M., Independent domination in chordal graphs. Oper. Res. Lett., 1:134–138,1981/82.

[9] Glebov, N.I., and Kostochka, A.V., On the independent domination number of graphs with given minimum degree. Discrete Math., 188:261–266, 1998.

[10] Goddard, W., and Henning, M.A., Nordhaus-Gaddum bounds for independent domination. Discrete Math., 268:299–302, 2003.

[11] Harary, F., and Livingston. M., Independent domination in hypercubes. Appl. Math. Lett., 6:27–28, 1993.

[12] Harary, F., and Livingston, M. Characterization of trees with equal domination and independent domination numbers. Congr. Numer., 55:121–150, 1986.

[13] Graham, R.L., Pollack, H.O., On the addressing problem for loop switching Bell Syst. Tech. J. (1971), pp. 2495–2519.

[14] Haviland., J., On minimum maximal independent sets of a graph. Discrete Math., 94:95–101, 1991.

[15] Haviland., J., Upper bounds for independent domination in regular graphs. Discrete Math., 307:2643–2646, 2007.

[16] Rosenfeld, M., Independent sets in regular graphs. Israel J. Math., 2:262–272, 1964.
[17] Southey, J., and Henning, M.A., Domination versus independent domination in cubic graphs. To appear in Discrete Math.
[18] Barefoot, C. A., Hamiltonian connectivity of the Halin graphs. Congr. Numer. 58(1987)93–102.
[19] Bondy, J. A., Pancyclic graphs: recent results. In: Infinite and Finite Sets. Colloq. Math. Soc. Ja'nos Bolyai 10, North-Holland, Amsterdam, 1975, pp. 181–187.
[20] Bondy, J. A., and Lova'sz, L., Lengths of cycles in Halin graphs. J. Graph Theory 9(1985) 397–410.

CHAPTER 8

An almost deterministic algorithm for finding the metric dimension and BIGS index of honeycomb torus networks

Libin Bright J.M.[1] and Simon Raj F.[1,a]

[1]Department of Mathematics for Excellence, Saveetha School of Engineering, Saveetha Institute of Medical and Technical Sciences, Saveetha University, Chennai, Tamil Nadu, India

Email: [a]simonrajf.sse@saveetha.com

Abstract

A set $W \subseteq V$ is the resolving set, where, for any two unique vertices u and v of $V \setminus W$, a vertex $w \in W$ exists for which $d(u,w) \neq d(v,w)$. The minimum resolving set W is referred as the metric bases and its cardinality is metric dimension. Honeycomb torus (HCT) networks are extensively used in high-performance computing environments, including supercomputers and large-scale clusters. They enhance parallel processing by linking numerous processors, enabling simultaneous execution crucial for accelerating scientific simulations and computational tasks. In this research, the metric dimension of HCT networks is computed. Also, an almost deterministic algorithm called BIGS (Bharati-Indra-George-Simon) Algorithm is introduced, which extracts all the metric bases and the total number of metric bases of a graph is defined as the BIGS index.

Keywords: Metric basis, metric dimension, honeycomb torus networks, BIGS index

1. Introduction

Graph Metric dimension has been originated from the field of combinatorial optimization. Early in the 1970s, Melter and Harary (1976) independently introduced the idea of metric dimension. They opened the door for more study in this field by examining its characteristics and uses in Graph Theory. This concept is crucial in various applications, such as network navigation, where it aids in efficiently pinpointing locations and simplifying routing algorithms. In robotics, metric dimensions enhance pathfinding and navigation by offering a minimal set of reference points for precise location determination. Furthermore, in chemistry, this concept facilitates the structural analysis of chemical compounds by differentiating between distinct molecular structures.In computational complexity theory, the phrases NP-hard and NP-complete are utilized to categorize problems based on their level of difficulty. Garey and Johnson (1990) provided a comprehensive manual for understanding theory of NP-completeness. The smallest number of vertices needed for which the distance vectors from these vertices to all other vertices in the graph are unique is the metric dimension. It is considered as NP-hard which indicates that, for a variety of graphs, there is no known polynomial-time solution to tackle this issue. Our goal is to determine the minimum resolving set of HCT networks of dimension n. We introduce an almost deterministic algorithm called BIGS Algorithm, which extracts all the metric bases of a graph and the total metric bases is defined as BIGS index.

2. Literature review

The metric dimension concept has been investigated for a variety of networks. Stojmenovic (1997) examined the Honeycomb network's topological characteristics. Approximation strategies for locating resolving sets in trees, grid graphs and non-planar graphs were presented by Khuller *et al.* (1996). The Hamiltonicity of some

DOI: 10.1201/9781003606611-8

generalized honeycomb tori was investigated by Cho *et al.* (2003). Metric dimension problem of honeycomb networks was examined by Paul *et al.* (2008). Sidra Bukhari *et al.* (2024) investigated the metric dimension of Honeycomb rhombic torus. Mudassar *et al.* (2023) studied the resolvability parameters of Honeycomb rectangular torus. Simon *et al.* proposed an almost deterministic algorithm to determine the metric bases of honeycomb, enhanced honeycomb networks of dimension one to six (2023, 2024) and various Hamming graphs (2024). Simon *et al.* (2024) determined the metric dimension of Hex Derived Networks of type one.

3. Honeycomb Torus Networks

The Torus is one of the most common topologies for integrating processors in high performance multicomputers. The honeycomb (HC) network of dimension one is a hexagon. Six hexagons are added to the boundary of the honeycomb network of dimension one to construct the honeycomb network of dimension two. If n indicates the total count of hexagonal layers extending from centre to the outer edge of the honeycomb mesh, then, in general, $HC(n)$ is constructed by incorporating a hexagonal layer to the outer edge of $HC(n-1)$. The honeycomb torus (HCT) is constructed by connecting pairs of vertices of degree two of the honeycomb mesh. The optimal choice for achieving vertex and edge symmetry can be found in pairs of vertices that show mirror symmetry in relation to three lines and traverses the centre of the hexagons and are perpendicular to each of three line orientations. The Honeycomb Torus network of dimension n, denoted by $HCT(n)$ has a total of $6n^2$ vertices and $9n^2$ edges respectively. The diameter is $2n$.

4. BIGS Algorithm

We introduce an almost deterministic algorithm called BIGS Algorithm, which extracts all possible metric bases of a graph and we define the total metric bases of G as BIGS index or BIGS number.

This algorithm involves the following steps:

- Calculate the distance matrix of the given graph G of size N and diameter d.
- If G is a path, then the columns of pendent vertices are unique, therefore the BIGS number of any path is two as there are only two pendent vertices possible in a path.
- Suppose that G is not a path, then we compare the uniqueness of the row vectors (C_{11}, C_{21}) with other row vectors (C_{1k}, C_{2k}), for $2 \le k \le n$. Repeat the same process for all row vectors in column 1 and column 2 that are below it. If (C_{1k}, C_{2k}) row vectors are unique then the corresponding vertex set {1, 2} is a metric basis. If (C_{1k}, C_{2k}) row vectors are not unique, then we repeat this process for other possible pairs of columns.
- If all possible pairs of columns are not unique, then we try the same process with three column comparisons.
- If all possible three column's row vectors are not unique, then we increase the number of columns and repeat the same process.
- Continuing this way, all metric bases can be extracted on comparing all possible $N-d$ columns, where N denotes the total vertices and d is the diameter of the graph G.
- The process of studying the uniqueness of row vectors using minimum number of columns in the distance matrix $D(G)$ is known as BIGS Algorithm. Here we may need to compare a maximum all possible $N-d$ columns in $D(G)$ to extract all Metric Bases of the graph G.
- We obtain the distance matrix of a graph using Matrix Laboratory (MATLAB) software and compute the BIGS number using Java Script.

HCT networks of dimension two and three are shown in Figures 8.1 and 8.2.

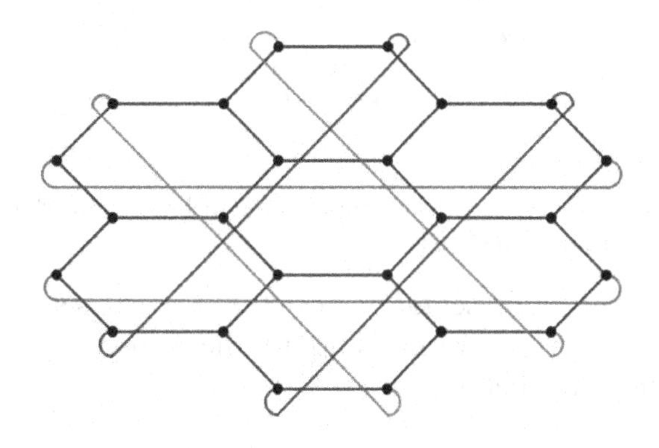

Figure 8.1: HCT network of dimension 2, $HCT(2)$

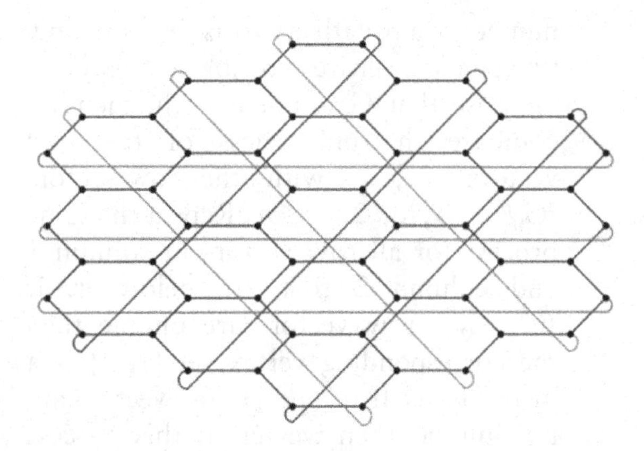

Figure 8.2: HCT network of dimension 3, $HCT(3)$

5. Results

Theorem 1:

The resolving set of Honeycomb Torus Network (HCT) of dimension $n \geq 2$ contains four elements.

Proof:

Let $W = \{c_{n,\,2n-1}, a_{n,\,1}, a'_{n,\,1}, c'_{n,\,2n-1}\}$ be the ordered vertex subset of $HCT(n)$. The symmetric representation of HCT(2) is shown in Figure 8.3.

We claim that the resolving set of $HCT(n)$ is $W = \{c_{n,\,2n-1}, a_{n,\,1}, a'_{n,\,1}, c'_{n,\,2n-1}\}$, which is of cardinality four.

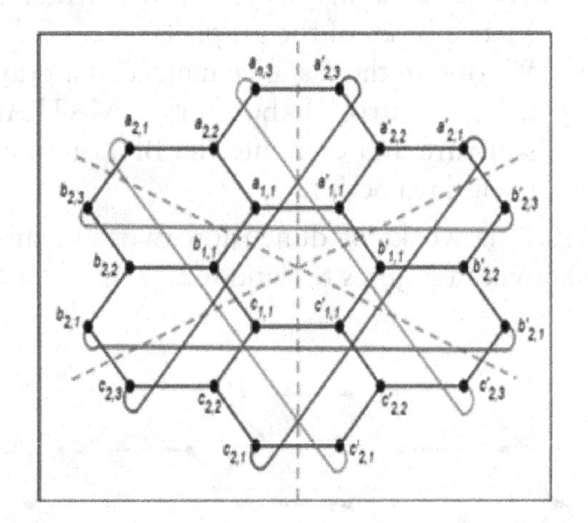

Figure 8.3: Symmetric representation of HCT network of dimension 2, $HCT(2)$

The representation of the vertex set of $HCT(n)$ corresponding to the subset W are indicated below.

For $i = 1, 2, 3, \ldots, n$ and $j = 1, 2, 3, \ldots, 2i - 1$, we have

$$r(b_{i,j} | W) = (2n + j - 2i,\ 2n - j,\ 2n - j + 1,\ 2n + j - 2i + 1)$$

Symmetrically,

$$r(b'_{i,j} | W) = (2n + j - 2i + 1,\ 2n - j + 1,\ 2n - j,\ 2n + j - 2i)$$

For $i = 1, 2, 3, \ldots, n$ and $j = 2i - 1$, we have

$$r(a_{i,j} | W) = (2n - j + 1,\ 2n - 2,\ 2n - 1,\ 2n - j)$$

$$r(c_{i,j} | W) = (2n - 1,\ 2n - j,\ 2n - j + 1,\ 2n - 2)$$

Symmetrically,

$$r(a'_{i,j} | W) = (2n - j,\ 2n - 1,\ 2n - 2,\ 2n - j + 1)$$

$$r(c'_{i,j} | W) = (2n - 2,\ 2n - j + 1,\ 2n - j,\ 2n - 1)$$

For $i = 2, 3, \ldots, n$ and $j = 1, 2, 3, \ldots, 2i - 2$, we have

$$r(a_{i,j} | W) = (2n - j + 1,\ 2n + j - 2i - 1,\ 2n + j - 2i + 2,\ 2n - j)$$

$$r(c_{i,j} | W) = (2n + j - 2i + 2,\ 2n - j,\ 2n - j + 1,\ 2n + j - 2i - 1)$$

Symmetrically,

$$r(a'_{i,j} | W) = (2n - j,\ 2n + j - 2i + 2,\ 2n + j - 2i - 1,\ 2n - j + 1)$$

$$r(c'_{i,j} | W) = (2n + j - 2i - 1,\ 2n - j + 1,\ 2n - j,\ 2n + j - 2i + 2)$$

The position vectors of the vertex set of $HCT(n)$ for $n \geq 2$ are unique.

Thus, the resolving set W of HCT network of dimension $n \geq 2$, has cardinality four.

The results obtained for dimensions two to six are mentioned in the following Table 8.1.

Table 8.1: Metric dimension and BIGS index of HCT networks of dimension two to six

HCT Dimensions	Diameter (Communication Delay)	Metric Dimension	BIGS Index	Time taken for extracting 1st metric basis (sec)
2	4	4	36	<1
3	6	4	243	<1
4	8	4	1728	1
5	10	4	6300	6
6	12	4	25596	13

(Source: Author's compilation)

6. Conclusion

This study reveals that the resolving set of Honeycomb Torus networks of dimension n has four elements. Thus, the metric dimension "of" HCT networks of dimension n is found to be four. All possible metric bases in all such networks from dimension two to six are computed. Metric bases help in designing and optimizing these interconnection networks by ensuring that the communication paths are minimized and that every node can be uniquely identified based on a small set of reference nodes. We work towards finding the metric dimension and all possible metric bases of Extended HCT networks.

References

[1] Abigail, M. V., Libin Bright, J. M., Simon Raj, F., and George, A. (2023). A deterministic algorithm for finding all metric bases of Honeycomb network and Enhanced Honeycomb network of dimension two with O(2^n) time complexity. Int. Conf. Comput. Eng. (ICCE), 308.

[2] Abigail, M. V., and Simon Raj, F. (2024). An algorithmic approach for determining metric dimension of HDN1. Int. Conf. Math., Comput. Eng. Sci. Malaysia, 25.

[3] Abigail, M. V., and Simon Raj, F. (2024). All possible positions of minimum number of satellites for honeycomb mesh and enhanced honeycomb networks of dimension two to six using BIGS Algorithm. Int. Conf. Math., Comput. Eng. Sci. Malaysia, 23.

[4] Bukhari, S., Jamil, M. K., Azeem, M., and Swaray, S. (2024). Honeycomb rhombic torus vertex-edge based resolvability parameters and its application in robot navigation. IEEE 12(1).

[5] Cho, H., and Hsu, L. Y. (2003). Generalized honeycomb torus. Inform. Process. Lett. 86(2):185–190.

[6] Dong, Q., Zhao, Q., and An, Y. (2015). The hamiltonicity of generalized honeycomb torus networks. Inform. Process. Lett. 115(2):104–111.

[7] Garey, M. R., and Johnson, D. S. (1990). Computers and Intractability: A Guide to the Theory of NP-Completeness. W. H. Freeman & Co.

[8] Harary, F., and Melter, R. A. (1976). On the metric dimension of a graph. Ars Comb. 2:191–195.

[9] Khuller, S., Raghavachari, B., and Rosenfeld, A. (1996). Landmarks in graphs. Discrete Appl. Math. 70(3):217–229.

[10] Libin Bright, J. M., and Simon Raj, F. (2024). An almost deterministic algorithm for finding all possible metric bases for Hamming Graphs $H_{3,4}$, $H_{4,3}$, $H_{2,6}$ and $H_{6,2}$. Proc. Int. Conf. Math., Comput. Eng. Sci. Malaysia, 24.

[11] Manuel, P., Rajan, B., Rajasingh, I., and Monica, C. (2008). On minimum metric dimension of honeycomb networks. J. Discret. Algorithms 6(1):20–27.

[12] Nazar, M., Azeem, M., and Jamil, M. K. (2023). Localization of honeycomb rectangular torus. Mol. Phys. 122(1):1–11.

[13] Stojmenovic, I. (1997). Honeycomb networks: Topological properties and communication algorithms. IEEE Trans. Parallel Distrib. Syst. 8(10):1036–1042.

CHAPTER 9

Advanced approximation techniques for fractional Van der Pol oscillators using Lucas wavelets

R. Rajaraman

[1]Department of Mathematics, Saveetha Engineering College, Chennai, Tamil Nadu
Email: rajaramanr@saveetha.ac.in

Abstract

In this study, the Lucas wavelet method (LWM) was effectively employed to approximate the analysis of the Van der Pol equation of fractional order. This equation generalizes the conventional Van der Pol oscillator by incorporating fractional calculus to model memory and hereditary properties in dynamical systems. The modified Riemann-Liouville fractional derivative was used to enhance its computational efficiency. The effectiveness of the LWM was validated by comparing its results with those obtained from the Runge-Kutta fourth-order method, demonstrating its superior accuracy and reliability. The results highlight the potential of LWM for solving complex fractional differential equations in various scientific and engineering applications.

Keywords: Lucas wavelet method, nonlinear oscillators, Van der Pol damped nonlinear oscillator of fractional order, modified Riemann-Liouville derivative, operational matrices of fractional derivative

1. Introduction

Van der Pol oscillator of fractional order is a variation where the derivatives in the equation are replaced with fractional derivatives. This additional complexity allows fractional Van der Pol oscillators to capture memory effects in real-world systems better, leading to more complex behaviours like chaos and multistability compared to their integer-order counterparts. The study of these oscillators of fractional order is an active area of research due to their potential for more accurate modelling in various fields. Numerous researchers [1–6] have explored the analysis of fractional-order Van der Pol oscillators using various semi-analytical methods.

Van der Pol employed this type of nonlinear oscillator in the 1920s to investigate oscillations in vacuum tube circuits, which are the components of early radios. The standard form is expressed as a second-order nonlinear differential equation of the form

$$\frac{d^2\phi}{d\tau^2}+\alpha\left(\phi^2-1\right)\frac{d\phi}{d\tau}+\phi=0 \tag{1}$$

x is the displacement or the position variable? t is the time variable. a is a parameter that controls the nonlinearity and the damping of the system. $\frac{d^2\phi}{d\tau^2}$ represents the acceleration with respect to time.is the nonlinear damping term. For small values of a, the system behaves similarly to a simple harmonic oscillator with slight damping. As a increases, the nonlinearity becomes more pronounced, and the system can exhibit inherent oscillations. This is the unforced Van der Pol oscillator describes a system with nonlinear damping and natural oscillations without any external driving force.

Van der Pol introduced a modified version of the aforementioned equation by adding a periodic forcing term

DOI: 10.1201/9781003606611-9

$$\frac{d^2\phi}{d\tau^2} + \alpha\left(\phi^2 - 1\right)\frac{d\phi}{d\tau} + \phi = Asin\omega\tau \quad (2)$$

ω is the oscillating frequency, A is the strength of the damping force. This is the forced Van der Pol oscillator includes an external driving force that influences the system's dynamics, adding complexity to the behaviour of the oscillator. The system's response depends on the frequency and amplitude of the external force.

The beginning conditions are given as

$$\phi(0) = a, \phi'(0) = 0 \quad (3)$$

It may be useful to investigate changes to the basic Van der Pol equation in which the derivative of the dependent variable x happens to a fractional power.

Mickens [4] investigated the following fractional Van der Pol oscillator equation

$$\frac{d^2\phi}{d\tau^2} + \alpha\left(\phi^2 - 1\right)\left(\frac{d\phi}{d\tau}\right)^{1/3} + \phi = 0 \quad (4)$$

The forced Van der Pol oscillator with fractional derivative is given by

$$\frac{d^2\phi}{d\tau^2} + \alpha\left(\phi^2 - 1\right)\left(\frac{d\phi}{d\tau}\right)^{1/3} + \phi = Asin\omega\tau$$

The term $\left(\frac{d\phi}{d\tau}\right)^{1/3}$ means the system exhibits fractional-order dynamics. Fractional-order differential equations describe systems with memory and hereditary properties, which can be used to model more complex and realistic physical phenomena than integer-order models. Due to the memory effect, the system's response to perturbations is more complex. It can exhibit intricate oscillatory patterns, including long-range temporal correlations and potentially more complex limit cycles or chaotic behaviour. Many real-world systems, such as viscoelastic materials, biological tissues, and anomalous diffusion processes, exhibit behaviours that are better captured by fractional-order models due to their memory and hereditary properties. Hereditary properties refer to the system's behaviour being influenced by its entire history. This is a key characteristic of systems modelled by Fractional Differential Equations (FDEs). These properties are crucial in materials and processes where the current state is significantly affected by all previous states, not just the immediate past.

Wavelet methods are widely recognized for their effectiveness in solving boundary value problems (BVPs). Their ability to handle substantial computational complexity has made them increasingly popular in this field [7–9]. Recent advancements in this domain highlight several significant contributions. In order to solve nonlinear and fractional differential equations with higher computational complexity, Rajaraman and Hariharan [10–13] introduced efficient wavelet-based spectrum approaches. Their work demonstrates the potential of wavelet techniques in enhancing the accuracy and efficiency of solving complex mathematical problems. Rajaraman [14] demonstrated the efficacy of mathematical analysis using wavelet techniques in studying the behaviour of immobilized enzymes within porous catalysts governed by nonlinear Michaelis–Menten kinetics, showcasing the potential of wavelet methods in handling nonlinear biochemical reactions. Additionally, Rajaraman [15] introduced an efficient wavelet algorithm for the fractional view analysis of the Bagley-Torvik equation, further establishing the utility of wavelets in addressing fractional differential equations. Furthermore, Faheem et al. [9] have developed a highly effective high-resolution method utilizing Hermite wavelets for solving space-time-fractional partial differential equations (STFPDE). This method offers significant improvements in computational efficiency and solution accuracy for fractional Partial Differential Equations (PDEs), which are essential in modelling various physical and engineering processes. Ankit Kumar [16] introduced a new technique using a derivative operational matrix based on the Lucas wavelets basis. This method effectively addresses Lane-Emden type differential equations of second order.

Wavelet methods are increasingly being used to address problems with multi-scale features and varying degrees of smoothness. The adaptive nature of wavelets allows for localized analysis and efficient representation of functions, making them applicable to a broad range of applications. The development of new wavelet bases and improved algorithms continues to expand the applicability of wavelet methods in scientific computing, providing reliable tools for tackling complex BVPs and advancing the field of numerical analysis. The fundamental principle of LWM is to use the operational matrices of integral or derivative to convert differential equations to a system of algebraic equations.

2. Investigation of Wavelets

2.1 Lucas Wavelets

Lucas wavelets defined in the interval [0 2) is given by

$$\psi_{p,q}(t)=\begin{cases}2^{\frac{k+1}{2}}\widetilde{L_q^*}\left(2^{k+1}it-\widehat{pi}\right), \frac{\hat{p}}{2^{k+1}}\le t\le\frac{\hat{p}+2}{2^{k+1}}\\ 0, \textit{otherwise},\end{cases} \quad (6)$$

$$\text{Where } \widetilde{L_q^*}=\begin{cases}\frac{1}{\sqrt{\pi}}, & q=0\\ \sqrt{\frac{2}{\pi}}L_q^*(t), & q>0\end{cases} \quad (7)$$

2.2 Operational Matrices of Derivatives of Lucas Wavelets

$$D=\begin{bmatrix}0&0&0\\ i\sqrt{2}&0&0\\ 0&4i&0\end{bmatrix}, D^2=\begin{bmatrix}0&0&0\\ 0&0&0\\ -4\sqrt{2}&0&0\end{bmatrix}$$

Then the Lucas wavelet matrix is

$$\psi(t)=\frac{2}{\sqrt{\pi}}\begin{bmatrix}\sqrt{2}\\ 2it-2i\\ -4t^2+8t-2\end{bmatrix}$$

2.3 Lucas Wavelet Operational Matrices of Fractional Derivatives

By applying the modified Riemann-Liouville fractional derivative to the Lucas wavelet matrix $\psi(t)$ the operational matrices for fractional derivatives are derived as follows:

$$D^{\frac{1}{2}}\psi(t)=\frac{2}{\sqrt{\pi}}\begin{pmatrix}0.8t^{-\frac{1}{2}}\\ 2.26it^{\frac{1}{2}}-1.13it^{-\frac{1}{2}}\\ -6.02t^{\frac{3}{2}}+9.03t^{\frac{1}{2}}-1.13t^{-\frac{1}{2}}\end{pmatrix}$$

$$D^{\frac{1}{3}}\psi(t)=\frac{2}{\sqrt{\pi}}\begin{pmatrix}1.04t^{-\frac{1}{3}}\\ 2.22it^{\frac{2}{3}}-1.48it^{-\frac{1}{3}}\\ -5.32t^{\frac{5}{3}}+8.88t^{\frac{2}{3}}-1.48t^{-\frac{1}{3}}\end{pmatrix}$$

3. Numerical Experiments

3.1 Unforced Fractional Van der Pol Oscillator

Case 1 At $a = 0.01$, the Eq. (4) becomes

$$\frac{d^2\phi}{d\tau^2}+0.01\left(\phi^2-1\right)\left(\frac{d\phi}{d\tau}\right)^{1/3}+\phi=0 \quad (8)$$

under initial conditions

$$\phi(0)=a, \phi'(0)=0 \quad (9)$$

Eq. (8) with the starting conditions is solved by applying the LWM to reach the required solution x(t)

The stated connection coefficients are

$$\mathrm{E}=\frac{\sqrt{\pi}}{2}\left(\mathrm{E}_{0,0},\mathrm{E}_{0,1},\mathrm{E}_{0,2}\right)^T=\frac{\sqrt{\pi}}{2}\left(\mathrm{E}_0,\mathrm{E}_1,\mathrm{E}_2\right)^T \quad (10)$$

then the Lucas wavelet scheme is stated as

$$E^T D^2\Psi(t)+0.01\left(E^T(\Psi(t))^2-1\right)E^T D^{\frac{1}{3}}\Psi(t) + E^T\Psi(t)=0 \quad (11)$$

Selecting the collocation point x=0.5 and employing the operational matrices of fractional derivatives in the Lucas wavelet framework, we obtain the following algebraic equation

$$-8E_2+0.01\left(\left(\sqrt{2}E_0-iE_1+E_2\right)^2-1\right)(1.31E_0-0.47iE_1+2.04E_2)+\left(\sqrt{2}E_0-iE_1+E_2\right)=0 \quad (12)$$

Using boundary conditions, we gain

$$\sqrt{2}E_0-2iE_1-2E_2=1 \quad (13)$$

$$2iE_1+8E_2=0 \quad (14)$$

On solving these simultaneous algebraic equations

$$E_0=0.2372,\ E_1=0.4438i,\ E_2=0.110\S$$

We get the Lucas wavelet solution by utilizing the above connection coefficients as

$$\varphi(\tau)=-0.4436\tau^2+1 \quad (15)$$

Case 2 For a = 1

The Lucas wavelet solution is

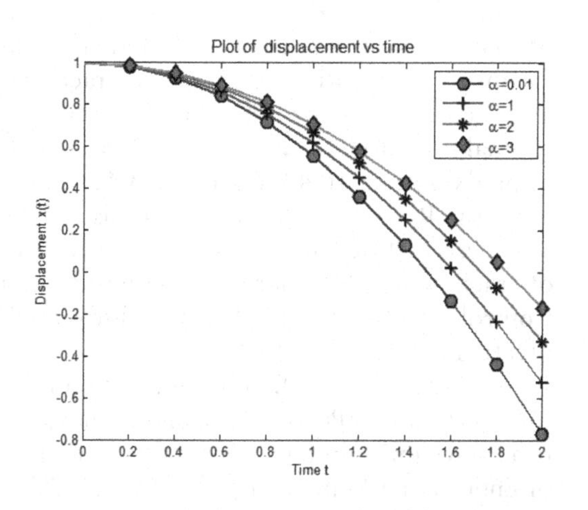

Figure 9.1: Plot of displacement vs time in an unforced fractional Van der Pol oscillator

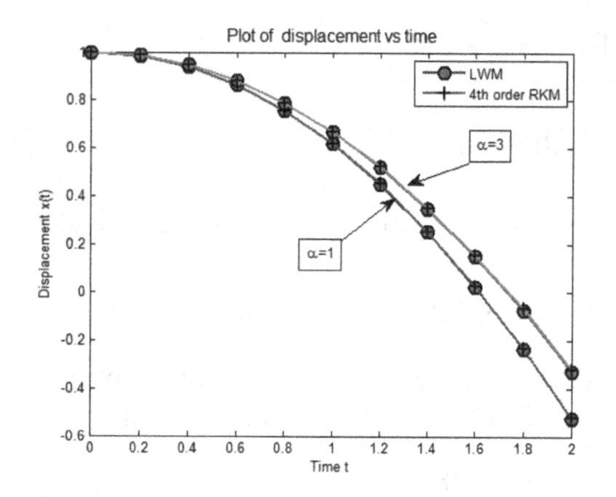

Figure 9.2: Comparison of LWM and 4th order RKM for unforced fractional Van der Pol oscillator

$$\varphi(\tau)=-0.382\tau^2+1 \quad (16)$$

Case 3 For a = 2

The Lucas wavelet solution is

$$\varphi(\tau)=-0.3324\tau^2+1 \quad (17)$$

Case 4 For a = 3,

The Lucas wavelet solution is

$$\varphi(\tau)=-0.2928\tau^2+1 \quad (18)$$

3.2 Forced Fractional Van der Pol Oscillator

Case 1 At $\alpha=1, A=1.2, \omega=0.5$ the Eq. (5) becomes

$$\frac{d^2\phi}{d\tau^2}+\alpha\left(\phi^2-1\right)\left(\frac{d\phi}{d\tau}\right)^{1/3}+\phi=1.2\sin(0.5)\tau \quad (19)$$

under initial conditions

$$\phi(0)=a, \phi'(0)=0 \quad (20)$$

Selecting the collocation point x=0.5 and employing the operational matrices of fractional derivatives in the Lucas wavelet framework, we obtain the following algebraic equation.

$$-8E_2+\left(\left(\sqrt{2}E_0-iE_1+E_2\right)^2-1\right)(1.31E_0-0.47iE_1+2.04E_2)+\left(\sqrt{2}E_0-iE_1+E_2\right)=1.2\sin(0.5) \quad (21)$$

Using boundary conditions, we gain

$$\sqrt{2}E_0 - 2iE_1 - 2E_2 = 1 \tag{22}$$

$$2iE_1 + 8E_2 = 0 \tag{23}$$

On solving these simultaneous algebraic equations

$$E_0 = 0.5402,\ E_1 = 0.1589i,\ E_2 = 0.0397$$

We get the Lucas wavelet solution by utilizing the above connection coefficients as

$$\varphi(\tau) = -0.1588\tau^2 + 1 \tag{24}$$

Case 2 For $a = 2$
The Lucas wavelet solution is

$$\varphi(\tau) = -0.1368\tau^2 + 1 \tag{25}$$

Case 3 For $a = 3$
The Lucas wavelet solution is

$$\varphi(\tau) = -0.1196\tau^2 + 1 \tag{26}$$

Case 4 For $a = 4$,
The Lucas wavelet solution is

$$\varphi(\tau) = -0.1064\tau^2 + 1 \tag{27}$$

4. Conclusion

In conclusion, this work demonstrates the efficacy of the LWM in approximating solutions for the Van der Pol equation of fractional order. By integrating fractional calculus, this method effectively models the memory and hereditary properties inherent in dynamical systems.

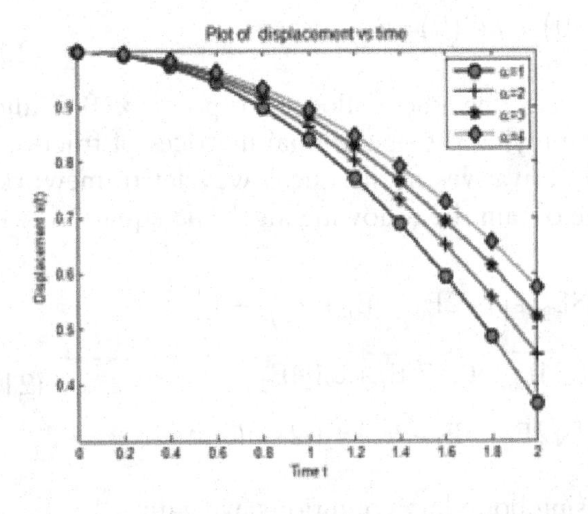

Figure 9.3: Plot of displacement vs time in a forced fractional Van der Pol oscillator

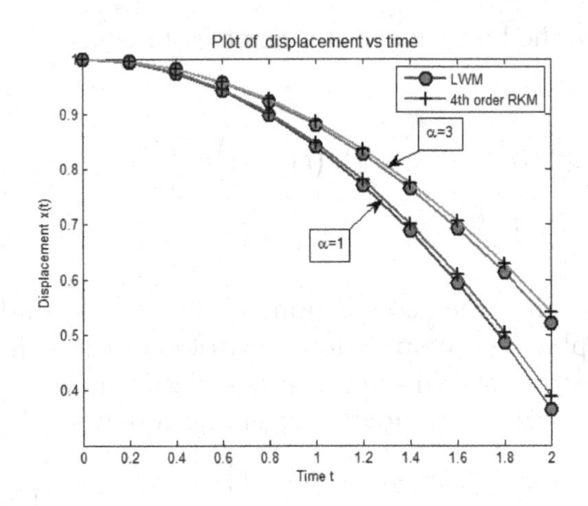

Figure 9.4: Comparison of LWM and 4th order RKM for forced fractional Van der Pol oscillator

Utilizing the modified Riemann-Liouville fractional derivative has further enhanced computational efficiency. The comparative analysis with the Runge-Kutta fourth-order method highlights the superior accuracy and reliability of LWM. These results highlight the efficacy of the Lucas wavelet approach in addressing intricate fractional differential equations, establishing it as a valuable asset for various scientific and engineering fields.

References

[1] Kavyanpoor, M., Shokrollahi, S. (2017). Challenge on solutions of fractional Van Der Pol oscillator by using the differential transform method. Chaos, Solitons & Fractals, 98, 44–45. https://doi.org/10.1016/j.chaos.2017.03.028.

[2] Mishra, V. et al. (2016). Study of fractional order Van der Pol equation. Journal of King Saud University – Science, 28, 55–60.

[3] Nofal, T.A. et al. (2013). Application of Homotopy Perturbation Method and Parameter Expanding Method to Fractional Van der Pol Damped Nonlinear Oscillator. Journal of Modern Physics, 4, 1490–1494. http://dx.doi.org/10.4236/jmp.2013.411179.

[4] Barbosa, R.S. et al. (2007). Analysis of the Van der Pol Oscillator containing derivatives of fractional order. Journal of Vibration and Control, 13(9–10), 1291–1301. https://doi.org/10.1177/1077546307077463.

[5] Juárez, G. et al. (2023). Hopf bifurcation for a fractional van der Pol oscillator and applications to aerodynamics: implications in flutter. Journal of Engineering Mathematics,139:1. https://doi.org/10.1007/s10665-023-10258-7.

[6] Ruth Isabels et al. Evaluating and Ranking Metaverse Platforms Using Intuitionistic Trapezoidal Fuzzy VIKOR MCDM: Incorporating Score and Accuracy Functions for Comprehensive Assessment, Decision Making: Applications in Management and Engineering, 7, No 1 (2024), 54–78.

[7] Doha, E. H., Abd-Elhameed, W. M., Youssri, Y. H. 2013. Second kind Chebyshev operational matrix algorithm for solving differential equations of Lane-Emden type. New Astronomy. 23–24: 113–117.

[8] Angadi, L. M. 2021. Numerical Solution of Singular Boundary Value Problems by Hermite Wavelet Based Galerkin Method. Annals of Pure and Applied Mathematics. 23(2): 101–110.

[9] Faheem, M., Khan, A., Raza, A. 2022. A high resolution Hermite wavelet technique for solving space–time-fractional partial differential equations. Mathematics and Computers in Simulation. 194(C): 588–609.

[10] Rajaraman, R., Hariharan, G. (2015). An efficient wavelet based spectral method to singular boundary value problems. J. Math. Chem, 53, 2095–2113. doi:10.1007/s10910-015-0536-0.

[11] Rajaraman, R., Hariharan, G. (2014). An efficient wavelet based approximation method to gene propagation model arising in population biology. Journal of Membrane Biology. 247, 561–570. doi:10.1007/s00232-014-9672-x.

[12] Rajaraman, R., Hariharan, G. (2023). Estimation of roll damping parameters using Hermite wavelets: An operational matrix of derivative approach. Ocean Engineering, 2023, 283, 115031. https://doi.org/10.1016/j.oceaneng.2023.115031.

[13] Rajaraman, R., Hariharan, G. (2023). A New Wavelet Collocation Algorithm for Solving a Nonlinear Boundary Value Problem of the Human Corneal Shape. Nonlinear Dynamics, Psychology, and Life Sciences, 27(4), 381–395.

[14] Rajaraman, R. (2024.) Waveletbased mathematical analysis of immobilized enzymes in porous catalysts under nonlinear Michaelis–Menten kinetics, Journal of Mathematical Chemistry, 62, 425–460. https://doi.org/10.1007/s10910-023-01548-7.

[15] Rajaraman, R. (2024). An efficient wavelet algorithm for the fractional view analysis of Bagley-Torvik equation. International Journal of Applied Mathematics, 37(1), 29–36. doi:http://dx.doi.org/10.12732/ijam.v37i1.3.

[16] Ankit Kumar (2022). A Computational Derivative Operational Matrix Technique for Solving Second-Order Lane-Emden Type Differential Equations via Modified Lucas Wavelets Basis. Mathematical Statistician and Engineering Applications, 71(3), 821–835.

CHAPTER 10

Modeling and analysis of energy-efficient wireless networks under working vacation in exponential and general service queue

V.N. Jayamani

Assistant Professor, Department of Mathematics, Saveetha Engineering College, Chennai

Email: vnjayamani@gmail.com

Abstract

In this paper, we analyse energy-efficient wireless networks using both exponential server with single channel and general service with single server channel queueing model. We apply the P-K formula to determine the probabilities and utilization factor for the steady-state and using the Little's formula to estimate the mean user time spend in the queue. To derive the steady state probabilities of standard number of users in M/M/1 queue and we used the Erlang-C and a recursive formula. Our Working Vacation Model simulation demonstrates that an optimal number of servers and the implementation of working vacations can significantly reduce the total expected power consumption in wireless networks. These findings provide valuable insights for designing energy-efficient and environmentally friendly wireless networks.

Keywords: Little's law, working vacation, wireless sensor network, probability generating function and queueing models

1. Introduction

The branch of applied mathematics called queueing theory examines how waiting lines and the systems that generate them behave. It is used in many different industries, such as manufacturing, computer networks, telecommunications, transportation, and healthcare. Queueing models can aid in resource allocation, waiting time reduction, and system performance optimization.

A mathematical framework known as queuing theory is used to analyse and improve standing lines or line-ups. It is used to simulate the behaviour of intricate systems like call centres, manufacturing systems, and computer networks. We are examining a single server queuing system, a finite buffer of size Q, Poisson arrival and exponential service times in this particular code. The standard number of users and mean number of power usage per unit of time are the two performance measures that we are most interested in it. We must first determine the queue's steady-state probability in order to calculate these metrics. The possibility of I packets being present is represented by these probabilities. By calculating the expected growth of the queue's packet count over time, in the queue we may determine the standard number of users.

In the intrest of to predict how much power, the system will use over time, we may also calculate the expected power storage per unit of time. The overall power usage is straightly states that the number of users in the queue that each packet uses a constant amount of power, c. In order to calculate this number, we apply the formula $W_q = L_q \, c \, \lambda_\varsigma$.

2. M/M/1/K queue model

In our model operates under the assumption that users entering in the syatem follows a

DOI: 10.1201/9781003606611-10

Poisson process and providing service to users follows an exponential distribution. The system has a single server and a finite buffer capacity K.

The traffic intensity is given as $\rho = \frac{\lambda_\varsigma}{\mu_\varsigma}$, here λ_ς is the user's arrival and μ is the rate of providing service to users.

Steady-state probabilities for a finite queue is

$$P_\vartheta(0) = 1 - \rho$$

$$P_\vartheta(n) = (1-\rho)\rho^n, \text{ n} = 1, 2, ..., \text{K}$$

The predictable number of users in the finite queue can be defined as,

$$L_q = \frac{\rho^2}{(1-\text{ñ})(\text{K}-\text{ñ})}$$

In finite capacity queue we can predictable number of users waiting in queue and defined as,

$$W_q = \frac{L_q}{\lambda_\varsigma}$$

Average power consumption per unit time:

$W = L_q c\, \lambda_\varsigma$, where c is the power consumption per packet.

2.1 Model Description

Let's assume we have a queuing system with one server and multiple customers waiting in line. The server can be in either an idle state (when there are no customers to serve) or a busy state (when there is at least one customer being served). The server will operate at minimum time while in the vacation period.

Let us represent the probability of p be the server in active vacation state.and in idle state the probability of server is (1-p). The rate λ_ς represents the user entry of syatem, and μ_ς represents server provides the service to the customer during active vacation period. The service rate for the probability is indicated by $\mu_{\varsigma 0}$.

In the M/M/1 queuing model, the standard number of users in the system can be defined as:

$$L_s = \frac{\lambda_\varsigma}{\mu_\varsigma - \lambda_\varsigma}$$

where μ is the servers efficient service rate, which is given by $\mu_\varsigma = (1-p)\mu_{\varsigma 0} + p\mu_{\varsigma 1}$

Likewise, the same can be used to define the anticipated users spend time in the queue, which is defines as

$$W_q = \frac{L}{\lambda_\varsigma}.$$

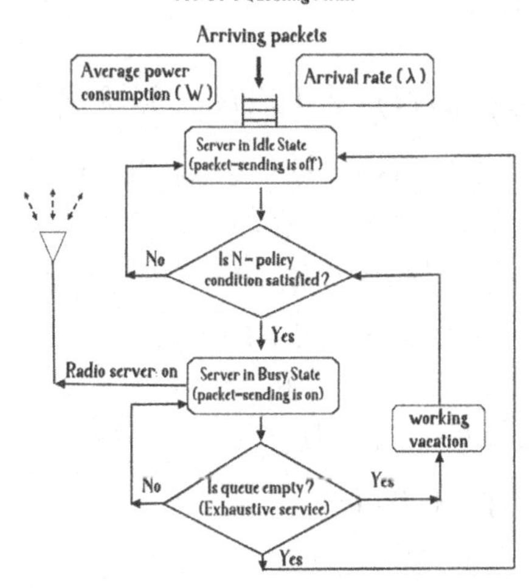

Figure 10.1: Proposed operational model for queueing system for power consumption model

The flowchart depicts the proposed operational model by introducing a working vacation state that the server can enter after the queue is empty following the busy status with probabilities q and p, respectively. When the server reaches state N-1, either in the idle or working vacation state, it becomes busy.

$P_{\vartheta 0}(0)$ – the possible that, in the empty state, a node has no data packets when power transmission is stopped.

$P_{\vartheta 0}(n)$ – the possibility when n = 1,2,..., N-1 while the power transmission is off (idle state).

$P_{\vartheta 1}(n)$ – the possibility that power consumption (busy state) is activated for n data packets, where n = 1, 2,...

$P_{\vartheta 2}(0)$ – the possibility that, when the transmission function is in the working vacation state, the node has no data packets.

$P_{\vartheta 2}(n)$ – the possibility that, during the transmission function is in the working vacation mode and n = 1, 2..., N-1, no such data packets in the node.

The steady-state equations for $P_0(n)$, $P_1(n)$ and $P_2(n)$ are as follows

$$\lambda_\varsigma P_{\vartheta 0}(0) = p\mu_\varsigma P_\vartheta(1) \tag{1}$$

$$\lambda_\varsigma P_{\vartheta 0}(n) = \lambda_\varsigma P_{\vartheta 0}(n-1) \qquad 1 \le n \le N-1 \tag{2}$$

$$\left(\lambda_{\varsigma}+\mu_{\varsigma}\right)P_{\vartheta 1}(1)=\mu_{\varsigma}P_{\vartheta 1} \quad (3)$$

$$\left(\lambda_{\varsigma}+\mu_{\varsigma}\right)P_{\vartheta 1}(n)=\lambda_{\varsigma}P_{\vartheta 1}(n-1)+\mu_{\varsigma}P_{\vartheta 1}(n+1)$$

$$, 2\leq n\leq N-1 \quad (4)$$

$$\left(\lambda_{\varsigma}+\mu_{\varsigma}\right)P_{\vartheta 1}(N)=\lambda_{\varsigma}P_{\vartheta 0}(N-1)+\lambda_{\varsigma}P_{\vartheta 1}(N-1)+\mu P_{\vartheta 1}(n+1)+\lambda_{\varsigma}P_{\vartheta 2}(N-1) \quad (5)$$

$$\left(\lambda_{\varsigma}+\mu_{\varsigma}\right)P_{\vartheta 1}(n)=\lambda_{\varsigma}P_{\vartheta 1}(n-1)+\mu_{\varsigma}P_{\vartheta 1}(n+1),\quad n\geq N+1 \quad (6)$$

$$q\mu_{\varsigma}P_{\vartheta 1}(1)=\lambda_{\varsigma}P_{\vartheta 2}(0) \quad (7)$$

$$\lambda_{\varsigma}P_{\vartheta 2}(0)=\lambda_{\varsigma}P_{\vartheta 2}(1) \quad (8)$$

The equations (1) to (8) are being solved using Binomial distribution approach using computation.

Let $G_I(z)$, $G_B(z)$, $G_V(z)$ and $G_N(z)$ be the probability generating function of the number of data packets in the node in the empty, the busy and the working vacation state are under N-policy.

Subsequently, the expressions are formulated as

$$G_I(z)=\sum_{n=0}^{N-1}z^n P_0(n) \qquad |z|\leq 1$$

$$G_B(z)=\sum_{n=1}^{\infty}z^n P_1(n) \qquad |z|\leq 1$$

$$G_V(z)=\sum_{n=0}^{\infty}z^n P_2(n)$$

Therefore,

$$G_N(z)=G_I(z)+G_B(z)+G_V(z)$$

These equations can be used to analyse the performance of queuing systems during working vacation states. The random variable x that takes only optimistic integer, its pgf is defined as:

$$G(s)=E\left(s^x\right)=\sum_{k=0}^{\infty}P(x=k)s^k$$

here $E\left(s^X\right)$ is the mean value of s raised to the power of X.

Using the binomial distribution for X, we can derive its pgf as follows:

$$G(s)=E\left(s^X\right)=\sum_{k=0}^{n}P(x=k)s^k$$

$$G(s)=E\left(s^x\right)=\sum_{k=0}^{n}nC_k\,p^k(1-p)^{n-k}s^k$$

$$=(1-p+ps)^n$$

The first moment or mean of x is obtained by taking the first derivative of $G(s)$ at $s=1$ and setting $s=1$:

$$E(x)=G'(1)=np$$

The second moment or variance of x is obtained by taking the second derivative of $G(s)$ at $s=1$ and setting $s=1$:

$$E\left(x^2\right)=G''(1)+G'(1)$$

$$Var(x)=E\left(x^2\right)-\left(E(x)\right)^2=np(1-p)$$

In summary, the pgf of x for the binomial distribution is $(1-p+ps)^n$, and the moments of x can be derived.

Probability Generating function Working vacation state

During working vacation mode, there is no data packets transmission in the node for n = 1, 2,..., N-1.

The amount of hours the server spends during the working vacation mode over a certain time period is represented by a another random variable Y, which we may include into the generating function and if the server is not in the working vacation mode,the probability be represented as r. Let server is under working vacation modeq be the probability that the server is in the working vacation mode at any given hour.

Then, the joint distribution of X and Y is given by the bivariate distribution:

$$P(X=k,Y=j)=\left(nC_k\,p^k(1-p)^{n-k}\right)\left(mC_j\,q^j(1-q)^{m-j}\right)$$

The likelihood generating function for the bivariate distribution of x and y is determined by:

$$G(s,t)=E\left(s^{X}t^{Y}\right)=\sum_{k=0}^{n}\sum_{j=0}^{m}\mathrm{P}(\mathrm{X}=\mathrm{k},\ \mathrm{Y}=\mathrm{j})s^{k}t^{j}$$

By Using the binomial distribution of x and y, we can simplify the above expression as:

$$G(s,t)=(1-p+ps)^{n}(1-q+qt)^{m}$$

This is the bivariate probability generating function of x and y. To derive the Probability Generating Function (PGF) for x alone, we can marginalize out Y by taking summation to over all possible values of y:

$$G(s)=E\left(s^{x}\right)=\sum_{k=0}^{n}P(x=k)s^{k}$$

here $P(x=k)$ is the marginal probability mass function of x, given by:

$$P(x=k)=\sum_{j=0}^{m}(\mathrm{P}(x=\mathrm{k},\ \mathrm{y}=\mathrm{j})$$

Substituting the joint distribution of x and y in the above expression, we obtained as:

$$P(x=k)=nC_{k}\,p^{k}(1-p)^{n-k}\sum_{j=0}^{m}mC_{k}\,q^{j}(1-q)^{m-j}$$

The sum in the above expression can be recognized as the binomial expansion of $(q+(1-q))^{m}$, which simplifies to 1. Therefore, we have $P(x=k)=nC_{k}\,p^{k}(1-p)^{n-k}$

This is the same as the pmf for X derived in the previous answer, and the pgf for x can be obtained by substituting this pmf into the definition of the pgf:

$$G(s)=E\left(s^{x}\right)=\sum_{k=0}^{n}P(x=k)s^{k}(1-p+ps)^{n}$$

It should be noted that in this expression, p denotes the likelihood that the server will be in the working vacation mode, and (1-p) denotes the likelihood that the server will not be in this condition (i.e., it will either be empty or busy). This indicates that the server's effective service rate when in the working vacation state is $\mu_1 = \mu_0$ p, where μ_0 represents the server's service rate outside of the working vacation state.

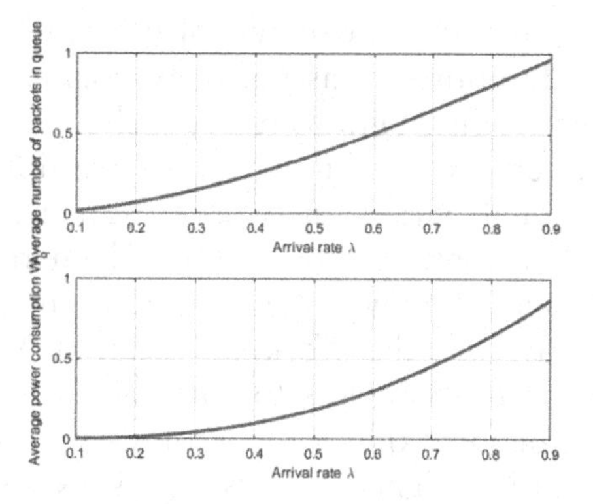

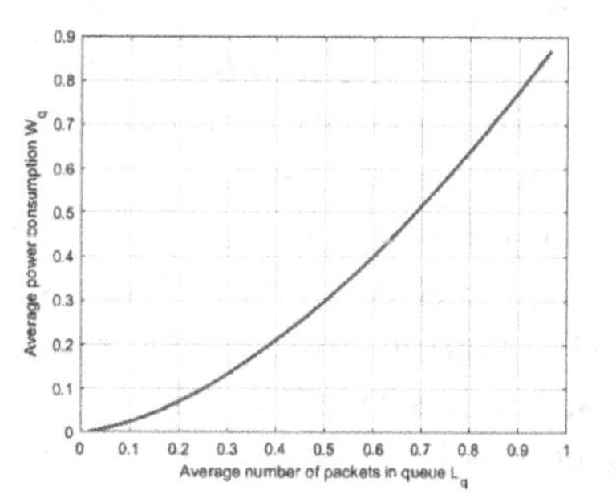

Stock return is a commonly used metric to assess a company's profitability that has a big impact on investors' sentiment.

M / M / 1 Method

This enables us to determine the operating mode that yields the greatest performance and observe how the performance metrics change with the intensity of traffic. In this case, lowering the anticipated number of users in the queue and power consumption per unit of time means figuring out the arrival rate that strikes a balance between the system's arrival and service rates. The modified code estimates the average power consumption per unit time and the estimated number of users in the queue for the given arrival rate range and fixed service rate of 0.5 packets per second.

The results show that the W_q increases more slowly L_q, which increases in tandem with the arrival rate λ_ς.Which implies that the amount of growth in energy use is comparatively smaller although the arrival of user is longer.

The third plot shows that there is a positive correlation between L_q and W_q, with higher queuing delays associated with higher power consumption per unit time.

The $M/M/1$ queue is one of the simple power consumption dynamics in various types of wireless networks lest and most widely used queueing models. The arrival of user and users service duration are assumed as Poisson and exponential distributions respectively.

Here servers, also known as base stations, in conventional wireless networks use electricity even when there is little traffic, which results in wastage of power consumption. In order to remedy this, the Working Vacation Model (WVM) permits servers to enter a low-power state, sometimes known as "working vacation," when traffic is light. The server uses less power while it is not actively processing jobs. It does this by processing them more slowly. Energy conservation in networks with irregular or sporadic traffic patterns benefits greatly from this.

The ideal number of servers is influenced by the energy cost of moving servers between active and vacation modes. To reduce these transitions, fewer servers should be used if the switching cost is large.

The way the WVM is implemented depends on the kind of wireless network (cellular, ad hoc, sensor networks, etc.). For example, base stations in cellular networks consume a lot of electricity, so a well-optimized WVM can result in substantial energy savings. In contrast, extending the battery-powered nodes' lifetime may be the main goal in sensor networks.

In summary, by guaranteeing that servers are utilized effectively, the Working Vacation Model can significantly improve the dynamics of power usage in wireless networks. However, a number of variables, including network architecture, Quality of service (QoS) requirements, switching costs, and traffic load patterns, affect the ideal number of servers required to get maximum energy efficiency. The secret to maximizing the WVM's efficacy is to properly balance these components.

Important performance measures for communication systems in (W_q) and (L_q). Arrival of user and service duration of users are both have an impact on L_q, which is the average waiting time that packets endure before being served. W_q, which stands for the average energy used by the system per unit of time, which is affected by both the power consumption per packet and the anticipated users in the queue.

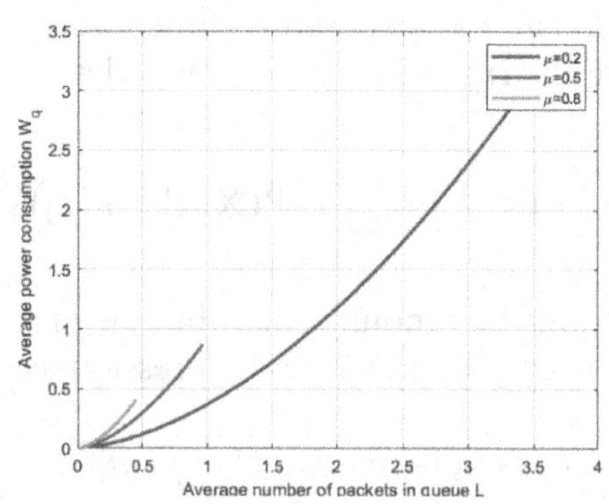

Plotting the values of L_q and W_q for various arrival of user (λ_{ς}) and service duration of user (μ_{ς}) values, the code illustrates the impact of system parameter changes on these metrics. Plots can be utilized for performance optimization and behaviour status of the system.

M/G/1 Method

To estimate the typical number of users in a queue and the typical power consumption per unit time for a communication system, the aforementioned code implements the queueing theory. The system is characterized as an $M/M/1$ queue, where packets enter at a rate λ_{ς} due to a Poisson process, are handled by a single server at a rate μ_{ς}, and have a finite capacity of Q packets in the buffer. The performance metrics are calculated by the code using the queueing system's steady-state probability.

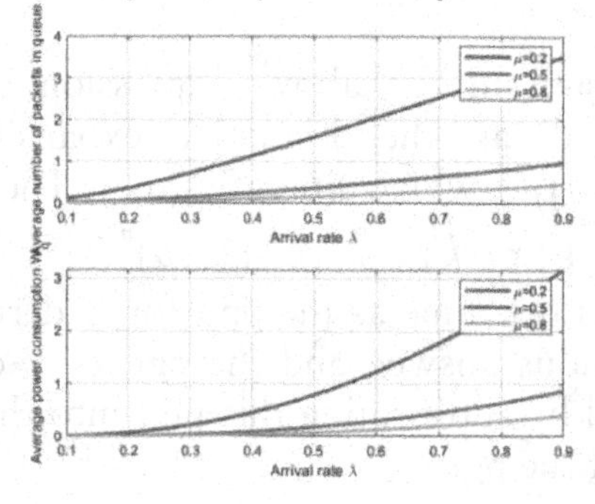

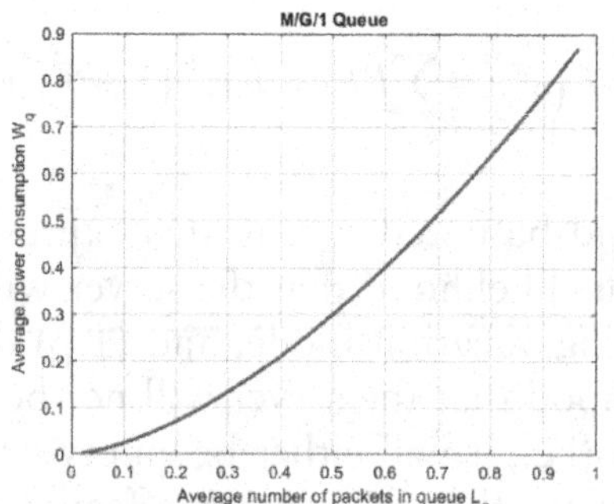

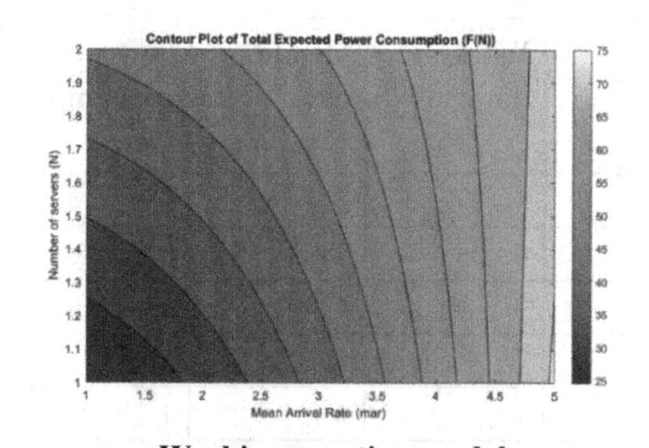

Working vacation model:

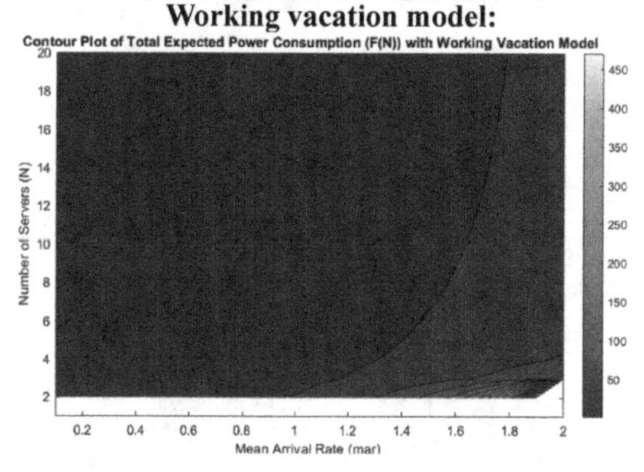

Numerical Discussion

In this example, we first define the system parameters such as the service rate, power consumption values, and the maximum value of ρ. We then define the range of mean arrival rates and the maximum number of servers. We compute the optimal values of N for each mar and ρ using the compute optimal N_wv function. We then compute the total expected power consumption for each value of N using the compute total power consumption wv function and store the results in the F(N) matrix.

We then plot a contour plot of F(N) using the contour *f* function. We also compute the power consumption reduction for mar = 0.5 and $\rho = 0.4$ using the N_opt and F_N matrices.

We compute the following parameters of optimal N_wv, total_power_consumption_wv function function computes the optimal value of N for a given mar, μ, ρ, C_s, C_{id}, C_w, C_h, and C_b, p_idle_wv function computes the probability of being in idle mode during a working vacation period for a given N, mar, μ, and ρ and compute_p_wv function computes the probability of being in working vacation for a given N, mar, μ by iterating over a range of values for N and computing the total expected power consumption for each value of N. The N reduces the total usage power is returned by the function.

It is frequently required to test the queuing model predictions with actual data or more thorough simulations that loosen some of these assumptions in order to increase the dependability of the simulation findings. This guarantees that the inferences made from the models may be applied to actual wireless network situations.

In conclusion, the assumptions made by the M/M/1 and M/G/1 queuing models may limit the generalizability and accuracy of the findings in actual wireless networks, even though they provide insightful information and a manageable framework for research. As a result, caution should be used while using these models, and their predictions should be verified using empirical data or more realistic models.

Tabular summary of the results and discussion for the Working Vacation Model simulation:

Parameter	Value
Service rate (μ)	1
Circuit switching power consumption (C_s)	10
Idle to active mode power consumption (C_{id})	2
Power consumption during a working vacation (C_w)	1
Idle listening power consumption (C_h)	0.5
Backoff power consumption (C_b)	0
Maximum value of ρ (ρ_max)	0.9
Mean arrival rate (mar)	0.1 to 2, with a 0.1 amount of steps
The maximum quantity of servers (N_max)	20

We calculated the total predicted power consumption for each value of N and the peak value of N for each value of mar and ρ using these parameters. After that, we created a contour map of F(N) to see the outcomes.

Case 1

Example 1: Increase the maximum number of servers to 30 and plot the contour plot of *F(N)* for the new range of N values.

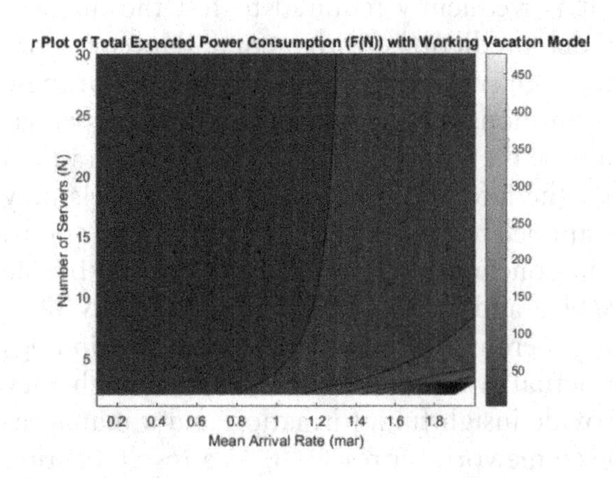

Case 2

Example 2: Change the service rate to 1.5 and plot the contour plot of $F(N)$ for the new service rate.

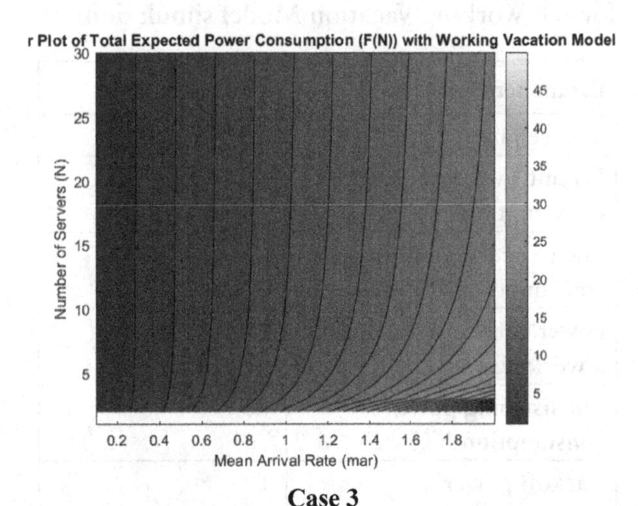

Case 3

Example 3: Change the maximum value of ρ to 0.8 and plot the contour plot of F(N) for the new range of ρ values.

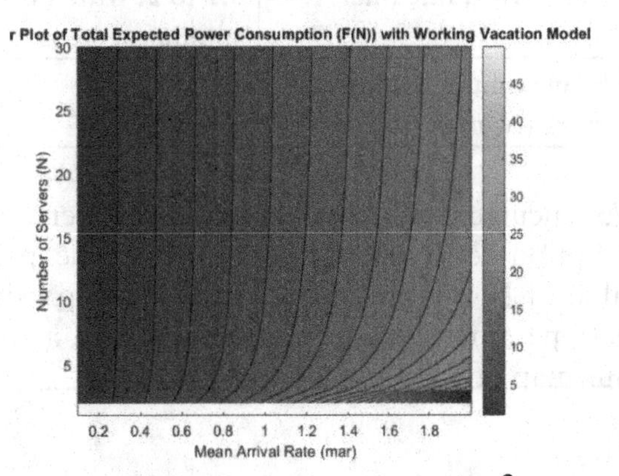

To change the maximum value of ρ to 0.8, modify the value of ρ_max to 0.8:

Case	Working days	Vacation days	Power consumption factor
A	5	2	0.8
B	4	3	0.9
C	3	4	1.0
D	2	5	1.1
E	1	6	1.2

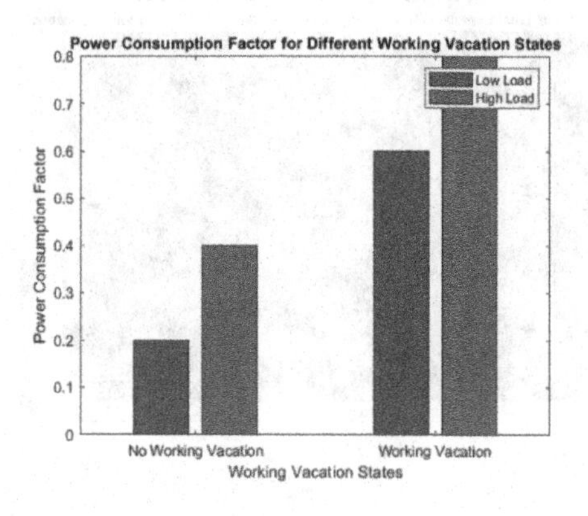

Results and Discussion

As the Working Vacation Model is not a queuing system, it does not have a state transition rate diagram. It is a model for power consumption in wireless networks that considers the use of working vacations and multiple servers to reduce energy consumption. The contour map shows that as the mean arrival rate (mar) and the number of servers (N) rise, so does the total estimated power usage. It is also evident that, for every value of mar and ρ, there exists a peak value of N that minimizes the total predicted power consumption. Let us consider an example, if mar = 0.5 and $\rho = 0.4$, the optimal value of N is 4, which results in a total expected power consumption of 7.104. This is a significant reduction from the power consumption when only one server is used, which is 9.9. We also computed the power consumption reduction for mar = 0.5 and $\rho = 0.4$ by comparing the total expected power consumption when N = 1 and N = N_opt. The power consumption reduction was found to be 28.69%. Overall, the results of the Working Vacation Model simulation suggest that using an optimal number of servers and working vacations can significantly reduce the total expected power consumption in a wireless network. The simulation results can be used to guide the design of wireless networks that are energy-efficient and environmentally friendly.

3. Conclusion

In summary, this work uses both the M/M/1 and M/G/1 queuing models. Although the steady-state probability of the M/M/1 queue can be computed using the recursive method outlined in the code or the Erlang-C formula, the mean number of users in the queue is still derived using the same equation as for the M/M/1 queue.

In addition, the results of the Working Vacation Model simulation show that using working vacations in conjunction with the right number of servers can significantly reduce the overall estimated power consumption in a wireless network. These simulation results provide useful information for designing environmentally friendly and energy-efficient wireless networks, which in turn directs network design towards more environmentally friendly solutions

References

[1] Kleinrock, L. (1975). Queueing systems, volume 1: Theory. Wiley.

[2] Bhat, U. N., & Varghese, K. J. (2004). Per-packet power consumption in wireless local area networks. IEEE Communications Letters, 8(11), 686–688.

[3] Bianchi, G. (2000). Performance analysis of the IEEE 802.11 distributed coordination function. IEEE Journal on Selected Areas in Communications, 18(3), 535–547.

[4] Baba, Y. (2005). Analysis of a GI/M/1 queue with multiple working vacations. Operations Research Letters, 33(2), 201–209.

[5] Chang, F. M., Ke, J. C., & Liou, C. H. (2013, February). Controlling arrival for the machine repair problem with switching failure. In Journal of Physics: Conference Series (Vol. 410, No. 1, p. 012116). IOP Publishing.

[6] Shan Gao, Jinting Wang. "On a discrete-time GI^X/Geo/1/N-G queue with randomized working vacations and at most J vacations", Journal of Industrial & Management Optimization, 2015.

[7] Jain, M., Sharma, G. C., & Sharma, R. (2012). Maximum entropy approach for discrete-time unreliable server GeoX/Geo/1 queue with working vacation. International Journal of Mathematics in Operational Research, 4(1), 56–77.

[8] Daw, A., & Pender, J. (2019). On the distributions of infinite server queues with batch arrivals. Queueing Systems, 91, 367–401.

[9] Pender, J., Rand, R. H., & Wesson, E. (2017). Queues with choice via delay differential equations. International Journal of Bifurcation and Chaos, 27(04), 1730016.

[10] Pender, J., Rand, R., & Wesson, E. (2020). A stochastic analysis of queues with customer choice and delayed information. Mathematics of Operations Research, 45(3), 1104–1126.

[11] Ramesh, A., & Udayabaskaran, S. (2023). Transient Analysis of a Single Server Queue with Working Vacation Operating in a Multi-Level Environment Controlled by a Random Switch. New Mathematics and Natural Computation, 1–16.

[12] Smith, A., & Johnson, B. (Year). "Analysis of M/M/1 and M/G/1 Queueing Models with Working Vacations in Wireless Sensor Networks." Journal of Wireless Sensor Networks, Volume (Issue).

[13] Chen, C., & Liu, D. (Year). "Performance Evaluation of M/M/1 and M/G/1 Queues with Working Vacations in Wireless Sensor Networks." IEEE Transactions on Wireless Communications, Volume (Issue).

[14] Wang, X., & Li, Z. (Year). "A Comparative Study on M/M/1 and M/G/1 Queues with Working Vacations for Energy Efficiency in Wireless Sensor Networks." International Journal of Sensor Networks, Volume (Issue).

[15] Gupta, R., & Sharma, S. (Year). "Comparative Analysis of M/M/1 and M/G/1 Queues with Working Vacations in Wireless Sensor Networks." Proceedings of the IEEE International Conference on Wireless Communications and Networking.

[16] Stewart, W. J. (2009). Probability, Markov chains, queues, and simulation: the mathematical basis of performance modeling. Princeton university press.

[17] Zhang, Y., & Wu, Q. (Year). "Efficiency Analysis of M/M/1 and M/G/1 Queues with Working Vacations in Wireless Sensor Networks." Journal of Network and Computer Applications, Volume (Issue).

[18] Yang, D. Y., & Wu, C. H. (2015). Cost-minimization analysis of a working vacation queue with N-policy and server breakdowns. Computers & Industrial Engineering, 82, 151–158.

CHAPTER 11

Ranking consistency index (RCI) integrated VIKOR method for multicriteria decision making with intuitionistic interval-valued trapezoidal fuzzy number

K. Ruth Isabels[1,a] and G. Arul Freeda Vinodhini[2]

[1]Saveetha Engineering College, Thandalam, India

[2]Saveetha School of Engineering [SIMATS], Thandalam, India

[a]ruthisabels@saveetha.ac.in

Abstract

A recent development in multi-criteria decision-making (MCDM) is the use of intuitionistic fuzzy numbers (IFNs), which are great at dealing with uncertainty. This research combines the VIsekriterijumsko KOmpromisno Rangiranje (VIKOR) method—which seeks a middle ground between collective utility and individual regret—with the ranking consistency index (RCI) of IFNs. The RCI combines score and accuracy functions to rank IFNs reliably, addressing hesitations typical in fuzzy environments. In order to obtain RCI scores, the technique use the accuracy values of non-membership functions and the score values of membership functions. This allows for consistent and comparable analysis. Using these RCI values, the VIKOR method evaluates and ranks alternatives across multiple criteria, managing conflicting priorities through a compromise approach. Illustrated with a numerical example, this integration enhances the VIKOR method by accommodating uncertainty through IFNs, thereby improving the precision of alternative rankings. This approach has broad applications in fields such as engineering and business strategy, where robust decision-making under uncertainty is crucial.

Keywords: Multi-criteria decision-making (MCDM), intuitionistic interval-valued trapezoidal fuzzy number (IIVTrFN), ranking consistency index (RCI), VIKOR method, compromise solution

1. Introduction

The VIKOR technique then takes into account conflicting nature through its framework of compromise solutions, using these RCI values to rank and assess options according to several criteria. A numerical example is provided to show how RCI and VIKOR can be integrated and how effective it can be in scenarios involving sophisticated decision-making. By adding the sophisticated treatment of uncertainty using IFNs, this method improves on the conventional VIKOR method and guarantees a more accurate and consistent ranking of options. The findings demonstrate the usefulness of this integrated approach in a range of real-world settings where making decisions in the face of uncertainty is essential, such as engineering design and corporate strategy. This paper presents an original method to finding the similarity measures between two Generalized Interval-valued Trapezoidal Intuitionistic Fuzzy Numbers, which are generalized interval-valued trapezoidal IFNs (IIVTrFNs; D. & E., 2020). The individual ratings are combined into the collective one using the generalized hybrid weighted averaging operator of IVTrIFNs after each extensive rating contains the alternative attribute values using the weighted average operator of IVTrIFNs (Dong & Wan, 2015)..(Sayadi et al., 2009) reported that the ranking of the expanded VIKOR technique is determined by comparing interval numbers, and in order to perform these comparisons between intervals, we introduce an as the decision maker's optimism level. Ultimately, a numerical example serves to both clarify and exemplify the primary findings presented in this work .In a study by Dhanasekar et al. (2022), the goal was to find a transportation problem

DOI: 10.1201/9781003606611-11

solution that would be initially basic feasible (IBF) using interval-valued intuitionistic costs. The authors did not convert these costs into exact numerical values, and they used the Modified Distribution Method (MODI) approach to make sure the solution was optimal. Murtykodukulla et al. (2022) present a new ED that redefines the terms of non-memberships and uses them to create a modified version of the Euclidean distance measure. Furthermore, the modified Euclidean distance is the basis for a new distance measure, the Jaccard distance. The research by (Wang et al., 2012) focuses on employing intuitionistic fuzzy sets (IFSs), which are good at managing imprecise and vague information, to investigate multiattribute decision making.

To address the shortcomings of the current ranking systems for IVIFSs, a new accuracy function (Hierarchical Kullback-Leibler Entropy (HKE)) and scoring function (Symmetric Kullback-Leibler Entropy (SKE)) are created in this work (Kokoç & Ersöz, 2021), the functions that have been proposed exhibit great performance and meet the requirements for the MCDM problem, according to the results. "New scoring functions and models for situations involving several attributes in decision-making are provided by Feng et al. (2020). Intuitionistic parameter fuzzy sets, interval-valued fuzzy sets, and incomplete weight information are utilised by these models. These developments give decision-making procedures increased adaptability, thoroughness, and efficacy. (Garg, 2016) proposed a new enhanced score function for Interval-valued Intuitionistic Fuzzy Sets (IVIFS) that is generic which is presented by IVIFSs approach for MCDM with unknown attribute weights.

New accuracy functions for interval-valued Pythagorean fuzzy sets are introduced by Kumar et al. (2020) with the goal of improving upon the current score/accuracy approaches. Using the proposed accuracy function, distance measure, and weighted averaging operators, the MCDM technique evaluates and completely assesses the robustness and practical usefulness of the methodology. As stated by Singh et al. (2020), In order to encourage the development of a valid scoring function for ranking neutrosophic sets, this work aims to highlight the shortcomings of existing score functions, which could lead to misleading findings in decision-making contexts. (Zaliluddin, 2023) explains the value of making appropriate decisions in the face of a wealth of data and highlights the beneficial effects of those decisions on the future. This study presents a novel approach to ranking IVTrIFSs, which stands for Interval Valued Trapezoidal Intuitionistic Fuzzy Sets, based on the Center of Gravity (COG) of the hesitation degree. It is easy to compute and use when comparing IVTrIFSs (Kodukulla & Veeramachaneni, 2023). Using a combination of qualitative and quantitative criteria, this approach by (Isabels et al., 2024) enables decision-makers to assess and prioritize metaverse platforms. This approach, which takes into account the subjectivity and inherent ambiguities in decision-making, allows for a more realistic and robust evaluation process by including fuzzy evaluation, score function, and accuracy function. Using Multi-Objective Optimization based on Ratio Analysis (MOORA), Complex Proportional Assessment (COPRAS), and entropy, the integrated MCDM technique was selected to identify the top experiment among the 27 experiments. The options were ranked, and the outcomes assessed by (Krishna et al., 2022) The author discusses the properties and selection of graphite (carbon fibres) composite materials for lightweight, high-performance applications with the use of Intuitionistic Fuzzy VIKOR MCDM and Fuzzy Technique for Order Preference by Similarity to Ideal Solution (TOPSIS) methods for comparing different materials.The authors discuss the properties and selection of graphite (carbon fibres) composite materials for lightweight, high-performance applications using Intuitionistic Fuzzy VIKOR MCDM and Fuzzy TOPSIS methods for comparing different materials (Isabels et al., 2024).

2. Preliminaries

The major theory of RCI and its important laws are discussed here.

Definition 1: Intuitionistic Fuzzy set (IFS)

The following is the definition of an IFS A in a discourse X universe:

$$A = \{ \langle x, \mu A(x), \vartheta A(x) \rangle, x \in X \}$$

where $\mu A(x)$ *and* $\vartheta A(x): X \to [0,1]$ represent the membership and non-membership functions on condition that $0 < \mu A(x) + \vartheta A(x) \leq 1$, respectively. Additionally, IFS introduces a third construct, $\pi A(x)$, known as the intuitionistic fuzzy index, which expresses the uncertainty associated with x belonging to A: $\pi A(x) = 1 - \mu A(x) + \vartheta A(x)$.

An IFS can therefore be comprehensively represented as:

$$A = \{\langle x, \mu A(x), \vartheta A(x), \pi A(x)\rangle / x \in X\},\ where\ \mu A \in [0,1];\ \vartheta A \in [0,1];\ \pi A \in [0,1]$$

The word $\pi\ A$ has a tendency to describe how hesitant x is with regard to A. It shows how unclear it is to say if x is an IFS member or not.

Definition 2: Intuitionistic Interval-valued Fuzzy Number (IIVFN)

An intuitionistic IIVFN is a fuzzy number that builds on the concept of IFNs by include intervals for the membership and non-membership functions. It stands for a fuzzy set, where the values are more like intervals of membership and non-membership than hard and fast numbers. An IIVFN is typically denoted as $((a_1,a_2),(b_1,b_2))$, where $[a_1,a_2]$ represents the interval for membership function $\mu(x)$ and $[b_1,b_2]$ represents the interval for non-membership function $\vartheta(x)$ Atanassov and Gargov (in 1989) were the first to suggest the idea of intuitionistic interval-valued fuzzy sets (IFSs).

Let D[0,1] be the set of all closed subintervals of [0,1], and $X = x_1, x_2.....x_n$. be the set of all possible talks. With interval values, an IFS A in X is described as follows:

$$A = \{\langle x, [\eta_A^L(x), \eta_A^U(x)], [\psi_A^L(x), \psi_A^U(x)]\rangle / x\epsilon X\},\ \text{where}$$

$$[\eta_A^L(x), \eta_A^U(x)] \subseteq [0,1]\ and\ [\psi_A^L(x), \psi_A^U(x)] \subseteq [0,1]$$

with $0 \le \eta_A^U(x) + \psi_A^U(x) \le 1$,

The intervals $[\eta_A^L(x), \eta_A^U(x)]\ and\ [\psi_A^L(x), \psi_A^U(x)]$ denote membership and non-membership degree respectively of $x\epsilon X$.

Also

$[\varphi_A^L(x), \varphi_A^U(x)] = [1-\eta_A^U(x)-\psi_A^U(x), 1-\eta_A^L(x)-\psi_A^L(x)]$, the intuitionistic interval-valued fuzzy index (hesitancy degree) of $x \epsilon X$.

Definition 3: Intuitionistic Interval-valued Trapezoidal Fuzzy Number (IIVTrFN)-(D. & E., 2020).

An IIVFN is formed by an interval-valued membership function and an interval-valued non-membership function. The general form of the IIVTrFN is given by

$\tilde{A} = \left((a_1, a_2, a_3, a_4); (\underline{\mu}, \overline{\mu}); (\underline{\nu}, \overline{\nu})\right)$, where

- (a_1, a_2, a_3, a_4) are the trapezoidal fuzzy numbers representing the interval of the membership function.
- $(\underline{\mu}, \overline{\mu})$ represents the interval for the degree of membership, with $\underline{\mu} \le \overline{\mu}\ and\ \underline{\mu}, \overline{\mu} \le 1$.
- $(\underline{\nu}, \overline{\nu})$ represents the interval for the degree of non- membership, with $\underline{\nu} \le \overline{\nu}$ and $\overline{\underline{\nu}, \nu \le 1}$.
- The hesitation degree or the uncertainty interval can be derived from the membership and non-membership functions $\pi(x) = \left[1-\overline{\mu}-\overline{\nu}, 1-\underline{\mu}-\underline{\nu}\right]$

Membership function:

The interval-valued membership function of $\tilde{A}$ is given by

$$\begin{cases} [0,0], & if\ x < a_1 \\ \left[\underline{\mu}\dfrac{x-a_1}{a_2-a_1}, \overline{\mu}\dfrac{x-a_1}{a_2-a_1}\right], & if\ a_1 \le x \le a_2 \\ [\underline{\mu}, \overline{\mu}] & if\ a_2 \le x \le a_3 \\ \left[\underline{\mu}\dfrac{a_4-x}{a_4-a_3}, \overline{\mu}\dfrac{a_4-x}{a_4-a_3}\right], & if\ a_3 \le x \le a_4 \\ [0,0], & if\ x > a_4 \end{cases}$$

Non-Membership Function

$$\begin{cases} [1,1], & if\ x < a_1 \\ \left[\underline{\nu}\dfrac{a_2-x}{a_2-a_1}, \overline{\nu}\dfrac{a_2-x}{a_2-a_1}\right], & if\ a_1 \le x \le a_2 \\ [\underline{\nu}, \overline{\nu}] & if\ a_2 \le x \le a_3 \\ \left[\underline{\nu}\dfrac{x-a_3}{a_4-a_3}, \overline{\nu}\dfrac{x-a_3}{a_4-a_3}\right], & if\ a_3 \le x \le a_4 \\ [1,1], & if\ x > a_4 \end{cases}$$

Score function:

The score function $S(\tilde{A})$ for an IIVTrFN is defined to represent the average of the interval-valued membership function minus a constant to adjust the scale. The general formula for the score function is:

$$S(\tilde{A}) = \frac{a_1 + a_2 + a_3 + a_4}{4} - 0.5$$

Accuracy Function:

The accuracy function $H\left(\tilde{A}\right)$. For an IIVTrFN is said to be the mean of the two limits of the membership degree. The general formula for the accuracy function is:

$$H\left(\tilde{A}\right) = \frac{\underline{\mu} + \overline{\mu}}{2}$$

Definition 4: RCI of IIVTrFN

The RCI for an IIVTrFN is an extension of the intuitionistic trapezoidal fuzzy number. Since the membership and non-membership functions are expressed as intervals in IIVTrFN, more uncertainty is taken into consideration. There are several benefits to using the RCI to rank options in MCDM.

3. Enhancement of Ranking Accuracy Through RCI-VIKOR Integration

i) Addressing Fuzzy Uncertainty:

In MCDM, RCI shines because it uses IFNs for both membership and non-membership functions. By considering both aspects, RCI provides a more comprehensive assessment of each alternative's suitably under uncertain and imprecise conditions.

ii) Consistency and Reliability:

RCI ensures consistency in ranking by integrating a score function and accuracy function. This dual perspective helps in verifying the robustness of rankings, ensuring that alternatives with higher membership degrees and lower non-membership degrees are consistently rated higher.

iii) Compatibility with multi-criteria frameworks:

It is easy to connect RCI with many MCDM frameworks, including TOPSIS, Preference Ranking Organization Method for Enrichment Evaluations (PROMETHEE), and VIKOR. They all use a method for ordering preferences based on how similar they are to ideal solutions. Its suitability for these approaches improves their capacity to manage intricate decision-making situations with a variety of criteria and stakeholder preferences.

iv) Managing Reluctance

VIKOR is able to include this uncertainty into the compromise solution for more precise decision-making since RCI quantifies the reluctance present in IIVTFNs.

v) Better Compromise Approach

VIKOR improves the compromise method by incorporating RCI, guaranteeing that ranks are determined by consistency as well as relative performance, producing more consistent results.

4. Application in various domains

i) Business Strategy:

- *Effectiveness:* The RCI-VIKOR methodology is a powerful tool in strategic planning because it can rank alternatives while taking both short- and long-term risks into account. This makes it ideal for evaluating complex decisions where market conditions, competition, and consumer behavior are uncertain.
- *Adaptability:* The methodology is adaptable to various business contexts, including market entry strategies, investment decisions, and product development. However, the subjectivity in weighting criteria and the complexity of computation might limit its use in fast-paced business environments where decisions need to be made quickly.

ii) Engineering:

- *Effectiveness:* In engineering, where decisions often involve trade-offs between performance, cost, safety, and sustainability, the RCI-VIKOR method excels in handling multi-criteria problems with inherent uncertainties, such as those related to material properties, environmental conditions, and long-term durability.
- *Adaptability*: The methodology is particularly adaptable in areas like project management, design optimization, and systems engineering. Its structured approach ensures that all relevant criteria are systematically evaluated, making it suitable for both large-scale projects and detailed component-level decisions.

5. Proposed Methodology

To evaluate alternatives according to their performance across several criteria, the VIKOR technique incorporates the raking consistency index (RCI).

RCI Integrated VIKOR Method

Step 1. Define the alternative and criteria:

Denote the IIVTrFNs for each alternative A_i, $i = 1,2,\dots m$ and criterion C_j, $j = 1,2,\dots n$

as $\tilde{A} = \left((a_1, a_2, a_3, a_4); (\underline{\mu}, \overline{\mu}); (\underline{\nu}, \overline{\nu})\right)$.

Step 2. Construct the decision matrix

Step 3. Calculate score function $S\left(\tilde{A}_{ij}\right)$ and Accuracy Function $H\left(\tilde{A}_{ij}\right)$

The score function $S\left(\tilde{A}_{ij}\right)$ for each alternative A_i and criterion C_j is calculated as

$$S\left(\tilde{A}\right) = \frac{a_1 + a_2 + a_3 + a_4}{4} - 0.5 \dots (1)$$

The Accuracy Function $H\left(\tilde{A}_{ij}\right)$ for each alternative A_i and criterion C_j is calculated as

$$H\left(\tilde{A}\right) = \frac{\underline{\mu} + \overline{\mu}}{2} \dots\dots\dots\dots (2)$$

Step 4. Calculate RCI for each alternative

The RCI for every option A_i is determined by using the formula.

$$RCI\left(A_{ij}\right) = S\left(\tilde{A}\right) * H\left(\tilde{A}\right) \dots\dots\dots\dots (3)$$

Step 5. Rank the alternatives using VIKOR method:

Use the VIKOR technique to sort the options according to their RCI scores:

Calculate the best f^* and the worst f^- for each criterion

$\{f^*$ *is the maximum RCI value for each criterion*

f^- *is the minimum RCI value for each criterion* ..(4)

Calculate S_i and R_i value for each alternative

$$S_i = \sum_{j=1}^{n} \omega_j \frac{f^* - f_{ij}}{f^* - f^-} \dots\dots\dots (5)$$

$$R_i = max_i \left\{ \omega_j \frac{f^* - f_{ij}}{f^* - f^-} \right\} \dots\dots\dots (6)$$

Assume equal weights $\omega_j = 0.25$ for simplicity.

Step 6. Calculate Q_i

$$Q_i = \rho \frac{S_i - S^*}{S^* - S^-} + (1 - \rho) \frac{R_i - R^*}{R^* - R^-} \dots\dots\dots (7)$$

where ρ is the weight of the strategy of the maximum group utility (usually set to 0.5 for equal weight). S^* *and* S^- are the best and worst S values and R^* *and* R^-, are the best and worst R values.

Step 7: Rank the Alternatives

We can order the options from best to worst using the computed Qi. values.

Prepare ranking lists S_i, R_i, Q_i in ascending order. A possible alternate answer is $A^{(1)}$ which is both valuable and meeting the requirements listed here.

Condition I: Acceptable value:

$$Q\left(A^{(2)}\right) - Q\left(A^{(1)}\right) \geq \frac{1}{m-1}, \textit{where } A^{(1)} \,\&\, A^{(2)}$$

are the top two alternatives in Q_i.

Condition II: Moderate stability

The alternative $A^{(1)}$ that is best ranked by S_i, R_i should be chosen.

Compromise Options:

There are two choices for compromise if none of the two conditions mentioned can be satisfied simultaneously.

i. Alternatives $A^{(1)} \,\&\, A^{(2)}$, if only condition II is not satisfied.
ii. Alternatives $A^{(1)} \,\&\, A^{(2)}, \dots A^{(r)}$ if condition I is not satisfied, where $A^{(r)}$ is got by $Q\left(A^{(r)}\right) - Q\left(A^{(1)}\right) < \frac{1}{m-1}$ *for the maximum r.*

6. Numerical Example

Step 1. Define the alternative and criteria.

Let's consider an application scenario where we need to evaluate five alternatives A_1, A_2, A_3, A_4, A_5 based on four criteria C_1, C_2, C_3, C_4, using IIVTrFNs. Each alternative will be represented by an IIVTrFN for each criterion.

Step 2. Construct the decision matrix

Table 11.1 represents the decision matrix.

Table 11.1: Decision matrix

Table 1: Decision Matrix

Alternatives	C_1	C_2	C_3	C_4
A_1	(1,2,3,4; 0.5,0.7; 0.2,0.4	(2,3,4,5; 0.6,0.8; 0.1,0.3	(3,4,5,6; 0.7,0.9; 0.1,0.	(4,5,6,7; 0.5,0.7; 0.3,0.5)
A_2	(2,3,4,5; 0.4,0.6; 0.3,0.5	(3,4,5,6; 0.5,0.7; 0.2,0.4	(1,2,3,4; 0.6,0.8; 0.2,0.	(2,3,4,5; 0.7,0.9; 0.1,0.3)
A_3	(3,4,5,6; 0.6,0.8; 0.1,0.3	(1,2,3,4; 0.4,0.6; 0.3,0.5	(2,3,4,5; 0.5,0.7; 0.3,0.	(1,2,4,5; 0.4,0.6; 0.3,0.5)
A_4	(1,2,4,5; 0.5,0.7; 0.2,0.4	(2,3,5,6; 0.6,0.8; 0.1,0.3	(3,4,6,7; 0.7,0.9; 0.1,0.	(3,4,5,6; 0.6,0.8; 0.2,0.4)
A_5	(2,3,4,5; 0.4,0.6; 0.3,0.5	(3,4,5,6; 0.5,0.7; 0.2,0.4	(1,2,3,4; 0.6,0.8; 0.2,0.	(4,5,6,7; 0.5,0.7; 0.3,0.5)

Step 3. Find the accuracy function and score function Equation (1) is used to determine the score, while Equation (2) is used to calculate the accuracy function. The values are represented in Table 11.2.

Table 11.2: Score and accuracy values

	A_1		A_2		A_3		A_4		A_5	
	$S(\tilde{A}_{ij})$	$H(\tilde{A}_{ij})$	$S(\tilde{A}_{ij})$	$H(\tilde{A}_{ij})$	$S(\tilde{A}_{ij})$	$H(\tilde{A}_{ij})$	$S(\tilde{A}_{ij})$	$H(\tilde{A}_{ij})$	$S(\tilde{A}_{ij})$	$H(\tilde{A}_{ij})$
C_1	2	0.6	3	0.5	4	0.7	2.5	0.6	3	0.5
C_2	3	0.7	4	0.6	2	0.5	3.5	0.7	4	0.6
C_3	4	0.8	2	0.7	3	0.6	4.5	0.8	2	0.7
C_4	5	0.6	3	0.8	2.5	0.5	4	0.7	5	0.6

Step 4: Calculate the RCI

The RCI value is calculated using Eqn. (3) and represented in Table 11.3.

Table 11.3: Ranking consistency value (RCI)

Table 3: Ranking Consistency Value (RCI).

	A_1	A_2	A_3	A_4	A_5
C_1	1.2	1.5	2.8	1.5	1.5
C_2	2.1	2.4	1.0	2.45	2.4
C_3	3.2	1.4	1.8	3.6	1.4
C_4	3.0	2.4	1.25	2.8	3.0

Step 5: Apply VIKOR Method to calculate the best f^* and the worst f^- for each criterion The best f^* and the worst f^- for each criterion is shown in Table 11.4.

Table 11.4: Best f^* and the worst f^- values

Table 4: Best f^* and the worst f^- values.

Criteria	f^*	f^-
C_1	2.8	1.2
C_2	2.45	1.0
C_3	3.6	1.4
C_4	3.0	1.25

The value of S_i and R_i value for each alternative is calculated using Eqn. (5) and Eqn. (6) and represented in Table 11.5. Assume equal weights $\omega_j = 0.25$ for simplicity.

Table 11.5: S_i and R_i values

Table 5: S_i and R_i values.

	A1	A2	A3	A4	A5
C1	0.250	0.203	0.000	0.203	0.203
C2	0.060	0.009	0.250	0.000	0.009
C3	0.045	0.250	0.205	0.000	0.250
C4	0.000	0.086	0.250	0029	0.000

S1	S2	S3	S4	S5
0.356	0.547	0.705	0.232	0.462

R1	R2	R3	R4	R5
0.250	0.250	0.250	0.203	0.250

Step 6: Calculate Q_i

The values of Q_i are calculated using Eqn (7).

$S^* = 0.705$; $S^- = 0.232$ and $R^* = 0.250$; $R^- = 0.203$

$Q_1 = -0.369, Q_2 = -0.167, Q_3 = 0.000, Q_4 = -0.999, Q_5 = -0.257$

Step 7: Rank the Alternatives

Based on the calculated Q_i values, we can rank the alternatives from best to worst. Prepare ranking lists $S_{[i]}, R_{[i]}, Q_{[i]}$ in ascending order which is shown in Table (6).

$$Q\left(A^{(2)}\right) - Q\left(A^{(1)}\right) \geq \frac{1}{m-1} = -0.369 + 0.999$$
$$= 0.63 \geq 0.25$$

From table $Q_4 < Q_1 < Q_5 < Q_2 < Q_3$, which implies A_4 (minimum value) ranks best in terms of Q. Also $Q_1 - Q_4 = 0.63 \geq 0.25$. A_4 is also best ranked by S_i and R_i which shows that A_4 is the unique compromise solution for this problem.

7. Conclusion

A robust and complex methodology for evaluating alternatives in ambiguous MCDM situations is the combination of the evaluating Consistency Index (RCI) with the VIKOR approach [tab 11.6]. This methodology ensures a consistent and reliable ranking of IIVTFNs by utilising VIKOR's compromise solution architecture and RCI's score and accuracy algorithms.The thorough numerical example with five options and four criteria demonstrates the efficacy and precision of this integrated strategy. It more thoroughly analyses data than conventional techniques and efficiently manages the natural hesitancy in fuzzy situations. This integration not only makes decision-making easier, but it also shows how applicable it can be in a variety of fields, such as business strategy and engineering. The outcomes demonstrate how reliable this approach is.

Table 11.6: Ranking list $S_{[i]}, R_{[i]}, Q_{[i]}$

$S_{[i]}$	$R_{[i]}$	$Q_{[i]}$
0.232 (A4)	0.203 (A4)	-0.999 (A4)
0.356 (A1)	0.250	-0.369 (A1)
0.462 (A5)	0.250	-0.257 (A5)
0.547 (A2)	0.250	-0.167 (A2)
0.705 (A3)	0.250	0.000 (A3)

References

[1] Atanassov, K., & Gargov, G. (1989). Interval valued intuitionistic fuzzy sets. *Fuzzy Sets and Systems*, *31*(3), 343–349. https://doi.org/10.1016/0165-0114(89)90205-4.

[2] D., S. D., & E., F. H. (2020). Similarity measures on generalized interval-valued trapezoidal intuitionistic fuzzy number. *Malaya Journal of Matematik*, *S*(1), 313–318. https://doi.org/10.26637/mjm0s20/0059.

[3] Dhanasekar, S., Rani, J. J., & Annamalai, M. (2022). Transportation Problem for Interval-Valued Trapezoidal Intuitionistic Fuzzy Numbers. *International Journal of Fuzzy Logic and Intelligent Systems*, 22(2), 155–168. https://doi.org/10.5391/IJFIS.2022.22.2.155.

[4] Dong, J.-Y., & Wan, S.-P. (2015). Interval-valued trapezoidal intuitionistic fuzzy generalized aggregation operators and application to multi-attribute group decision making. In *Scientia Iranica E* (Issue 6). www.scientiairanica.com.

[5] Feng, F., Zheng, Y., Alcantud, J. C. R., & Wang, Q. (2020). Minkowski weighted score functions of intuitionistic fuzzy values. *Mathematics*, *8*(7). https://doi.org/10.3390/math8071143.

[6] Garg, H. (2016). A new generalized improved score function of interval-valued intuitionistic

fuzzy sets and applications in expert systems. *Applied Soft Computing Journal*, *38*, 988–999. https://doi.org/10.1016/j.asoc.2015.10.040.

[7] Isabels, R., Vinodhini, A. F., & Anandan, V. (2024). Evaluating and Ranking Metaverse Platforms Using Intuitionistic Trapezoidal Fuzzy VIKOR MCDM: Incorporating Score and Accuracy Functions for Comprehensive Assessment. *Decision Making: Applications in Management and Engineering*, *7*(1), 54–78. https://doi.org/10.31181/dmame712024858.

[8] Kodukulla, S. N. M., & Veeramachaneni, S. (2023). A METHOD TO RANK INTERVAL VALUED TRAPEZOIDAL INTUITIONISTIC FUZZY SETS AND ITS APPLICATION TO ASSIGNMENT PROBLEM. *Journal of Theoretical and Applied Information Technology*, *31*, 20. www.jatit.org.

[9] Kokoç, M., & Ersöz, S. (2021). New Ranking Functions for Interval-Valued Intuitionistic Fuzzy Sets and Their Application to Multi-Criteria Decision-Making Problem. *Cybernetics and Information Technologies*, *21*(1), 3–18. https://doi.org/10.2478/cait-2021-0001.

[10] Krishna, M., Kumar, S. D., Ezilarasan, C., Sudarsan, P. V., Anandan, V., Palani, S., & Jayaseelan, V. (2022). Application of MOORA & COPRAS integrated with entropy method for multi-criteria decision making in dry turning process of Nimonic C263. *Manufacturing Review*, *9*. https://doi.org/10.1051/mfreview/2022014.

[11] Kumar, T., Bajaj, R. K., & Ansari, M. D. (2020). On accuracy function and distance measures of interval-valued Pythagorean fuzzy sets with application to decision making. *Scientia Iranica*, *27*(4), 2127–2139. https://doi.org/10.24200/sci.2019.51579.2260.

[12] Murtykodukulla, S. N., Sireesha, V., & Anusha, V. (2022). *Distance Measure Approaches to Rank Interval-Valued Trapezoidal Intuitionistic Fuzzy Sets*. *71*(4). http://philstat.org.ph.

[13] Sayadi, M. K., Heydari, M., & Shahanaghi, K. (2009). Extension of VIKOR method for decision making problem with interval numbers. *Applied Mathematical Modelling*, *33*(5), 2257–2262. https://doi.org/10.1016/j.apm.2008.06.002.

[14] Singh, A., Bhat, S., & Bhat, S. A. (2020). *Posted on Authorea*. https://doi.org/10.22541/au.159612891.19134659.

[15] Wang, W., Liu, X., & Qin, Y. (2012). Interval-valued intuitionistic fuzzy aggregation operators. *Journal of Systems Engineering and Electronics*, *23*(4), 574–580. https://doi.org/10.1109/JSEE.2012.00071.

[16] Zaliluddin, D. (2023). Bibliometric Analysis of "Accuracy of Multi Criteria Decision Making (MCDM) of Assistance Recipients with Fuzzy Logic Algorithm." In *West Science Interdisciplinary Studies* (Vol. 01, Issue 07).

[17] Isabels, K. R., Vinodhini, G. A. F., & Anandan, V. (2024). Implementation of FMCDM Techniques to Select the Best Composite Material-Comparative Ranking. International Journal of Vehicle Structures and Systems, 16(2), 149–155. https://doi.org/10.4273/ijvss.16.2.03.

CHAPTER 12

Application of ABC-VED Analysis to Control the Inventory of Sri-Amman Textile Mill PVT LTD

A Case Study

Kanagajothi D.[1,a] **and Premila S.C.**[2,b]

[1]Associate Professor, Department of Mathematics, Vel Tech Rangarajan Dr.Sagunthala R&D Institute of Science and Technology, Chennai, India

[2]Assistant Professor, Department of Mathematics, Saveetha Engineering College, Thandalam, India

EMAIL: [a]kanagajothi82@gmail.com, [b]premilac@saveetha.ac.in

Abstract

An enterprise administration and oversight of its inventory system are integral parts. Inventory is seen as one of a projects most important resource. Inventory executives should be proactive, precise and efficient. Each firm needs inventory to guarantee that the creative engagement runs smoothly. Sri-Amman Textile Mill pvt ltd, a startup has conducted research to learn more about the inventory and the control system that will be used. Through Always Better Control (ABC) analysis, Vital, Essential and Desirable (VED) analysis and Economic Order Quantity (EOQ), variance analysis, Inventory turnover ratio and Inventory conversion period was examined and the control system was put into action. A study of inventory management at Sri-Amman Textile Mill pvt ltd is undertaken in order to know the inventory performance and position of the company and to know the strength and weakness and to assess the profitability of the company. The variation of the prices of raw materials are also analyzing with their effects on the overall working of the unit. This is a descriptive research in which the already available information is critically evaluated. This study describes the inventory management tools to be adopted by Sri-Amman Textile Mill for efficient inventory management.

Keywords: Inventory systems, ABC analysis, VED analysis, EOQ, Variance analysis, Inventory turnover ratio, Inventory conversion period.

1. Introduction

The three core functional subsystems of an organization are the departments of materials, manufacturing, and marketing. The materials subsystem gathers the input, the manufacturing subsystem transforms it, and the marketing subsystem sells it. The three working sub-systems' needs can be met by the other sub-systems, such as personnel control and finance. As these three operational subsystems work in a chain, it is easy to see how much heat will be produced at their interfaces.

Additionally, if the output of one subsystem serves as the input for the other two, all three depend on the other one to exist. At these interfaces, inventory, which is basically a store of commodities, acts as a buffer to facilitate the easy movement of materials from one subsystem to another. The lines can be balanced if there are sufficient stocks. Since the supplies for the other sub-systems can be obtained from inventories, even in the event of a brief failure in one, the other sub-system need not suffer. As a result, inventory serves as a bank and separates different operational phases.

Inventory management has experienced a lot of changes recently. The market for inventory software has been expanding, as more businesses are choosing to use inventory applications rather

DOI: 10.1201/9781003606611-12

than sticking with antiquated and unreliable tools like spreadsheets. The e-commerce business sector, which anticipates reaching $4 trillion by the end of 2020, and the rising demand for goods are mostly to blame for this. Businesses are investing in contemporary inventory management strategies in order to take advantage of this expansion as well as to increase process efficiency and optimize return on investment. In businesses such as grocery stores, fabrication, parts management and others that involve a lot of stock, a company's primary operations are centered on the raw materials and final commodities it produces.

A stock ought to be easy to locate when and where it's needed.

Knowing when to restock inventory, what totals to develop or acquire, what expenses to incur, when to sell, and at what price are some of the tough considerations.

2. Literature review

Minor companies will frequently physically inspect the s tock to ascertain how much has to be replenished and how much is reordered. Different inventory management strategies are acceptable, depending on the industry. ABC, method to classify (XYZ), Scare, Difficult and easy (SDE), MNG, Fast, Slow and Non-moving (FSN), High, Medium and Low (HML), VED Government, Ordinary, Local and Foreign (GOLF) and SOS analysis are a few of the several methods of inventory control.

Beyond hesitation, one of the areas required for any nation's economic growth is the industrial sector.

The manufacturing sector is thought to be especially significant since it witnessed significant technological advancements, along with the corresponding development of machinery and equipment and the introduction of contemporary production techniques (Al-Msary et al. 2022). Keeping track of inventory is essential since it protects an organizations most valuable asset, which if not handled effectively can have a detrimental impact on a company's performance since it consumes money that might be used for other business operations (Kumari 2020). Due to the rapid advancement of production techniques and equipment as well as the advent of automation, product quality has improved which has increased demand and sales and greatly influenced the production of a wide variety of goods and their distribution in the markets. As a result many organizations, particularly those in the industrial sector have purchased significant amounts of raw materials, replacement parts and other products. (Rahman et al. 2022).

The goods or amenities that come from the manufacturing procedure constitute the outputs, and the unprocessed ingredients or skills needed to finish the fabrication cycle are the production inputs. (Ayat 2017) the ABC-VED matrix was applied to the inventory items in this study, and the results showed that doing so helped determine the proper quantities to be retained in the inventory as well as decrease buildup and spoilage. They came to the conclusion (Al-Najjar et al. 2020) that the Pharmacy drug stocks are mostly under control thanks to the ABC-VED matrix inventory analysis.

This examination's superior use of inventory models benefits the hospital's health service delivery system, patient-hospital interaction, and the allotment of sufficient financial resources.

Inventory analysis helps cut down on waste, which in turn reduces processing expenses and damages to improve product quality and speed of data storage (Mor and others, 2021). According to (1) Abdolazimi et al. (2021), inventory is defined as all tangible assets, which include things that are now being produced and kept for sale as well as those that are actively being used to create goods and services.

By integrating both the ABC and VED analyses and using the ABC, VED matrix, three unique master groups with various characteristics are produced, which are used to categorise inventory items (Andawaningtyas 2020, (Bulkan & Ceylan 2017;) (Guner Goren & Dagdeviren 2017; Karim & Andawaningtyas, 2020). (Durmus & Dugral 2021). By cross-tabulating the results of the ABC-VED analysis, three groupings of master categories are created.

It should be noted that the first letter in this category denotes where the item falls within the ABC analysis's categories, while the second letter denotes where the item falls within the VED analysis's categories.

When calculating inventory costs, (Fadel and Mohammed 2021) took into account things like employee pay, gasoline, building upkeep, power, etc.

VED analysis was conducted using 1540 inventory.Items were investigated independently in comparison to the literature and a few additional criteria, such as FHC prescription rates and pharmacy customer profiles.

Fahriati et al.'s categorization of the inventory was made in 2021. Simple frequency categorization of the components that go towards making the product cost and utility In order to make it simple to apply stringent control to certain categories based on their value and amount of consumption, this analysis divides each item of inventory into one of the three A, B, and C groups by Pandya, B., & Thakkar, H. (2016). According to the study's findings, ABC-VED analysis should be used to manage resources more effectively and efficiently, to decrease wastage and to address medicine inventory shortages in the healthcare facilities it analysed.

While other researchers have linked the ABC approach, other authors have carried out comparable studies in various places, including Kumar et al. (2016) and Hazrati et al. (2018). Taddele et al. (2019) used the ABC-VED matrix in their study to analyse the inventory at Arbaminch Hospital (Ethiopia). The study's sample included 218 medications, and it was conducted from 2013 to 2015. By using ABC Analysis, VED Analysis, and EOQ, (Annie Rose Nirmala 2022) analysed the inventory management and implemented the control system.

Adopting the right inventory management techniques may reduce the negative impacts on pharmacy economies from unanticipated events like price changes, seasonal demand fluctuations, and taking advantage of bulk discounts Yilmaz, F. (2018). Muhammad Ayat (2017) suggested using an inventory system. Management strategies for effectively managing spare parts inventory management, effective product distribution and buying decisions, and close monitoring of products that fall within key categories. These academic Tools aid management in preventing out-of-stock situations. Additionally, remove dead stock without obstructing large amount of money. Additionally, these scientific instruments identify the Categories of goods require strict management oversight. These methods lessen the amount of cash that is locked up on stockpile of replacement components and lessen disruption period. Ebru DURSA (2022) discussed the knowledge and awareness of pharmacists about the importance of inventory control.

3. Objective of the Study

- For the chosen substance items, comprehend and quantify the amount of economic order.
- Utilizing ABC, VED, Ratio, and variance analyses to evaluate its inventory management techniques.
- To evaluate the inventory management practices of SRI- AMMAN TEXTILE MILLS.

4. Methodology and Model Specifications

4.1. ABC Analysis

ABC system of classification is used to classify and control the inventories. Establishing the form and level of control needed for each substance is helpful.

Group A items are those that require significant investment, and there should be more stringent and thorough inventory control there, inventory which involves relatively small investment but fairly large numbers of items are included under group C which needs minimum attention. Group B stands in between and hence it demands less attention than A but more than C group.

The various items of inventory being maintained in SRI-AMMAN TEXTILE MILL PVT LTD but I chosen selected 10 materials are thus classified and presented in the following tables on an yearly basis from 2014–2015 to 2017–2018.

A - Goods: In general, just 5–10% of the total items are responsible for 70–75% of the funds invested in commodities.

B - Goods: It frequently covers 10–15 percent of the total goods and 10–15 percent of the overall material expenses..

C - Goods: These are cheap (representing about 5–10 percent of the overall annual expenditure on materials), frequent (representing up to 90 percent of the total goods), and therefore inconsequential (do not require tight management) items.

Interpretation

From the tab.12.1, it is observed that most materials have been classified in the same category consistently for five years while for some materials there have been some changes in their classifications.

Table 12.1: The ABC analysis for the five year consolidated report.

S.NO	MATERIAL NAME	2014 – 2015	2015 – 2016	2016 – 2017	2017 – 2018	2018 – 2019
1	Cotton	A	A	A	A	A
2	Fiber Waste	C	C	C	C	B
3	Polyester	B	B	B	B	B
4	Viscose	C	C	C	C	C
5	Paper cone	B	B	B	C	C
6	Green color powder	C	C	C	C	C
7	Bearing	C	C	C	C	C
8	V-belt	C	C	C	C	B
9	Polythene	C	C	C	C	C
10	Ring Traveler	C	C	C	C	C

Cotton has been classified as category A consistently followed by polyester in category B and green color powder, viscose, bearing, polythene, ring traveler in category C without any changes over the five years period. But fiber waste was categorized as B in the last year (2018–2019) and Paper cone was categorized as B in the first three year (2014–2015 to 2016–2017) and then v-belt was categorized as B in the last year (2018–2019).

Table 12.2: The raw material consumption (Cotton).

Year	Amount of raw material (at closing stock)	Raw material consumed (Cotton)	% of Cotton consumption value
2014	943.15	18905.23	90.05%
2015	889.94	18890.88	88.34%
2016	911.73	17607.23	85.46%
2017	862.38	20801.19	86.58%
2018	110.32	22431.15	76.34%

Interpretation

The above graph shows consumption of raw material cotton. In the ABC analysis the raw material cotton constitute the major value in the chart.

Table 12.3: The VED analysis.

Category	% of Items	Total Money Value	% of The Total Money Value
V	40%	191962	84.11%
E	40%	320166	10.84%
D	20%	412630	5.04%
Total	100%	380626	100%

4.2. VED Analysis

Each supply product in the VED method (vital, essential, and desirable) is categorized as either vital, essential, or desirable depending on how important it is to the provision of health services. The essential items are kept in medium amounts, the critical items in large amounts, and the desirable items in modest amounts. Essential and vital commodities are always stocked, resulting in minimal disturbance to the services provided to the public.

Three distinct groups are employed in VED analysis to classify the inventory according to its functional importance (Table 12.3).

- Vital
- Essential
- Desirable

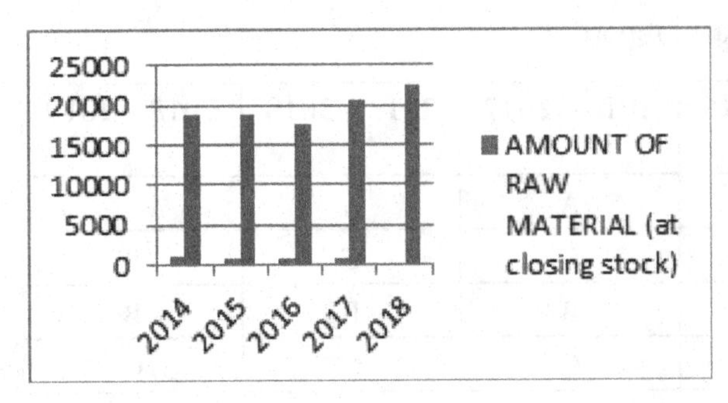

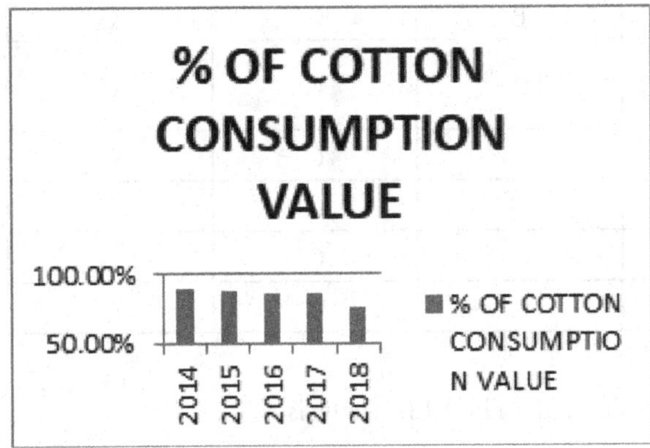

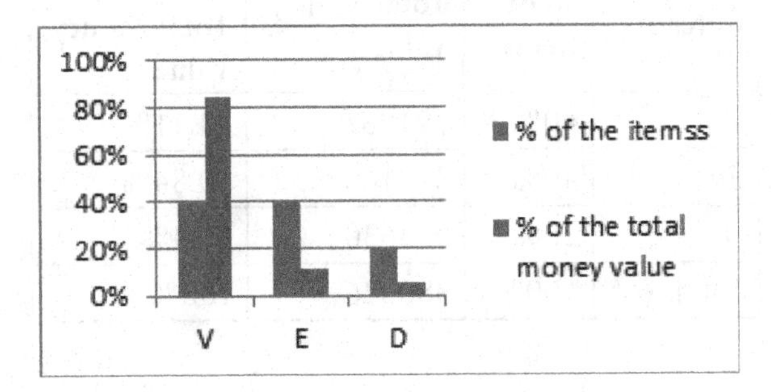

Interpretation:

From the above VED analysis, it is understood that 40% of the inventory items constituting 84.11% money value and the desirable items in modest amounts. Essential and vital commodities are always stocked, resulting in is classified under group 'V'. 40% of the items constituting 10.84% money value is classified as group 'E'. 20% of the items constituting 5.04% money value are classified as group 'D'.

4.3. Economic order quantity

The amount of merchandise that is most cost effective to order is known as the economic order of quantity. EOQ refers to the order size that maximizes economy while buying any item and, in the end, helps to maintain the material at the lowest possible cost and optimal level. One should evaluate the costs associated with economic order number two.

The various items of inventory being maintained in SRI-AMMAN TEXTILE MILL PVT LTD but I chosen selected 4 cotton materials. They are V797, S-4, MCU-5 and DCH. V797 cotton material is used for manufacturing bed sheet and towel. S-4 cotton material is used for manufacturing baniyan. MCU-5 cotton material is used for manufacturing saree. DCH cotton material is used for cloth this material gives over strength to material. It is highest rate cotton material.

Table 12.4: The EOQ for the five year consolidated report

Year	V797 (In Units)	S-4 (In Units)	MCU-5 (In Units)	DCH (In Units)
2014–2015	567	774	1218	996
2015–2016	435	772	1575	1032
2016–2017	557	623	19	1029
2017–2018	451	55	1861	113
2018–2019	366	597	1606	700

Interpretation

V797, S-4, MCU-5 and DCH. V797 cotton material is used for manufacturing bed sheet and towel. S-4 cotton material is used for manufacturing baniyan. mechanical condition unknown (MCU)-5 cotton material is used for manufacturing saree. Deal-Contigent-Hedging (DCH) cotton material is used for cloth this material gives over strength to material. It is highest rate cotton material. From this calculation MCU-5 cotton material of EOQ is high will compare to other material order.

4.4. Material cost variance

It can be determined by multiplying the total number of units acquired by the distinction between the actual and projected cost of acquiring supplies. It is employed to identify situations when a company might be overpaying for raw materials.

$$MCV = (SP * SQ - AP * AQ)$$

4.5. Material price variance

It is established bymultiplying the total number of units acquired by the distinction between the real and projected expenses for acquiring supplies. It is employed to identify situations when a company might be exceeding its budget for essential supplies.

$$MPV = AQ\ (SP - AP)$$

4.6. Material usage variance

The material usage variance is the difference between the actual quantities of materials used and the standard quantity of materials allowed for the actual output multiplied by the standard price. There can be many reasons for material usage variance including the use of defective or sub standard products, pilferage, wastage, the difference in material quality etc.

Material usage variance (MUV) = SP (SQ – AQ)

4.7. Deviation in Material Mix

It is the distinction between the real material costs used in a production process as budgeted and as actual. This variance can be used to determine whether a product can be made with a lower cost combination of ingredients by isolating the aggregate unit cost of each item and excluding all other variables. The idea only makes sense when it is feasible to change the composition of the materials used without lowering the final product's quality below a certain threshold.

digital material management (DMM) = SP (recommended storage quantit (RSQ) – AQ)

4.8. Material Yield Variance

The material yield variance is calculated by multiplying the variance of what actually happens and the normal manufacturing process by the cost per unit of the usual output.

It calculates how a variation in the production yield from the norm affects the cost of materials.

To provide more analysis of the material consumption variation, it is utilized in conjunction with the material mix variance.

$$MYV = SP\ (SQ - RSQ)$$

Interpretation

The material cost variance has been in the favorable position for the last five year. The material price variance has been in the favorable position for the last five year. The material usage variance has been in the favorable position for the last five year. In the year 2014–2015 The material mix variance has been in the adverse position and then the year 2015–2016 to 2018–2019 "MCU-5" raw materials in the adverse position because of the actual quantity of material consumed more than RSQ. The material yield variance has been in the favorable position for the last five year.

5. Investigation of the Turnover of Inventory Rate

5.1. Turnover of Inventory Rate

One type of accounting turnover rate that is associated with finance is the inventory turnover rate.

This ratio calculates how many goods, on average, are sold throughout the course of time. Its objective is to gauge the inventory's liquidity.

Converting the inventory turnover ratio into an average day to sell the goods in terms of days is a common variation of the ratio.

The formula: Cost of goods sold / average value of Inventory.

Table 12.5: The variance analysis for the five year consolidated report

Variance Analysis	2014–2015	2015–2016	2016–2017	2017–2018	2018–2019
Material cost variance	F	F	F	F	F
Material price variance	F	F	F	F	F
Material usage variance	F	F	F	F	F
Deviation in Material Mix	A	A	A	A	A
Material yield variance	F	F	F	F	F

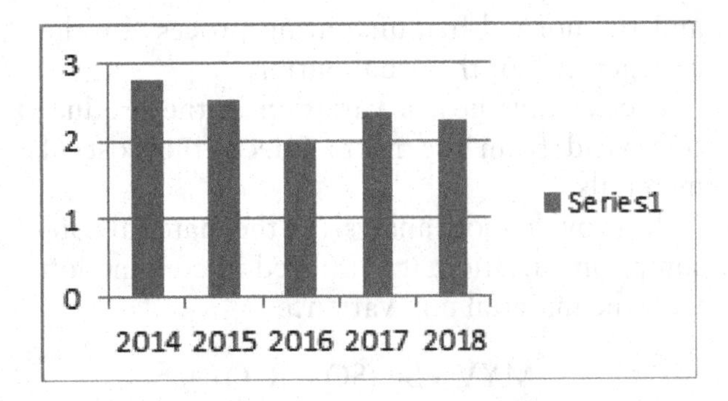

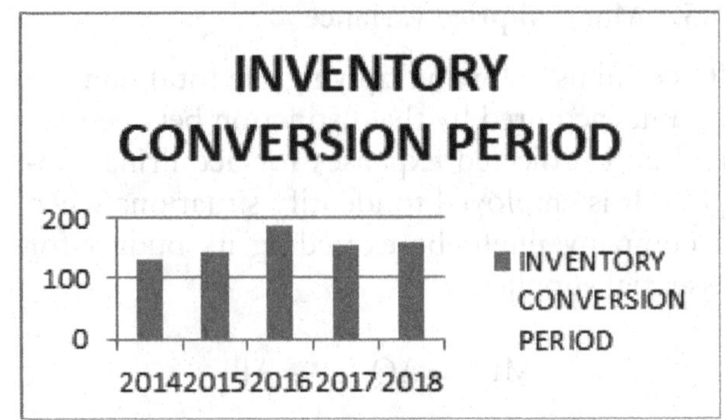

Table 12.6: Turnover of inventory rate

Year	Cost of goods sold	Average value of inventory	Turnover of inventory rate
2014	253511.60	90877.59	2.79
2015	232032.17	91654.67	2.53
2016	176674.28	90084.04	1.96
2017	210836.28	88706.27	2.37
2018	224311.56	98282.86	2.28

6. The period of inventory Transition

The period of time from the day materials are obtained to the date a good or service is sold is known as the period of inventory transition.

The formula: No.of. Days in a Year/inventory turnover ratio (Table 12.7).

Table 12.7: The period of Inventory Transition

Year	No. of. days	Inventory turnover ratio	Inventory conversion period
2014	365	2.79	130
2015	365	2.53	144
2016	366	1.96	186
2017	365	2.37	154
2018	365	2.28	160

Interpretation

It is clear from the above table that the stock turnover ratio in the year 2014–2015 is 2.74, which is recorded high in the study period of five years. It was 2.53 in the year 2015–2016. It was much low, 1.96 in the year 2016–2017. An excellent inventory turnover ratio indicates strong sales.A substantial amount of inventory is implied by a poor turnover of assets ratio. It is concluded that in the year 2014–2015, it was high.

Interpretation: The lowest inventory conversion period had shown 130 days during the year 2014–2015. The highest inventory conversion period had shown 186 days during the year 2016–2017. So the inventory conversion period of the company was not satisfactory

7. Conclusions

7.1 ABC analysis

ABC analysis reveals that most materials have been classified in the same category consistently for five years while for some materials there have been some changes in their classifications.

Cotton has been classified as category A consistently followed by polyester in category B and green color powder, viscose, bearing, polythene, ring traveler in category C without any changes over the five years period.

But fiber waste was categorized as B in the last year (2018–2019) and Paper cone was categorized as B in the first three year (2014–2015 to 2016–2017) and then v-belt was categorized as B in the last year (2018–2019).

7.2. VED analysis and economic order quantity

VED analysis reveals that 40% of the inventory items constituting 5.04% money value are classified under group 'V'. 40% of the items constituting 84.11% money value are classified as group 'E'. 20% of the items constituting 10.84% money value are classified as group 'D'.

In the MCU-5 cotton material of EOQ is high will compare to other material order.

7.3. Variance analysis

The material cost variance has been in the favorable position for the last five year. The material price variance has been in the favorable position for the last five year.

The material usage variance has been in the favorable position for the last five year.

In the year 2014–2015 The material mix variance has been in the adverse position and then the year 2015–2016 to 2018–2019 "MCU-5" raw materials in the adverse position because of the actual quantity of material consumed more than RSQ.

The material yield variance has been in the favorable position for the last five year.

7.4. Ratio analysis

The stock turnover ratio reveals in the year 2014–2015 is 2.74, which is recorded high in the study period of five years. It was 2.53 in the year 2015–2016. It was much low, 1.96 in the year 2016–2017.

An excellent inventory turnover ratio indicates strong sales.

A substantial amount of inventory is implied by a poor turnover of assets ratio. It is concluded that in the year 2014–2015, it was high.

The lowest inventory conversion period had shown 130 days during the year 2014–2015. The highest inventory conversion period had shown 186 days during the year 2016–2017. So the inventory conversion period of the company was not satisfactory.

The present study is an attempt to analyze the inventory management in SRI-AMMAN TEXTILE MILL PVT LTD. The study focuses on different aspect such as ABC analysis, VED analysis, economic order quantity, variance analysis and ratio analysis. The findings made in the study will help the company to have proper control over inventory management.

References

[1] Abdolazimi, O., Shishebori, D., Goodarzian, F., Ghasemi, P., & Appolloni, A. (2021) Design a new mathematical model based on ABC analysis of an inventory control problem: a real case study, 2309–2335.

[2] Al-Msary, A. J. K., Nasrawi, B. K. M., & Alisawi H. G. J. (2022). Analyzing and evaluating economic indicators and occupational safety to raise performance efficiency in the industrial company: Applied research in Babylon Cement Factory, 8(1), 1–19.

[3] Al-Najjar, S. M., Jawad, M. K., & Saber, O. A. (2020). Application of ABC-VED Matrix Analysis to Control the Inventory of a Central Pharmacy in a Public Hospital: A Case Study , 9(1), 1328–1336.

[4] Andawaningtyas, K., & Karim, C. (2020). Analysis of grouping ABC -VED and predicting the number of requests, 1–12.

[5] Ayat, M. (2017). A Case Study of Spare Parts Inventory of Printing Industry using ABC and VED analysis, 5, 1–7.

[6] Ceylan, Z. & Bulkan, S. (2017). Drug Inventory Management of a Pharmacy using ABC and VED Analysis, 2(1), 13–18.

[7] Durmus, A., & Dugral, E. (2021). Inventory Management with ABC and VED Analysis in Hospitals During the Covid-19 Pandemic Process, 368–377.

[8] Fadel, S. S., & Mohammed M. J. (2021). Design a supply chain model for Baghdad Soft Drinks Company, 27(128), 202–218.

[9] GunerGoren, H., & Dagdeviren, O. (2017). An Excel-Based Inventory Control System Based on ABC and VED Analyses for Pharmacy: A Case Study, 2, 11–17.

[10] Kumari , M. (2020). Inventory Management Issues in Indian Steel Industry: A Qualitative Study, 15, 14–25.

[11] Mor, R. S., Kumar, D., Yadav, S., & Jaiswal, S. K. (2021). Achieving cost efficiency through increased inventory leanness: Evidence from manufacturing industry, 27(1) ,42–49.

[12] Pandya, B., & Thakkar, H. (2016). A Review on Inventory Management Control Techniques: ABC-XYZ Analysis, 2(3), 82–86.

[13] Rahman, A., Muda, I., Caroline, C., Panjaitan, P.D & Situmorang, R.Y. (2022). The panel Data regression: Relationship of the Exports, Imports and intake of Oil Reserves in Southwest Asia the Country Contribution to State Revenue (The Implementation of International Trade Theory), 7(3), 1–13.

[14] Hazrati, E., Paknejad, B., Azarashk, A. and Taheri, M., (2018), ABC and VED Analysis of Imam Reza Educational Hospital Pharmacy, Annual Military Health Science Research., Vol. 16, No. 3, pp. 1–5.

[15] Kumar Y., Lilhare A., Sahu, A., Lal, B. and Khaperde Y., (2016), ABC Analysis for Inventory ManagementCase Study of Sponge Iron Plant, International Journal for Research in Applied Science & Engineering, Vol.4, No. 4, pp. 32–36.

[16] Taddele, W., Wondimagegn, A., Asaro, A., Sorato, M., Gedayi, G. and Hailesilase, A., (2019), ABC-VEN Matrix Analysis of the Pharmacy Store in a Secondary Level Health Care Facility in Arbaminch Town, Southern Ethiopia, Journal of Young Pharmacists, Vol. 11, No. 2, pp. 182–185. DOI: 10.5530/jyp.2019.11.38.

[17] Annie Rose Nirmala D., Vijila Kannan, Thanalakshmi M., Joe Patrick Gnanaraj S., Appadurai M., (2022),"Inventory management and control system using ABC and VED anlaysis", materials today Proceedings, Volume 60, pp:922–925.

[18] Yilmaz, F. (2018). The drug inventories evaluation of healthcare facilities using ABC and VED analyses. Istanbul Journal of Pharmacy, 48, 43–48.

[19] Muhammad Ayat, (2017), "A Case Study of Spare Parts Inventory of printing Industry using ABC and VED analysis", Merit Research Journal of Business and Management (ISSN: 2408-7041) Vol. 5(1) pp. 001–007.

[20] Ebru DURSA* , Miray ARSLAN, (2022), "ABC, VED, and ABC-VED Matrix Analyses for Inventory Management in Community Pharmacies: A Case Study", FABAD J. Pharm. Sci., 47, 3, 293–300.

[21] Kalaiarasi K., Mary Henrietta H., Sumathi, A. Stanley Raj M., The Economic order quantity in a fuzzy environment for a periodic inventory model with variable demand, Iraqi Journal of Computer Science and Mathematics. 3(1):102–107 (2022).

[22] Kalaiarasi K., Mary Henrietta H., Sumathi M., Stanley Raj A., The optimality representation of a finite-life inventory model with exiguous defectives with Hessian matrix approach, Nonlinear Studies. 30(2): 393–398 (2023)

CHAPTER 13

Integration of particle swarm optimization with COPRAS method for optimal material selection

N. Kavitha[1,a], G. Uthra[2] and V. Anandan[1]

[1]Department of Mathematics, Saveetha Engineering College, Chennai, India
[2]PG and Research Department of Mathematics, Pachaiyappa's College, Chennai, India
[a]sakkavi238@gmail.com

Abstract

Material selection is an important task in engineering and manufacturing, involving conflicting criteria that need to be examined for its sustainability, safety, and performance. Traditional multi criteria decision-making (MCDM) methods use fixed weights, it may not capture the reality of the environment as there is a chance to influence the environment by the decision-makers. This work proposes an integration of particle swarm optimization (PSO) with the complex proportional assessment (COPRAS) method to enhance the flexibility and precision of material selection process. Using PSO, the weights of criteria can be optimized based on its importance and requirement in real-time situations. The integrated algorithm is utilized for a material selection problem in manufacturing. The optimized criterion weights significantly enhance the ranking accuracy of the modified COPRAS method. The results are further compared with the conventional method to prove the need of the integrated approach. The hybrid PSO-COPRAS technique facilitates to make accurate material selection decision-making in engineering applications.

Keywords: COPRAS, PSO, MCDM, material selection

1. Introduction

Material selection is a critical process in manufacturing and engineering, as the choice of materials significantly impact the performance, cost, durability, and overall quality of the product. The material selection process involves evaluating and selecting the most suitable materials based on the conflicting criteria, such as cost, quality, durability, efficiency, and innovation. Given the complexity and multifaceted nature of these criteria, effective material selection necessitates robust decision-making methodologies to balance the various factors involved.

1.1 Multi Criteria Decision-Making

MCDM has emerged as an essential tool for addressing complex decision problems that involve multiple conflicting criteria [3, 7, 8]. MCDM methods provide a structured framework for evaluating various alternatives and identifying the optimal choice by considering all relevant criteria simultaneously. In the recent decades, MCDM has seen significant growth and development, with numerous methods being proposed and refined to enhance decision-making processes. Among these methods, the COPRAS method has garnered attention for its effectiveness and simplicity.

1.2 The COPRAS Method

The COPRAS method [1] is an MCDM approach that evaluates and ranks alternatives based on a set of criteria. It calculates the relative importance of all available options by considering both the beneficial and non-beneficial criteria, providing a clear and rational ranking of options. Recent modifications to the COPRAS method have aimed at improving its accuracy and adaptability [6]. For instance, researchers have explored various weighting

DOI: 10.1201/9781003606611-13

schemes, dynamic adjustments, and integration with other optimization techniques to enhance decision-making outcomes.

In the manufacturing sector, COPRAS has been widely used for material selection, process optimization, supplier evaluation, and product design. The method's ability to handle multiple criteria and provide a comprehensive ranking makes it particularly suitable for material selection, where various attributes need to be considered simultaneously. Studies have shown that COPRAS can effectively identify the best materials by balancing different criteria, leading to improved product performance and cost efficiency.

1.3 Particle Swarm Optimization

PSO is a metaheuristic optimization method replicates the social behaviour of birds flocking or fish schooling, found to be useful for solving complex optimization problems due to its simplicity and effectiveness [4, 10, 11]. PSO has been successfully applied in various domains, including engineering, economics, and artificial intelligence. Recent literature highlights the increasing use of PSO in manufacturing for optimizing production processes, resource allocation, and quality control [2, 16, 18].

1.3 Background

Material selection is a significant part of manufacturing, directly affecting product cost, performance, and sustainability. In complex decision-making problems involving multiple conflicting criteria such as cost, quality, and durability, it often requires suitable methods like the COPRAS. While COPRAS effectively ranks alternatives depending on fixed criteria weights, it lacks adaptability in dynamic environments where these weights might vary. Recent studies have focused on improving COPRAS through integration with optimization algorithms like PSO, which dynamically change the weights considerably. But, the integrated PSO and COPRAS in material selection remains under-explored. This study addresses this gap by integrating PSO with a modified COPRAS method, aiming to optimize material selection processes and contribute to more effective and nuanced decision-making in manufacturing.

1.4 Research Gap

The proposed method integrates the meta-heuristic algorithm PSO with the MCDM technique COPRAS to determine the best alternative under the conflicting criteria [13, 19]. The PSO algorithm is proposed to determine the weight of the criteria depending on its importance, making sure that that the criteria conditions remain open to existing situations. The integrated technique helps to enhance the flexibility and precision in material selection process, found to be suitable for complex decision-making situations [20].

2. Integrated PSO and COPRAS Method

The integrated PSO-COPRAS method is utilized to obtain the optimal alternative in the material selection problem under conflicting criteria. The optimal criterion weights help for effective decision-making. The superiority and efficiency of the integrated method based on its adaptability and precision for better optimization in manufacturing is addressed in this study [14].

This research supports for the development of MCDM techniques by the novel combination of PSO and COPRAS, to address the limitations and provide a comprehensive solution for material selection in dynamic situations [9, 12, 15]. The two-stage proposed algorithm, is explained as follows.

2.1 Stage 1: PSO Method

PSO is an optimization algorithm which replicates the social behaviour of birds flocking or fish schooling.

1. Set the swarm:

- Set the number of particles n
- Fix a search space dimensions d
- Initialize the position x_i of each particle i randomly within the search space.
- Initialize the velocity v_i of each particle i randomly.
- Initialize particle's favourite location p_i to its initial position.
- Initialize the global favourite location g to the best position among all particles.

2. Fitness Estimation:
   - For every particle i, evaluate the fitness (x_i) of its current position.

3. Update individual and global best positions:
   - For each particle i:
     If $(x_i) < f(p_i)$
     Update p_i to x_i.
   - If $f(p_i) < f(g)$:
     Update g to p_i.

4. Update velocity and position:
   - For each particle i:
     Update the velocity v_i using the formula:
     $v_i = w \cdot v_i + c_1 \cdot r_1 \cdot (p_i - x_i) + c_2 \cdot r_2 \cdot (g - x_i)$

   where, w refers inertia weight; c_1 and c_2 indicates coefficients of cognitive and acceleration; r_1 and r_2 refers random numbers in [0, 1].

   Update the position *xi* using the formula:
   $x_i = x_i + v_i$

5. Iteration:
   - Repeat steps 2 to 4 until the criterion converges.

6. Return the global best position g and its fitness $f(g)$ to determine the criterion weights.

2.2 Stage 2: Modified COPRAS Method

1. The initial matrix is normalized to reduce the values in [0,1] for each criterion.

$$n_{ij} = \frac{x_{ij}}{\sqrt{\sum x_{ij}^2}} \quad \text{(For benefit criteria)} \qquad (1)$$

$$n_{ij} = 1 - \frac{x_{ij}}{\sqrt{\sum x_{ij}^2}} \quad \text{(For cost criteria)} \qquad (2)$$

2. The weighted normalization matrix is obtained using the criterion weights obtained from PSO

$$w_{ij} = n_{ij} * w_j; \qquad (3)$$

$$i \cdot j = 1, 2, \ldots m \text{ and } n \text{ respectively}$$

3. Calculate the aggregation value for the 'k' benefit criteria $P_{ij} = \sum_{i=1}^{k} w_{ij}$ (4)

4. Calculate the aggregation value for the 'm-k' cost criteria $R_{ij} = \sum_{i=k+1}^{m} w_{ij}$ (5)

5. Determine the relative weight

$$Q_{ij} = P_{ij} + \frac{\sum_{i=1}^{m} R_{ij}}{R_{ij} \sum_{i=1}^{m} \frac{1}{R_{ij}}} \qquad (6)$$

6. The maximum Q_i measure directs the best alternative.

3. Numerical Example

Consider a material selection problem with 6 alternatives and 5 criteria. The values are represented as initial matrix in Table 13.1. The conflicting criteria are cost, quality, durability, efficiency and innovation.

Table 13.1: Initial matrix

	Cost	Quality	Durability	Efficiency	Innovation
M1	690	3.1	9	7	4
M2	590	3.9	7	6	10
M3	600	3.6	8	8	7
M4	620	3.8	7	10	6
M5	700	2.8	10	4	6
M6	650	4.0	6	9	8

The PSO algorithm was iterated for 100 runs, optimizing the weights for each criterion to enhance the decision-making process. The optimal weights are obtained using PSO as shown in Table 13.2. The PSO weights obtained in Table 13.2 supports the decision-maker in making an unbiased estimation about the criteria against each material.

Table 13.2: PSO weight

Criteria	PSO Weight
Cost	0.18
Quality	0.22
Durability	0.20
Efficiency	0.19
Innovation	0.21

The values in the initial matrix are different as criteria is different with different data range. Hence the normalization is necessary to reduce all values in [0,1]. Normalize the initial matrix values in Table 13.1 using (1) and (2) as illustrated in Table 13.3.

Table 13.3: Normalized matrix

	Cost	Quality	Durability	Efficiency	Innovation
M1	0.8429	0.775	0.9	0.7	0.4
M2	0.7214	0.975	0.7	0.6	1.0
M3	0.7286	0.900	0.8	0.8	0.7
M4	0.7714	0.950	0.7	1.0	0.6
M5	0.8429	0.700	1.0	0.4	0.6
M6	0.7714	1.000	0.6	0.9	0.8

Obtain the weighted Normalized matrix using (3) and the criterion weights from Table 13.2 as shown in Table 13.4. The PSO weight represents the importance of the criteria providing reasonable relationship among the criteria.

Table 13.4: Weighted normalized matrix

	Cost	Quality	Durability	Efficiency	Innovation
M1	0.0843	0.155	0.27	0.175	0.06
M2	0.0721	0.195	0.21	0.175	0.15
M3	0.0729	0.180	0.24	0.150	0.105
M4	0.0771	0.190	0.21	0.200	0.09
M5	0.0843	0.140	0.20	0.100	0.09
M6	0..0771	0.200	0.18	0.225	0.12

Determine the Regret and Utility Measure using (4), (5) and (6) as shown in Table 13.5. The regret and utility measure indicates the closeness coefficient of each material, which is used to rank the importance of the material.

Table 13.5: Regret and utility measure

	Ri	Ui	Rank
M1	0.7443	0.9110	5
M2	0.7771	0.9510	4
M3	0.7979	0.9764	3
M4	0.8171	0.9990	1
M5	0.7143	0.8736	6
M6	0.8021	0.9816	2

The results indicate that M4 is the best alternative based on the multiple criteria, followed closely by M6 and M3. The integration of PSO has allowed for dynamic optimization of weights, leading to a more refined decision-making process.

4. Results and Discussion

The integration of PSO with the COPRAS method was utilized to determine the best alternative in the material selection problem, involving six materials evaluated across five criteria. The optimal criterion weights were obtained as follows: Cost: 0.18; Quality: 0.22; Durability: 0.20; Efficiency: 0.19; Innovation: 0.21

Using the PSO criterion weights, the COPRAS method was used to rank the alternatives. The determination of the relative importance of each material was done based on the positive and negative ideal measures. The final rankings based on the utility measure is as follows:

M3 > M6 > M1 > M4 > M2 > M5

4.1 Integrated COPRAS vs. Traditional COPRAS

The traditional COPRAS method was applied to the same problem with equal criterion weights and the rankings was obtained as given below:

M6 > M3 > M1 > M4 > M5 > M2

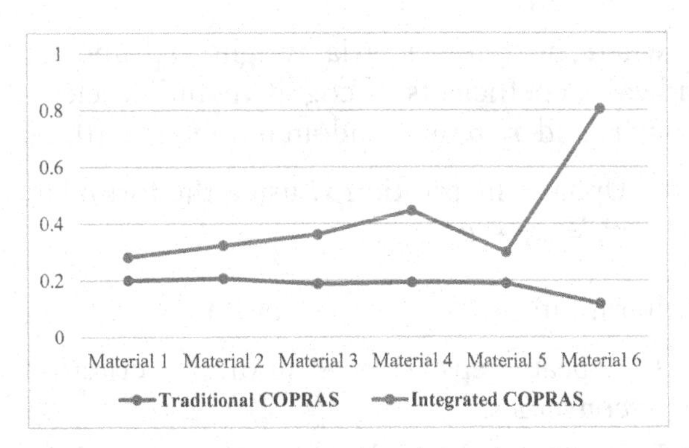

Figure 13.1: Comparison of traditional and integrated COPRAS

To assess the correlation between the traditional and proposed COPRAS methods, the comparison is done based on the COPRAS grade and represented graphically. Figure 14.1 shows the comparison between the integrated and traditional COPRAS methods based on the utility measure of each material.

The traditional COPRAS method found out that M6 as the better choice, but the integrated PSO-COPRAS method, ranked M3 as the best alternative. This nonconformity is because of the effect of dynamic weight optimization predicted using PSO, which offers better accuracy of the proposed technique in addressing the differences among the alternatives.

4.2 Symmetric Mean Absolute Percentage Error (SMAPE) Analysis

The SMAPE was calculated to determine the correlation between the traditional and integrated COPRAS methods. The SMAPE score was 14.26%, representing a relatively low SMAPE error and realistic agreement. The integrated PSO-COPRAS method offers a distinct ranking and maintains better consistency with the traditional method.

The integrated PSO-COPRAS technique exhibited its adaptability by providing distinct rankings compared to the traditional COPRAS method. This adaptability is very much necessary for material selection, where decision criteria can change depending on the requirements and other external factors. The minimal SMAPE value indicates the accuracy of the proposed method, representing its effectiveness in predicting the material selection outcomes [5, 17]. The SMAPE test is a validation test to confirm the reliability of the integrated PSO-COPRAS method.

4.3 Discussion

The integrated PSO-COPRAS technique significantly improves the decision-making in the context of material selection. The optimized criterion weights replicate the dynamic nature of decision-making environment. The proposed method addresses the limitations in the traditional COPRAS method, as it depends on fixed weights and may even influence the decision-making as per the decision-makers wish.

The low SMAPE value refers that the integrated technique provides a better correlation against the traditional method and improves the sensitivity and adaptability.

The proposed PSO-COPRAS method acts as a strong and flexible tool for material selection in manufacturing, significant improvements in decision-making accuracy and precision addressing all the research gaps. The research proves the efficiency of integrating metaheuristic optimization technique and MCDM method to improve its effectiveness in complex decision-making situations.

5. Conclusion

1. The work has confirmed the significant efficiency of integrating PSO and the COPRAS method to improve the material selection process in manufacturing. The traditional COPRAS method, though effective, depends on fixed weights depending on the decision-makers choice. The use of dynamic weight determination optimization using PSO, an accurate mechanism is established for decision-making.
2. The criterion weights obtained from PSO indicates the importance of criteria, supports in getting context-sensitive rankings. The SMAPE analysis confirmed a less error rate of 14.26%, highlighting a strong correlation with the traditional method and enhancing sensitivity and adaptability.
3. The results confirms that M3 was recognized as the optimal alterative using the integrated PSO-COPRAS method, whereas M6 was considered as the best in traditional method. This deviation highlights the ability of the proposed approach to capture distinct changes between the materials, offering robust and flexible tool.
4. The integrated PSO-COPRAS method signifies an essential advancement in MCDM techniques, particularly in the field of material selection for manufacturing. The proposed method improves the decision-making accuracy and adaptability, offering a useful tool for solving complex and dynamic decision-making problems. This research offers the groundwork for further exploration and combination of metaheuristic optimization techniques with MCDM methods.

References

[1] Alinezhad, Alireza, Javad Khalili, Alireza Alinezhad, and Javad Khalili. "COPRAS method." New methods and applications in multiple attribute decision making (Madm) (2019): 87–91.

[2] Anandan, V, Naresh Babu, M, Vetrivel Sezhian, M, Cagri Vakkas Yildirim, Dinesh Babu, M, Influence of graphene nanofluid on various environmental factors during turning of M42 steel, Journal of Manufacturing Processes, 68 (2021), 90–103.

[3] Anandan, V, and G. Uthra. "Defuzzification by area of region and decision making using Hurwicz criteria for fuzzy numbers." Applied Mathematical Sciences 8, no. 63 (2014): 3145–3154.

[4] Bharathi Raja, S., and N. Baskar. "Particle swarm optimization technique for determining optimal machining parameters of different work piece materials in turning operation."

The International Journal of Advanced Manufacturing Technology 54 (2011): 445–463.

[5] Benioff, Paul. "The computer as a physical system: A microscopic quantum mechanical Hamiltonian model of computers as represented by Turing machines." Journal of statistical physics 22 (1980): 563–591.

[6] Chatterjee, Prasenjit, and Shankar Chakraborty. "Materials selection using COPRAS and COPRAS-G methods." International Journal of Materials and Structural Integrity 6, no. 2–4 (2012): 111–133.

[7] Duc Trung, Do. "A combination method for multi-criteria decision making problem in turning process." Manufacturing review 8 (2021): 26.

[8] Ehrgott, Matthias, Kathrin Klamroth, and Christian Schwehm. "An MCDM approach to portfolio optimization." European Journal of Operational Research 155, no. 3 (2004): 752–770.

[9] Feng, Ruiliang, Jingchao Jiang, Zhichao Sun, Atul Thakur, and Xiangzhi Wei. "A hybrid of genetic algorithm and particle swarm optimization for reducing material waste in extrusion-based additive manufacturing." Rapid Prototyping Journal 27, no. 10 (2021): 1872–1885.

[10] Gad, Ahmed G. "Particle swarm optimization algorithm and its applications: a systematic review." Archives of computational methods in engineering 29, no. 5 (2022): 2531–2561.

[11] Juneja, Mudita, and S. K. Nagar. "Particle swarm optimization algorithm and its parameters: A review." In 2016 International Conference on Control, Computing, Communication and Materials (ICCCCM), pp. 1–5. IEEE, 2016.

[12] Kornyshova, Elena, and Camille Salinesi. "MCDM techniques selection approaches: state of the art." In 2007 IEEE symposium on computational intelligence in multi-criteria decision-making, pp. 22–29. IEEE, 2007.

[13] Krishna, Munuswamy, Sathuvachari Devarajan Kumar, Chakaravarthy Ezilarasan, Perumalsamy Vishnu Sudarsan, Viswanathan Anandan, Sivaprakasam Palani, and Veerasundram Jayaseelan. "Application of MOORA & COPRAS integrated with entropy method for multi-criteria decision making in dry turning process of Nimonic C263." Manufacturing Review 9 (2022): 20.

[14] Maity, Saikat Ranjan, Prasenjit Chatterjee, and Shankar Chakraborty. "Cutting tool material selection using grey complex proportional assessment method." Materials & Design (1980–2015) 36 (2012): 372–378.

[15] Onut, Semih, Selin Soner Kara, and Tugba Efendigil. "A hybrid fuzzy MCDM approach to machine tool selection." Journal of intelligent manufacturing 19 (2008): 443–453.

[16] Shami, Tareq M., Ayman A. El-Saleh, Mohammed Alswaitti, Qasem Al-Tashi, Mhd Amen Summakieh, and Seyedali Mirjalili. "Particle swarm optimization: A comprehensive survey." IEEE Access 10 (2022): 10031–10061.

[17] Sidik, Aryo De Wibowo Muhammad, Ilyas Aminuddin, Cahya Laxa Eka Putra, Edwinanto, Apriditia Karisma, and Yufriana Imamulhak. "Enhancing Material Selection Efficiency: A Multi-Criteria Decision-Making Approach." In 2nd International Conference on Consumer Technology and Engineering Innovation (ICONTENTION 2023), pp. 80–85. Atlantis Press, 2024.

[18] Son, Pham Vu Hong, Nguyen Huynh Chi Duy, and Pham Ton Dat. "Optimization of construction material cost through logistics planning model of dragonfly algorithm—particle swarm optimization." KSCE Journal of Civil Engineering 25, no. 7 (2021): 2350–2359.

[19] Yazdani, Morteza, Ali Jahan, and Edmundas Kazimieras Zavadskas. "Analysis in material selection: influence of normalization tools on COPRAS-G." (2017).

[20] Yazdani-Chamzini, Abdolreza, Mohammad Majid Fouladgar, Edmundas Kazimieras Zavadskas, and S. Hamzeh Haji Moini. "Selecting the optimal renewable energy using multi criteria decision making." *Journal of Business Economics and Management* 14, no. 5 (2013): 957–978.

CHAPTER 14

The Estimation of Optimal Order Quantity in an Inventory Model utilizing Fuzzy Logic

H. Mary Henrietta

[1]Department of Mathematics, Saveetha Engineering College, Chennai, Tamil Nadu, India
Email: henriettamaths@gmail.com

Abstract

Operations research is one among the most prominent areas of mathematics. Beginning in the early 19th century, the study of inventory models evolved into the study of economic order quantity (EOQ). In this study, the ideal order quantity was calculated for an inventory model's overall cost. The two problems of when to order and what to order are mostly answered by inventory. A fuzzy inventory model is a type of inventory model that uses fuzzy logic to deal with uncertainty in the demand and other factors that affect inventory levels. Fuzzy logic is a mathematical approach to dealing with uncertainty that allows for the use of imprecise or vague information. This makes fuzzy inventory models well-suited for situations where the demand for a product is difficult to predict, such as for seasonal products or products with a short shelf life. Fuzzy inventory models are typically more complex than traditional inventory models, but they can provide more accurate results. This is because they can take into account the full range of possible values for the demand and other factors, rather than just the most likely values. In this paper, we fuzzied the lead-time and the demand parameter using trapezoidal fuzzy numbers and compared the results. The results of the sensitivity analysis between the crisp and fuzzy numbers were tabulated.

Keywords: Trapezoidal fuzzy number, economic order quantity (EOQ), defuzzification, fuzzy, inventory model

1. Introduction

The most important aspect of any corporate operation is inventory management. It entails regulating the flow of materials and products effectively from the perspective of origination to the perspective of consumption. Businesses that effectively manage their inventories can save expenses, improve customer satisfaction, and increase profits. It is more crucial than ever for businesses to understand inventory management in the fast-paced, cutthroat business environment of today. In order to optimize inventory levels, reduce expenses, and increase profits, this presentation will give a general overview of the important ideas and strategies that can be applied. We sincerely hope that you enjoy and learn from this presentation. The EOQ formula helps businesses decide on the right number of orders to place in order to reduce their costs associated with holding inventory. The method determines the most cost-effective quantity to order by taking into consideration variables including ordering costs, holding costs, and demand rates. One real-world example of the benefits of using EOQ is seen in the automotive industry. By implementing EOQ, car manufacturers can reduce their inventory costs by up to 20%, resulting in significant savings for the company. This is achieved through better management of raw materials and parts, reducing waste and excess inventory.

Traditional (crisp) inventory models are usually based on methods that make use of deterministic, defined parameters and assumptions. These models do not account for uncertainty or unpredictability since they presume that demand, lead times, and other factors are

DOI: 10.1201/9781003606611-14

known with certainty. In contrast, stochastic or probabilistic components are frequently included in contemporary inventory management approaches to take lead times, demand variations, and other uncertainties into consideration. Decision-makers can more effectively comprehend and control the risks related to stockouts, overstocking, and supply chain disruptions by taking uncertainty into account when creating inventory models.

(a) Enhanced Service Levels: Using probabilistic components makes it possible to estimate service levels and customer satisfaction more precisely, which improves customer loyalty and retention. (b) Optimized Inventory Policies: With the use of contemporary techniques, inventory policies may be created that are more adaptable to shifts in market dynamics and demand trends, which lowers holding costs and boosts overall productivity. (c) Improved Resource Allocation: Decision-makers can allocate resources more efficiently and guarantee optimal inventory levels while minimizing expenses and optimizing revenue by taking uncertainty into consideration. (d) Adaptability to Dynamic contexts: More flexibility and adaptation to changing conditions is offered by modern approaches, while traditional models may find it difficult to adjust to dynamic and uncertain environments. In today's complicated and uncertain business climate, adopting contemporary inventory management strategies helps firms satisfy consumer demand more effectively, cut costs, and make smarter decisions.

This paper investigates an inventory model and determines the EOQ in crisp and fuzzy by using trapezoidal fuzzy numbers. Subsequently verifying the outputs in numerical analysis.

2. Literature review

Zadeh (1965) introduced fuzzy sets for the first time. Significant advances have since been made in the study of fuzzy logic and its applications. Harris (1913) first presented the EOQ model. Fuzzy sets were first used in operations research studies by Zimmermann (1983). Parameters such as stockouts and demand were treated as crisp quantities (Park, 1987) interpretation of fuzzy sets in EOQ, whereas order cost and carrying cost were treated considered in fuzzy sense. Chen (1996) investigated a fuzzy backorder inventory model using the function principle. It can be seen from the literature analysis, it is observed that demand parameter was often kept constant, whereas Roy (1997) explored an EOQ model in a fuzzy based environment while taking demand-dependent cost into account. For fuzzification, Jing (2000) applied signed distance and fuzzy ranking-numbers. Jing (2003) investigated a back-order inventory model utilizing signed-distance and centroid defuzzification techniques. Rosenblatt (1986) investigated the economic production-quantity model and sampled goods of mediocre quality. Additionally, investigations on items of variable quality were considered by fuzzy EOQ models by Wang (2007), Salameh (2000), and Hung (2004). The supply chain expense examined in a fuzzy EOQ model in 1996, according to Vujosevic, 1996. Wagner, 1958 introduced the effective economic-lot size models. Kalaiarasi et. al (2023) discussed a defective inventory model by examining the optimization using Hessian matrix approach. Isabels et.al (2024) applied trapezoidal intuitionistic fuzzy number for decision making psychological behavior problem. Sivan (2023) considered an inventory system with consistent demand and applied triangular fuzzy number for optimal order quantity.

3. Methodology

There are several methodologies applied to fabricate the impact of uncertain parameters, few to be mentioned are sensitivity analysis, Monte-Carlo simulation, scenario analysis, variance decomposition, Taguchi methods and response surface methodology. Here, in this study, trapezoidal fuzzy-numbers were utilized for the fuzzification of the EOQ for the total cost. The defuzzification technique signed-distance method were applied. Sensitivity analysis is used to perform a numerical comparison of the fuzzy and crisp values between the two fuzzy numbers.

4. The Inventory Model

The inventory comprises of several parameters that are involved in framing the final total inventory cost. The optimum value is derived for the expenditure function

Variables engaged

$Q\rightarrow$ the order-quantity
$I(\alpha)\rightarrow$ the outlay required for the outlay to reduce the lost-revenue fraction
$\alpha\rightarrow$ the annual apportioned expense of capital investment
$S\rightarrow$ stockout on an average
$P\rightarrow$ Security factor
$h\rightarrow$ storage cost/unit/year
$E\rightarrow$ Demand
$R\rightarrow$ Lead-time weekly
$E(X - W)\rightarrow$ insufficiency quantity at the cycle-end on an average
$C(l - t)\rightarrow$ Crashing expense in lead-time

The annual combined expenses are

$$T_C = \alpha I(\alpha) + \frac{1}{Q}\left[S + C(l-t)\right] + h\left[P - ER + \frac{EQ}{2} + \alpha E(X-W)\right] \quad -(1)$$

Differentiating the above equation partially w.r.t 'Q'

$$\frac{\partial T_C}{\partial Q} = -\frac{1}{Q^2}\left[S + C(l-t)\right] + \frac{hE}{2} \quad -(2)$$

$$\frac{\partial T_C}{\partial Q} = 0$$

$$\frac{1}{Q^2}\left[S + C(l-t)\right] = \frac{hE}{2} \quad -(3)$$

we derive the optimal quantity as

$$Q = \sqrt{\frac{2(S + C(l-t)}{hE}} \quad -(4)$$

4.1 Fuzzification technique

The demand and stockout expected parameters are subjected to trapezoidal fuzzy numbers followed by signed-distance defuzzification method, (S_1, S_2, S_3, S_4) and (E_1, E_2, E_3, E_4)

$$\tilde{T}_c = \alpha I(\alpha) + \frac{1}{Q}\left[(S_1, S_2, S_3, S_4) + C(l-t)\right] + h\left[\begin{array}{l} P - (E_1, E_2, E_3, E_4)R + \frac{(E_1, E_2, E_3, E_4)Q}{2} \\ +\alpha E(X-W) \end{array}\right] \quad -(5)$$

$$\begin{aligned}\tilde{T}_c = {} & \alpha I(\alpha) + \frac{1}{Q}[S_1 + C(l-t)] \\ & + h\left[P - E_1R + \frac{E_1Q}{2} + \alpha E(X - W)\right], \alpha I(\alpha) + \frac{1}{Q}[S_2 + C(l-t)] \\ & + h\left[P - E_2R + \frac{E_2Q}{2} + \alpha E(X - W)\right], \alpha I(\alpha) + \frac{1}{Q}[S_3 + C(l-t)] \\ & + h\left[P - E_3R + \frac{E_3Q}{2} + \alpha E(X - W)\right], \frac{1}{Q}[S_4 + C(l-t)] \\ & + h\left[P - E_4R + \frac{E_4Q}{2} + \alpha E(X - W)\right] \end{aligned} \quad -(6)$$

$$Q^* = \sqrt{\frac{2(S_1 + S_2 + S_3 + S_4) + C(l-t)}{h(E_1 + E_2 + E_3 + E_4)}} \quad -(7)$$

5. Numerical Analysis and Discussion

e parameters $S = 70$, $C = 2.5$, $h = 5$, $l = 13$, $E = 10$, $t = 10$ was estimated. The sensitivity examination for the trapezoidal fuzzy numbers is shown in Table 14.1, and values were contrasted with crisp values. It is clear that the fuzzy values remain the same after defuzzification using the signed distance defuzzification approach and fuzzy trapezoidal numbers. Between the crisp and fuzzy numbers, there are very tiny differences. The graph in Figure 14.1 contrasts the sharp and fuzzy values over both table values. The outcomes demonstrate that the fuzzification method has no appreciable impact on the parameter values. This means that the sensitivity of the model to the parameter uncertainty can be examined using the fuzzification method.

6. Conclusion

In a fuzzy setting, the EOQ for a supply model was created using trapezoidal fuzzy numbers. It was done to implement the signed-distance defuzzification. The findings reveal a decreasing tendency in the trapezoidal fuzzy outputs. Other fuzzy numbers and defuzzification methods can be used to review and compare the outcomes. The distinction between crisp and fuzzy values is minimal. The post optimal analysis is contrasted in Figure 14.1. The variations of fuzzy

Table 14.1: Sensitivity analysis applying trapezoidal fuzzy numbers

Sensitivity Variations	S (TrFN)	E (TrFN)	Crisp Values Q	Fuzzy Values Q^*
-50%	$35(15,25,45,55)$	$5(3,4,6,7)$	1.760682	1.695582
-25%	$52.5(32.5,42.5,62.5,72.5)$	$7.5(5.5,6.5,8.5,9.5)$	1.732051	1.688194
No Variations	$70(50,60,80,90)$	$10(8,9,11,12)$	1.717556	1.684488
$+25\%$	$87.5(67.5,77.5,97.5,107.5)$	12.5 $(10.5,11.5,13.5,14.5)$	1.708801	1.68226
$+50\%$	$105(85,95,115,125)$	$15(13,14,16,17)$	1.702939	1.680774

(*Source:* Author's compilation)

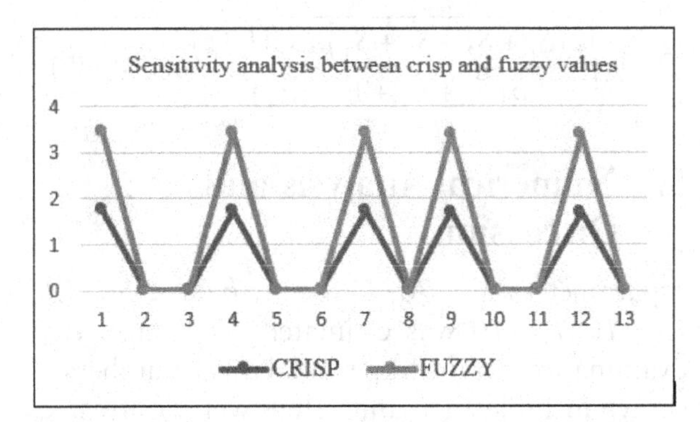

Figure 14.1: Differentiation between the crisp & fuzzy outputs

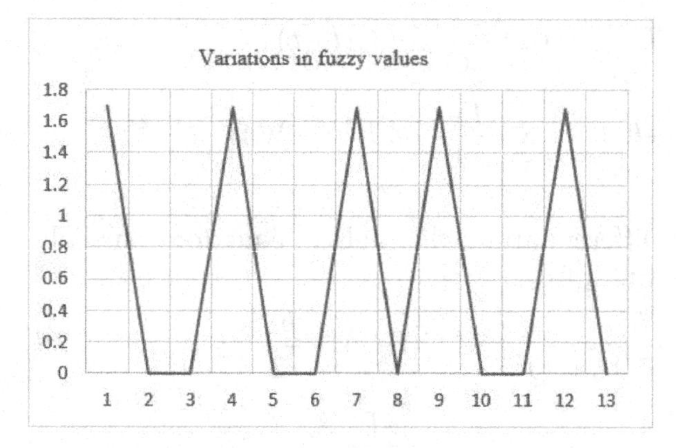

Figure 14.2: Variations among the fuzzy values (Refer Table 14.1)

(*Source:* Author's compilation)

values from Table 14.1 are depicted in Figure 14.2. Fuzzy values are also less than crisp values. In conclusion, fuzzy numbers provide new opportunities for handling uncertainty and adding subjective insights into decision-making processes, even if modern probabilistic techniques and classic crisp inventory models continue to be fundamental to inventory management. Organizations may create more resilient and adaptable inventory management systems that better fit the intricacies of day-to-day operations by embracing a wide range of approaches.

References

Harris F (1913) Operations and cost. AW Shaw Co. Chicago.

Zadeh L.A(1965). Fuzzy sets. Information Control 8, 338–353.

Zimmerman H.J (1983). Using fuzzy sets in operational Research. European Journal of Operational Research 13, pp.201–206.

Park K.S. (1987). Fuzzy Set Theoretical Interpretation of economic order quantity. IEEE Trans. Systems Man. Cybernet *SMC* 17, pp.1082–1084.

Chen S H, Wang C, Arthur R. (1996). Backorder fuzzy inventory model under function principle. Information sciences 95(1–2), pp.71–79.

Roy, T.K., Maiti, M. (1997). A fuzzy EOQ model with demand-dependent unit cost under limited storage capacity. European Journal of Operational Research 99(2), pp.425–432.

Jing-Shang Yao, Kweimei Wu. (2000). Ranking fuzzy numbers-based decomposition principle and signed distance. Fuzzy sets and systems 116(2), pp.275–288.

Jing-Shang Yao, Chiang J. (2003). Inventory without back order with fuzzy total cost and fuzzy storing cost defuzzied by centroid and signed distance. European Journal of Operational Research 148(2), 401–409.

Hung-Chi Chang (2004). An application of fuzzy sets theory to the EOQ model with imperfect quality. Computers and Operations Research 31(12), pp. 2079–2092.

Rosenblatt M J, H L Lee. (1986) Economic production cycles with imperfect production processes. IIE Transactions 18, pp.48–55.

Wang X, Tang W, Zhao R. (2007). Random fuzzy EOQ model with imperfect quality items. Fuzzy Optimization and decision making 6(2), pp.139–153.

Salameh M.K, H.L Lee. (2000). Economic production quantity model for items with imperfect quality. International Journal of Production Economics 64(1), pp.59–64.

Vujosevic M, Petrovi D, R Petrovic. (1996). EOQ formula when inventory cost is fuzzy. International Journal of Production Economics 45(1–3), pp.499–504.

Donaldson W A. (1977). Inventory replenishment policy for a linear trend in demand- An analytical solution. Operational Research Quarterly 28, pp.663–670.

Wagner H M, Whitin T M. (1958). Dynamic version of the Economic lot size model. Management Science 5(1), pp.89–96.

Kalaiarasi K, Mary Henrietta H, Sumathi M, Stanley Raj A. (2023). The optimality representation of a finite-life inventory model with exiguous defectives with Hessian matrix approach, Nonlinear Studies. 30(2), 393–398.

Isabels R, AF Vinodhini, A Viswanathan. (2024). Evaluating and Ranking Metaverse Platforms Using Intuitionistic Trapezoidal Fuzzy VIKOR MCDM: Incorporating Score and Accuracy Functions for Comprehensive Assessment, Decision Making: Applications in Management and Engineering. 7(1), 54–78.

Sivan V, Thirugnanansambandam K, Sivasankar N. (2023). A Fuzzy innovative ordering plan using stock dependent holding cost of inspection with shortages in time reliability demand using TFN, Reliability: Theory & Applications. 18 4(76), pp 921–939.

CHAPTER 15

Significance of vertex antimagic labelings in cloud computing

J. Joy Priscilla[1,*] K. Thirusangu[2]

[1]Department of Mathematics, Saveetha Engineering College, Chennai, India
[2]Department of Mathematics, SIVET College, Chennai, India
Email: *joypriscilla16@gmail.com

Abstract

The work focuses on the research of various VATL (vertex antimagictotal labelings) applied to Generalized Petersen graphs, which are simple graphs with v vertices and e edges. A VATL of the graph G is a one-to-one mapping from the set of edges and vertices of G onto the integers 1 to v+e, ensuring that each vertex has a distinct weight. The significance of these labelings is explored in the context of enhancing security in cloud computing.

Keywords: Generalized Petersen graphs, cloud computing, vertex antimagic total labeling

1. INTRODUCTION

A cloud is commonly defined as a specialized IT environment designed to remotely provide scalable and metered IT resources. The term "cloud" emerged as a metaphorical representation of the Internet, which enables remote access to a decentralized set of IT resources. In various specifications and established frameworks based on the web, the cloud symbol is commonly used to represent the Internet. Computing, on the other hand, refers to a purposeful task that involves creating a step-by-step mathematical sequence, known as an algorithm, to be executed by computers. It is also a branch of engineering science that systematically studies algorithmic processes used for expressing and transforming information. Computing encompasses the design and development of software and hardware systems to structure, process, and manage information in order to facilitate scientific research.

Cloud computing involves accessing and storing data over the Internet instead of relying on a computer's local hard drive. Instead of utilizing their own hardware and software, organizations can avail themselves of cloud computing services provided by external companies and accessed via the Internet. This allows organizations to consume virtual machines, caches, or applications as utilities, like how electricity is used, without the need to build and maintain in-house computing infrastructures. Consequently, all programs and processing tasks are carried out in a virtual environment that is not physically visible. The Internet serves as the cloud, and systems act as the intermediary medium for accessing the data available within it.

1.1 Security In Cloud Computing

Though the cloud may be versatile and cost-efficient, there is always a challenge in maintaining the safety and seclusion of the data of cloud consumers and in the confidentiality of the processing strategies of the cloud providers which becomes an obstruction sometimes. Some of the safety issues in cloud computing are discussed in [2] and [4].

Cloud Computing undergoes these security issues because the mandatory services are mostly outsourced to a third party, which becomes difficult to maintain data security and privacy. Security problems in cloud computing are kindred to those in conventional IT environments.

DOI: 10.1201/9781003606611-15

But the hazards will be different due to the following reasons:

(i) sharing of commitment between the cloud customer and the cloud service provider
(ii) the cloud service provider has the control of the technical planning and operations
(iii) interface(s) exists between the cloud customer and their cloud service providers
(iv) the rights to access the data that managers and legitimate administrators have with respect to data detained in cloud services.

Companies make use of the cloud in various service models namely SaaS, IaaS, PaaS and grouping patterns. The security problems are encountered by cloud providers, the companies delivering infrastructure as a service or software as a platform or via the cloud and also by the cloud consumers. However, both should be responsible in maintaining the system secure. The provider should guarantee that the framework and the data are secured, whereas the consumer must take care in strengthening their information using intense passwords and testament actions.

Cloud computing security is a collection of techniques, strategies and guidelines employed to safeguard the data, implementations, and the affiliated infrastructure of cloud computing. Identity and Access Management (IdAM) is one of the most important strategies to secure the data and to have the control over the access.

1.1 Identity And Access Management (IdAM)

The cloud provider must possess a safety system for providing and governing unique identities for their customers. This IdAM functionality should assist the resource access and powerful user application and service progress. Identity Management aids to secure access benefits not only for end users, but also for favored access by programmers, executives, and installers.

1.2 Federated Identity Management (FIM)

Companies may want to integrate identity over various applications to give single sign-on (SSO) and single sign-off to verify proper termination of the user. Especially organizations using distinct SaaS applications for ERP (Enterprise resource planning) and CRM (Customer relationship management) may need SSO, sign-off, and endorsement over these applications.

1.3 Service ID and API Keys

With the perspective of cloud native micro services, the identity and access of services need to be managed by the cloud service customers. This is exactly executed by using the service ID which is an identity that a service or an application make use of. The contriver can also develop and associate API keys with the service ID which is used to validate and access various services based on the agreement policy and authorizations set.

1.4 Personally Identifiable Information (PII)

The security of data needs imposing restrictions over the usage and convenience of PII based on strategies drawn by non-IT employees, specifically the Legal and Risk Management departments, that are compatible with statutes and laws. The PII otherwise called Sensitive Personal Information (SPI), refers to data that, either alone or in combination with other details, can be used to identify, locate, or distinguish an individual, or link them to a specific person within a system.

1.5 Role Of Unique Labeling

When such restrictions are implemented in various above-mentioned identifications, it is necessary to label the data with suitable labels, preserve it safely, and grant access only to authorized users. In order to protect resources, decrease the expense, and conserve efficiency, cloud providers frequently store data of many customers on the same server. Because of this, there exists a potential risk that the personal data of a user could be accessed by others, including potential adversaries. Hence the cloud service providers must safeguard and provide secure storage privacy to handle such threats. To handle these issues of data isolation and secured access it is just not sufficient to label the data but also to define its uniqueness which can be achieved by the concept of vertex anti magi labeling of graphs.

1.6 Graph Labeling In Cloud Computing

Generally, a network is a group of interconnected people or things or organizations. The easier way of visualizing and understanding is comparing that network with a graph where the people or the organizations are indicated as nodes or vertices of the graph and the relationship between them can be represented as the links or the edges of the graph. When cloud computing is considered, the graph vertices may act for the cloud consumers or the cloud providers and the graph links may constitute their inter-connection or the mutual communications. From the network we can study the properties of its structure, the role and behavior of individual organizations whether they are cloud providers or cloud customers. For instance, we can relate the following sample cloud computing technology figure to a collection of graphs.

In the late 1960s, a concept known as graph labeling emerged within the field of Graph theory. Graph labeling involves assigning values to edges or vertices (or both) of a graph, with respect to specific conditions. A comprehensive exploration of graph labelings is presented in a scholarly work referred to as [3]. This paper specifically focuses on the examination of VATL within a particular class of graphs known as Generalized Petersen graphs. These graphs possess a well-defined structure and can be readily utilized as models for both cloud providers and cloud customers in the context of Cloud Computing.

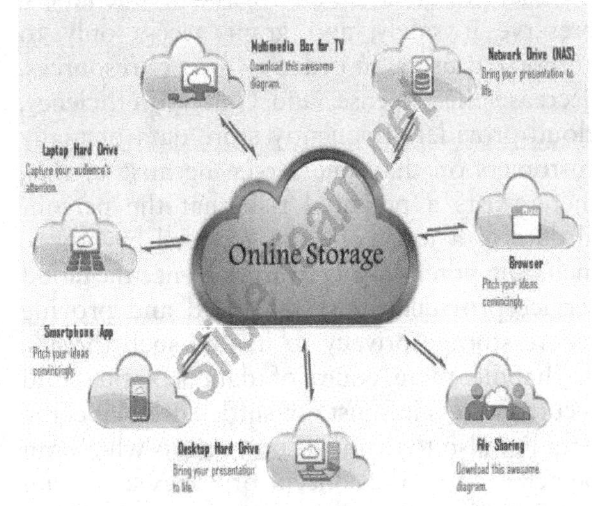

Figure 15.1: Cloud technology icons

1.7 Some Preliminary Definitions

Definition 1

For every vertex $v \in V$, its weight $w(v) = \sum_{u \in N(v)} f(uv) + f(v)$, $N(v)$ being the neighborhood of v.

Definition 2

Let G = (V, E) be a simple graph with e edges and v vertices. A VATL of G is a bijection, denoted as $f: V(G) \cup E(G) \rightarrow \{1, 2, ..., v + e\}$, satisfying the condition that for every vertex v in V, the weight assigned to v, denoted as w(v), is distinct. A graph is deemed to be vertex antimagic total if there exists a valid VATL for it.

Definition 3

A Generalized Peters en graph P(n,m) is a 3-regular graph with edges (u_i, v_i), (u_i, u_{i+1}) (v_i, v_{i+m}) and 2n vertices $u_0, u_1, \ldots u_{n-1}, v_0, v_1, \ldots v_{n-1}$ with $1 \leq m <, \frac{n}{2} n \geq 3$, where the subscripts are taken modulo n for all $i \in \{0,1,2,\ldots n-1\}$ is called. The standard Petersen graph is given by P(5,2). In [1] , [5] and [6], a few antimagic labelings of Generalized Petersen graphs are observes.

Vertex AntimagicLabelings of rP(n, m)

Theorem 1

When n is odd and $n \geq 5$, r P(n,2) is vertex antimagic.

Proof

Let $f: V(rP(n,2)) \cup E(rP(n,2)) \rightarrow \{1,2,\ldots 5nr\}$ be defined as follows:

For t = 0,1, 2…r-1 and i = nt, nt+1, nt+2, …, nt+(n-1),

$$f(u_i) = \{\frac{1}{2}(8n - i + 11nt), for\ i - nt \equiv 0\ (2),$$
$$\frac{1}{2}(7n - i + 11nt), for\ i - nt \equiv 1\ (2).$$

$$f(v_i) = \{\frac{1}{2}(10n - i + 11nt), for\ i - nt \equiv 0\ (2),$$
$$\frac{1}{2}(9n - i + 11nt), for\ i - nt \equiv 1\ (2).$$

$$f(u_i v_i) = \{\frac{1}{2}(2 + i + 9nt), for\ i - nt \equiv 0\ (2),$$
$$\frac{1}{2}(n + 2 + i + 9nt), for\ i - nt \equiv 1\ (2).$$

$f(u_iu_{i+1}) = \{ 2n + 5nt + 1, for\ \ i - nt = 0,$

$\frac{1}{2}(6n - i + 11nt + 2),$

$for\ \ i - nt \equiv 0(\ 2),$

$i - nt \neq 0, \frac{1}{2}(5n - i + 11nt + 2),$

$for\ \ i - nt \equiv 1(\ 2).$

Consider the following two subcases to label the edges $v^i v^{i+2}$

Case (i): n ≡ 1 (mod 4)

$f(v_iv_{i+2}) =$

$\{ n + 5nt + 1, for\ \ i - nt = 0,$

$\frac{1}{4}(5n - i + 21nt + 4),$

$for\ \ i - nt \equiv 1(\ 4), \frac{1}{4}(6n - i + 21nt + 4),$

$for\ \ i - nt \equiv 2(\ 4), \frac{1}{4}(7n - i + 21nt + 4),$

$for\ \ i - nt \equiv 3(\ 4), \frac{1}{4}(8n - i + 21nt + 4),$

$for\ \ i - nt \equiv 0(\ 4), i - nt \neq 0.$

Case (ii) : n ≡ 3 (mod 4)

$f(v_iv_{i+2}) =$

$\{ n + 5nt + 1, for\ \ i - nt = 0,$

$\frac{1}{4}(7n - i + 21nt + 4), for\ \ i - nt \equiv 1(\ 4),$

$\frac{1}{4}(6n - i + 21nt + 4), for\ \ i - nt \equiv 2(\ 4),$

$\frac{1}{4}(5n - i + 21nt + 4), for\ \ i - nt \equiv 3(\ 4),$

$\frac{1}{4}(8n - i + 21nt + 4), for\ \ i - nt \equiv 0(\ 4),$

$i - nt \neq 0.$

Under this labeling, the weights of u_i 's and v_i 's are derived as follows:

For t = 0,1, 2...r-1 and i = nt, nt+1, nt+2, ...,nt+(n-1),

$v_i) = \{ \frac{1}{4}(30n + 80nt + 14),\ \ for\ \ i - nt = 0,$

$\frac{1}{4}(30n - i + 81nt + 12),\ \ for\ \ i - nt = 2,$

$\frac{1}{4}(32n - 2i + 82nt + 14),$

$for\ \ i - nt \equiv 1(4)3(4),$

$\frac{1}{4}(30n - 2i + 82nt + 14),\ \ for\ \ i - nt \equiv 2(4);$

$i - nt < 2,\ \ \frac{1}{4}(34n - 2i + 82nt + 14),$

$for\ \ i - nt \equiv 2(4);\ \ i - nt > 2,$

$\frac{1}{4}(34n - 2i + 82nt + 14),$

$for\ \ i - nt \equiv 0(4),\ \ i - nt \neq 0.$

$w(u_i) = \{ \frac{1}{2}(17n + 40nt + 7), for\ \ i - nt = 0,$

$\frac{1}{2}(17n + 40nt + 5), for\ \ i - nt = 1,$

$\frac{1}{2}(19n - 2i + 42nt + 7),$

$for\ \ i - nt \neq 0,\ \ i - nt \neq 1.$

Since the vertex weights are distinct for all vertices, r P(n,2) is vertex antimagic

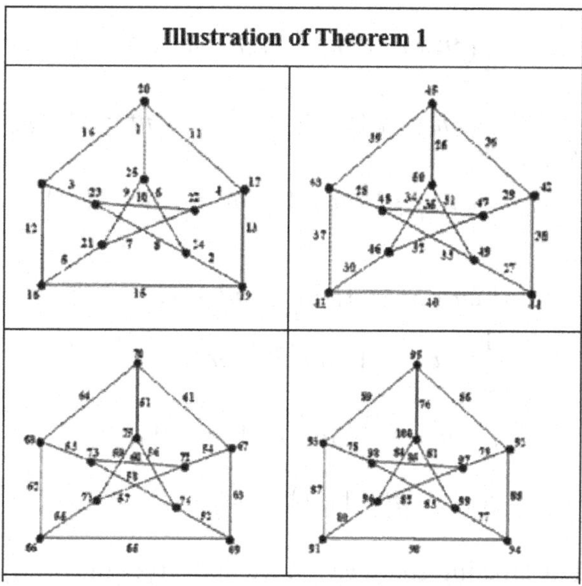

Figure 15.2: VATL of 4 P(5,2)

Theorem 2

For n odd, n ≥ 9, r P(n,4) is vertex antimagic.

Proof

Let the labeling f: V(rP(n,4)) ∪ E(rP(n,4)) → {1,2,... 5nr} be defined as given below:

We consider two possible cases.

Case 1: If n ≡ 1 (mod4)

For t = 0,1, 2…r-1and i = nt, nt+1, nt+2, …, nt+(n-1) ,

$f(u_i) = \{\frac{1}{4}(16n-i+21nt), for\ i-nt \equiv 0(4),$
$\frac{1}{4}(13n-i+21nt), for\ i-nt \equiv 1(4),$
$\frac{1}{4}(14n-i+21nt), for\ i-nt \equiv 2(4),$
$\frac{1}{4}(15n-i+21nt), for\ i-nt \equiv 3(4).$

$f(v_i) = \{\frac{1}{4}(20n-i+21nt), for\ i-nt \equiv 0(4),$
$\frac{1}{4}(17n-i+21nt), for\ i-nt \equiv 1(4),$
$\frac{1}{4}(18n-i+21nt), for\ i-nt \equiv 2(4),$
$\frac{1}{4}(19n-i+21nt), for\ i-nt \equiv 3(4).$

$f(u_iv_i) = \{\frac{1}{4}(i+19nt+4), for\ i-nt \equiv 0(4),$
$\frac{1}{4}(3n+i+19nt+4), for\ i-nt \equiv 1(4),$
$\frac{1}{4}(2n+i+19nt+4), for\ i-nt \equiv 2(4),$
$\frac{1}{4}(n+i+19nt+4), for\ i-nt \equiv 3(4).$

$f(u_iu_{i+1}) = \{2n+5nt+1, for\ i-nt=0,$
$\frac{1}{2}(6n-i+11nt+2), for\ i-nt \equiv 0(2),$
$i-nt \neq 0, \frac{1}{2}(5n-i+11nt+2),$
$for\ i-nt \equiv 1(2).$

For labeling edges v_iv_{i+4} , we have the following subcases:

Case (i) n ≡ 1 (mod8)

$f(v_iv_{i+4}) = \{\frac{1}{8}(9n-i+41nt+8),$
$for\ i-nt \equiv 1(8),\ \frac{1}{8}(10n-i+41nt+8),$
$for\ i-nt \equiv 2(8),\quad for\ i-nt \equiv 3(8),$
$\frac{1}{8}(12n-i+41nt+8),\quad for\ i-nt \equiv 4(8),$
$\frac{1}{8}(13n-i+41nt+8),\quad for\ i-nt \equiv 5(8),$
$\frac{1}{8}(14n-i+41nt+8),\quad for\ i-nt \equiv 6(8),$
$\frac{1}{8}(15n-i+41nt+8),\quad for\ i-nt \equiv 7(8),$
$\frac{1}{8}(16n-i+41nt+8),\quad for\ i-nt \equiv 0(8),$
$i-nt \neq 0.$

$f(v_iv_{i+4})$ = n+5nt+1 fori-nt = 0 .

Case (ii) n ≡ 5 (mod8)

$f(v_iv_{i+4}) =$

$\{\frac{1}{8}(13n-i+41nt+8), for\ i-nt \equiv 1(8),$
$\frac{1}{8}(10n-i+41nt+8),\ for\ i-nt \equiv 2(8),$
$\frac{1}{8}(15n-i+41nt+8),\ for\ i-nt \equiv 3(8),$
$\frac{1}{8}(12n-i+41nt+8),\ for\ i-nt \equiv 4(8),$
$\frac{1}{8}(9n-i+41nt+8),\ for\ i-nt \equiv 5(8),$
$\frac{1}{8}(14n-i+41nt+8),$
$for\ i-nt \equiv 6(8), \frac{1}{8}(11n-i+41nt+8),$
$for\ i-nt \equiv 7(8), \frac{1}{8}(16n-i+41nt+8),$
$for\ i-nt \equiv 0(8),\ i-nt \neq 0.$

$f(v_iv_{i+4})$ = n+5nt+1 where i-nt = 0 .

The vertex weights under this labeling are derived and given below:

For t = 0,1, 2…r-1 and i = nt, nt+1, nt+2, …, nt+(n-1),

$w(u_i) = \{\frac{1}{2}(17n+40nt+7),$
$for\ i-nt=0,\ \frac{1}{2}(17n+40nt+5),$
$for\ i-nt=1,\ \frac{1}{2}(19n-2i+42nt+7),$
$for\ i-nt \neq 0,\ i-nt \neq 1.$

w(v_i) = {$\frac{1}{4}(30n+80nt+14)$,
$for\ i-nt=0$,
$\frac{1}{4}(30n+80nt+10)$,
$for\ i-nt=4$,
$\frac{1}{4}(34n-i+81nt+14),\ for\ i-nt\equiv 0\,(4)$,
$i-nt\neq 0, i-nt\neq 4,\ \frac{1}{4}(31n-i+81nt+14)$,
$for\ i-nt\equiv 1\,(4)$,
$\frac{1}{4}(32n-i+81nt+14),\ for\ i-nt\equiv 2\,(4)$,
$\frac{1}{4}(33n-i+81nt+14),\ for\ i-nt\equiv 3\,(4)$.

We observe that w(u_i) ≠ w(u_j); w(v_i) ≠ w(vj) and w(ui) ≠ w(vj) for all i, j.

Hence r P(n,4) is vertex antimagic.

Case 2: If n ≡ 3 (mod4)

For t = 0,1, 2…r-1 and i = nt, nt+1, nt+2, …, nt+(n-1),

f(u_i) = {$\frac{1}{4}(16n-i+21nt), for\ i-nt\equiv 0\,(4)$,
$\frac{1}{4}(15n-i+21nt), for\ i-nt\equiv 1\,(4)$,
$\frac{1}{4}(14n-i+21nt), for\ i-nt\equiv 2\,(4)$,
$\frac{1}{4}(13n-i+21nt), for\ i-nt\equiv 3\,(4)$.

f(v_i) = {$\frac{1}{4}(20n-i+21nt), for\ i-nt\equiv 0\,(4)$,
$\frac{1}{4}(19n-i+21nt), for\ i-nt\equiv 1\,(4)$,
$\frac{1}{4}(18n-i+21nt), for\ i-nt\equiv 2\,(4)$,
$\frac{1}{4}(17n-i+21nt), for\ i-nt\equiv 3\,(4)$.

u_iv_i) = {$\frac{1}{4}(i+19nt+4), for\ i-nt\equiv 0\,(4)$,
$\frac{1}{4}(n+i+19nt+4), for\ i-nt\equiv 1\,(4)$,
$\frac{1}{4}(2n+i+19nt+4), for\ i-nt\equiv 2\,(4)$,
$\frac{1}{4}(3n+i+19nt+4), for\ i-nt\equiv 3\,(4)$.

f(u_iu_{i+1}) = {$2n+5nt+1,\ for\ i-nt=0$,
$\frac{1}{2}(6n-i+11nt+2)$,
$for\ i-nt\equiv 0\,(2)$,
$i-nt\neq 0, \frac{1}{2}(5n-i+11nt+2)$,
$for\ i-nt\equiv 1\,(2)$.

For labeling edges v_iv_{i+4}, we have the following subcases:

Case (i) n ≡ 3 (mod8)

f(v_iv_{i+4}) = {$\frac{1}{8}(11n-i+41nt+8)$,
$for\ i-nt\equiv 1\,(8),\ \frac{1}{8}(14n-i+41nt+8)$,
$for\ i-nt\equiv 2\,(8),\ \frac{1}{8}(9n-i+41nt+8)$,
$for\ i-nt\equiv 3\,(8),\ \frac{1}{8}(12n-i+41nt+8)$,
$for\ i-nt\equiv 4\,(8),\ \frac{1}{8}(15n-i+41nt+8)$,
$for\ i-nt\equiv 5\,(8),\ \frac{1}{8}(10n-i+41nt+8)$,
$for\ i-nt\equiv 6\,(8),\ \frac{1}{8}(13n-i+41nt+8)$,
$for\ i-nt\equiv 7\,(8),\ \frac{1}{8}(16n-i+41nt+8)$,
$for\ i-nt\equiv 0\,(8),\ i-nt\neq 0$.

f (v_iv_{i+4}) = n+5nt+1 fori-nt = 0 .

Case (ii) n ≡ 7 (mod8)

f(v_iv_{i+4}) = {$\frac{1}{8}(15n-i+41nt+8)$,
$for\ i-nt\equiv 1\,(8),\ \frac{1}{8}(14n-i+41nt+8)$,
$for\ i-nt\equiv 2\,(8),\ \frac{1}{8}(13n-i+41nt+8)$,
$for\ i-nt\equiv 3\,(8),\ \frac{1}{8}(12n-i+41nt+8)$,
$for\ i-nt\equiv 4\,(8),\ \frac{1}{8}(11n-i+41nt+8)$,
$for\ i-nt\equiv 5\,(8),\ \frac{1}{8}(10n-i+41nt+8)$,
$for\ i-nt\equiv 6\,(8),\ \frac{1}{8}(9n-i+41nt+8)$,
$for\ i-nt\equiv 7\,(8),\ \frac{1}{8}(16n-i+41nt+8)$,

$for\ \ i-nt \equiv 0\,(\,8),\ \ i-nt \neq 0.$

f ($v_i v_{i+4}$) = n+5nt+1 where i-nt = 0.

The vertex weights under this labeling are derived and given below:

For t = 0,1, 2…r-1 and i = nt, nt+1, nt+2, …, nt+(n-1),

w(u_i) = { $\frac{1}{2}(17n+40nt+7),$

$for\ \ i-nt=0,\ \ \frac{1}{2}(17n+40nt+5),$

$for\ \ i-nt=1,\ \ \ \frac{1}{2}(19n-2i+42nt+7),$

$for\ \ i-nt \neq 0,\ \ \ i-nt \neq 1.$

w(v_i) = { $\frac{1}{4}(30n+80nt+14),$

$for\ \ i-nt=0,\ \ \frac{1}{4}(30n+80nt+10),$

$for\ \ i-nt=4,\ \ \frac{1}{4}(34n-i+81nt+14),$

$for\ \ i-nt \equiv 0\,(\,4),\ i-nt \neq 0, i-nt \neq 4,$

$\frac{1}{4}(33n-i+81nt+14),\ \ for\, i-nt \equiv 1\,(4),$

$\frac{1}{4}(32n-i+81nt+14),\ \ for\, i-nt \equiv 2\,(4),$

$\frac{1}{4}(31n-i+81nt+14),\ \ for\ \ i-nt \equiv 3\,(4).$

We observe that w(u_i) ≠ w(u_j) ; w(v_i) ≠ w(vj) and w(u_i) ≠ w(vj) for all i, j .

Hence r P(n,4) is vertex antimagic.

Illustration of Theorem 2

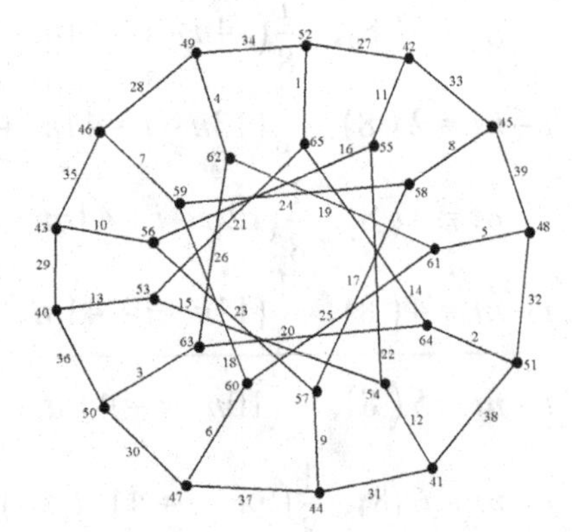

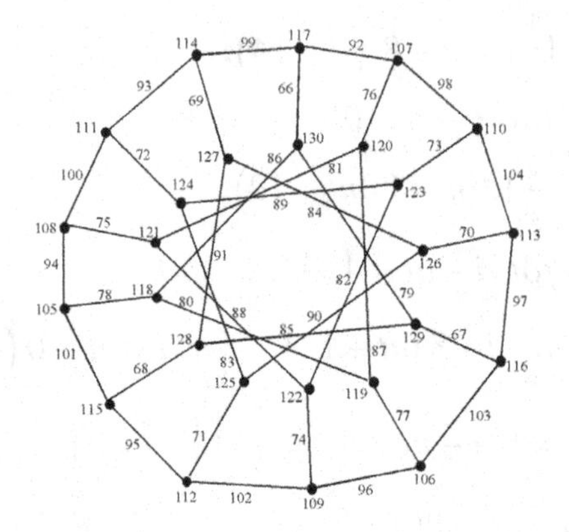

Figure 15.3: VATL of 2 P(13,4)

2. CONCLUSION

This study introduces several vertex antimagic-labelings for Generalized Petersen graphs. These labeling constructions can be made use to define unique identification numbers wherever needed in cloud computing strategies. As these graphs can be imitated as the group of cloud providers or the cloud customers in Cloud Computing, we can restrict each organization to a particular pattern of attributes there by providing a secured atmosphere for both cloud providers and cloud customers

REFERENCES

[1] Anak Agung Gede Ngurah, Edy Tri Baskoro, "On magic and antimagic total labeling of generalized Petersen graph", Utilitas Mathematica 63(2003), 97–107.

[2] Jinpeng Wei, Xiaolan Zhang, Glenn Ammons, Vasanth Bala,and Peng Ning, Managing security of virtual machine images in a cloud environment. In CCSW '09: Proceedings of the ACM workshop on Cloud computing security.

[3] Joseph A.Gallian, A Dynamic survey of Graph labeling, Electronic Journal of Combinatorics, 2014.

[4] P. Mell and T. Grance. Effectively and securely using the cloud computing paradigm. National Institute of Standards and Technology. October 7, 2009.

[5] Mirka Miller, Martin Baca, "Antimagic valuations of generalized Petersen Graphs", Australasian Journal of Combinatorics 22 (2000), 135–139.

[6] K.A.Sugeng, M.Miller and M.Baca, "Super Antimagic total labeling of graphs", Utilitas Mathematica 76 (2008) , 161–171.

CHAPTER 16

Analyzing the effectiveness of fuzzy in solving first-order differential equations using Laplace transform

S. Sindu Devi[1,a] H. Mary Henrietta[2,b,*] K. Ruth Isabels[3,c]

[1]Department of Mathematics, Faculty of Engineering and Technology, SRM Institute of Science & Technology, Ramapuram, Chennai, India
[2]Department of Mathematics, Saveetha Engineering, College (Autonomous), Chennai, India
[3]Department of Mathematics, Saveetha Engineering, College (Autonomous), Chennai, India
Email: [a]sindudes@srmist.edu.in [b*]henriettamaths@gmail.com [c]ruthisabels@saveetha.ac.in

Abstract

This research proposes a generalized concept of discrimination using the Laplace transformation in solving ambiguous differential equations (FDEs). Obscure sizes are expressed in parametric form. It uses the terms I gh variation and (ii) the variation. Numerical illustrations, including graphs, are provided for solving fuzzy differential equations of linear first order to exhibit an analytical solution of our proposed technology.

Keyword: generalized-differentiability, fuzzy number, Fuzzy initial value problem (IVF), fuzzy Laplace transform method.

1. Introduction

In modern engineering, precise problems of electrodynamics, general electricity, classical mechanics, electrical maneuverability, and general relativity are described using dim differential equations (FDE). These equations can be subdivided using FDE. First-order fuzzy differential equations (FODEs) represent a natural progression of classical differential equations into the field of fuzzy mathematics, where fuzzy numbers are used to represent the imprecision and uncertainty present in real-world occurrences. A FODE may incorporate fuzzy values for the unknown function, its derivative, or the parameters involved. Fuzzy sets like as triangular, trapezoidal, or pentagonal fuzzy numbers are commonly used to represent fuzzy values. Finding a fuzzy function that, under specific initial conditions, satisfies a specified fuzzy differential equation is the aim. This type of equation is particularly helpful in systems for which the dynamics are not well understood, enabling more adaptable and reliable models in disciplines such as biology, economics, and control theory. It is frequently necessary to use specialized numerical techniques or modify conventional differential equation methods in order to solve FODEs.

Laplace transformation is applied to solve large-scale differential equations, which convert linear differential equations into algebraic expressions. The fuzzy Laplace Transform method is applied to analyze first-order linear fuzzy initial value problems analytically under strongly normalized variation. The ambiguous Laplace transformation thus transforms the task of solving the blur differential equation into a computative problem. Computational

DOI: 10.1201/9781003606611-16

calculus is a branch of applied mathematics dealing with transitions from calculus functioning to computative performance. It reviews necessary definitions and facts about obscure sets and obscure Laplace transformation methods, as well as novel results used to solve problems and show similar conclusions to those obtained in one sense. [1] Seashore et al. Explain the absolute values of the fuzzy-value functions, as well as the useful theory for the fuzzy Laplace variant of the second derivative and some of their properties. At the end of them, under certain conditions, blurred-value functions include blurred Laplace transformers, which are employed to solve first-order differential equations in a vague sense based on h-differentiation. Under the strong generalized heterogeneous theory, Allavirans et al. [2] The analytical solution approach for many obscure differential equations used the blur Laplace transformation. Describing the absolute values of the blur-value functions, as well as the useful theory for the vague Laplace variant of the second derivative and some of their properties. At the end of them, under certain conditions, blurred-value functions include blurred Laplace transformers, which are applied to first-order differential equation solutions in a vague sense based on h-differentiation. Kalaiarasi [9] employed fuzzy sense in inventory models.

Preliminaries

Definition 2.1. A fuzzy set $\tilde{a}$ efined on the set of real numbers R is said to be a fuzzy number if its membership function $\tilde{a}: R \to [0,1]$ as the following:

i $\tilde{a}$ convex, i.e.,

$\tilde{a}\{\lambda x_1 + (1-\lambda)x_2\} \geq \min\{\tilde{a}x_1, \tilde{a}x_2\}$,

for all $x_1, x_2 \in R$ *and* $\lambda \in [0,1]$.

ii $\tilde{a}$. is normal i.e., there exists an $x \in R$ such that $\tilde{a}(x) = 1$.

iii $\tilde{a}$. is piecewise continuous."

Definition 2.2. A triangular fuzzy number is denoted as $\tilde{a} = (a_1, a_2, a_3)$ and is defined by the membership function

$$\tilde{a} = \begin{cases} 0, & x \leq q, \\ \dfrac{x-a_1}{a_2-a_1}, & a_1 \leq x \leq a_2 \\ \dfrac{a_3-x}{a_3-a_2}, & a_2 \leq x \leq a_3 \\ 0, & x \geq a_3 \end{cases}$$

Fuzzy Derivative:

Definition 2.3. {Hukugara Derivative} :Consider a fuzzy mapping $F:(a,b) \to R$ *and* $t_0 \in (a,b)$. We say that F is differentiable at $t_0 \in (a,b)$ if there exists an element $F'(t_0) \in R$, such that for all $h > 0$ sufficiently small $\exists\, F(t_0+h) \ominus F(t_0),\ F(t_0) \ominus F(t_0-h)$ and the limits (in the metric D)

$$\lim_{h \to 0_+} \frac{F(t_0+h) \ominus F(t_0)}{h} = \lim_{h \to 0_-} \frac{F(t_0) \ominus F(t_0-h)}{h}$$

xists and are equal to $F'(t_0)$.

Fuzzy Initial Value Problem.

An initial value problem is a system of ordinary differential equations together with the initial conditions. Consider a function of nth order fuzzy differential equations with initial conditions are

$$\tilde{y}^n(t) = \tilde{f}(t, y(t), y'(t), \ldots, y^{n-1}(t))$$

$$\tilde{y}(t_0) = y_0, \ldots, \tilde{y}^{n-1}(t_0) = y_0$$

By using the extension principle, the membership functions are

$$[\tilde{f}(t,\tilde{y})]^{\alpha} = \tilde{f}(t, [\tilde{y}]^{\alpha}) = \tilde{f}\left(t, [\underline{y}_{\alpha}, \overline{y}_{\alpha}]\right) =$$

$$\left(\min \tilde{f}\left(t, [\underline{y}_{\alpha}, \overline{y}_{\alpha}]\right), \max \tilde{f}\left(t, [\underline{y}_{\alpha}, \overline{y}_{\alpha}]\right)\right)$$

2. Basic Concepts of Fuzzy Laplace Transform

Definition 3.1. The fuzzy-valued function's fuzzy Laplace transform $\tilde{f}(t)$ circumscribed as

$$\tilde{F}(s) = L(\tilde{f}(t)) = \int_0^{\infty} e^{-st} \tilde{f}(t)\,dt = \lim_{h \to \infty} \int_0^{\infty} e^{-st} \tilde{f}(t)\,dt$$

Whenever the limit exists. Where $L(\tilde{f}(t))$ enotes fuzzy-valued function's fuzzy Laplace transform $\tilde{f}(t)$, and the integral is the fuzzy integral of Riemann in error and generate a novel feature with fuzzy values

$$\tilde{F}(s) = L(\tilde{f}(t))$$

Considering about the fuzzy-valued function $\tilde{f}$, followed by the upper and lower fuzzy This function's Laplace transform, based on the fuzzy-valued function's upper and lower bounds $\tilde{f}$, is as follows:

$$\tilde{F}(s,\alpha) = L(\tilde{f}(t,\alpha)) = \left[L(\underline{f}(t,\alpha)), L(\overline{f}(t,\alpha)\right] where$$

$$L(\underline{f}(t,\alpha)) = \int_0^\infty e^{-st}\underline{f}(t,\alpha)dt$$

$$= \lim_{h\to\infty}\int_0^\infty e^{-st}\underline{f}(t,\alpha)dt, \quad 0 \le \alpha \le 1.$$

$$L(\overline{f}(t,\alpha) = \int_0^\infty e^{-st}\overline{f}(t,\alpha)dt$$

$$= \lim_{h\to\infty}\int_0^\infty e^{-st}\overline{f}(t,\alpha)dt, \quad 0 \le \alpha \le 1.$$

Definition 3.2 When $\tilde{f}(t)$ is continuous and $\tilde{F}(s,\alpha) = L(\underline{f}(t,\alpha)), (\overline{f}(t,\alpha)$ we have $(\underline{f}(t,\alpha)), (\overline{f}(t,\alpha) = L^{-1}\tilde{F}(s,\alpha)$ and L^{-1} is called the inverse Laplace transform, $\tilde{f}$ is known as the inverse Laplace transform of $\tilde{F}(s,\alpha)$.

Theorem 3.3 [7]. Let $\tilde{f}$ be an integrable fuzzy valued function, and it is the primal of $\tilde{f}'(t) on\ [0,\infty)$.

Then

$L[\tilde{f}'(t)] = sL[\tilde{f}] \ominus f'(0)$ where $\tilde{f}$ is differentiable

$L[\tilde{f}'(t)] = (-\tilde{f}(0) \ominus (-sL[\tilde{f}(t)])$ where $\tilde{f}$ is differentiable

3. Example I

Examine an electrical circuit including an alternating current source (RL circuit). Circuits with combining an inductor (I) and a resistor (R) are known as RL circuits (R). Both the resistor and inductor are connected in a sequence. A battery with emf ε, is also included, as well as a switch, initially set to "no battery". The circuit is devoid of current at first. As time goes by, it is due to the inductor's response to change that it is change in current is equal to RL circuit. The voltage throughout the resistor and inductor fluctuates, however the loop rule remains consistently upheld are environmental conditions, unknown circuit results, and ambiguous element values.

$$\tilde{f}'(t) = -f(t) + \sin(t),\ t \in [0,1],$$

$$f(0;\alpha) = (0.96, 1, 1.01),$$

Consequently, the fuzzy notion enables us to understand the issue in the following way:

$$\tilde{f}'(t) = -\frac{I}{R\tilde{f}(t)} + u(t), \qquad t \in [0,1]$$

$$\tilde{f}'(0,\alpha) = (0.96 + 0.04\,\alpha, 1.01 - 0.01\alpha), \qquad \alpha \in [0,1]$$

R represents the circuit resistance, while I is a coefficient associated with the solenoid. Presume $u(t) = \sin(t),\ I = 1\Omega\ and\ R = 1H.$

Therefore, the above equation can be rewritten as:

Case (i): Multiply both sides by Laplace Transform $L(f^1) = -L(f(t)) + L(\sin(t))$

Let us consider $\tilde{y}(t)$ is (i) gh differentiable then, use

$$sL[\underline{f}(t)] \ominus (\underline{f}(0)) = L[\underline{f}(t)]$$

$$sL[\overline{f}(t)] \ominus \overline{f}(0) = L[\underline{f}(t)]$$

Then we get the membership function as

$\tilde{f}(t;\alpha) = -1/2(0.05 - 0.05\alpha)e^t + 1/2$ 2.97+0.03α $e^{-t} + 1/2$) (sin(t) – cos(t)),

$\tilde{f}(t;\alpha) = 1/2(0.05 - 0.05\alpha)e^t + 1/2$ 2.97+0.03α $e^{-t} + 1/2$) in(t) – cos(t)),

Case (ii): In this instance, we consider $\tilde{f}(t)$ is (ii) gh – differentiable

$$-sL\left[\underline{f}(t)\right]\ominus\left(\underline{f}(0)\right) = L\left[\underline{f}(t)\right],$$

$$-sL[\bar{f}(t)]\ominus\bar{f}(0) = L[\underline{f}(t)]$$

Then we get the membership function of the solution as

$\tilde{f}(t;\alpha)=1/2\left(\sin(t)-\cos(t)\right)+e^{t}$ 46+0.04α),

$\tilde{f}(t;\alpha)=1/2\left(\sin(t)-\cos(t)\right)+e^{t}$ 51+0.01α),

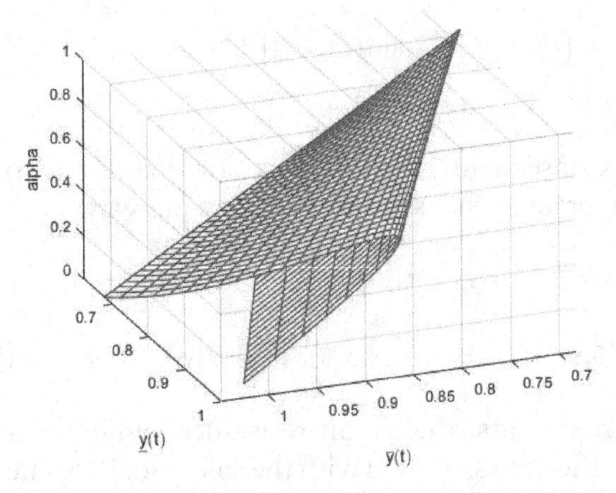

Figure 16.1: Membership function of fuzzy Laplace transforms under (i) –gh differentiability.

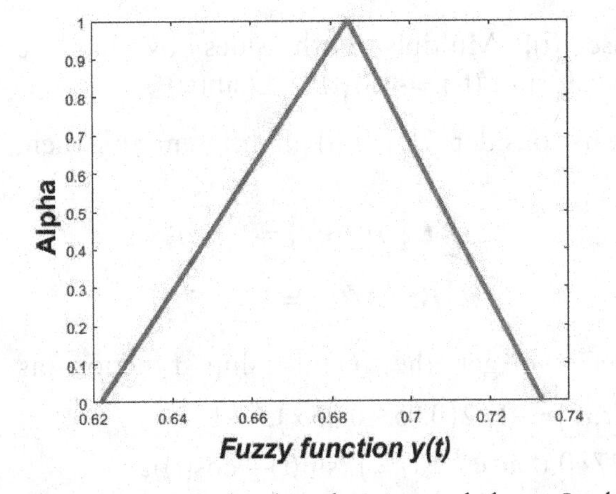

Figure 16.2: Membership functions of fuzzy Laplace transform t = 0.8 under (i) –gh differentiability.

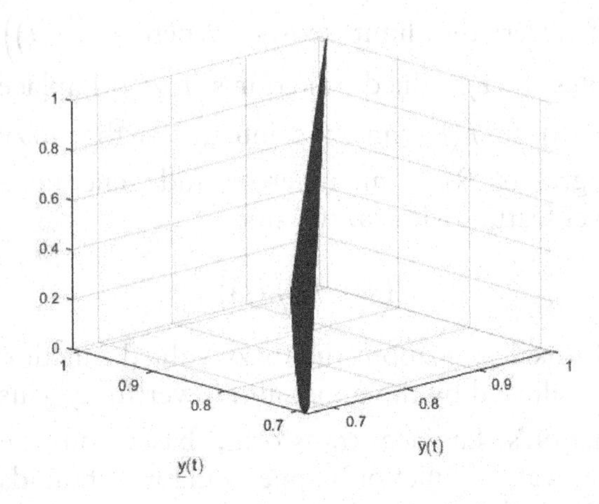

Figure 16.3: Membership functions of fuzzy Laplace transform for t = 0 to 1 under (i) –gh differentiability

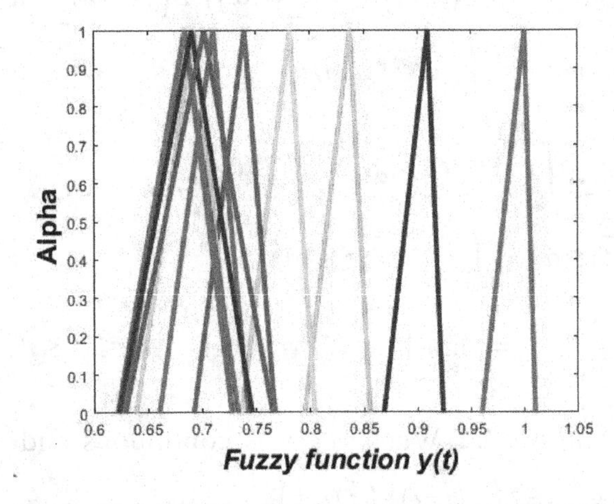

Figure 16.4: Membership function at the various alpha level sets and for t = 1 to 2 under (ii) –gh differentiability.

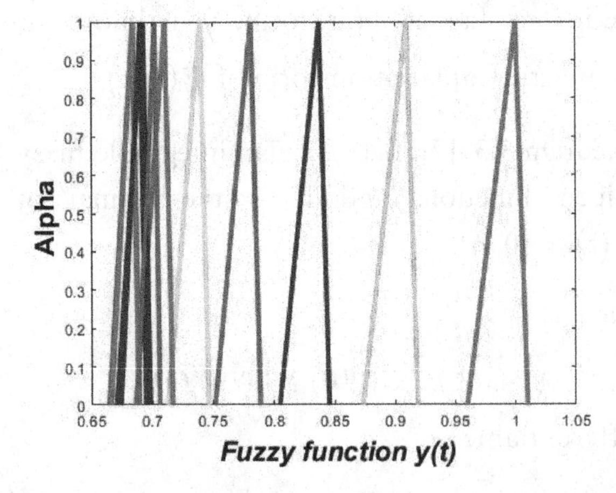

Figure 16.5: Membership functions of fuzzy Laplace transform for ᵗ = 0.8 under (ii) –gh differentiability.

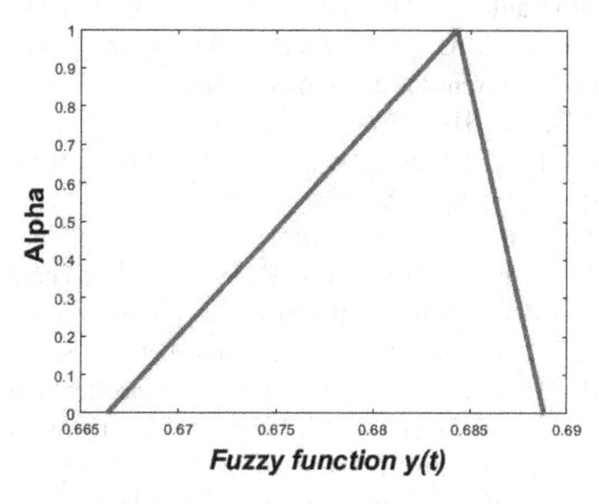

Figure 16.6: Membership functions of fuzzy Laplace transform for t = 0.8 under (ii) –gh differentiability.

Example 2: We consider a fuzzy initial value problem

$$\tilde{f}'(t)=(1-t)\tilde{f}(t), \quad t\in[0,2]$$

$$\tilde{f}(0;\alpha)=(0,1,2),$$

The below figure shows the analytic solution of (i)-gh and (ii)-gh differentiability

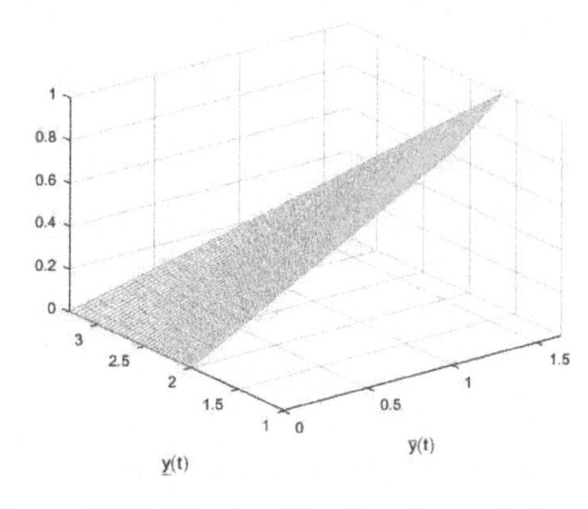

Figure 16.7: Membership function of fuzzy Laplace transforms for t = 0.8 under (i) – GH

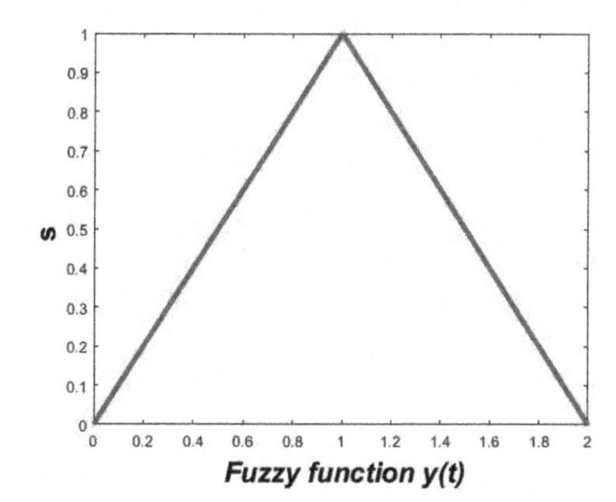

Figure 16.8: Membership function of fuzzy Laplace transforms for t = 0.8 under (i) – GH

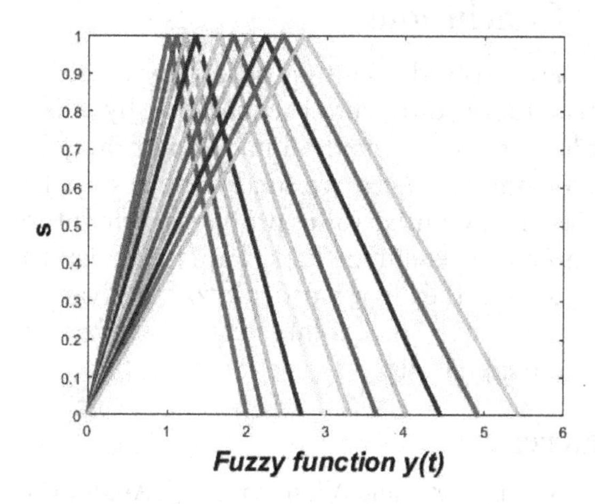

Figure 16.9: Membership functions for t = 0 to 1 under (i) – gh

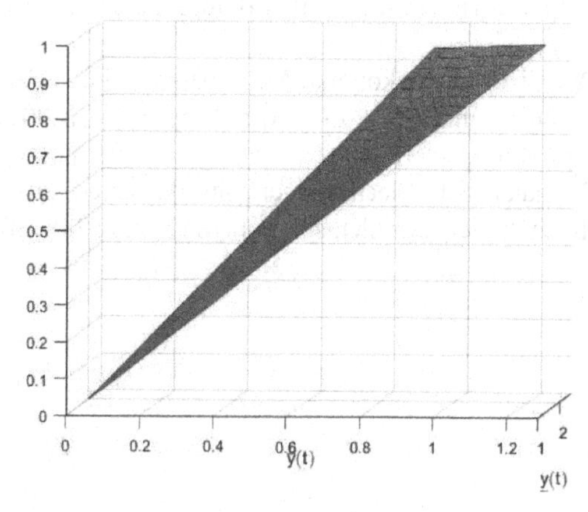

Figure 16.10: Membership function for various level set and t = 1 to 2 under (ii) – gh.

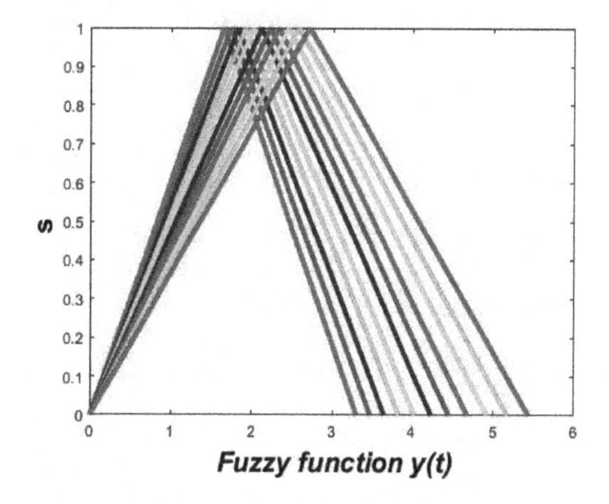

Figure 16.11: Membership functions of fuzzy Laplace transform t = 1 to 2 under (ii) – GH.

4. Conclusions

We have solved the first-order linear generalized fuzzy differential equations by fuzzy Laplace transform method involving triangular fuzzy numbers. Then we used these results for solving fuzzy initial value problems. For future research, we extend the fuzzy Laplace transform method for the higher-order fuzzy derivatives and apply it to solve a large class of fuzzy differential equations.

References

[1] Salahshour, S and Allanviranloo.T, Application of Fuzzy Laplace Transforms, Soft Computing, 17(1), (2013), pp;145–158.

[2] Allanviranloo, T, and Barkhordari, M, Fuzzy Laplace Transforms, Soft Computing, 14, (2010), pp; 235–243.

[3] Yazdi.H.S, Pakdaman.M, Effati. S, Fuzzy circuit analysis, Int.j.Appl.Eng.res.3 (2008), pp: 1061- 1071.

[4] Yamakawa.T, Electronic circuits dedicated to the fuzzy logic controller, sci. Iran.18 (2011) 528–538.

[5] Allanviranloo.T, Abbasbandy.S, salahshour.S and Hakimzadeh.A., A new method for solving fuzzy linear differential equations, computing, 92(2), (2011), pp:181–197.

[6] Zadeh.L.A., Fuzzy sets, Information and control, 8(3), (1965), pp:338-353.

[7] Isabels R, AF Vinodhini, A Viswanathan. (2024). Evaluating and Ranking Metaverse Platforms Using Intuitionistic Trapezoidal Fuzzy VIKOR MCDM: Incorporating Score and Accuracy Functions for Comprehensive Assessment, Decision Making: Applications in Management and Engineering. 7(1), 54–78.

[8] Sivan V, Thirugnanansambandam K, Sivasankar N. (2023). A Fuzzy innovative ordering plan using stock dependent holding cost of inspection with shortages in time reliability demand using TFN, Reliability: Theory & Applications. 18 4(76), pp 921–939.

[9] K. Kalaiarasi, H. Mary Henrietta, M. Sumathi, A. Stanley Raj, The Economic order quantity in a fuzzy environment for a periodic inventory model with variable demand, Iraqi Journal of Computer Science and Mathematics. 3(1):102–107 (2022).

CHAPTER 17

Electrical Study of Mn Doped NiO Nanoparticles

C. Saveetha[1, a]

[1]Department of Physics, Saveetha Engineering College, Chennai, India
Email: [a]saveethaphysics@gmail.com

Abstract

In this work, NiO doped with Mn nanoparticles were synthesized for varied composition using chemical co-precipitation technique. Its electrical and dielectric properties were studied using impedance spectroscopy. As frequency is increased, an increase in activation energy of the samples with Mn doping concentration and a decrease in dielectric constant as well as dielectric loss is reported.

Keywords: Nanoparticles, Mn doped NiO, impedance spectroscopy, dielectric constant, DC conductivity

1. Introduction

NiO is employed for electronic devices and as functional layer material for making chemical sensors. Electrochemical materials offer dynamic control of the output radiant energy and have a major role in energy efficient smart windows in order to reduce the lighting and cooling cost of buildings. It is also used in making electrical ceramics such as thermistors and varistors.

Nickel (II) oxide, has a face centered cubic structure with a = 4.1684 Å. NiO is usually non-stoichiometric compound and has a complex band structure [1]. NiO is attractive due to its chemical stability, electrical and magnetic properties. It is categorized as a Mott-Hubbard insulator having electrical conductivity at room temperature to be less than 10–13 ohm–1cm–1. The holes of intermediate mass are responsible for electrical conduction [2].

NiO in bulk form behaves as an insulator having ~4.0 eV as band gap [3] with antiferromagnetic property at room temperature, possessing a Neel temperature of 250 °C [4]. In comparison with NiO single crystals, NiO nanoparticles exhibit an enhanced electrical conductivity of the order of six to eight times. Doping of Mn within the NiO lattices improves the electrical property significantly.

2. Sample Preparation

Stoichiometric samples of $Ni_{0.99}$Mn0.01O [NMO1], $Ni_{0.98}Mn_{0.02}O$ [NMO2] and $Ni_{0.97}Mn_{0.03O}$ [NMO3] were synthesized by the co-precipitation method. The starting materials, $NiCl_2.6H_2O$ and $MnCl_2.H_2O$ were taken in appropriate proportion and weighed. The mixture was then dissolved in double distilled water. Aqueous solution of NaOH was introduced into the stoichiometric solution gradually for about 30 min with a continuous stirring at room temperature in order to reach a pH value of 8. The precipitates were allowed to settle for 24 h and then washed thoroughly with distilled water several times and filtered to remove the water-soluble compounds. The samples were dried, mildly ground using agate mortar and pestle. After this treatment, the samples were annealed at various temperatures ranging from 600 °C to 800 °C in air atmosphere for 4 h.

3. The DC Conductivity Studies

Compressed pellets having 1.88 mm thickness and 8 mm diameter was considered for carrying out complex impedance measurements in air. Before measuring the impedance of the samples, the pellets were initially dried at 100 °C for 30 min. The measurement of impedance was

DOI: 10.1201/9781003606611-17

performed in a wide range of frequency from 1 Hz to 10 MHz. The measurement was performed from room temperature to 270 °C.

Figures 17.1 and 17.2 illustrate the complex impedance plot of samples NMO1 and NMO2 at various temperatures. The spectra show only one semicircle. This clearly verifies that the grain boundary mainly contributes to the electrical conductivity.

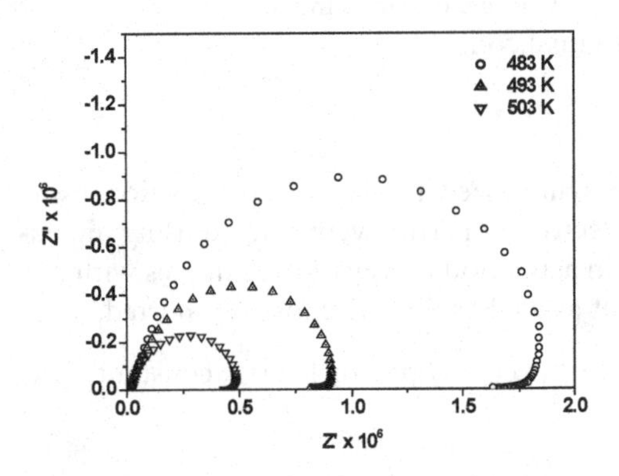

Figure 17.1: Complex impedance spectra of $Ni_{0.99}Mn_{0.01}O$

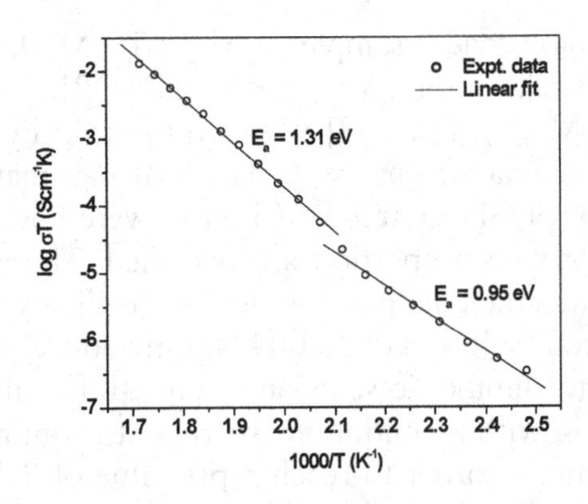

Figure 17.2: Complex impedance spectra of $Ni_{0.98}Mn_{0.02}O$

By considering the geometrical dimensions of the sample and by measuring the grain boundary resistance (R_{gb}) value, the dc conductivity of the as prepared sample was determined. Figures 17.3 and 17.4 display the Arrhenius plots of electrical conductivity in the temperature ranging from 423 K and 603 K. It is found that as temperature increases, electrical conductivity also increases in accordance with the semiconducting behavior of NiO in NMO1 and NMO2 samples. The activation energy was calculated by fitting DC component of electrical conductivity to Arrhenius relation,

$$\sigma T = \sigma_0 \exp\left[-E_a/k_B T\right]$$

where,

σ_0 – pre-exponential factor,
E_a – activation energy for DC component of electrical conductivity,
k_B – Boltzmann constant.

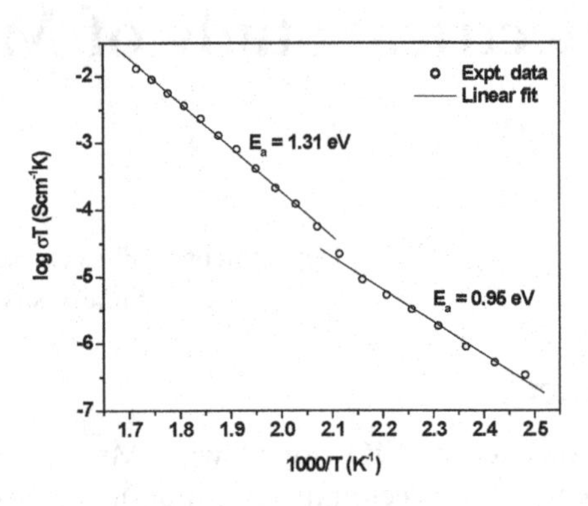

Figure 17.3: Arrhenius plot of $Ni_{0.99}Mn_{0.01}O$ using DC conductivity

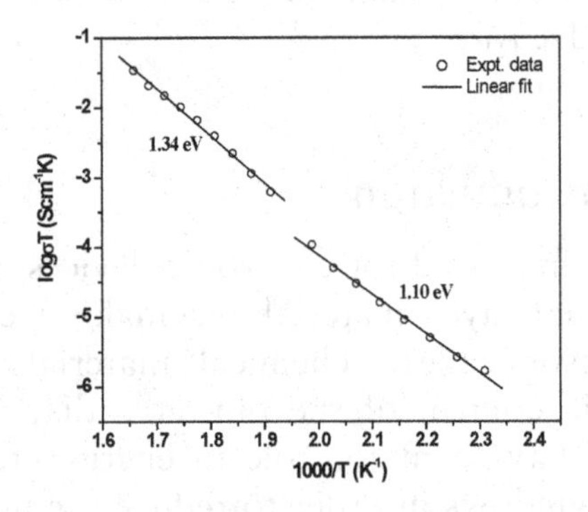

Figure 17.4: Arrhenius plot of $Ni_{0.98}Mn_{0.02}O$ using DC conductivity

The changes in slope of the fitted lines suggest the change in conduction mechanism over the concerned temperature range. The activation energies (E_a) of Ni0.99Mn0.01O and Ni0.98Mn0.02O were obtained at high temperature and low temperature regions as 1.31 eV, 0.95 eV and 1.34 eV, 1.10 eV respectively. In bulk NiO, the activation energy was calculated as ~0.6 eV in the temperature range from 200 K to 1000 K [5]. V. Biju et al., have observed a decrease in activation energy of NiO when it is prepared in nano form. Even though the samples prepared in this work are in nanoform, a significant increase in activation energy (~1 eV) is observed when compared to that reported for pure NiO in both bulk and nanoform. However, the reason for the activation energy to be increased in the synthesized sample is the doping effect of Mn in NiO as observed in Fe and Cr doped NiO [6, 7]. This is supported by the observation of increase in activation energies in both NMO1 and NMO2 samples with an increase in Mn concentration.

4. Effect of Dielectric Constant (ε′) and dielectric loss (ε″) Upon Frequency

Figures 17.5 and 17.6 present the dielectric constant (ε′) and dielectric loss (ε″) variation of $Ni_{0.98}Mn_{0.02}O$ with respect to frequency at different temperatures. From the figure, it is inferred that as frequency increases, dielectric values decrease. $Ni_{0.98}Mn_{0.02}O$ shows high values of ε′ and ε″ at lower frequencies and relatively low values at higher frequencies. The dielectric values also increase with temperature.

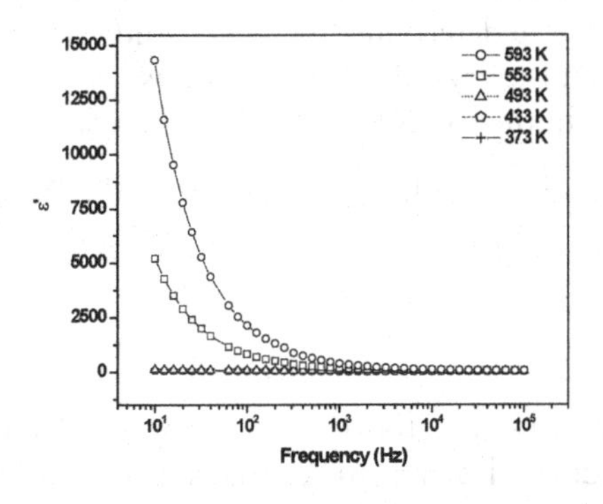

Figure 17.5: Variation of ε′ with frequency of $Ni_{0.98}Mn_{0.02}O$ at different temperatures

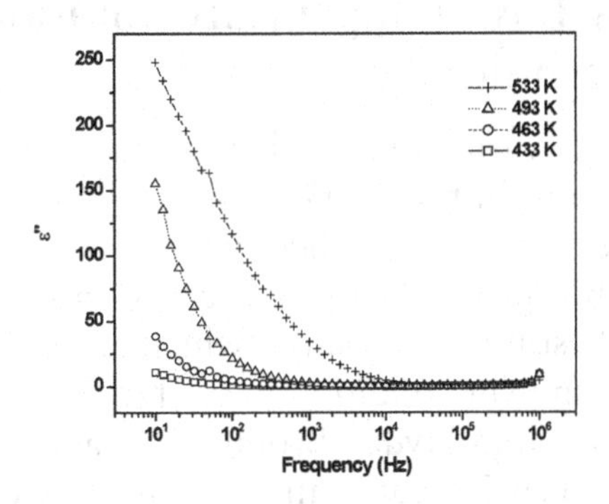

Figure 17.6: Variation of ε″ with frequency of $N_{0.98}Mn_{0.02}O$ at different temperatures

At lower frequencies, all the four polarization mechanisms, namely electronic, dipolar, interfacial, and atomic polarizations contribute to the dielectric properties of a material. Hence high dielectric values are expected at lower frequencies when compared to those at higher frequencies where only the electronic and atomic polarizations contribute [8]. It is also notable that there is no much variation in dielectric values and their behavior with respect to temperature, as a function of Mn doping.

5. Conclusion

Mn doped NiO were successfully synthesized for varied chemical composition by employing chemical co-precipitation technique and characterized with impedance spectroscopy. It is found that the activation energies of the samples increase with Mn doping concentration. Power law dispersion holds good with a plateau region of independent frequency that extends towards higher frequency values with an increase in temperature. As frequency increases, there is a decrease in dielectric constant and dielectric loss. The increase in ε′ and ε″ values are significant for lower frequencies with elevated temperatures.

References

[1] D.C. Onwudiwe, N.H. Seheri, L. Hlungwani, H. Ferjani, R. Rikhotso-Mbungela, J. Mol. Struct. 13179, 139084 (2024)

[2] F. Wrobel, H. Park, C. Sohn, H.-W. Hsiao, J.-M. Zuo, H. Shin, H.N. Lee, P. Ganesh, A. Benali, P.R.C. Kent, O. Heinonen, A. Bhattacharya, Phys. Rev. B 101, 195128 (2022)

[3] S. Mohseni Meybodi, S.A. Hosseini, M. Rezaee, S.K. Sadrnezhaad, D. Mohammadyani, Ultrason. Sonochem. 19(4), 841 (2012)

[4] A. Mandziak, G.D. Soria, J.E. Prieto, P. Prieto, C. Granados-Miralles, A. Quesada, M. Foerster, L. Aballe, J. de la Figuera, Nat Sci Rep. 9, 13584 (2019)

[5] V. Biju, M. Abdul Khadar, Mater. Res. Bull. 36, 21–33 (2001)

[6] Y.H. Lin, J. Wang, J. Cai, M. Ying, R. Zhao, M. Li, C.W. Nan, Phys. Rev. B 73, 193308 (2006)

[7] J. Nowotny, M. Rekas, Solid State Ion. 12, 253 (1984)

[8] L.L Hench, J.K. West, *Principles of Electronic Ceramics* (Wiley, New York, 1990), p. 205

CHAPTER 18

Growth and Characterization of a Single Crystal: BisGlycine HydroBromide

C. Saveetha[1, a] and L. Sangeetha[1, b]

[1]Department of Physics, Saveetha Engineering College, Chennai India
Email: [a]saveethaphysics@gmail.com, [b]sangeethabalaji0305@gmail.com

Abstract

A perfect and transparent single crystals of bisglycine hydrobromide (BGHB) were grown by solution method. Good and transparent BisGlycine Hydrobromide crystal is thus grown is subjected to different characterization which includes Etching study, X-Ray Diffraction (XRD) study and high resolution XRD (HRXRD) study. Crystal unit cell dimensions were determined by XRD study and the structure is found to be orthorhombic. HRXRD confirms that the crystalline perfection is very good. Etching studies indicate that the etch pit density is very high in the case of dilute acid as an etchant.

Keywords: Single crystal, bisglycine hydrobromide, etching studies, crystal growth, XRD, crystalline purity

1. Introduction

The inorganic acids and salts in combination with amino acid compounds behaves as a suitable material for optical second harmonic generation since it has the added benefits of inorganic salts along with the organic amino acid. The salts of amino acids L-arginine [1] and L-histidine [2] are reported to have high efficiency in second harmonic conversion.

Compounds formed by the combination of mineral acids along with Glycine and its methylated analogues shows interesting properties like ferroelastic, ferroelectric or antiferroelectric behavior. Two kinds of ionic groups of glycine are present in the ferroelectric triglycine sulphate, [3] namely glycinium ions and zwitter ions. The ferroelectric behavior of such type of crystal is decided by the O-H–O hydrogen bonds between the ions.

The paraelectric and ferroelectric structure of the glycine molecules has zwitterionic configuration and the other is monoprotonated [4]. Diglycine hydrochloride [5], and Diglycine hydrobromide [6] has the same structure.

Attempts are made for growing BGHB, a single crystal at room temperature by solution method.

2. Growth of BGHB Single Crystals by Low Temperature Solution Technique

There are various methods available for growing crystals. In this work, growth from solution is chosen rather than other methods namely melt and vapor growth. This is because this method works best for compounds, which are not sensitive to ambient condition in the laboratory.

Glycine and Hydro Bromide (Analar Grade) are taken in 2:1 ratio and dissolved in doubly purified water. The below mentioned reaction is likely to take place and the expected glycine compound is obtained.

$$2\,(NH_2CH_2COOH) + HBr\ (NH_3^{+}\text{-}CH_2COO^{-})\ (NH_3^{+}CH_2COOH)Br^{-} \qquad \text{I} \qquad \text{II}$$

I indicate glycine in the form of zwitterion and II indicate glycinium bromide which is a form

DOI: 10.1201/9781003606611-18

of glycinium ion. Transparent crystals of size 12 × 5 × 5 mm^3 crystallized within three weeks. Using the method of Floatation, the crystal density was determined. The density of BGHB was determined as 1.88(3) g/cm^3. This agrees with the density, calculated from crystallographic data [6]. Seed crystals are formed due to spontaneous nucleation. For the growth of crystals, good quality transparent seed crystals are used. When the crystal is allowed to grow for a longer time in solution, inclusions appear in the crystal, reducing the quality. The photographic picture of as grown crystal of BGHB is given below in Figure 18.1.

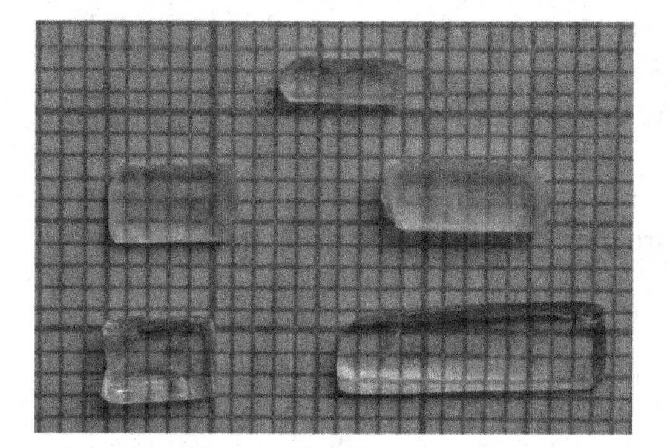

Figure 18.1: Photograph of as grown BGHB crystal

3. Single Crystal XRD

Enraf Nonius CAD4 single crystal X-ray diffractometer which is fitted with monochromatic MoK_α beam is used to carry out the single crystal XRD analysis. Leica polarizing microscope was used to analyze the single crystallinity since the crystal was transparent. A required size of the crystal was placed on the X-ray goniometer. The crystal was adjusted to have the optical center at the sphere of confusion. Nearly twenty five reflections were taken through various planes of the crystal. Short vectors method is used for indexing of reflections which was then corrected by least squares structure refinement process. The parameters of the unit cell were found by this way was transformed to correct Bravais lattice.

From single crystal XRD study, the parameters of the unit cell were calculated as a = 5.373(4) angstrom, b = 8.153(3) angstrom and then c = 18.268(5) angstrom. Unit cell volume = 800.367(5) $Å^3$ and the glancing angles are $\alpha = \beta = \gamma = 90°$.

The crystal structure was drawn using the data and the structure is orthorhombic having space group $P2_12_12_1$. The values which are calculated matches with the already stated value [5, 6].

BGHB crystal was placed over the X-ray goniometer and reflections from particular planes were taken for indexing the crystal structure. The morphology (Figure 18.2) study shows the prominent faces of the crystal. On seeing the morphology of the crystal, it is observed that the crystal growth was faster along the axis of the crystal. (Lattice parameter 'a' is lesser than 'b' and 'c').

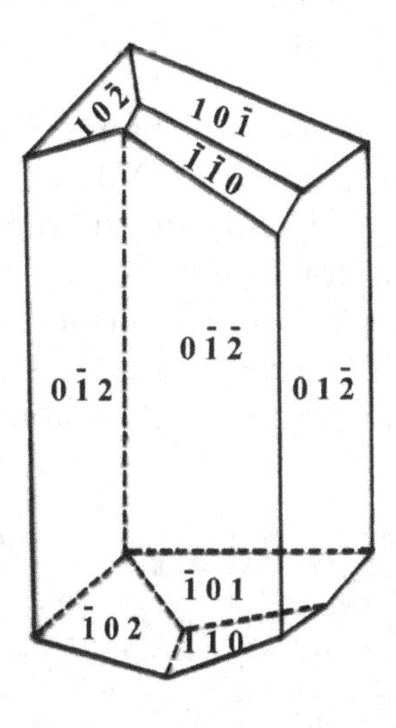

Figure 18.2: Morphology of solution grown single crystals of BGHB

4. Multi crystal XRD

A multi crystal X-ray diffractometer designed at National Physical Laboratory [7] was employed to get diffraction curves with high-resolution. A narrow beam of monochromatic $MoK\alpha_1$ radiation was used as the exploring X-ray beam.

BGHB crystal with well-polished surface and etched by using an etchant which is a mixture of water and acetone mixed in the ratio of 1:2 is taken and subjected to multi crystal XRD study. The Diffraction Curve (DC) for the specimen using (2 2 0) diffracting planes is shown in the Figure 18.3. From this, it is observed that the DC is sharp without any additional peaks which may arise due to structural grain boundaries [8] or the presence of some organic additives leads to extra peaks [9].

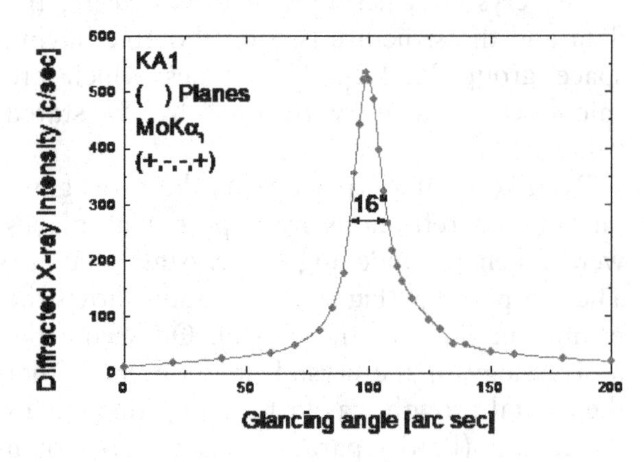

Figure 18.3: High resolution diffraction curve of BGHB crystal for (2 2 0) plane

From the curve, it is observed that the full width at half maximum (FWHM) is 16 arcsec which goes nearly close to the reported value [10]. Such a low value of FWHM implies that the perfection of the crystal is good. So the crystal is nearly a perfect single crystal devoid of any structural grain boundaries and dislocations.

5. Etching Studies

The crystal BGHB quality was studied by observing the surface morphology of the cut and polished wafers using Reichert MD 400E Ultra Microscope. The different types of etch patterns obtained for crystals confirm the anisotropic nature of the crystals. Water and 1N hydrochloric acid were used as etchants in the studies.

The BGHB crystal was subjected to etching studies. Etchants used for this study is water and 1N hydrochloric acid. The capacity of the solvent which reveals the etch pits is also get varied based on crystallographic orientation. Hence, the as grown crystal was cut into thin plates of thickness 2–3 mm using wet thread. Polished and transparent samples devoid of any visible additives or cracks were taken for etch pits analysis. To etch, the crystal surface was made to get dipped in the etchant for 10–30 s at room temperature and then using a filter paper the crystal is dried by simply wiping it. The etch patterns on the surface of the crystal were photographed using optical (Reichert MD 400E) Optical Ultra microscope and reflected light.

Figure 18.4: Surface of the crystal under optical microscope (×250)

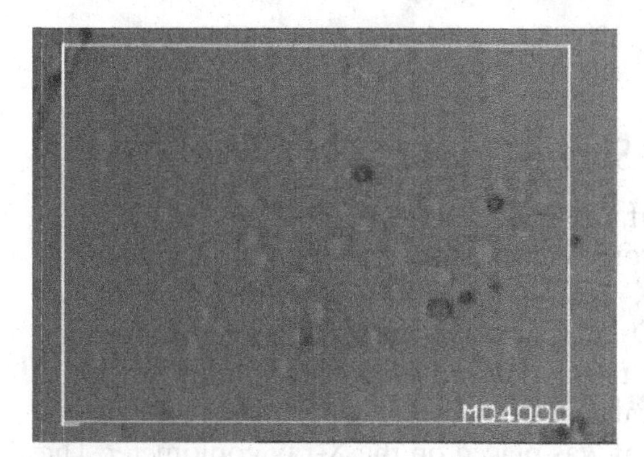

Figure 18.6: Growth hillocks and etch pits observed on (0-1-2)

Figure 18.5: Growth of hillocks observed on (0-1-2) when etched with water for 10 s (×400)

Figure 18.7: Growth hillocks, etch pits and grooves observed on (0-1-2) when etched with water for 30 s (×200)

Figure 18.8: Growth terraces and circular etch pits observed on (0-1-2) when etched with dilute hydrochloric acid for 10 s (×200)

Figure 18.10: Growth terraces and grooves observed on (0-1-2) when etched with dilute hydrochloric acid for 20 s (×200)

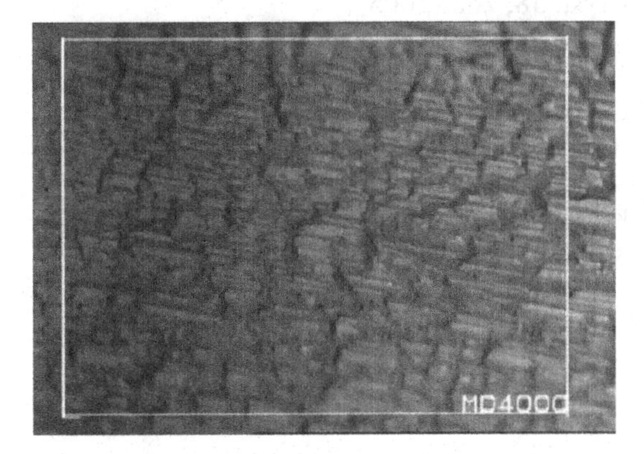

Figure 18.9: Growth terraces and grooves observed on (0-1-2) when etched with dilute hydrochloric acid for 20 s (×200)

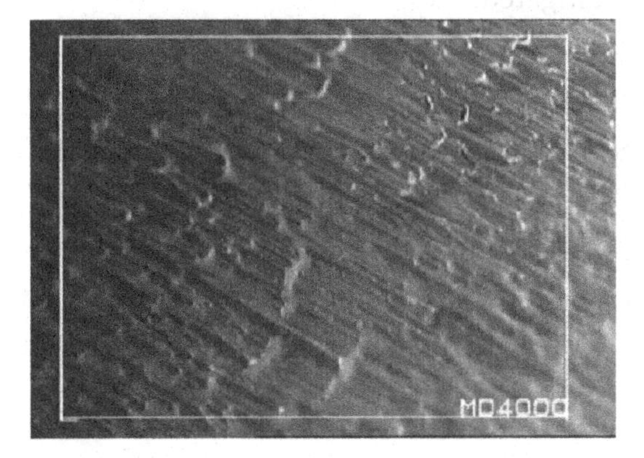

Figure 18.11: Growth macro steps and grooves observed on (0-1-2) when etched with dilute hydrochloric acid for 30 s (×200)

Figure 18.4 shows the as grown surface (0-1-2) of the crystal before etching. Structures commonly observed on this face are explained in relation to the growth condition. There are mild lines seen, which could be due to growth strain. The dark spots may be due to segregation of impurities on the exposed surface.

In Figures 18.5, 18.6, and 18.7, the growth features of the crystal under the action of the etchant as distilled water are shown. The pictures taken for progressively increasing time intervals clearly point out that as the etching time increases, the features get developed very well. In 10 s interval, only circular mound like pattern gets developed, and this changes over to well defined circular etch pits in the case of 20 s etching time. In fact the surface shows the formation of grooves also along with hillocks and etches pits when the etching time is increased to 30 s.

The circular etch pits formation is attributed to the following reason. In terms of crystal solubility when under some condition of etching, a transition stage from one morphology to another takes place, an etchant in which the crystal dissolves poorly (i.e., at a slow rate) produces circular etch pits [11].

In Figures 18.8, 18.9, 18.10, and 18.11, the etchant used is dilute hydrochloric acid. The reaction of this etchant is so much that even after 10 s, the circular etch pits as well as growth terraces with grooves and kinks show up. When the etching time is 20 s, the features are more prominent but the pits shrink in size. When the etching interval is made to 30 s, there are macro steps seen along with grooves.

From the etching study we can conclude that the growth of the BGHB crystal is due to the formation of macro steps and grooves which also modify to circular etch pits under condition of lower dissolution around the growth site, resulting in slow rate of growth.

6. Conclusion

The growth of BGHB crystals were done by solution method and the crystal is characterized by XRD study, HRXRD study and Etching study. The single crystal XRD analysis determines the dimensions of the unit cell and hence the structure was found to be orthorhombic. HRXRD confirms the crystalline perfection is very good. Etching studies indicate that the density of etch pit is very high if dilute acid is used as an etchant. All the properties make BGHB an ideal candidate for NLO applications. Hence the growth, morphology, crystalline perfection and etching behaviour of BGHB single crystals are investigated.

References

[1] S.N. Vijayan, M. Vij, H. Yadav, R. Kumar, D. Surd, B. Singh, S.A. Martin Britto Dhas, S. Verma, J. Phys. Chem. Solids **129**(1), 401 (2019)

[2] H.A. Petrosyana, H.A. Karapetyanb, M.Yu Antipinc, A.M. Petrosyana, J. Cryst. Growth **2759**(1–2), 103 (2005)

[3] M. Senthil Pandian, P. Ramasamy, B. Kumar, Mater. Res. Bull. **47**(6), 1289 (2012)

[4] M. Luisa Pita, R.A. Mosquera, Compounds, an MDPI journal **2**(4), 252 (2022)

[5] S. Natarajan, C. Muthukrishnan, S. Asath Bahadur, R.K. Rajaram, S.S Rajan, Z. Kristallogr. **195**, 265 (1992)

[6] M.J. Buerger, R. Barney, T. Hahn, Z Kristallogr. **108**, 130 (1956)

[7] K. Lal, G. Bhagavannarayana, J. Appl. Cryst. **22**, 209 (1989)

[8] G. Bhagavannarayana, R.V. Ananthamurthy, G.C. Budakoti, B. Kumar, K.S. Bartwal, J. Appl. Cryst. **38**, 768 (2005)

[9] S. Meenakshisundaram, S. Parthiban, N. Sarathi, R. Kalavathy, G. Bhagavannarayana, J. Cryst. Growth **293**, 376 (2006)

[10] B.W. Betterman, H. Cole, Rev. Mod. Phys. **36**, 681–717 (1964)

[11] K. Pandurangan, S. Suresh, J. Mater. **2014**, 362678 (2014)

CHAPTER 19

Luminescent properties of CaS and $Ca_xSr_{1-x}S$: Eu red emitting sulfide phosphors

S.Varalakshmi[1*, a], Arunachalam Lakshmanan[1], L.Sangeetha[1], M. Selvi[2]

[1]Department of Physics, Saveetha Engineering College, Thandalam, India
[2]Department of Electronics Communication and Engineering, Saveetha Engineering College, Thandalam, India
[a]varalakshmi2904@gmail.com

Abstract

Europium doped red emitting CaS and $Ca_{(1-x)}Sr_xS$ phosphors were synthesized through a solid state reaction at 1100 °C for 2 h in a CO reducing atmosphere. The resulting phosphor can be excited by visible spectrum wavelengths and emits a wide band of red light. The PL emission intensity of CaS:Eu was studied with different monovalent fluxes and by varying Ca/Sr concentrations. The presence of KCl flux was found to increase the photoluminescence (PL) intensity of the synthesized phosphor. The emission spectrum of the $Ca_XSr_{1-X}S$:Eu phosphor underwent a shift from 647 to 623 nm with Ca concentration x from 1 down to 0. Thermoluminescence studies indicate a prominent TL peak at 90 °C under UV irradiation, with additional shoulder peaks at 145 °C and 310 °C. On X-ray irradiation, the main TL glow peak appeared at 98 °C, with weaker TL peaks observed at 156 °C and 305 °C. CIE indicates that the emission colour of the CaS:Eu^{2+} phosphors were located in the red region. Hence this phosphor could be implanted in blue LED along with YAG:Ce as red phosphor to produce white light. Also, emission wavelength can be varied from 623 to 647 nm by adjusting Ca/Sr ratio.

1. Introduction

The exceptional brightness and durability of light-emitting diodes (LEDs) make them well-suited for display applications, while semiconductor laser diodes (LDs) are used in various devices, including optical communication systems and compact disc (CD) players. LEDs offer greater durability and lower power consumption compared to traditional incandescent bulbs or fluorescent lamps. LEDs are more efficient than Fluorescent Lamps (FL), eco friendly and based on electroluminescence and not on photoluminescence. II-VI compounds synthesized earlier have a smaller band gap and hence emit light in the green or red region only. III-V nitride based semiconductors possess a direct band gap, making them well-suited for blue light-emitting devices [1]. These semiconductors also enable the development of practical visible LEDs across the blue to yellow spectrum [2]. From blue LEDs, it is easy to get other colours such as green or red with the assistance luminescence phosphors. The superposition of three basic colours (Red, Blue and Green) is essential to get white light. In general, a LED made by InGaN semiconductor diode with blue emission and a YAG:Ce phosphor with yellow emission (λ_{exi} = 450 nm, λ_{emi} = 520–540 nm) is used in producing white light [3]. But, a low CRI in the range 60 to 70 is to be enhanced. Adding a small amount of red phosphor to Cerium doped YAG improved the color rendering index (CRI) to an acceptable level (>80) and enhanced light conversion efficiency. The development of red phosphor is essential to get white light with a better colour rendition in YAG:Ce coated blue LEDs. Several sulfide materials are capable of producing red emission so they are investigated as conversion phosphors [4]. CaS doped with different activators were investigated by Lehmann [5] by

DOI: 10.1201/9781003606611-19

firing mixtures of pure CaS and desired impurities in quartz container at about 1200 °C for 1 h to several hrs. CaS phosphors with different activators are very efficient luminescent materials with quantum efficiencies of Photoluminescence (PL) of about 80% or more and energy efficiencies of cathodoluminescence (CL) of about 15 to 20%. Also, the (Ca_x, Sr_{1-x}) S:Eu^{2+} phosphors were prepared by physical mixing of $CaSO_4$ and $SrSO_4$ with proper concentration in carbon reducing atmosphere for the reduction of sulphate to sulfide phase. It is reported that as the Sr/Ca ratio decreases, the lattice parameter of (Ca_x, Sr_{1-x})S:Eu^{2+} decreases and also influences the emission of the phosphor towards redshift [6].

Nanocrystalline SrS and CaS have been synthesized by Wang et al. [7] via solvo-thermal method using $SrCl_2.6H_2O$, $CaCl_2$, Sulfur and ethylenediamine as starting raw materials at low temperatures. Suitable amount of $SrCl_2$, $CaCl_2$ and sulfur were introduced into a 50 ml autoclave. Then, required quantity of cooled ethylenediamine was added. The autoclave was then heated to 170 °C–220 °C for 12 h. After cooling and vacuum desiccation at 60 °C for 4 h, grey-white powders were produced. Photoluminescence studies indicate that the CaS nanophosphor emits a yellow-green band at 550 nm, while SrS displays a blue band at 465 nm [7]. Using this method, alkaline sulfides with RE or transition metal ions as dopants at elevated temperatures upto 300 °C can be synthesized. Europium doped SrS and CaS were prepared by Van Haecke et al. [8] by solvo- thermal method which was earlier adopted by Wang et al. [7] .The phosphors displayed a broad PL emission band with emission peaks at 663 nm for CaS:Eu and 623 nm for SrS:Eu. In addition, thioglycerol was used as capping agent to control the size and shape of the materials. CaS:Eu powdered samples have been prepared using calcium carbonate, Eu_2O_3, nitric acid, ethanol and ammonium sulfate by co-precipitation method. Reduction of sulphate to sulfide is taking place in the temperature range 550 °C–1150 °C for 2 h under H_2 5 ml/min and N_2 95 ml/min. The phosphor showed red emission of wavelength 634 nm when excited at 465 nm. It was reported that PL emission intensity increased with firing temperature from 550 °C to 950 °C and showed a decline in intensity at 1150 °C due to thermal quenching [9]. Eu^{3+} doped SrS and CaS have been synthesized by solid state technique in carbon reducing atmosphere at 950 °C for 2 h by Rao et al. [10]. Reported that both CaS:Eu^{3+} and SrS:Eu^{3+} phosphors are suitable for wavelength-tunable red emission in 3-band white LEDs driven by a blue LED (460 nm). Rao et al. [10] developed a two band and three band white LEDs by coating phosphors on to a blue LED chip and investigated the optical properties. Kumar et al. [11] used a chemical co-precipitation method to produce CaS:Ce^{3+} nanophosphors with varying Ce^{3+} concentrations, analyzing their luminescence and decay properties. CaS and related phosphors are very attractive for fundamental research and for various applications such as television screens, fluorescent lamps and high pressure mercury lamps.

In the present study, Europium doped red emitting CaS and $Ca_xSr_{1-x}S$ phosphors were synthesized through solid state reaction method in CO reducing atmosphere. Photoluminescence properties of the phosphors with varying Ca/Sr ratio were investigated. The PL emission of the phosphor shifted from 647 to 623 nm when Ca molar concentration x was decreased from 1 to 0. PL emission intensity when excited with 310 nm (violet emission) and 470 nm (red emission) was maximum with Ca concentration of 1. Emission spectrum shifted towards shorter wavelength with decreasing Ca/Sr ratio. Higher luminescence efficiency is observed in red emitting CaS:Eu phosphor as a propitious choice for white LEDs. Sulphide phosphors are chemically unstable in humid conditions due to their conversion to oxides accompanied by the emanation of H_2S odour ($SrS + H_2O \rightarrow SrO + H_2S\uparrow$). Therefore, enough attention is needed to improve the storage stability of sulfide phosphors [12–14] . The synthesized CaS:Eu by solid state technique at 1100 °C for 2 h in a carbon reducing atmosphere are coated with polymer by simple wet mixing method. After it is dried, coated phosphors are stored in normal ambient conditions. PL intensities were measured for a period of 18 days to analyze the degradation in luminescence efficiency of coated phosphors.

2. Experimental Method

CaS:Eu and $Ca_xSr_{1-x}S$:Eu powdered samples were synthesized by solid state technique. Raw materials used were analar grade $CaSO_4$ and $SrSO_4$. Eu_2O_3 was added with specific doping concentration of 1 mol%. Different monovalent materials NaCl, LiCl, KCl and NaF were used as flux with 7 mol%. The samples were sintered in a carbon reducing atmosphere for 2 hrs. The temperature maintained in the furnace was 1100 °C. The sample was allowed to cool overnight. The sample turned orange and red in colour for CaSrS and CaS respectively. Again, the sintered samples were transferred into mortar, mildly crushed before taking the PL measurements. The PL excitation and emission spectra were obtained with a Jobin Yvon Spectrofluorometer (Model-Fluoromax 4C) . TL intensity and glow curves after UV (254 nm) and 100 kV X-ray irradiations as a function of radiation exposure were measured using a computer-controlled Nucleonix reader (Type: TL 10091) with a linear heating rate of 8 °C/s. For UV exposure, a 254 nm UV lamp was utilized, while X-ray irradiation was performed utilizing a Siemens Heliophos D 500 mA X-ray generator set to 100 kV and 100 mA.

3. Results and Discussion

3.1 Photoluminescence Studies

Figure 19.1a shows the appearance of CaS:Eu synthesized in the present work under fluorescent room light. The body colour of phosphor lumps looks red under fluorescent room light as a result of the absorption of room light by Eu^{2+} ions doped in CaS in the blue (250–500 nm with a peak absorption at 450 nm) region. Under blue LED light illumination, due to the 647 nm red emission from CaS:Eu^{2+} superposed with incident 450 nm blue LED light, the phosphor appears magenta in colour (Figure 19.1b). Figure 19.2 compares the PL intensities of violet emission (λ_{exi} = 310 nm, λ_{emi} = 380 nm) and red emission (λ_{exi} = 397 nm, λ_{emi} = 647 nm) of CaS:Eu with KCl synthesized (A) and CaS:Eu (commercial phosphor B). The phosphors exhibit violet emission when excited at 310 nm characteristics of Eu^{2+}. Intense red emission is observed at 647 nm when excited at 470 nm. The excitation spectrum of the phosphor aligns closely with emission spectrum of LED. Emission intensity of synthesized CaS:Eu is comparable with the commercial CaS:Eu^{2+}.

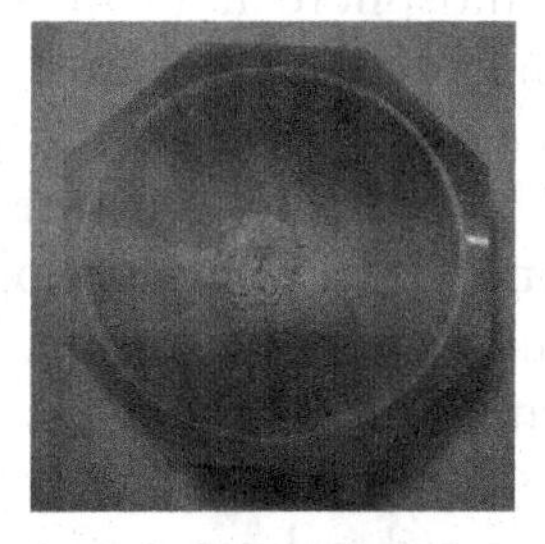

Figure 19.1: Appearance of CaS:Eu lumps before (a) and during (b) 450 nm blue LED exposure respectively

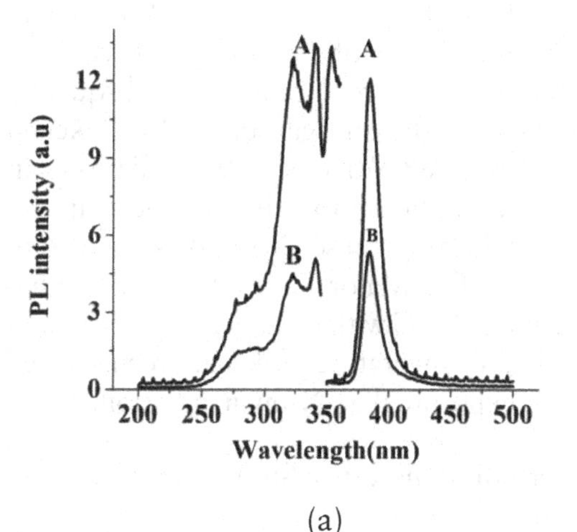

(a)

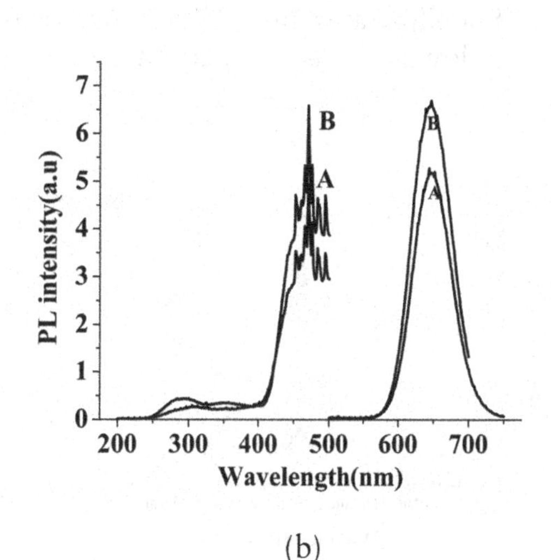

(b)

Figure 19.2: (a) PL intensities of violet emission (λ_{exi} = 310 nm, λ_{emi} = 375 nm) and (b) red emission (λ_{exi} = 397 nm, λ_{emi} = 647 nm) of synthesized CaS:Eu (A) and Commercial CaS:Eu (B)

3.2 Effect of Monovalent Flux

Figure 19.3 shows the PL intensity of CaS:Eu synthesized with different monovalent flux materials. KCl flux enhanced the intensity of red emission in CaS:Eu to the level of commercial CaS:Eu. NaCl and LiCl flux also showed high PL efficiency. Cl^- based monovalent fluxes enhanced the PL efficiency of the CaS:Eu phosphors. This is due to the fact that the ionic radii of Cl^- (181 pm) closely matched with that of S^{2-} (184 pm). The ionic radii of F^- (133 pm) is too small than S^{2-}. Hence, NaF flux yielded least PL intensity. KCl flux enhanced the red fluorescence more than that of NaCl and LiCl.

3.3 Thermoluminescence Studies on UV Exposure

Figure 19.4 compares the TL glow curves of synthesized CaS:Eu under different UV exposure time. TL sensitivities of CaS:Eu enhanced with UV exposure for up to 10 min and subsquently declined. CaS with band gap of 4.43 eV excited with 254 nm (4.88 eV) short UV irradiation can produce electron-hole traps which can get trapped. Subsequent heating cause these pairs to recombine to cause TL. Major TL peak on UV irradiation occurs at 90 °C with shoulders at 145 °C and 310 °C. High temperature peak is more visualized under UV exposure than X-ray irradiation.

3.4 Thermoluminescence Studies on X-ray Irradiation

Figure 19.5 illustrates the TL glow curve of CaS:Eu following X-ray irradiation dose with

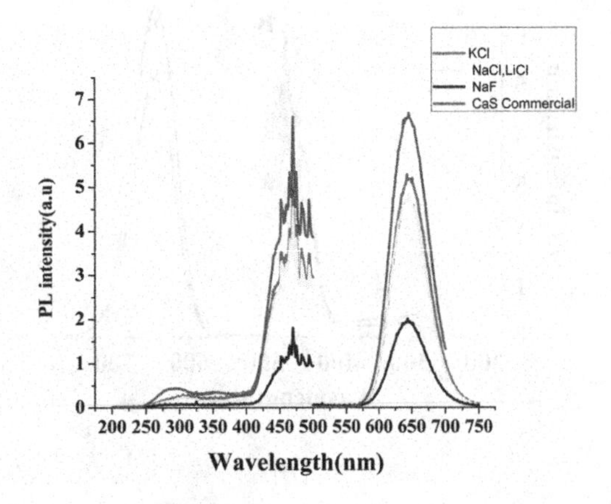

Figure 19.3: Comparison of PL excitation and emission spectrum (λ_{exi} = 397 nm, λ_{emi} = 647 nm) of synthesized CaS:Eu with different monovalent fluxes and Commercial CaS:Eu

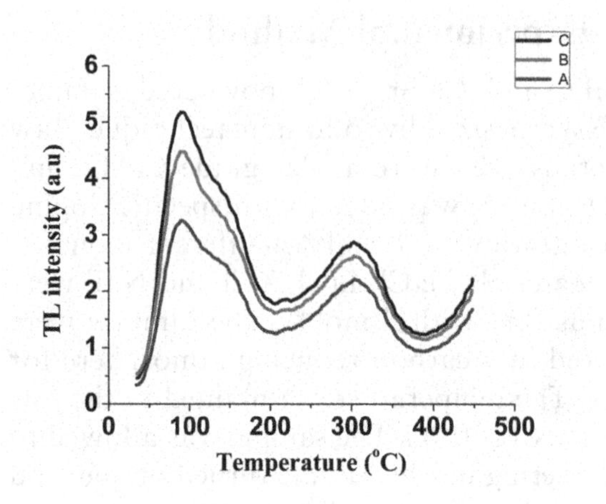

Figure 19.4: TL glow curves of synthesized CaS:Eu (a) under different UV exposure time 5 min (A), 10 min (B) and 15 min (C).

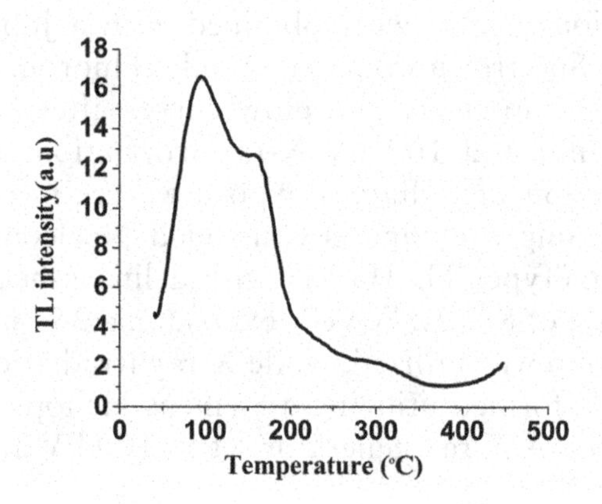

Figure 19.5: TL glow curve of CaS:Eu after X-ray irradiation with dose 0.51 Gy

0.51 Gy. The primary TL glow peak occurs at 98 °C on X-ray irradiation. An additional weak TL peak occurs at 156 °C and 305 °C.

3.5 Effect of Ca/Sr Ratio

Phosphors were also prepared by varying the molar ratio of Ca to Sr. Table 19.1 illustrates the effect of Ca/Sr ratio on the PL intensity of $Ca_xSr_{1-x}S$:Eu with NaCl flux synthesized through solid state reaction in CO reducing atmosphere at 1100 °C for 2 h. It also includes the peak excitation and emission wavelengths for CaSrS:Eu at various Ca/Sr ratios. The red emission intensity is same for SrS:Eu and CaS:Eu while violet emission is 1.5 times higher for the former. The wavelength of light emission of the phosphor shifts from 623 to 647 nm when Sr content is decreased. Thus, the red colour purity of CaS:Eu is far better than that of SrS:Eu.

Table 19.1: Effect of Ca/Sr ratio on the PL intensity of $Ca_xSr_{1-x}S:Eu$ with NaCl flux synthesized by solid state technique in CO reducing atmosphere at 1100 °C for 2 h

S.No Phosphor	Ca:Sr	PL intensity (arb. units)	
		Violet λ_{exi}-310 nm	Red λ_{exi}-470 nm
1 $Ca_1Sr_0S:Eu$	1:0	0.90×10^6	4.84×10^6
2 $Ca_{0.75}Sr_{0.25}S:Eu$	3:1	1.12×10^6	3.63×10^6
3 $Ca_{0.50}Sr_{0.50}S:Eu$	1:1	1.22×10^6	4.82×10^6
4 $Ca_{0.25}Sr_{0.75}S:Eu$	1:3	0.50×10^6	3.04×10^6
5 $Ca_0Sr_1S:Eu$	0:1	0.59×10^6	4.27×10^6

3.6 CIE Chromaticity Diagram

The CIE chromaticity coordinates (x, y) for the synthesized $CaS:Eu^{2+}$ phosphors were determined using CIE software. Figure 19.6 displays the CIE chromaticity coordinates with a dominant red wavelength of 647 nm for $CaS:Eu^{2+}$. These coordinates are positioned close to the edge of the CIE diagram, indicating that the phosphors demonstrate high color purity.The calculated chromaticity coordinates are (x - 0.6814, y - 0.3176), indicating that the emission color of the $CaS:Eu^{2+}$ phosphors is in the red region.

3.7 Storage Stability of CaS:Eu Phosphor

The PL sensitivity (red emission) of coated and uncoated CaS:Eu phosphors is listed in Table 19.2. The PL sensitivity slightly decreases with polymer coating due to the substance covered over the phosphors and small variation in intensity was observed.

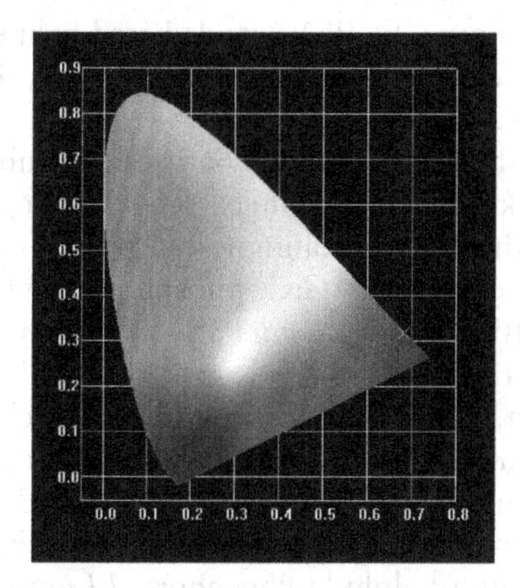

Figure 19.6: CIE chromaticity diagram of $CaS:Eu^{2+}$ prepared by solid-state method

Table 19.2: PL intensities of $CaS:Eu^{2+}$ with and without polymer coating

S.No Phosphor	PL intensity (arbitrary units) Red emission (λ_{exi} = 470 nm, λ_{emi} = 647 nm)
1 Uncoated	5.72×10^6
2 Coated with 0.3 ml of polymer	5.43×10^6

Table 19.3 shows the improved storage stability of PL intensity of polymer coated CaS:Eu in comparison with uncoated phosphor. Phosphor coated with polymer is stable for up to 10 days, beyond which a decrease in intensity was observed slowly. 29% of degradation of PL intensity was observed on 18 days of storage. Uncoated phosphor, however, shows nearly 48% degradation of PL intensity as shown in Figure 19.7. Coated polymer improved the stability to 41% when compared with uncoated phosphor. But further improvement in composition of the polymer and / or coating techniques is needed for long term moisture stability. For comparison, the storage stability of commercial $YGd(BO)_3$ red phosphor , LG chemical (E) is also shown.

4. Conclusion

The CaS:Eu red emitting phosphor was synthesized using a simple solid state technique in CO reducing atmosphere. Most of the synthesis methods for sulphide phosphors described in literature use highly toxic H_2S and hence avoided in this work. The synthesized phosphor

Table 19. 3: Storage stability of $CaS:Eu^{2+}$ with and without polymer coating

Phosphor	Relative PL intensity on storage period (Days)					
	0	1	5	10	15	18
Uncoated	1	0.95	0.92	0.88	0.61	0.52
Coated with 0.3 ml of polymer	1	0.99	0.99	0.96	0.85	0.71
$YGd(BO)_3$ red Phosphor	1	1.03	0.96	1.03	0.98	0.99

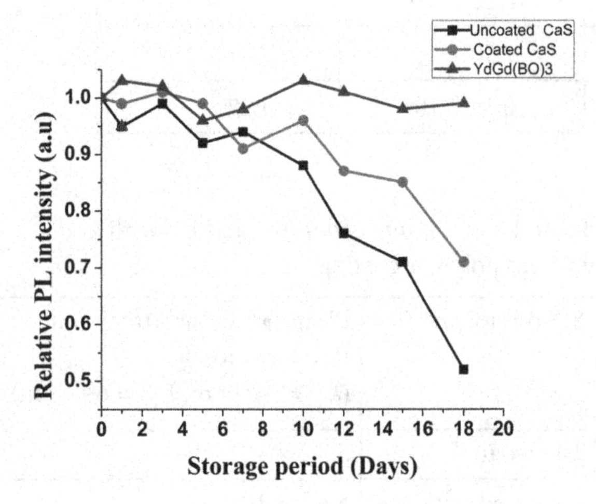

Figure 19.7: PL degradation of uncoated $CaS:Eu^{2+}$ phosphor (■) and those coated with 0.3 ml of polymer (●) as a function of period of storage in room temperature ambience. For comparison, the storage stability of commercial $YGd(BO)_3$ red phosphor, LG chemical (▶) is also shown

can be excited by visible light and emits a single broad band of red light. The PL emission intensity of CaS:Eu was analyzed with different monovalent fluxes, showing that the use of KCl flux resulted in increased PL intensity. Thermoluminescence studies indicate a prominent TL peak (T_{max}) at 90 °C under UV irradiation, with additional shoulder peaks at 145 °C and 310 °C. On X-ray irradiation, the major TL glow peak appears at 98 °C, with additional weaker TL peaks observed at 156 °C and 305 °C. CIE indicates that the emission colour of the $CaS:Eu^{2+}$ phosphors were positioned in the red region and hence a commercial version of this phosphor has been implanted in blue LED along with YAG:Ce to emit a white light. In $CaSrS:Eu^{2+}$, emission wavelength can be varied from 623 to 647 nm by adjusting Ca/Sr ratio. Storage stability of sulphide phosphor under atmospheric condition is a prime requirement for all applications. Polymer coating achieved by a simple wet mixing method slightly improved the storage stability of CaS:Eu in atmospheric conditions. 0.3 ml polymer coated $CaS:Eu^{2+}$ phosphor was stable for up to 12 days, beyond which the intensity gradually decreased. 29% of PL intensity degraded on 18 days of storage in polymer coated CaS:Eu. Further improvement with different coating techniques in improving the storage stability for a long term is in progress. The commercial $CaS:Eu^{2+}$ is quite stable for several months due to improved encapsulation.

5. Acknowledgments

The authors extend their gratitude to the Management of Saveetha Engineering College for their financial support in conducting this research and to Mr. D. Albert Raja from Saveetha Institute of Medical and Technical sciences for providing the X-ray irradiation facility.

References

[1] Nakamura, S., The roles of structural imperfections in InGaN-based blue light-emitting diodes and laser diodes. *Science*, 1998, 281, 956–961.

[2] Nakamura, S., III-V nitride based light emitting devices. *solid state commun.*, 1997, 102 (2), 237–248.

[3] Lakshmanan, A.R., Satheesh Kumar, R., Sivakumar, V., Thomas, P.C. and Jose M.T., Synthesis, photoluminescence and thermal quenching of YAG:Ce phosphor for white light emitting diodes. *Indian J. Pure Appl Phys*, 2011, 49, 303 -307.

[4] Smet, P., Moreels, I., Hens, Z. and Poelman, D., Luminescence in Sulfides: A Rich History and a Bright Future. *Materials*, 2010, 3, 2834–2883.

[5] Lehmann, W., Activators and Co-activators in Calcium Sulphide Phosphors. *J.Lumin.*, 1972, 5, 87–107.

[6] Hu, Y., Zhuang, W., Ye, H., Zhang Ying Fang, S. and Huang, X., Preparation and luminescent properties of (Ca_{1-x},Sr_x)S:Eu^{2+} red emitting phosphor for white LED. *J. Lumin*, 2005, **111**,139–145.

[7] Wang, C., Tang, K., Yang, Q., An, Q., Hai, B., Shen, B. and Qian, Y., Blue-light emission of nanocrystalline CaS and SrS synthesized via a solvothermal route. *Chem. Phys. Lett.*, 2002, **351**, 385–390.

[8] Van Haecke, J.E., Smet, P.F., De Keyser, K. and Poelman, D., Single crystal CaS:Eu and SrS:Eu Luminescent particles obtained by solvothermal synthesis. *J. Electrochem. Soc.*, 2007, **154** (9) 278–282.

[9] Kim, K.N., Kim, J.M., Choi, K.J., Park, J.K. and Kim, C.H., Synthesis, characterization and luminescent properties of CaS:Eu phosphor'. *J. Am. Ceram. Soc.*, 2006, **89** (11) 3413–3416.

[10] Rao, C.A., Poornachandra Rao, N.V. and Murthy, K.V.R., Synthesis, characterization and photoluminescence properties of CaS:Eu^{3+} and SrS:Eu^{3+} phosphors for white LED. *Advance Physics letter*, 2015, **2** (4) 4–10.

[11] Kumar, V., Pitale, S.S., Mishra, V., Nagpure, I.M., Biggs, M.M., Ntwaeaborwa, O.M. and Swart, H.C., Luminescence investigations of Ce^{3+} doped CaS nanophosphor. *J. Alloys Compd.*, 2010, **492**, 8–12.

[12] So-Ra Gang, Kim, D., Kim, S.M., Hwang, N. and Lee, K.C., Improvement in the moisture stability of CaS:Eu phosphor applied in light-emitting diodes by titania surface coating. *Microelectron Reliab*, 2012, **52** , 2174–2179.

[13] Lee, J.S., Unithrattil, S., Kim, S., Lee, I.J., Lee, H. and Im, W.B., Robust moisture and thermally stable phosphor glass plate for highly unstable sulphide phosphors in high-power white light-emitting diodes. *Opt. Lett.*, 2013, **38**, 3298–3300.

[14] Guo, C., Chu, B. and Su, Q., Improving the stability of alkaline earth sulfide-based phosphors. *Appl. Surf. Sci.*, 2004, **225**, 198–203.

CHAPTER 20

Nano Boron Carbide Reinforcement in Al6061 Metal Matrix Composite

A comprehensive study of microstructural changes and mechanical enhancements

K. Venkitachalapathy,[1*,a] I. Manivannan,[2] S. Periandy,[1] S. Suresh, [4*]M. Priya[4]

[1]Department of Physics, Kanchi Mamuniver Government Institute for Post Graduate Studies and Research (An Autonomous institution, Affiliated to Pondicherry University, Puducherry) Airport Road, opposite to TAC, Lawspet, Puducherry India
[2]Department of Mechanical Engineering, Motilal Nehru Government Polytechnic College, Puducherry, India
[3]Department of Physics, Saveetha Engineering College (Autonomous), Thandalam, Chennai Tamil Nadu, India
Email: [a]pathyphysics@gmail.com, [b]sureshthanjai2003@gmail.com

Abstract

Nano reinforcements significantly enhance the properties of Al6061 composites compared to standalone counterparts. A hybrid approach of stir casting and ultrasonic cavitation incorporated diverse weight fractions of Nano Boron carbide as reinforcements, with varying weight percentages (0.6, 1.2, 1.8 wt.%). Microstructural and mechanical properties of the fabricated metal matrix composite were investigated. Homogeneous distribution of Nano Boron Carbide particles in Al6061-1.8wt%B4C composites was confirmed through examination with FESEM and EDS analysis. The interfacial analysis of the metal matrix composite was conducted using the XRD technique. The composites' hardness was measured using Brinell and Micro-Vickers systems, and the density and porosity were studied based on these values resulting in a 36.09% increase in specific hardness for Brinell and 38.50% for Micro-Vickers in this composite. The Universal Testing Machine was used to assess the tensile, flexural, and compression properties of the Al6061/1.8wt% B4C composite. The results showed an 11.47% increase in specific uniaxial tensile strength, a 39.30% increase in specific compression strength, and a 40.03% increase in specific flexural strength. Additionally, the impact energy of this composite material was measured at 12 J.

Keywords: Metal matrix nano composite, Al6061, Boron carbide, Ultra sonification, Stir casting.

1. Introduction

Securing high-quality materials faces substantial challenges due to the limited availability of abundant natural or monolithic resources across the globe. As the world's attention shifts towards greener technologies, manufacturing companies have increasingly integrated Metal Matrix Composites (MMCs) into their processes. This adaptation allows for the creation of materials that are not only lighter but also more efficient, catering to the requirements of fuel-efficient automobiles and aircraft. Amid the growing global emphasis on sustainability in technology, manufacturing enterprises have adopted MMCs as a strategy to produce materials that offer enhanced efficiency and reduced weight. This strategic alignment meets the demands of fuel-efficient automobiles and aircraft effectively [1,2]. Aluminum metal matrix composites (AMMCs) offer a solution as they fall under the category of advanced lightweight materials centered around aluminum, addressing these challenges. The characteristics of AMMCs can be customized to suit the requirements of various industrial uses through the judicious selection of matrix, reinforcement, and processing method [3, 4]. The aviation and automotive industries are increasingly intrigued by the AA 6XXX aluminum alloy series, where silicon and magnesium serve

DOI: 10.1201/9781003606611-20

as the primary alloying components. This series of alloys presents notable benefits, including an exceptional strength-to-weight ratio, superior formability, weldability, corrosion and wear resistance, and cost-effectiveness. These attributes make it a compelling option for the production of lightweight vehicles. Among the 6XXX series, AA 6061 stands out as a widely favored alloy, frequently employed as the matrix material in various AMMCs due to its capacity for composite strength adjustment through appropriate heat treatments [5, 6]. AMMCs can be reinforced with Particulate, Whiskers, Continuous fibers, and Mono-filaments. Particulate reinforcement provides advantages over other types, incorporating ceramics, glasses, metals, and amorphous materials [3]. Ceramic reinforcements like Al2O3, TiO2, SiC, TiC, TiB2, B4C etc., are widely used in high-performance materials[7]. Generally, these Ceramic particulate reinforcements usually have equiaxed shapes and aspect ratio less than about 5 [3]. Notably, Boron carbide (B4C), a black solid with a metallic lustre, stands out as one of the earth's hardest ceramic substances. With its elevated melting point, impressive hardness, favourable mechanical properties, low specific weight, exceptional chemical resistance, and substantial neutron absorption cross-section (BxC, x > 4), B4C finds practical utility in cutting-edge domains like fast-breeder technology, lightweight armour development, and high-temperature thermoelectric conversion [5,8]. Composite fabrication techniques can be broadly categorized into liquid phase processing using infiltration technique (molten metal infiltration [9], compo/stir casting[10–14] [15–20][21][22] squeeze casting [23–25], friction stir process[26], and laser based additive manufacturing[26–28]) and solid-state processing methods (powder metallurgy [10, 29–37]). In the former, the discontinuous reinforcement phase is introduced into the continuous metal matrix phase while in a liquid state, and the molten metal is subsequently cast into desired shapes using conventional casting techniques. Conversely, solid-state processing involves the production of MMCs below the melting temperature of the matrix material. Another approach worth mentioning is reactive processing, also known as in-situ processing, where the reinforcement is synthesized within the matrix through chemical reactions involving different elements [38].

The size of the particulates incorporated as reinforcements, along with the volume fraction of these reinforcements within the metal matrix, play a pivotal role in elevating the mechanical and other material properties of composites. Notably, when contrasted with composites reinforced by micro-sized particles, those bolstered by Nano-sized particles demonstrate remarkable enhancements in terms of strength, ductility, wear resistance, and their ability to perform effectively at elevated temperatures [39]. Unlike incorporating micro-sized reinforcement particles into metal matrix composites, the inclusion of Nano-sized particles presents certain challenges that need to be addressed. These challenges include: (1) achieving an even distribution of Nano-sized particles within the matrix and (2) establishing a strong interfacial bonding [39]. Attaining a uniform dispersion of Nano-sized particles offers significant advantages when integrating them with various base materials. This results in enhanced nucleation and grain growth compared to the use of micro-sized particles [4, 40].

Numerous synthesis processes have been explored and outlined above, yet achieving efficient mass production in the synthesis and fabrication of MMNCs remains a significant challenge. The achievement of a homogeneous and uniformly distributed nanoparticle dispersion within a metal matrix, while employing innovative synthesis processes, has garnered substantial interest among material scientists. Despite the adoption of various synthesis techniques, ranging from solid-state mixing to powder metallurgy, attaining a consistent dispersion of nanoparticles throughout the matrix has proven elusive [41].

After extensive research and deliberation, the author has embraced a novel instrumentation technique: ultrasonic cavitation-based stir casting processing, as a suitable approach for material fabrication. The ultrasonic cavitation technique offers a remarkable advantage, being the most promising method for manufacturing MMNCs with an almost uniform distribution of nanoparticles [41].

Having taken into account all pertinent aspects, the author subsequently undertook the present study to create Metal Matrix Nano Composites (MMNCs) of Al6061-nB_4C through the utilization of Ultrasonic Cavitation assisted stir casting, employing varying weight percentages of B_4C (0.6, 1.2, 1.8wt%). The primary objective was to comprehensively analyze these composites, assessing both their microstructural features and mechanical properties.

2. Materials and methods:

2.1. Materials details

Al6061 metal pieces measuring dia. 25.4mm x length 3050 mm were sourced from PMC corporation in Bangalore, India, whereas Graphite (≤ 20µm, 99.9%) and Boron carbide (~300nm, 99.9%) were procured from US Research Nanomaterials, INC. USA to be used as reinforcement materials. The chemical composition of Al6061 is presented in Table 20.1, while the physical properties of Al6061 and Boron carbide are listed in Table 20.2.

2.2 Fabrication of composition

In this study, the composite material was fabricated using the experimental setup illustrated in Figure 20.1. Al6061 materials were received in the specified dimensions, cut into small pieces as per the desired weight percentage, and then cleaned thoroughly before use. The Al6061 core material was melted to form a vortex of molten metal by placing it in a crucible and heating it up to 750°C. To prevent oxidation of the melt, a controlled argon gas atmosphere was maintained throughout the process, while a PID (Proportional Integral Derivative) temperature controller was used to keep the required heat. Boron Carbide Nano particles (B_4C) of 0.6wt.% were pre-heated to 350^{0}C and slowly introduced onto the molten metal, which was then allowed to solidify. The resulting semi-solid state composite mix was subsequently re-heated to 750°C to ensure uniform distribution of the reinforcements.

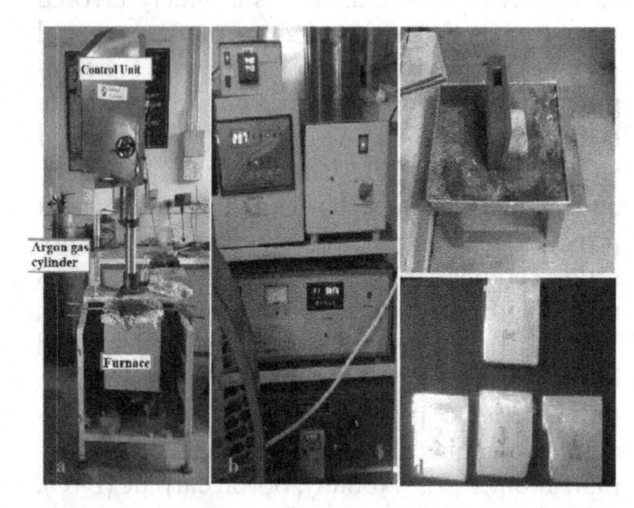

Figure 20.1: Experimental setup of composite material fabrication

The formation of micro-volatile hot void spots in the molten composite mix was caused by the cavitation effect produced by the titanium alloy ultrasonic probe, which was connected to an ultrasonic generator (Johnson Platonic Pvt., Ltd, India) with a 2.5 KW, 25 KHz capability. These hot void spots had temperatures of up

Table 20.1: The chemical composition of Al6061

Element Al6061	Mg	Si	Cu	Mn	Cr	Ti	Zn	Al
Weight %	0.88	0.7	0.29	0.33	0.006	0.02	0.003	99

Table 20.2: Physical properties of Al6061 and Boron carbide

Material	Density	Thermal conductivity	Coefficient of Thermal expansion	Melting temperature	Modulus of Elasticity	Specific Heat Capacity
Notation and unit	$\rho\ (g/cm^3)$	$K(W/mK)$	$CTE(mm\times10^{-6}/m)$	T_m	$E\ (GPa)$	J/Kg K)
Al6061	2.71	180	23	620	70	896
B_4C	2.52	32-40	3-6	2350	350-550	1100

to ~5000K and pressures of up to ~1000 atm, and moved rapidly with velocities on the order of ~10^7 m/s [42]. The composite sample was created by pouring a metal matrix mixture into a preheated rectangular die at temperatures of up to 400^{0}C using the top pouring method, and the same procedure was used to prepare other composite samples with 1.2% and 1.8% variations of Boron Carbide mixed into the matrix. The prepared specimens were of standard dimensions for various Micro-structural and Mechanical examinations. The detailed specifications of the Al6061/nB_4C can be found in Table 20.3.

Table 20.3: Detailed specifications of the Al6061/nB_4C

Composite Specimen	Part of the Components with weight percentage
Sample 1	Al6061
Sample 2	Al6061/0.6% nB_4C
Sample 3	Al6061/1.2% nB_4C
Sample 4	Al6061/1.8% nB_4C

3. Properties of Composites, Results and Discussion

3.1 Microstructure

The specimens underwent a series of steps including grinding, polishing, and etching, before being analysed with an FESEM and optical microscope. Wet grinding was carried out using SiC emery paper with grit sizes ranging from P120 to P2000, in the presence of water. Polishing was achieved using a diamond paste (1 μm) on a specialized cloth with a lubricant. The samples were then etched in a solution containing 99% water and 1% HF, rinsed with water and alcohol, and dried thoroughly. Figure 20.2(a) exhibits FESEM images depicting Al6061 metal matrix composites, while Figures 20.2(b), 3(a), 3(b), 4(a), and 4(b) present composites characterized by different weight percentages of Nano B_4C fine particles embedded within the Al6061 matrix.

In Figure 20.2(a), an even and smooth surface is evident. Moving to Figure 20.2(b), the Al6061-0.6wt% nB_4C composite showcases a uniform dispersion of Nano B_4C fine particles within the matrix. Figure 20.3(a) demonstrates a similarly even and smooth surface as seen in the previous figure. Transitioning to Figure 20.3(b), the Al6061-1.2wt%nB_4C composite reveals a consistent dispersion of Nano B4C fine particles throughout the matrix. The exceptional uniformity in Nano B_4C fine particle distribution is observable in Figures 20.4(a) and 20.4(b), specifically in the composite containing 1.8wt% of Nano Boron Carbide. Shifting attention to Figure 20.4(a), the surface of the composite material reveals minor cracks and clustered particles, attributed to an elevated concentration of Nano Boron Carbide fine particles. Notably, Figure 20.4(b) emphasizes the presence of a high-density reinforcement distribution pattern.

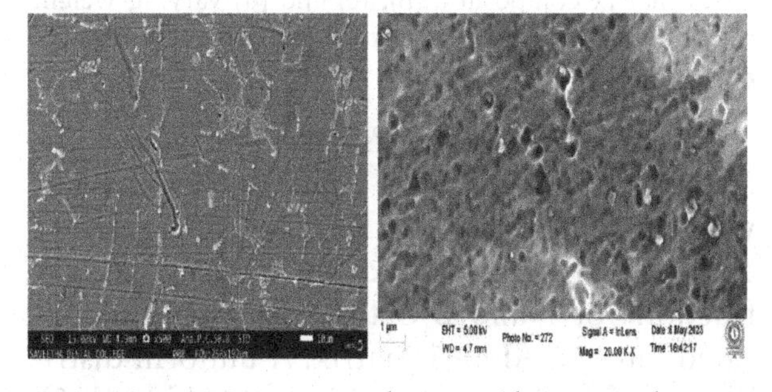

Figure 20.2: FESEM images depicting Al6061 metal matrix composites

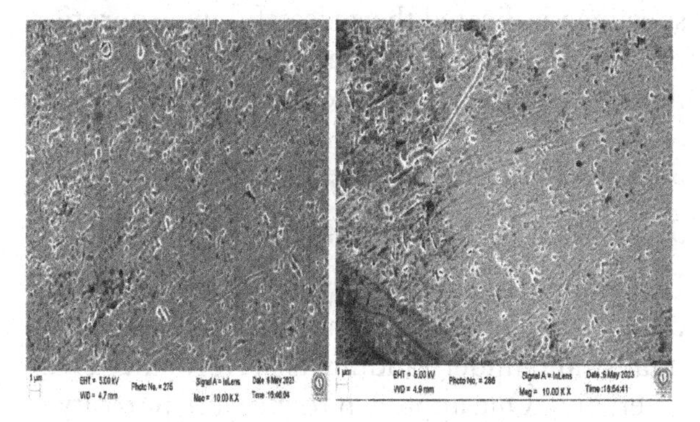

Figure 20.3: Different weight percentages of Nano B_4C fine particles embedded within the Al6061 matrix

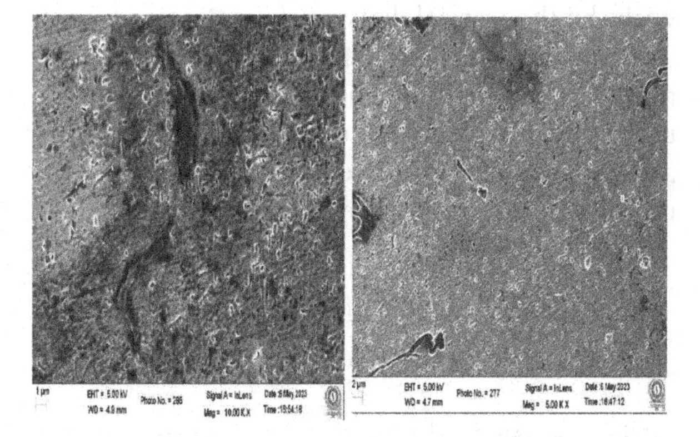

Figure 20.4: Different weight percentages of Nano B4C fine particles embedded within the Al6061 matrix

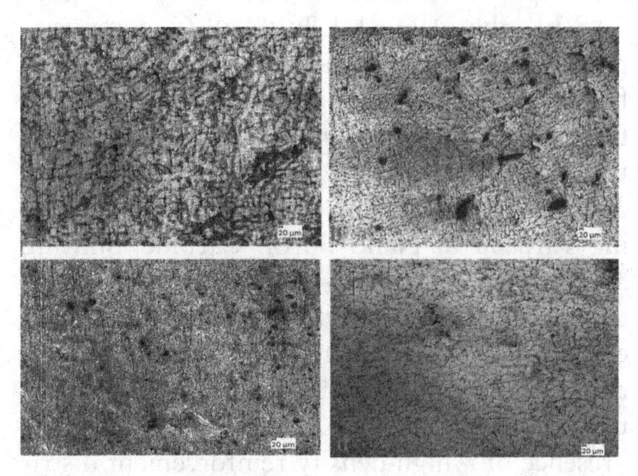

Figure 20.5: (a) displays optical images illustrating Al6061 metal matrix composites, (b), (c) and (d) varying weight fractions of Nano B_4C fine particles incorporated into the Al6061 matrix

Figure 20.5(a) displays optical images illustrating Al6061 metal matrix composites, whereas Figure 20.5(b), 5(c), and 5(d) portray composites characterized by varying weight fractions of Nano B_4C fine particles incorporated into the Al6061 matrix. A uniform distribution of isolated Nano Boron Carbide (B_4C) particles in the Al6061-1.8wt%B_4C composite specimen indicated the successful incorporation of the reinforcement material into the metal matrix, as observed by the author.

3.2 Mechanical Properties

3.2.1 Porosity and Density

Determining the porosity of a metal matrix composite material is crucial for assessing its quality and understanding its performance characteristics. One approach to measure porosity is through density-based calculations. By comparing the actual density of the composite material with the theoretical density of a solid composite without any porosity, the porosity can be estimated. This method relies on accurately measuring the mass and volume of the composite, allowing for quantitative evaluation of the volume fraction occupied by voids or pores.

$$\rho_{density\ ratio} = \left(\frac{\rho_{exp}}{\rho_{the}} \right) \quad (1)$$

$$Porosity(\%) = \left[1 - \rho_{density\ ratio} \right] \times 100 \quad (2)$$

Archimedes' principle is used to measure the experimental density of aluminium composites by assessing the buoyant force when the sample is submerged in a fluid, typically water. The process begins by weighing the dry sample in air, known as the weight in air(g). Next, the apparent weight of the fully submerged sample is measured, referred to as the weight in water (g). The difference between these two weights represents the buoyant force, equivalent to the weight of the displaced fluid. This difference allows for the calculation of the sample's volume. The experimental density, denoted as is then calculated by dividing the sample's weight in air by its volume, as shown in Equation (3).

$$\rho_{\mathrm{MMNC(exp)}} = \frac{m}{V} = \frac{Weight\ in\ air\,(g)}{Weight\ in\ air\,(g) - Weight\ in\ water\,(g)} \times density\ of\ water\left(g/cm^3\right) \quad (3)$$

The theoretical density can also be calculated using the following equation (4).

$$\rho_{MMNC(the)} = \left[\left(\frac{w_{Al6061}}{\rho_{Al6061}} \right) + \left(\frac{w_{B4C}}{\rho_{B4C}} \right) \right]^{-1} \quad (4)$$

Where V_{MMNC}, V_{Al6061} and V_{B4C} are volume of Metal Matrix Nano Composite, Al6061 metal matrix and reinforcement; W_{Al6061} and W_{B4C} are mass fraction of metal matrix and reinforcement; $\rho_{\mathrm{MMNC(the)}}$, ρ_{Al6061} and ρ_{B4C} are theoretical density of Metal Matrix Nano Composite, Al6061 metal matrix and reinforcement respectively.

Utilizing the formulas described earlier, the experimental and theoretical densities, density ratio, and porosity values were computed. These results are outlined in Table 20.4.

Figure 20.6 (a) provides a graphical representation of the theoretical and measured densities for each composite. Furthermore, Figure 20.6(b) illustrates the porosity percentage of the composites.

Figure 20.6(a) demonstrates a noticeable trend wherein the measured density values of composites Sample 1, 2, 3, and 4 consistently remain lower than their respective theoretical values. This trend offers valuable insights into the crystalline formation of the Nano reinforcement material B_4C, showcasing a linear decrease up to 1.8wt% nB_4C.

Table 20.4: The experimental and theoretical densities, density ratio, and porosity values of the four samples

Composite Specimen	Theoretical density $\rho_{MMNC(the)}$ (g/cm^3)	Experimental density $\rho_{MMNC(exp)}$ (g/cm^3)	Density Ratio	Porosity (%)
Sample 1	2.710	2.700	0.9963	0.3690
Sample 2	2.709	2.692	0.9938	0.6246
Sample 3	2.708	2.688	0.9927	0.7328
Sample 4	2.707	2.678	0.9894	1.0627

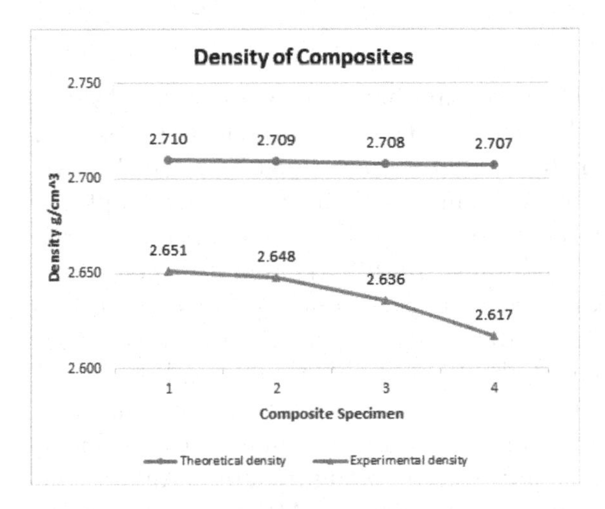

Figure 20. 6 (a) Graphical representation of the theoretical and experimental densities for each composite

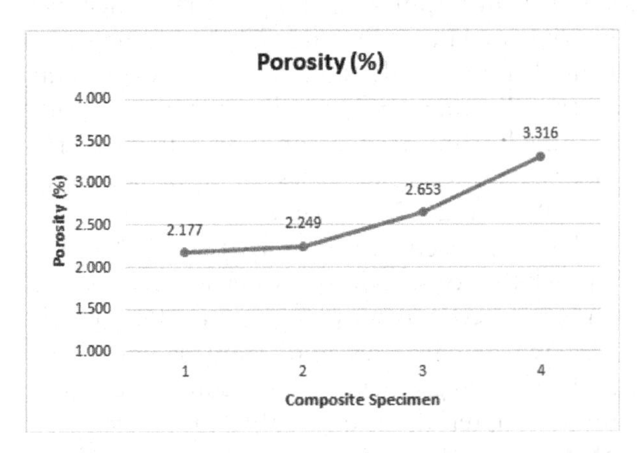

Figure 20.6 (b) Graphical representation of porosity percentage of the composites.

The Al6061 alloy, in its molten state, adopts the FCC crystal structure, whereas the reinforcement material Nano B_4C exhibits a Rhombohedral crystal structure. As the concentration of B_4C increases, the lattice strain between the FCC structure of the Al6061 base material and the B_4C reinforcement also undergoes a proportional increase. The addition of 0.6wt%-nB_4C to the base matrix results in a slight increase, as observed by the author, and this trend persists in a linear fashion up to 1.8wt%-nB_4C, as indicated in Figure 20.6(b). With an increasing content of B_4C, the strain on the FCC lattice of Al6061, caused by the presence of B_4C reinforcement, demonstrates a progressive increment. The experimental findings show a gradual rise in strain as the proportion of B_4C reinforcement is increased in the Al6061 matrix, as depicted in Figure 20.6(b).

3.2.2 Hardness

Two metric systems, Micro-Vickers and Brinell, were utilized to assess the hardness of the specimen. The hardness of the composite system was determined using a Shimadzu HMV Vickers micro-hardness tester, employing a diamond indenter with a 136-degree angle. For the Vickers micro-hardness (HV) measurement, an applied load of 0.5 kg was exerted for 10 seconds. Formula (5) was applied to calculate the micro-hardness value.

$$HV = \frac{2P\,sin\left(\frac{\alpha}{2}\right)}{D^2} kgf/mm^2 \quad (5)$$

Where P is applied load measured in kgf; α is Angle between two opposite faces of diamond indenter measured in degree; D is Mean diagonal length of the indentation measured in mm.

The Brinell Hardness system, following ASTM E10 standards, utilized the AKB-3000 testing machine for hardness measurements. A 10mm steel ball was employed as the indenter, applying a 1000 kgf load for 15 seconds. The

Table 20.5: The Hardness and specific hardness value of composite specimen

Composite	Hardness		Experimental Density $\rho_{MMNC(exp)}$ (g/cm^3)	Specific Hardness $\left(\frac{Hardness}{\tilde{n}_{mmnc(exp)}}\right)$	
	Vickers	Brinell		Vickers	Brinell
	HV	BHN		HV	BHN
Specimen 1	92.3	55.32	2.700	34.185	20.489
Specimen 2	111.2	65.35	2.692	41.308	24.276
Specimen 3	115.6	68.24	2.688	43.006	25.387
Specimen 4	120.1	70.64	2.678	44.847	26.378

diameter of the resulting impression was then measured using a low-power microscope, and the obtained values were used in formula (6) to calculate the Brinell Hardness Number (BHN).

$$BHN = \frac{P}{\frac{\pi D}{2}\left[D - \sqrt{D^2 - d^2}\right]} kgf/mm^2 \quad (6)$$

Where P is applied load measured in kgf; D is diameter of the steel ball in mm; d is diameter of the indentation in mm. The precision of composite material hardness measurement, particularly on surfaces with multiple phases and inhomogeneity, relies heavily on the indenter's depth and area. To achieve higher accuracy, both metric systems were utilized. The hardness and specific hardness data (Vickers and Brinell) of the composites are presented in Table 20.5, and visual representations of the specific hardness values can be found in Figure 20.7.

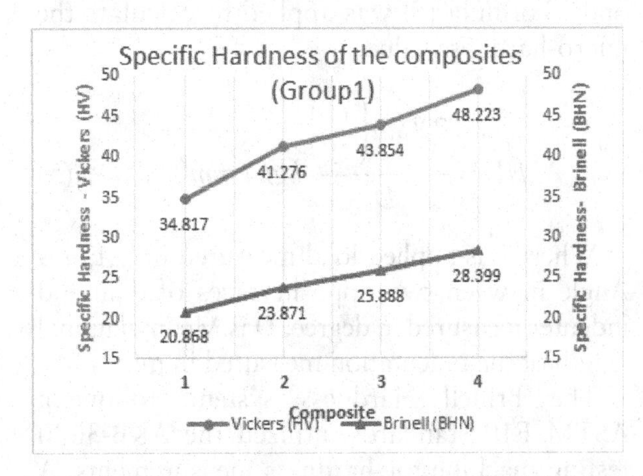

Figure 20.7: Specific hardness of Composite Vickers HV and Brinell BHN

The material's hardness was evaluated using both Vickers HV and Brinell BHN scales, which resulted in variations between the two systems. Figure 20.7 illustrates the graphical representation, revealing that specific hardness values for both Vickers and Brinell scales exhibit an increasing trend up to 1.8wt%nB_4C. The hardness of B_4C is notably higher than that of the base metal matrix, and an initial increase in B_4C concentration contributes to enhancing the composite material's hardness, as demonstrated in Figure 20.7. Among the composite specimens, Al6061-1.8wt%nB_4C displayed the maximum hardness value, showcasing approximately 30.1% increase in Vickers HV scale and 27.6% increase in Brinell BHN scale compared to the metal matrix. Furthermore, specific hardness values obtained from Vickers and Brinell measurements exhibited increases of approximately 31.19% and 28.74%, respectively.

The increased hardness of the Al6061-1.8 wt% B4C composite is primarily due to the higher concentration of individual B4C particles, which enhances load transfer capability and results in greater hardness compared to the matrix material. Additionally, the refinement of the microstructure, influenced by ultrasonic cavitation, contributes to the increased hardness. The cavitation process disrupts the formation of large dendrites or grains by promoting the nucleation of smaller grains. The collapse of the bubbles introduces shear forces that break up dendritic structures and refine the grain size, further enhancing the hardness of the composite

Table 20.6: Hardness variation of metal matrix composites in Percentage compared to Al6061 metal matrix.

Composite	Hardness variation in %		Specific Hardness Variation in %	
	Vickers	Brinell	Vickers	Brinell
	HV	BHN	HV	BHN
Specimen 1	-	-	-	-
Specimen 2	20.48	18.13	20.63	18.48
Specimen 3	25.24	23.35	25.80	23.90
Specimen 4	30.12	27.69	31.19	28.74

Table 20.7: Yield Strength, Ultimate Tensile Strength, Specific Tensile Yield, Specific UTS and Elongation variation of Composite specimen

Composites	Yield strength	UTS $MPa\cdot$	Experimental density $\rho_{MMNC(exp)}$ (g/cm^3)	Specific Yield Strength $(MPa/g/cm^3)$	Specific UTS $(MPa/g/cm^3)$	Elongation in %
Specimen 1	80.965	124.770	2.700	29.987	46.211	5.17
Specimen 2	90.230	138.321	2.692	33.518	51.382	3.72
Specimen 3	96.303	143.235	2.688	35.725	53.287	3.45
Specimen 4	114.706	153.464	2.678	42.833	57.305	3.31

3.2.3 Uniaxial Tensile, Flexural and Compression test.

A single universal testing machine (TUE-CN-200) with a maximum capacity of 200 kN and elongation capability of 200 mm was employed to conduct uniaxial tensile, flexural, and compression tests on the composite systems. The uniaxial tensile tests followed the ASTM E8/E8M standard, employing a controlled strain rate of 0.01 mm/s. Compression tests adhered to the ASTM E9-89a (Reapproved 2000) standard, while flexural tests were carried out in accordance with the ASTM D790 standard, all using the same machine. The table labelled as Table 20.7 presents the Specific Tensile Yield strength (), Specific Ultimate Tensile Strength (), and Elongation values for all the composites. These values are graphically interpreted in Figure 20.8(a) and Figure 20.8(b). Additionally, Figure 20.9(a) displays the specific Compressive strength of the composite materials, while Figure 20.9(b) illustrates their specific Flexural strength.

The composite material exhibits higher yield strength and ultimate tensile strength (UTS) than the base Al6061 metal matrix. Additionally, the specific compressive and flexural strengths of the composites are also higher than the pure matrix. Among all weight fractions of composites, 1.8wt%-nB_4C exhibits the highest values. This is consistent with the earlier observation that increasing the reinforcement concentration leads to an increase in hardness of the material. The Specific Tensile Yield strength and Ultimate Tensile Strength values are found to increase up to a concentration of 1.8wt%-nB_4C as shown in Figure 20.8(a). Based on the study of elongation percentage, as the hardness and tensile strength of the material increase, the elongation percentage is expected to decrease. This was observed in the composite material 1.8wt%-nB_4C. The yield and tensile strength of particulate-reinforced metal matrix composites (MMCs), particularly those with Nano-sized reinforcements are predominantly influenced by various strengthening

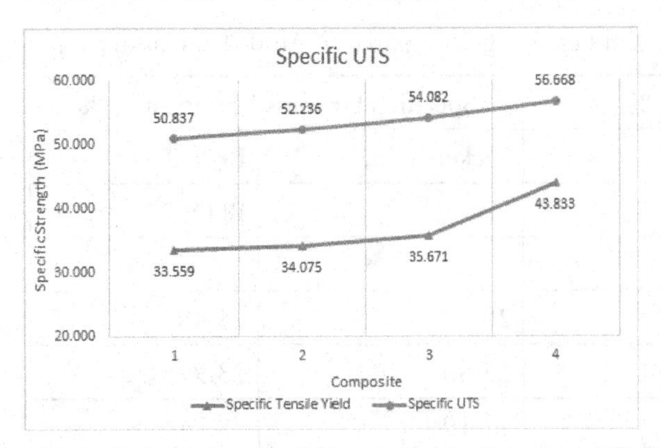

Figure 20.8: (a) Specific Ultimate Tensile Strength of four nanocomposite sample

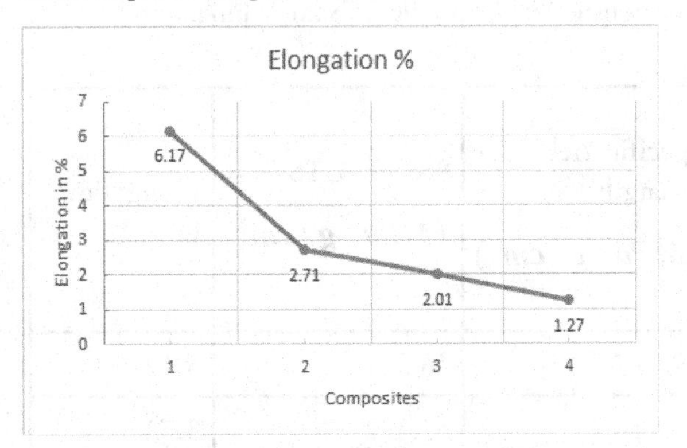

Figure 20.8: (b) Elongation of four nanocomposite sample

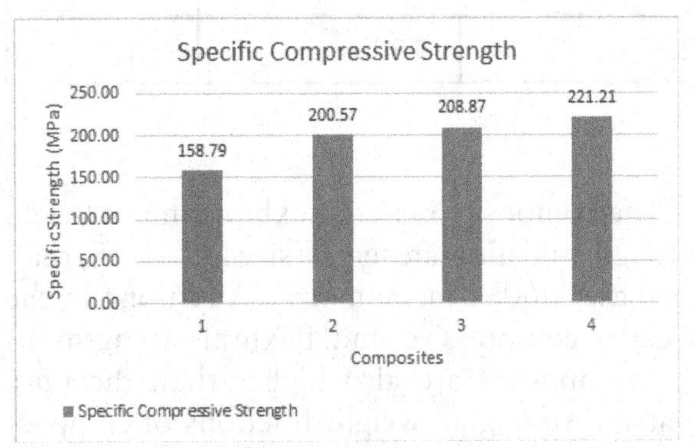

Figure 20.9: (a) Specific Compressive Strength of four nanocomposite sample

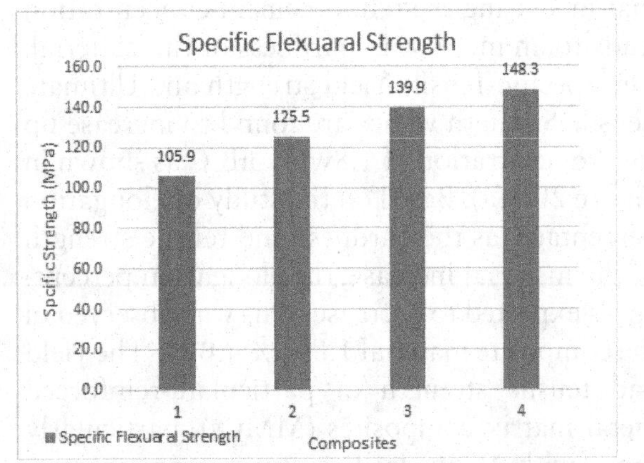

Figure 20.9: (b) Specific Flexuaral Strength of four nanocomposite sample

mechanisms, such as load transfer, grain size (Hall-Petch strengthening), Orowan strengthening, thermal dislocation strengthening.

In Al6061 composites, the higher strength value is achieved by incorporating stiff and hard B_4C particulate reinforcement into the soft and compliant metal matrix, which enables load transfer from the weaker matrix to higher stiffness reinforcements when subjected to external load. The higher stiffness of reinforcements can impede the movement of dislocations, resulting in an increase in strength but a decrease in ductility. Grain size of the reinforcement can have a significant impact on the strength of composites due to the ability of grain boundaries to hinder dislocation motion. Nano-sized B_4C reinforcement particles act as pins for intersecting dislocations, inducing dislocation bending around the particles (referred to as Orowan loops) when subjected to external loads. The significant variation in Coefficient of Thermal Expansion (CTE) and Elastic Modulus between the A16061 metal matrix and B_4C reinforcement can lead to uneven cooling during solidification from high processing temperatures. Consequently, this non-uniform cooling creates residual strains within the composite material, impeding dislocation motion under external loading. Furthermore, porosity within the matrix and the clustering of particles contribute to restricted strength and ductility in the composites.

3.2.4 Impact energy

The Charpy impact test (ASTM E23) was conducted to assess the impact energy of the specimen. Each specimen was precisely machined into a standard square cross-section of 10mm × 10mm and a length of 55mm. A U-notch, 5mm deep and with a 2mm root diameter, was meticulously crafted along the midpoint of the specimen's length. The specimen was positioned horizontally as a simply supported beam, with the notch facing upwards. To initiate the test, a 21.3kg hammer was released from a 140^0 drop angle, reaching a velocity of 5.308m/s upon impact behind the notch. Consequently, the specimen underwent bending and fracture, experiencing high strain rates during this process. The Charpy impact test setup initially held a potential energy of 300J. The energy absorbed by the specimen during fracture was determined by calculating the difference

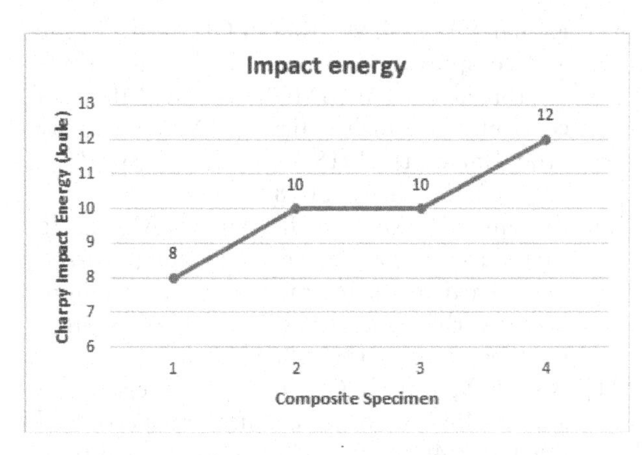

Figure 20.10: Impact energy of four nanocomposite sample

between the potential energies of the Charpy pendulum before and after striking the specimen. This difference was obtained by observing the position of the pointer on the scaling dial, with a precision of 2J as the least count. The graphical representation of impact energy values for both the base matrix and the composites is illustrated in Figure 20.10. It appears that Nano B_4C reinforcing particles reduced the ductility of the composite specimens, and composite structures behave stiffer with increasing volume fraction. As a result, with increasing particle volume fraction, metal matrix less deforms plastically to dissipate impact energy and reinforcing particle provide more rigidity to composite structure by increasing load bearing capacity. Meanwhile, this also makes the structure more brittle.

4. Conclusion

The Al6061 metal matrix composites (MMCs) containing NanoB_4C particles at weight percentages (0.6%, 1.2%, and 1.8%) were synthesized using the ultrasonic cavitation-assisted method within the stir casting route. The addition of varying weight percentages of Boron Carbide aimed to enhance the mechanical properties of Al6061. However, during the fabrication process, certain imperfections such as fine defects, voids, and particle agglomerations were observed within the samples. Successful integration and uniform dispersion of B_4C particles into the Al6061 matrix were achieved at a concentration of 1.8wt% nB4C, facilitated by a well-suited ultra-sonication setup. The experimental findings from this investigation can be summarized as follows.

The Al6061-1.8wt%nB_4C composite demonstrated a notable enhancement in specific hardness values, both in terms of Vickers and Brinell measurements, with increases of approximately 31.19% and 28.74%, respectively, when compared to the properties of the base matrix. Moreover, the specific ultimate tensile strength (UTS) exhibited a substantial improvement, with Al6061-1.8wt%nB_4C showing an increase of nearly 23.00% compared to the base matrix material. The hard ceramic Nano-particles of B_4C, approximately 300nm in size, function as obstacles to the dislocation movement within the material, thereby contributing to the reinforcement of the softer base matrix.

Declarations

Funding: None

Conflict of Interest/Competing Interests: None

Availability of data and material Data and material will be available upon request.

Authors' Contribution Venkitachalapathy K: Conceptualization, Data Curation, writing – Original Draft, Writing – Review and Editing (First Author & Corresponding author)
Manivannan I: Writing – Review and Editing, Supervision (Co-Author)
Periandy S: Writing – Review and Editing, Supervision (Co-Author)

References

[1] Trinh SN. Processing and properties of metal matrix composites. Washington University, Open Scholarship [Internet]. https://openscholarship.wustl.edu/mems500/10

[2] Stojanović B, Ivanović L. Primjena aluminijskih hibridnih kompozita u automobilskoj industriji. Teh Vjesn. 2015;22:247–51.

[3] Surappa MK. Aluminium matrix composites: challenges and opportunities. Sadhana. 2003;28:319–34. Department of Metallurgy, Indian Institute of Science, Bangalore.

[4] Singh N, Belokar RM, Walia RS. A critical review on advanced reinforcements and base materials on hybrid metal matrix composites. Silicon. 2022;14:335–58.

[5] Kareem A, Qudeiri JA, Abdudeen A, et al. A review on aa 6061 metal matrix composites produced by stir casting. Materials (Basel). 2021;14:1–22.

[6] Mukhopadhyay P. Alloy designation , processing , and use of AA6XXX. 2012;2012.

[7] Baron C, Springer H. Properties of particle phases for metal-matrix-composite design. Data Brief [Internet]. 2017;12:692–708. http://dx.doi.org/10.1016/j.dib.2017.04.038.

[8] Thevenot F. Boron carbide: A comprehensive review. J Eur Ceram Soc. 1990;6:205–25.

[9] Arslan G, Kalemtas A. Processing of silicon carbide – boron carbide – aluminium composites. J Eur Ceram Soc. 2009;29:473–80.

[10] Oñoro J, Salvador MD, Cambronero LEG. High-temperature mechanical properties of aluminium alloys reinforced with boron carbide particles. Mater Sci Eng A. 2009;499:421–6.

[11] Mohanty P, Mahapatra R, Padhi P, et al. The role of ultrasonic cavitation in refining the microstructure of aluminum based nanocomposites during the solidification process. Mater Sci Eng A [Internet]. 2019;1:1–22. http://dx.doi.org/10.1016/j.jallcom.2015.12.228.

[12] Reddy PS, Kesavan R, Vijaya Ramnath B. Investigation of mechanical properties of aluminium 6061-silicon carbide, boron carbide metal matrix composite. Silicon [Internet]. 2018;10:495–502. http://dx.doi.org/10.1007/s12633-016-9479-8.

[13] Canakci A, Arslan F, Yasar I. Pre-treatment process of B4C particles to improve incorporation into molten AA2014 alloy. J Mater Sci. 2007;42:9536–42.

[14] Park B, Lee D, Jo I, et al. Automated quantification of reinforcement dispersion in B4C/Al metal matrix composites. Compos Part B Eng [Internet]. 2020;181:107584. https://doi.org/10.1016/j.compositesb.2019.107584.

[15] Khare M, Gupta RK, Bhardwaj B. Dry sliding wear behaviour of Al 7075/Al2O3/B4C composites using mathematical modelling and statistical analysis. Mater Res Express. 2019;6:126512.

[16] Mahesh VP, Nair PS, Rajan TPD, et al. Processing of surface-treated boron carbide-reinforced aluminum matrix composites by liquid-metal stir-casting technique. J Compos Mater. 2011;45:2371–8.

[17] Ramanathan A, Krishnan PK, Muraliraja R. A review on the production of metal matrix composites through stir casting – furnace design, properties, challenges, and research opportunities. J Manuf Process. 2019;42:213–45.

[18] Gudipudi S, Selvaraj N, Chandra DTS, et al. A study on geometrical features of electric discharge machined channels on AA6061-4%B4C composites. Meas Control (United Kingdom). 2020;53:358–77.

[19] Kumar VA, Anil MP, Rajesh GL, et al. Tensile and compression behaviour of boron carbide reinforced 6061Al MMC's processed through conventional melt stirring. Mater Today Proc [Internet]. 2018;5:16141–5. https://doi.org/10.1016/j.matpr.2018.05.100.

[20] Ibrahim MF, Ammar HR, Samuel AM, et al. Metallurgical parameters controlling matrix/B4C particulate interaction in aluminium-boron carbide metal matrix composites. Int J Cast Met Res. 2013;26:364–73.

[21] Auradi V, Rajesh GL, Kori SA. Preparation and evaluation of mechanical properties of 6061Al-B 4Cp composites produced via two-stage melt stirring. Mater Manuf Process. 2014;29:194–200.

[22] Gudipudi S, Nagamuthu S, Subbu K, et al. Optimization of processing parameters for the Al + 10% B4C system obtained by mechanical alloying. Mater Today Proc [Internet]. 2019;8:1–22. http://dx.doi.org/10.1016/j.ultras.2017.06.023.

[23] Mazahery A, Shabani MO. Mechanical properties of squeeze-cast A356 composites reinforced with B 4C particulates. J Mater Eng Perform. 2012;21:247–52.

[24] Lee M, Park S, Jo I, et al. Analysis of metal matrix composite (MMC) applied armor system. Procedia Eng [Internet]. 2017;204:100–7. https://doi.org/10.1016/j.proeng.2017.09.761.

[25] Composites ABC, Pozdniakov AV, Lotfy A, et al. Author ' s accepted manuscript. Mater Sci Eng A [Internet]. 2017. http://dx.doi.org/10.1016/j.msea.2017.01.075.

[26] Mehta KM, Badheka VJ. Wear behavior of boron-carbide reinforced aluminum surface composites fabricated by Friction Stir Processing. Wear [Internet]. 2019;426–427:975–80. https://doi.org/10.1016/j.wear.2019.01.041.

[27] Shuai C, Cheng Y, Yang Y, et al. Laser additive manufacturing of Zn-2Al part for bone repair: formability, microstructure and properties. J Alloys Compd [Internet]. 2019;798:606–15. https://doi.org/10.1016/j.jallcom.2019.05.278.

[28] Yang Y, Guo X, He C, et al. Regulating degradation behavior by incorporating mesoporous silica for Mg bone implants. ACS Biomater Sci Eng. 2018;4:1046–54.

[29] Park J, Hong S, Lee M, et al. Enhancement in the microstructure and neutron shielding efficiency of sandwich type of 6061Al – B 4 C composite material via hot isostatic pressing. Nucl Eng Des [Internet]. 2015;282:1–7. http://dx.doi.org/10.1016/j.nucengdes.2014.10.020.

[30] Chen H, Wang W, Nie H, et al. SC. 2017.
[31] Kumar N, Gautam A, Sevak R, et al. Study of B4C/Al – Mg – Si composites as highly hard and corrosion-resistant materials for industrial applications. Trans Indian Inst Met [Internet]. 2019. https://doi.org/10.1007/s12666-019-01717-w.
[32] Mohanty RM, Balasubramanian K, Seshadri SK. Boron carbide-reinforced alumnium 1100 matrix composites: fabrication and properties. Mater Sci Eng A. 2008;498:42–52.
[33] Ghasali E, Alizadeh M, Ebadzadeh T. Mechanical and microstructure comparison between microwave and spark plasma sintering of Al-B4C composite. J Alloys Compd [Internet]. 2016;655:93–8. http://dx.doi.org/10.1016/j.jallcom.2015.09.024.
[34] Chen HS, Wang WX, Li YL, et al. The design, microstructure and tensile properties of B4C particulate reinforced 6061Al neutron absorber composites. J Alloys Compd 2015;632:23–29.
[35] Bai Q, Zhang L, Ke L, et al. The effects of surface chemical treatment on the corrosion behavior of an Al-B4C metal matrix composite in boric acid solutions at different temperatures. Corros Sci [Internet]. 2019;108356. https://doi.org/10.1016/j.corsci.2019.108356.
[36] Abenojar J, Velasco F, Martínez MA. Optimization of processing parameters for the Al + 10% B4C system obtained by mechanical alloying. J Mater Process Technol. 2007;184:441–6.
[37] Karabulut Ş, Karakoç H, Çıtak R. Influence of B4C particle reinforcement on mechanical and machining properties of Al6061/B4C composites. Compos Part B Eng [Internet]. 2016;101:87–98. http://dx.doi.org/10.1016/j.compositesb.2016.07.006.
[38] Sharma DK, Mahant D, Upadhyay G. Manufacturing of metal matrix composites: A state of review. Mater Today Proc [Internet]. 2019;26:506–19. https://doi.org/10.1016/j.matpr.2019.12.128.
[39] Zhou D, Qiu F, Wang H, et al. Manufacture of nano-sized particle-reinforced metal matrix composites: a review. Acta Metall Sin (English Lett). 2014;27:798–805.
[40] Raj R, Thakur DG. Effect of particle size and volume fraction on the strengthening mechanisms of boron carbide reinforced aluminum metal matrix composites. Proc Inst Mech Eng Part C J Mech Eng Sci. 2019;233:1345–56.
[41] Mohanty P, Mahapatra R, Padhi P, et al. Ultrasonic cavitation: an approach to synthesize uniformly dispersed metal matrix nanocomposites—a review. Nano-Struct Nano-Obj [Internet]. 2020;23:100475. https://doi.org/10.1016/j.nanoso.2020.100475.
[42] Suslick KS. Acoustic cavitation and its chemical consequences. Philos Trans R Soc Lond A. 1999;357:335–53.

CHAPTER 21

Synthesis and Characterization of SnO_2 and V_2O_5 thin Films for Solar Cell Applications

Prabakaran Natarajan,[1] Chandralingam Rameshkumar,[1*,a] Karupaiyyan Subalakshmi[2]

[1]Sathyabama Institute of Science and Technology, Chennai India

[2]Department of Physics, University of Madras, Guindy campus, Chennai India

Email: [a]crkpapers@gmail.com

Abstract

SnO_2 thin films doped with aluminium were effectively completed using the spin coating method under various deposition conditions. The synthesized films were investigated using various optical studies such as FTIR and UV. These studies are intended to verify whether the doping substance influences the optical characteristics of the thin films, specifically the generated energy band gap. We deduced from the results that the layer generated was more transparent at higher doping concentrations. Furthermore, the applied doping has an influence on optical characteristics of thin films, such as the band gap energy. A successful method to developing tin-doped vanadium oxide thin films is thermal evaporation. These films, containing varying weight percentages of SnO_2 (3% and 6%), were placed on substrates made of glass. The deposited films were characterized using UV-Vis spectroscopy. All of the thin film's amorphous properties were confirmed by X-ray diffraction (XRD). Using UV-V is spectroscopy, the film's optical properties (absorption, transmission, and band gap) were examined in the 300–800 nm range.

Keywords: Thin film, solar cell application, spin coating, thermal evaporation, vanadium, tin

1. Introduction

Industries are now the most important factor for a nation's growth. The industrial revolution has been initiated under the name Industry 4.0. Technology, especially artificial intelligence, automated machinery, and the internet, is the key factor that will drive this revolution.This development will rely solely on semiconductor materials (Prit Singh 2023). It is inseparable from the industrial revolution. In this regard, metal oxide materials are crucial to the development and performance of solar cells (Ricardo 2022).Oxide materials are typically transparent and act as electrodes in solar cell panels (Ikhmayies 2017), as well as in light-emitting diodes, transistors, Li-ion batteries, and gas sensors. Specifically, SnO_2 exhibits high reactivity, easy adsorption, low-temperature processing capability, excellent stability, and cost-effective production (Kaur et al. 2007). SnO_2 thin films are prepared using synthesis processes such as spray pyrolysis and chemical vapor deposition. To achieve the desired results, the substrate must be heated to over 400 °C (Jayaprakash and Mariappan 2021). However, after heating the substrate, some drawbacks were observed that need to be addressed.

After iron, titanium, and manganese,the fourth popular transition metal is Vanadium. Vanadium pentoxide (V_2O_5) was reduced with calcium metal in the year 1925 to yield high purity vanadium (around 99.7%).Investigating the properties of V_2O_5 is essential due to its potential application in thermochromics, electronics, electrochemical activity, and optics (Arbab and Mola 2016). V_2O_5 thin films have band gap energy of 2.2–2.3 eV in their bulk form, which makes them extremely stable. This is significant data to consider when evaluating the characteristics of V_2O_5.Several chemical and physical methods, such as physical vapour deposition (PVD) (Zhang et al. 2020), sputtering (Chang et al. 2020), hydrothermal (Guo et al. 2017), sol gel (Chae et al. 2006), spray pyrolysis

DOI: 10.1201/9781003606611-21

(Zhussupbekova et al. 2020), and spin coating, are able to develop V_2O_5thinfilms (Guo et al. 2021). The synthesis process and substrate used have significant effects on structural and morphological factors. V_2O_2 can be found in nanowire, nanotube, and nanosheet forms.

2. Experimental

Using Sol-gel spin coating, SnO_2 thinfilm was done with Al dopant concentration of 3% and 6% with the annealing temperature of 470 °C. $SnCl_2.2H_2O$ is the precursor material, C_2H_5OH is the solvent, and $AlCl_3$ is the dopant. The glass substrate (Khan et al. 2015) was cleaned by acetone, ethyl alcohol and deionised water (DI) in the ultrasonic path for 20 min. It was dried by the flow of N_2 and cleaned by UV–ozone for 10 min. The precursor solution $SnCl_2.2H_2O$ with 0.915 g was dissolved in 45 ml C_2H_5OH. The mixture was stirred for 15 min using hot plate magnetic stirrer to obtain pristine SnO_2 sample. Furthermore, $AlCl_3$ was dissolved in the main solution with weight percentages of 0.025 g (3%) and 0.042 g (6%). These solutions was stirred further 15 min vigorously to obtain dopant solutions. By retaining the solution for 24 h, the mixture was coated onto the cleaned glass substrate at a spin speed of 2500 rpm for 5 min.

In concentrated HCl, vanadium pentoxide powder was dissolved. When the solution was heated to the appropriate temperature, a green powder was obtained. Then, the powder was left in a water bath for 24 h. The powder was dissolved in 30 ml of 2-methoxyethanol with the addition of 10 drops of concentrated HCl. It was doped with SnO_2 solution at 3% and 6%, and the spin coating technique was used to produce a thin film on the glass substrate. (Marques Lameirinhas RA et al. 2022)

3. Results and Discussion

3.1 Optical Properties

3.1.1 Spectroscopy Characterization

The FTIR spectra of a pristine SnO_2 thin film formed on a glass substrate is displayed in Figure 21.1. The spectrum shows several absorption peaks, which confirm the formation of the material. The FTIR spectrum reported in the wave number range 2000–500 cm^{-1}. The vibrational bands are found between Sn and O in the band 830.52 cm^{-1} and 728.23 cm^{-1}. The presence of alkoxy C–O band found from the band 1077.45 cm^{-1}. Some noises have been found in the peak 599.92 cm^{-1} and it is happened due to deposition process. This spectrum analysis holds good agreement with the earlier research article (Benrabah et al. 2011; Saeedabad et al. 2018).

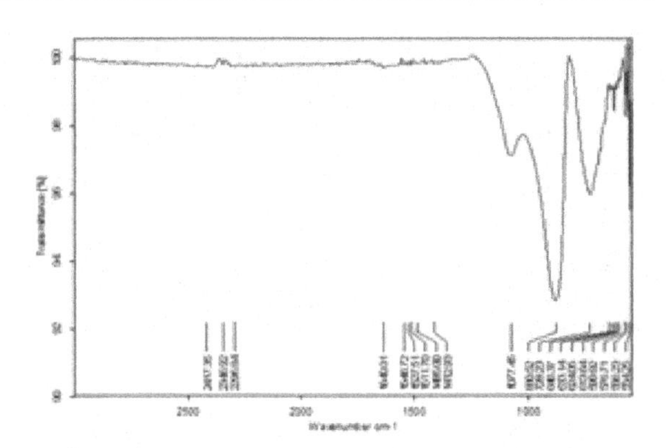

Figure 21.1: FTIR spectrum of SnO_2 sample

3.1.2 Optical Absorption and Transmittance

The thin films optical absorption and transmission properties have been investigated with UV–Vis spectroscopy. Based on these spectra, the highest transmittance was determined to be between 70% and 80%, with a cut-off wavelength of 287 nm. Due to the spectrum, the highest transmittance that can be seen ranges from 70% to 80%, with a cut-off frequency of 287 nm. The results clearly show that, compared to other CdTe, CdO, and ZnO thin films, the SnO_2 thin film transfers more energy. In solar cell technology, energy conversion depends entirely on the amount of solar radiation transmitted to the coating materials (Figures 21.2 and 21.3).

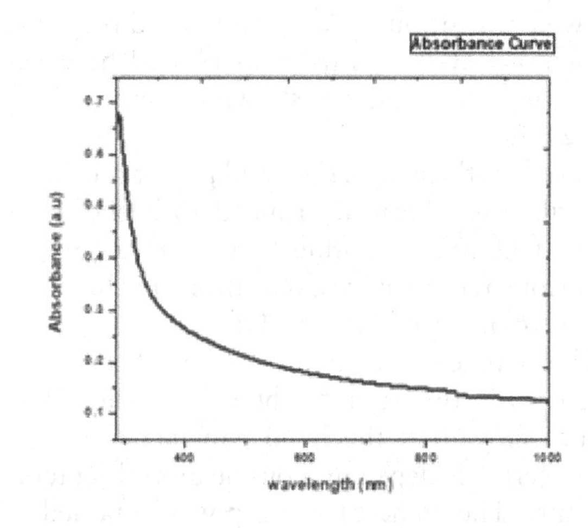

Figure 21.2: Absorption spectra of SnO_2 sample

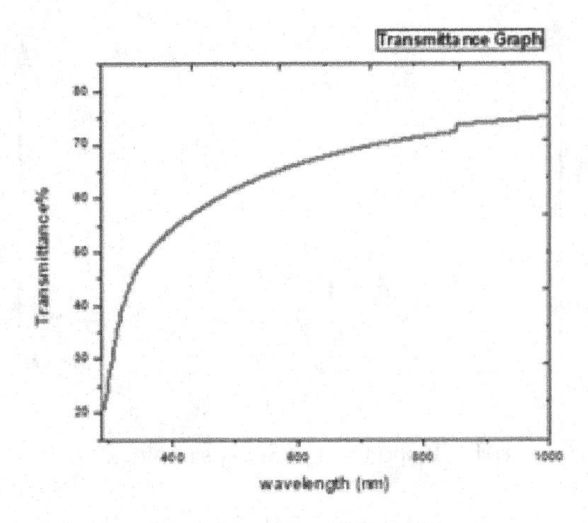

Figure 21.3: Transmittance spectra of SnO_2 sample

3.2 Band Gap Calculation

The Lambert-Beer-Bouguer Law may be used to express the energy that is transmitted light's intensity as

$$I = I_0 \exp(-\alpha d) \quad (1)$$

where d is the samples thickness and Io is the incident light's intensity.
Simplifying equation (1) and using the relationship to find the absorption coefficient

$$\alpha = (1/d)\ln(1/A) \quad (2)$$

Here, A represents the thin film's absorbance at a specific wavelength. In the end, Tauc's relationship produced the thin film's band gap (E_g), which is given by

$$(\alpha h\nu)1/n = A(h\nu - E_g)^n$$

The exponential n depends on the type of transition. For direct allowed n = 1/2, indirect allowed transition, n = 2, and for direct forbidden, n = 3/2. A graph was plotted between $(\alpha h\nu)1/n$ and $h\nu$ and it is shown in Figures 21.4 and 21.5.

Based on the graph, the band gap of the SnO_2 thin films has been determined to be between 3.3 and 3.7 eV. After annealing the same sample at various temperatures, deviations in the band gap were observed (Table 21.1).

The changes in the band gap have been compared with the results obtained earlier. The results show that the band gap increases and then decreases depending on the annealing temperature. The highest band gap was obtained at the annealing temperature of 470 °C. The doping

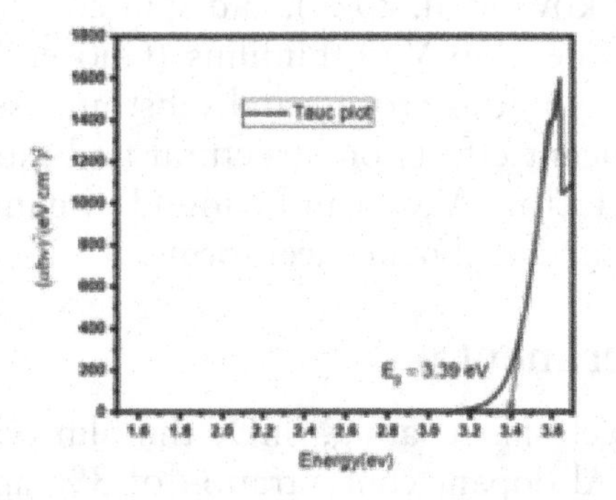

Figure 21.4: Band gap graph of SnO_2 sample

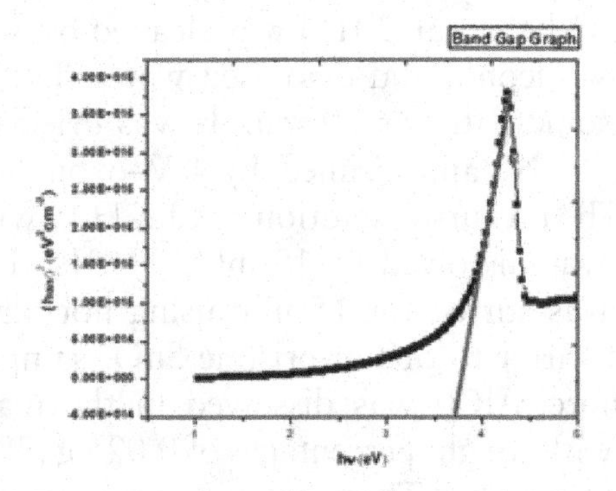

Figure 21.5: Band gap graph of V_2O_5 sample

of Al in SnO_2 shows a variance in the band gap, consistent with previous findings. Between doping agent concentrations ranging from 3% and 6%, the optical band gap is expected to be increasing from 3.2 eV to 3.32 eV by extending the graph's linear section onto the photon energy axis.

4. Conclusions

In this study, using the spin coating process, we successfully completed the synthesis and characterization of SnO_2 thin films for solar cell applications. We examined the electrical and optical characteristics of SnO_2 thin films. The stretching vibrational mode between Tin and Oxygen has been found in the wavebands at 883 cm^{-1} and 730 cm^{-1}, using FTIR graph. The optical and electrical properties of the SnO_2 thin film were used to prove the existence of tin oxide.

About 80% of the UV-visible spectrum is transmitted, which is found to be suitable for photovoltaic devices. The optical band gap value of tin oxide has been determined to be between 3.39 and 3.7 eV using Tauc's Plot. At

Table 21.1: Comparison of the energy band gap obtained with those of other researchers on tin oxide thin films at various annealing temperatures

Energy gap of SnO_2 (eV) (in the present study)	Method	Annealing temperature (°C)		
		370	470	570
	Spin coating	3.39	3.7	3.4
	RF Sputtering	3.00	3.39	2.95
	Chemical Bath Deposition	3.20	2.60	2.40
	Sol Gel Method	3.79	4.50	3.69
Energy Gap of Tin doped V_2O_5 (eV) – 3% (in the Present Study)	Spin Coating	3.6	3.8	3.72
Energy Gap of Tin doped V_2O_5 (eV) – 6% (in the Present Study)		3.25	3.28	3.3

varying concentrations (3% and 6%), the band gap energies of Sn-doped V_2O_5 thin films energy gap vary between 3.6 eV and 3.8 eV. This exhibits how tin oxide behaves as a semiconductor. Thus, the distortion associated with the increased band gap energy can also be linked to the minor rise in the optical band gap that has been observed. This can be explained by the generated samples enhanced transmittance and insulating properties.

References

[1] Arbab EAA, Mola GT (2016) V_2O_5 thin film deposition for application in organic solar cells. Appl Phys A 122:405

[2] Benrabah B, Bouaza A, Kadari A, Maaref MA (2011) Impedance studies of Sb doped SnO_2 thin film prepared by sol gel process. Superlattices Microstruct 50(6):591–600

[3] Chae B-G, Kim H-T, Yun S-J, Kim B-J, Lee Y-W, Youn D-H, Kang K-Y (2006) Highly oriented VO_2 thin films prepared by sol-gel deposition. Electrochem Solid State Lett 9(1):C12

[4] Chang J-Y, Chen Y-C, Wang C-M, Chen Y-W (2020) Electrochromic properties of Li-doped NiO films prepared by RF magnetron sputtering. Coatings 10(1):87

[5] Guo D, Zhao Z, Li J, Zhang J, Zhang R, Wang Z, Chen P, Zhao Y, Chen Z, Jin H (2017) Symmetric confined growth of superstructured vanadium dioxide nanonet with a regular geometrical pattern by a solution approach. Cryst Growth Des 17(11):5838–5844

[6] Guo X, Tan Y, Hu Y, Zafar Z, Liu J, Zou J (2021) High quality VO_2 thin films synthesized from V_2O_5 powder for sensitive near-infrared detection. Sci Rep 11(1):21749

[7] Ikhmayies SJ (2017) Transparent conducting oxides for solar cell applications. In: Sayigh A (eds) Mediterranean green buildings & renewable energy. Springer, Cham

[8] Jayaprakash RN, Mariappan R (2021) Effect of substrate temperature on the structural, optical and electricalproperties of Tin Oxide thin films. Chalcogenide Lett 18(4):991–200

[9] Kaur J, Roy SC, Bhatnagar MC (2007) Highly sensitive SnO_2 thin film NO_2 gas sensor operating at low temperature. Sens Actuators B Chem 123(2):1090

[10] Khan GR, Bilal Ahmad Bhat KA, Asokan K (2015) Role of substrate effects on the morphological, structural, electrical and thermoelectrical properties of V_2O_5 thin films. RSC Adv 5:52602–52611

[11] Marques Lameirinhas RA et al. (2022), Torres JPN, de Melo Cunha JP (2022) A photovoltaic technology review: history, fundamentals and applications. Energies 15:1823

[12] Mola GT et al. (2017), Arbab EA, Taleatu BA, Kaviyarasu K, Ahmad I, Maaza M (2017) Growth and characterization of V_2O_5 thin film on conductive electrode. J Microsc 265(2):214–221

[13] Prit Singh Z (2023) The power of semiconductor materials paving the way for technological advancements. Adv Mater Sci Res 6(4):67-69

[14] Ricardo C.L.A, D'Incau M, Leoni M, Malerba C, Mittiga A, P. ScardiStructural properties of RF-magnetron sputtered Cu_2O thin filmsThin Solid Films, 520 (2011), pp. 280-286.

[15] Saeedabad SH, Selopal GS, Rozati SM, Tavakoli Y, Sberveglieri G (2018) From transparent conducting material to gas-sensing application of SnO_2: Sb thin films. J Electron Mater 47(9):5165–5173
[16] Wei Z et al. (2021), Van Le Q, Peng W, Yang Y, Yang H, Gu H, Shiung Lam S, Sonne C (2021) A review on phytoremediation of contaminants in air, water and soil. J Hazard Mater 403:123658
[17] Zhang C, Koughia C, Güneş O, Luo J, Hossain N, Li Y, Cui X, Wen S-J, Wong R, Yang Q (2020) Synthesis, structure and optical properties of high-quality VO2 thin films grown on silicon, quartz and sapphire substrates by high temperature magnetron sputtering: properties through the transition temperature. J Alloys Compd 848:156323
[18] Zhussupbekova A, Caffrey D, Zhussupbekov K, Smith CM, Shvets IV, Fleischer K (2020) Low-cost, high-performance spray pyrolysis-grown amorphous zinc tin oxide: the challenge of a complex growth process. ACS Appl Mater Interfaces 12(41):46892–46899

CHAPTER 22

Trace Elements and Minerals in Rock Crystals

A Spectral Analysis

Andrew M. Appaji,* N. Joseph John

Department of Physics, Kamarajar Govt Arts College, Surandai, Tamil Nadu, India
*Manonmaniam Sundaranar University, Tirunelveli, Tamil Nadu, India.
Email: andrew.appaji@gmail.com

Abstract

Among the naturally occurring elements; Titanium, Magnesium, Sodium, Potassium, Calcium, Aluminium, Iron, Hydrogen, Silicon, and Oxygenconstitute the mass of Earth's crust by 99%. The rest of the 0.5% mass is covered by other elements called Trace Elements. In the formation of Earth's crust, the trace elements do not have any basic role. Still, their importance in health, toxicology, medicine, farming, economy, ecology, and other many vital areas are abundantly large compared to their disproportionate low value of abundance. Crystal analysis is done using spectroscopic methods, and the elements are analyzed along with the trace elements. This paper gives an account of trace elements found in rock crystals. Their qualitative behaviourand quantitative presence in the crystal samples are analyzed. EDS and powder XRD are taken for the samples. The results show the presence of trace elements along with other elements and minerals.

Keywords: EDS, XRD, trace elements, rock crystals

1. Introduction

1.1. Trace Elements

When the concentration of an element is lower in a given phase it is considered as a "Trace Element". At the same time, the same element may be at a higher concentration and constitute another phase's main part. For example, in common soils and rocks, Aluminium is a main element but in water and agriculture, it occurs as a trace element. Considering Mafic minerals, Fe is the main element but it comprisesa trace element in Quartz. Thus, in Science and Engineering, the Trace Element may be defined as a chemical element of which the concentration is less than 0.1% by mass in Earth's crust [22].

There is an acute need for trace elements in the production of many materials and commodities and to continue research. Also, there is a severe shortage of Trace Elements in the Engineering and scientific arena. This leads to finding good sources of Elements and Minerals to suffice this need. Rocks are such sources of abundant Elements and Minerals. They are natural, clean, and reliable. Finding their qualitative and quantitative presence using spectroscopy is a practical scientific method for this purpose (Figure 22.1).

Figure 22.1: Western Ghats Range of Rocks

DOI: 10.1201/9781003606611-22

Figures 22.2, 22.3, 22.4, 22.5, and 22.6 of the samples are as follows:

Figure 22.2: Sample 1

Figure 22.3: Sample 2

Figure 22.4: Sample 3

Figure 22.5: Sample 4

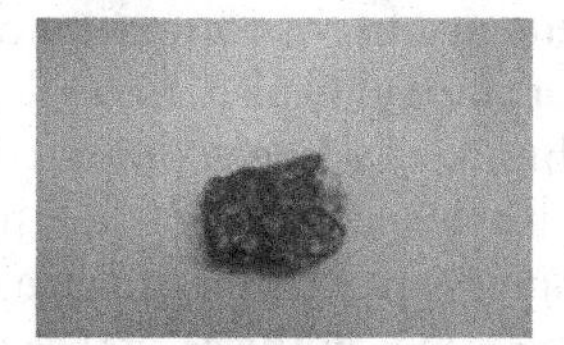

Figure 22.6: Sample 5

The research area's Geographical location is in the Western Ghats, covering a distance of around 100 km from Perunchilambu to Kumarakovil. The average height of the mountains ranges from 300 to 400 ft above sea level (Table 22.1).

Table 22.1: The presence of some of the trace elements in the materials and material productions

	As	Cd	Zn	Pb	Cu	Hg	Sb	Cr
Coatings and Paints	•	•	•	•	•	•	•	•
Processing of Crude oil	•		•	•				•
Production of Iron and steel	•	•	•	•	•	•	•	•
Metal and Nonferrous refinery	•	•	•	•	•	•	•	•
Production of Plastic		•	•	•				•
Batteries and Cells		•	•	•	•	•	•	•
Pesticides, herbicides, fungicides	•		•	•		•		•

2. Materials and Methods

2.1. Preparation of Samples

The collected rock samples are first cleaned with clear but not hot water. The samples are brushed to wipe out bacteria and any biological impurities on the sample. The cleaned samples are further cleaned with distilled water. The cleaned samples are dried in bright sunlight and are powdered with a hand mortar. Fine-sieved powders are collected without contaminations as samples for this study.

2.2. Instrumentation of EDS (Energy Dispersive X-Ray Spectroscopy)

The EDS instrument used in this analysis is "JEOL Model JED – 2300". It has resolutions 15 nm (Acc V 1.0 KV, WD 6 mm, SEI), 8 nm (Acc V 3.0 KV, WD 6 mm, SEI), 3 nm (Acc V 30 KV, WD 8 mm, SEI), It has a magnifying capacity of 5 × 300, 000 × (Either in low and high Vacuum Modes). The two image modes are namely BEI and SEI. The probe current is in the range of 1 pA to 1 mA.

2.3. Instrumentation of XRD (X-Ray Diffraction)

The Diffractometer has the following technical and operational parameters. The make/model is "Bruker AXS D8 Advance". It works on vertical configuration with theta/2 theta geometry. The predefined 435 mm, 500 mm and 600 mm are the measuring circle diameters. The angular range that can be used is 3–135°. 360° being its maximum angle 0.001° is the minimum addressable increment. 30°/s is the maximum angular speed. It has a Cu X-ray source with a wavelength of 1.5406 arcsec.

3. Spectroscopical Analysis of Samples-EDS

3.1. Characterization of the Sample-1

Figure 22.7 depicts the EDS spectrum of the sample-1. The peak with the energy 0.525 keV is the K_α excitation. This energy peak corresponds to the element Oxygen which is a light element. The excitation is at high intensity but with low energy. The peak with the energy of 1.739 keV is the element Silicon due to Kα excitation. The element is light but has high intensity and low energy. The peak with the energy of 1.694 keV represents the element **Rubidium** which is an Lα1 excitation. The element has high intensity with low energy.

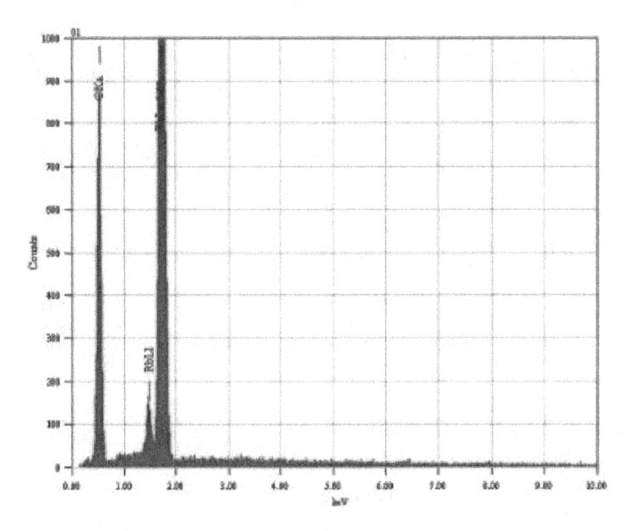

Figure 22.7: EDS of sample-1

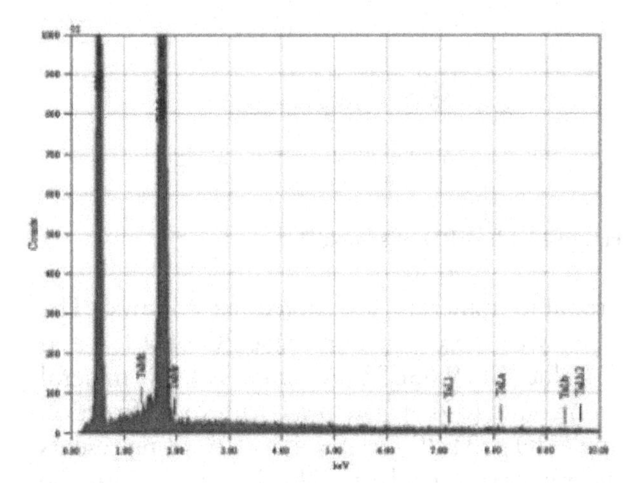

Figure 22.8 EDS of sample-2

3.2. Characterization of the Sample-2

Figure 22.8 reveals the EDS spectrum of sample-2. The peak with the energy of 0.525 keV is due to the Kα excitation. This peak corresponds to Oxygen, a light element with high intensity and low energy. The peak with the energy of 1.739 keV is the element Silicon which is due to the Kα excitation. This is a light element with high intensity and low energy. The peak with the energy of 1.709 keV is the element **Tantalum** which is a heavy element with high intensity and low energy. The excitation is Mα

3.3. Characterization of the Sample-3

Figure 22.9 is the display of the EDS spectrum of the sample-3. The sample has many elements and all are light elements. All the peaks are at

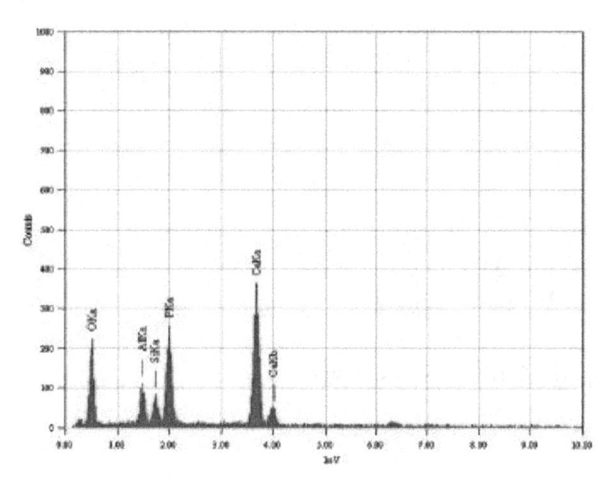

Figure 22.9: EDS of sample-3

low intensities. The peak with the energy of 0.525 keV is the element Oxygen which is a light element with low intensity. This peak is due to Kα excitation. The peak with the energy of 1.739 keV is the element Silicon which is due to Kα excitation. It is a light element with low intensity and low energy. The peak with the energy of 1.486 keV is the element Aluminium which is a light element with both low intensity and energy. The cause of this energy peak is due to Kα excitation. The peak with the energy of 3.69 keV is the element Calcium which is a light element with low energy and low intensity. This energy peak is due toKα excitation. The peak with the energy of 4.508 keV is the element **Titanium**. This is due to Kα excitation. The element is a light element with low intensity and low energy. The peak with the energy of 6.398 keV is due to the Kα excitation. The element is Iron, which is an intermediate element with low intensity and low energy.

3.4. Characterization of the Sample-4

Figure 22.10 represents the EDS spectrum of sample-5. The peak with the energy of 0.525 keV is the peak due to Kα excitation. The element Oxygen which is a light element with minimum intensity and low energy.

The peak with the energy of 1.922 keV is the element **Yttrium** which is intermediate with medium intensity and low energy. This is Lα excitation. The peak with the energy of 3.69 keV is the element Calcium. This excitation is a Kα excitation. Calcium element is a light element with medium intensity and low energy.

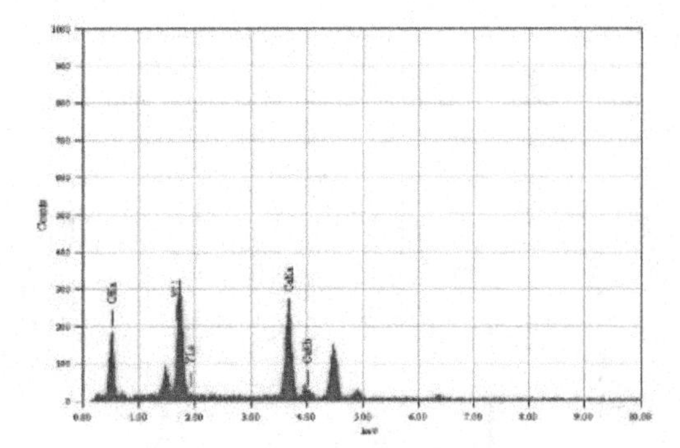

Figure 22.10: EDS of sample-4

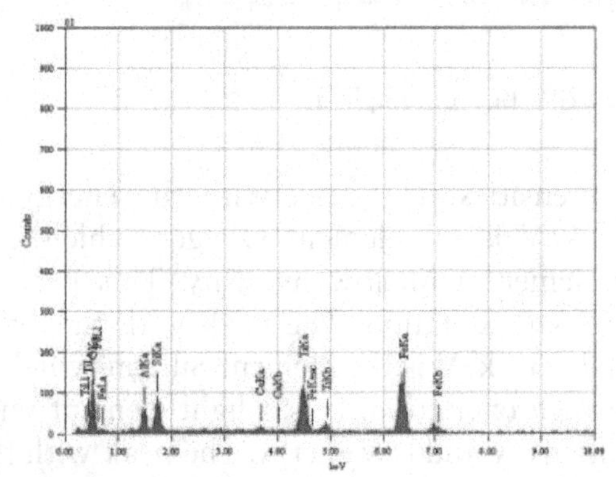

Figure 22.11: EDS of sample-5

3.5. Characterization of the Sample-5

Figure 22.11 is the EDS spectrum of the sample-5. The peak with the energy of 0.525 keV is Oxygen, a light element with moderate intensity and low energy. This peak is due to excitation. The peak with the energy of 1.739 keV is due to $K\alpha$ excitation. The element is Silicon which is a light element with minimum intensity and low energy. The peak with the energy of 1.468 keV is the element Aluminum which is a light element with low intensity and low energy. The results are due to the $K\alpha$ excitation. The peak with the energy of 3.69 keV is the element Calcium which is due to $K\alpha$ excitation. The element is a light element with high medium intensity and low energy. The peak with the energy of 2.013 keV is the element **Potassium** which is a light element with moderate low intensity and energy. This peak is due to $K\alpha$ excitation,

4. Spectroscopical Analysis of Samples-XRD

4.1. Characterization of the Sample-1

Sample-1 XRD pattern is shown in Figure 22.12. The graph has 2Θ values in degrees along the X-axis and intensity in cps along the Y-axis.

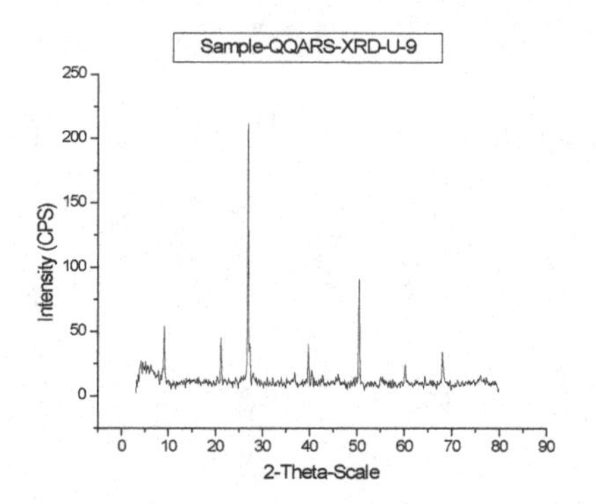

Figure 22.12: XRD of sample-1

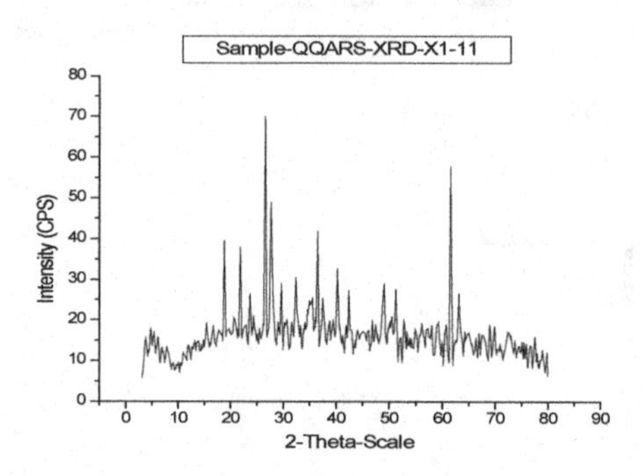

Figure 22.13: XRD of sample-2

Probable minerals are noted as follows. The peak with the diffraction angle 2Θ of 9.17° and d value of 9.63672 A° corresponds mineral Nontronite [8–10]. The peak with the diffraction angle 2Θ = 21.207 and = 4.18614 A° represents the mineral Cristobalite [1]. The peak with the diffraction angle 2Θ of 26.986° and d value of 3.30142 A° represents the mineral Magnetite [4]. The peak with the diffraction angle 2Θ = 50.447° and d = 1.80757 A° represents the mineral Quartz [5–7]. The peak with the diffraction angle 2Θ of 10.77° and d value of 8.20798A° corresponds to the mineral **Rubicline** [11, 12]. The peak with the diffraction angle 2Θ = 60.274° and d = 1.53425 A° corresponds to the mineral Moganite [5–7].

4.2. Characterization of the Sample-2

The XRD pattern of sample-2 is represented in Figure 22.13. The graph has 2Θ values in degrees along the X-axis and intensity in cps along the Y-axis. Probable minerals are noted as follows. The peak with the diffraction angle 2Θ = 21.151° and d = 4.19713 A° represents the mineral Cristobalite [1]. The peak with the diffraction angle 2Θ of 26.96° and d value of 3.30449 A° corresponds to the mineral Magnetite [4]. The peak with the diffraction angle 2Θ = 50.33°, d = 1.81151 A° represents the mineral Quartz [5–7]. The peak with the diffraction angle 2Θ of 60.271° and d value of 1.53434 A° corresponds to the mineral Moganite [5–7]. The diffraction angle in 2Θ = 68.342°, d value in 1.37146 A° corresponds to the mineral **Tantalite** [14, 15].

4.3. Characterization of the Sample-3

In Figure 22.14, the XRD pattern of sample-3 is shown. The graph has 2Θ values in degrees along the X-axis and intensity in cps along the Y-axis. Probable minerals are noted as follows. The peak with the diffraction angle of 2Θ 21.856°, d value of 4.06333 A° represents the mineral Cristobalite [1]. The peak with the diffraction angle 2Θ = 26.577° and d = 3.35131 A° corresponds to the mineral Sanidine [13]. The peak with the diffraction angle 2Θ of 27.651° and d value of 3.22351 A° represents the mineral Orthoclase [3]. The peak with the diffraction angle 2Θ = 32.39° and d = 2.76188 A° corresponds to the mineral Kaolinite [2–4]. The diffraction angle 2Θ = 36.512° and d = 2.45896 A° corresponds to the mineral **Rutile** (TiO_2) [19, 21]. The peak with the diffraction angle 2Θ of 42.374° and d value of 2.13137 A° represents the mineral Sanidine [13]. The peak with the diffraction angle 2Θ = 49.076°, d = 1.8548 A° corresponds to the mineral Hematite [13]. The peak with the diffraction angle 2Θ of 51.263° and d value of 1.78069 A° represents the mineral Quartz [5–7]. The peak with the diffraction angle 2Θ = 61.514° and d value of 1.50627 A° correspond to the mineral Moganite [5–7].

The peak with the diffraction angle 2Θ of 40.22° and d value of 2.45896 A° represents the mineral Quartz [5–7].

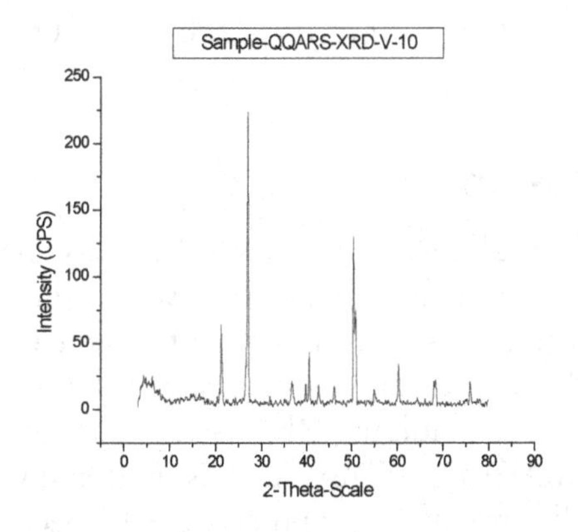

Figure 22.14: XRD of sample-3

4.4. Characterization of the Sample-4

Figure 22.15 displays the sample-4 XRD pattern. The graph has 2Θ values in degrees along the X-axis and intensity in cps along the Y-axis. Probable minerals are noted as follows. The peak with the diffraction angle 2Θ of 28.907° and d value of 3.08617 A° represents the mineral Calcite [18]. The peak with the diffraction angle 2Θ = 33.158° and d = 2.69959 A° corresponds to the mineral Kaolinite [2–4]. The peak with the diffraction angle 2Θ of 42.21°, and d value of 2.13926 A° represents the mineral Sanidine [13]. The peak with the diffraction angle 2Θ = 49.343° and d = 1.8454 A° corresponds to the mineral Quartz [5, 6]. The peak with the diffraction angle 2Θ = 58.837° and d = 1.56824 A° corresponds to the mineral **Yttrium oxide** (Y_2O_3) [20].

4.5. Characterization of the Sample-5

Figure 22.16 shows the XRD pattern of sample-5. The graph has 2Θ values in degrees along the X-axis and intensity in cps along the Y-axis. Probable minerals are noted as follows. The peak with the diffraction angle 2Θ of 8.967° and d value of 9.8538 A° represents the mineral Nontronite [8–10]. The peak

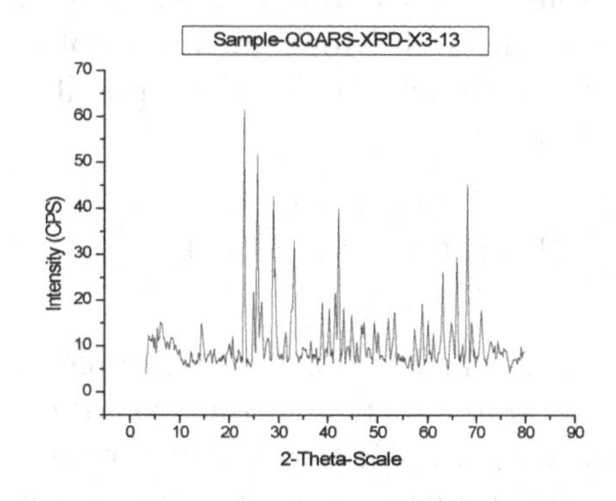

Figure 22.15: XRD of sample-4

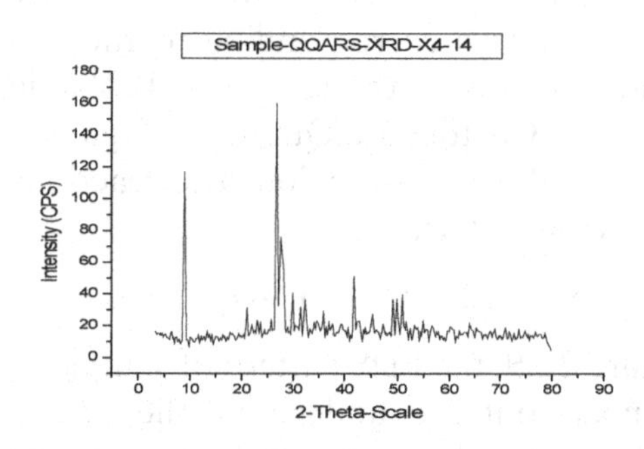

Figure 22.16 XRD of sample-5

with the diffraction angle 2Θ = 26.754° and d = 3.32945 A° corresponds to the mineral Sanidine [13]. The peak with the diffraction angle 2Θ of 29.879° and d value of 2.98799 A° represents the mineral Zeolite [16]. The peak with the diffraction angle 2Θ of 50.886° and d value of 1.79299 A° represents the mineral Quartz [5, 6]. The peak with the diffraction angle 2Θ = 31.383° and d = 2.848137 A° corresponds to the mineral **Hydrogen-Phosphate** (H_2PO_4) [17]. The peak with the diffraction angle 2Θ of 50.886° and d value of 1.79299 A° represents the mineral Quartz [5, 6].

5. Result and Discussions

The characteristics of EDS reveal the presence of several elements, some of which are trace elements. Further, the characteristics of XRD ascertain the presence of elements and trace elements in minerals.

From Figure 22.7, the EDS pattern of sample-1, we canlook atand understand the excitation levels of the elements. The elements present in Rock Sample-1 areSilicon and Rubidium. From Figure 22.12, the XRD pattern of the sample-1, we scrutinize the intensity of the diffracted X-rays corresponding to the elements and minerals. Nontronite, Cristobalite, Magnetite, Quartz, Rubicline and Moganite are probable minerals. Stoichiometric reactions possible within the sample 1 are:

$$Rb + O \rightarrow RbO_2$$
$$2Rb + H_2O \rightarrow 2RbOH + H_2$$
$$Si + O \rightarrow SiO$$

From Figure 22.8, the EDS pattern of sample-2, we may spot and understand the excitation levels of the elements. Tantalum, Silicon and Oxygen are the elements present in the Rock sample-2. From Figure 22.13, the XRD pattern of the sample-2, we observe the intensity of diffracted X-rays corresponding to the elements and minerals of the sample-2. Probable minerals are Cristobalite,Quartz, Moganite, and Tantalite. Possible stoichiometric reactions within the sample 2 are:

$$SiO_2 + 2\,H_2O \rightarrow H_4SiO_4$$

From Figure 22.9, the EDS pattern of sample-3, we may make out and understand the excitation levels of the elements. The names of the elements present in Rock Sample-3 as Oxygen, Silicon, Aluminium, Calcium, Titanium and Iron. From Figure 22.14, the XRD pattern of the sample-3, we see the intensity of diffracted X-rays corresponding to the elements and minerals of the sample-3. Probable minerals are Cristobalite, Sanidine, Othoclase, Kaolonite, Rutile, Hematite, Quartz, Moganite and Quartz. Possible stoichiometric reactions within the sample 3 are as follows:

$$Ti + 2\,H_2O \rightarrow TiO_2 + 2H_2$$
$$Ca + 2H_2O \rightarrow Ca(OH)_2 + H_2$$

From Figure 22.10, the EDS pattern of sample-4, we may see and understand the excitation levels of the elements. The names of the elements present in the Rock Sample-4 as Oxygen, Silicon, Aluminium, Calcium and Potassium. From Figure 22.15, the XRD pattern of the sample-4, we see the intensity of diffracted X-rays corresponding to the elements and minerals of the sample-4. Probable minerals are Calcite, Kaolinite, Sanidine, Quartz and Yttrium oxide. Possible stoichiometric reactions within the sample 4 are as follows:

$$4Y + 3O2 \rightarrow 2Y2O3$$
$$2Ca + O \rightarrow 2CaO$$

From Figure 22.11, the EDS pattern of sample-5, we may see and understand the excitation levels of the elements. The names of the elements present in the Rock Sample-5 as Oxygen, Silicon, Aluminium, Calcium and Potassium. From Figure 22.16, the XRD pattern of the sample-5, we see the intensity of diffracted X-rays corresponding to the elements and minerals of the sample-5. Probable minerals are Nontronite, Sanidine, Zeolite, Quartz, Hydrogen-Phosphate (H_2PO_4) and Quartz. Possible stoichiometric reactions within the sample 5 are as follows:

$$10Ca(OH)_2 + 6H_3PO_4 \rightarrow Ca_{10}(PO_4)2H_2 + 18H_2O$$
$$Si + O \rightarrow SiO$$

6. Conclusion

The Spectral Analysis of the Rock Crystals as affirmed above concludes that along with other elements and minerals, the presence of trace elements is evident. The analysis ofsample-1 shows the presence of Rubidium (Rb) in the mineral Rubicline ($RbAlSi_3O_8$). Rubidium is extremely rare, especially present in authenticated specimens. In the sample-2 Tantalum (Ta) is present in the mineral Tantalite ((Fe)(Ta_2O_6). Tantalum is a highly corrosion-resistant and blue-grey transition metal. It is hard, ductile and lustrous. It is used to produce strong and high-melting-point alloys. The analysisof sample-3 shows the presence of the element Titanium (Ti) in the mineral Rutile (TiO_2) and Ilmenite ($FeOTiO_2$) itis found in nature as an oxide. The materials made of Titanium are lustrous, low in density, high in strength and resistant to aqua corrosion. In sample-4 the element Yttrium (Y) is present in the mineral Yttrium oxide (Y_2O_3Yttrium is a moderately soft, ductile and silvery-white metal. In the air, it is fairly stable, although, above 450 ˚C, rapid oxidation may start. The analysis of sample-5 reveals the presence of the element Phosphorus (P). It exists in nature as a Hydrogen Phosphate mineral or as a Calcium hydroxyapatite $Ca_5(PO_4)_3(OH)$ mineral.

References

[1] Nabil M, Mahmoud KR, El-Shaer A, Nayber HA (2018) Preparation of crystalline silica (quartz, cristobalite, and tridymite) and amorphous silica powder (one step). J Phys Chem Solids. https://doi.org/10.1016/j.jpcs.2018.05.001

[2] Tironia A, Trezzaa MA, Irassara EF, Scianb AN (2012) Thermal treatment of kaolin: effect on the pozzolanic activity. Proc Mater Sci 1:343–350

[3] Marvila MT, Alexandre J, Zanelato EB, Azevedo ARG, Goulart MA, Cerqueira NA, Xavier GC, Espirito Santo TL (2018) Study of the use of kaolinite clay in mortars. In: 7th International Congress on Ceramics & 62°, June 2018.

[4] Dewi R, Agusnar H, Alfian Z, Tamrin (2018) Characterization of technical kaolin using XRF, SEM, XRD, FTIR and its potentials as industrial raw materials. J Phys Conf Ser 1116:042010

[5] Parthasarathy G, Kunwar AC, Srinivasan R (2001) Occurrence of moganite-rich chalcedony in Deccan flood basalts, Killari, Maharashtra, India. Eur J Mineral 13:127–134

[6] Paralı L, Garcia Guinea J, Kibar R, Cetin A, Can N (2011) Luminescence behaviour and Raman characterization of dendritic agate in the Dereyalak village (Eskis-ehir), Turkey. J Lumin 131:2317–2324

[7] Moxon T, Ríos S (2004) Moganite and water content as a function of age in agate:an XRD and thermogravimetric study. Eur J Mineral 16:269–278

[8] Baker LL, Strawn DG (2014) Temperature effects on the crystallinity of synthetic nontronite and implications for nontronite formation in Columbia River Basalts. Clays Clay Miner 62(2):89–101

[9] Delvaux B, Mestdagh MM, Vielvoye L, Herbillon AJ (1989) XRD, IR and ESR study of experimental alteration of Al-nontronite into mixedlayer. Clay Minerals 24:617–630

[10] Liu R, Xiao D, Guo Y, Wang Z, Liu J (2014) A novel photosensitized Fenton reaction catalyzed by sandwiched iron in synthetic nontronite. RSC Adv 4:12958–12963

[11] Liu Y, Tan Y, Wu J (2020) Rubidium doped nano-hydroxyapatite with cytocompatibility and antibacterial. J Asian Ceram So. https://doi.org/10.1080/21870764.2020.1865861

[12] Cigler AJ, Kaduk JA (202) Structures of dipotassium rubidium citrate monohydrate, $K_2RbC_6H_5O_7(H_2O)$, and potassium dirubidium citrate monohydrate, $KRb_2C_6H_5O_7(H_2O)$, from laboratory X-ray powder diffraction data and DFT calculations. Acta Crystallogr E Crystallogr Commun E76:1566–1571

[13] Morris RV, Rampe EB, Vaniman DT, Christoffersen R (2020) Hydrothermal precipitation of sanidine (adularia) having full Al,Si structural disorder and specular hematite at Maunakea Volcano (Hawai'i) and at Gale Crater (Mars). J Geophys Res Planets 125:e2019JE006324. https://doi.org/10.1029/2019JE006324, 1-22, Sanidine-Hematite-2020.

[14] Chennakesavulua K, Ramanjaneya Reddy G (2013) Synthesis, characterization of carbon microtube/tantalum oxide composites and their photocatalytic activity under visible irradiation. Royal Society of Chemistry, RSC Adv 5:56391-56400

[15] Jakubowicz J, Sopata M, Adamek G, Koper JK (2017) Properties of high-energy ball-milled and hot pressed nanocrystalline tantalum. GSTF J Eng Technol 4(3):124

[16] Kwakye-Awuah B, Von-Kiti E, Nkrumah I, Erdoo Ikyreve R, Radecka I, Williams C (2016) Parametric, equilibrium, and kinetic study of the removal of salt ions from Ghanaian seawater by adsorption onto zeolite X. Desalin Water Treat Balaban , 1–16. https://doi.org/10.1080/19443994.2015.1128361. Desalination Publications

[17] Cianflone E, Brouillet F, Grossin D, Soulié J, Josse C (2023) Toward smart biomimetic apatite-based bone scaffolds with spatially controlled ion substitutions. Nanomaterials 13:519. https://doi.org/10.3390/nano13030519

[18] Ignjatovi N, Jovanovi J, Suljovrujić E, Uskoković D (2003) Injectable polydimethylsiloxane–hydroxyapatite composite cement. Biomed Mater Eng 13:401–410

[19] El-Sherbiny S, Morsy F, Samir M, Fouad OA (2013) Synthesis, characterization and application of TiO_2 nanopowders as special paper coating pigment. Appl Nanosci. https://doi.org/10.1007/s13204-013-0196-y

[20] Tamrakar RK, Upadhyay K, Bisen DP (2014) Gamma ray induced thermoluminescence studies of yttrium (III) oxide nanopowders doped with gadolinium. J Rad Res Applied Sci 7:526. http://dx.doi.org/10.1016/j.jrras21. Phung Kim Phu, Nguyen Minh Thuan, Tran Nam Trung, In-sang Yang, Nguyen Van Minh (2010) Synthesis and characterization of $Rb_xMn[Fe(CN)_6]$ and $Mn_3[Cr(CN)_6]_2$. In: Proceedings of 10th IEEE, international conference on nanotechnology joint symposium with Nano Korea 2010, KINTEX, Korea, pp 17–20

[21] Phung Kim Phu, Nguyen MinhThuan, TranNam Trung, In-sang Yang, Nguyen Van Minh (2010) Synthesis and characterization of Rbx Mn $[Fe(CN)_6]$ and Mn_3 $[Cr(CN)_6]$ 2. In: Proceedings of 10th IEEE, international conference on nano technology joint symposium with Nano Korea 2010, KINTEX, Korea, pp17–20

[22] Navrátil T, Minařík L. Trace elements and contaminants. In: ©Encyclopedia of Life Support Systems (EOLSS), vol IV

CHAPTER 23

Design and implementation of a standalone photovoltaic system with zeta converter for efficient energy conversion and voltage regulation

Mettilda M.,[1*,a] Chrisona S.,[2] Anand S.,[1,b] and Kavitha S[3]

[1]Department of Chemistry, Saveetha Engineering College, Thandalam, Chennai, India
[2]Jeppiar Maamallan Engineering College, Chennai, India
[3]Department of Electrical and Electronics Engineering, Saveetha Engineering College, Thandalam, Chennai, India
Email: [a]mettilda@saveetha.ac.in, [b]anand.singaravelu@gmail.com

Abstract

This paper explores the design and implementation of a standalone photovoltaic (PV) system integrated with a Zeta converter for efficient voltage regulation in remote areas. The system focuses on converting solar energy into stable electrical power while overcoming challenges like fluctuating irradiance and varying load demands. A maximum power point tracking (MPPT) algorithm ensures optimal energy extraction, while the Zeta converter is employed to boost and regulate the output voltage. The Zeta converter's ability to minimize output voltage ripple and maintain continuous conduction mode makes it ideal for this application. The system's inverter stage incorporates space vector modulation (SVM) to improve power quality by reducing total harmonic distortion (THD) and efficiently utilizing the DC bus voltage. Simulation and hardware results confirm that a 12.7 V DC input from the solar panel can be boosted to 350 V DC, which is then converted to AC to power local loads. The system exhibits stable operation under varying conditions, with quick stabilization after load changes. In future a quadratic-boost converter with the Zeta converter for higher output voltages in employing advanced semiconductor technologies to can be integrated improve performance in high-speed applications. This design offers a reliable solution for off-grid energy needs in remote locations.

1. Introduction

Power Engineering deals with electrical devices like motors, generators and transformers. Part of this field solves obstacles of 3-phase AC power, supplied throughout world. The other major fraction of this field the concentrates on AC and DC power systems in aero planes and electric railway networks. A power engineer is considered to be an energy engineer who draws theoretical base from electrical engineering. The first work power station was built in Godalming in England in the year 1881 by two electricians, employing two waterwheels which produced an alternating current intermittently. The first steam-powered electric power station was developed in New York in 1882. Initially the power station powered 59 customers with, several generators. The station used direct current at a single voltage, which could not be transformed to higher voltage required for minimizing the power loss. The objective of this work is to construct the circuit and execute the Simulink output using MATLAB software by using zeta converter and space vector modulation technique. The following objectives are achieved in this work.

- The output voltage should be greater than the input.
- Output voltage using SVM technique is to be improved.
- To produce higher output quality and flexible control of output voltage.

DOI: 10.1201/9781003606611-23

2. Existing System

Low crossover frequency decreases the performance of DC–DC converter [1]. To enhance suppressing effect an inner current closed loop is applied. Low cross, over frequency results in poor dynamic performance. Load transient introduces a change in bus voltage. The large overshoot increases voltage stress of power switches, in power system but triggers the overvoltage protection, whereas large undershoot leads to distorted output voltage of DC–AC inverter.

Frequency of the Sub-Harmonic Component (SHC) is constant and can be suppressed by lowering the magnitude of loop gain of the DC–DC converter only instead of entire frequency domain. This could, be attained by insertion of a notch filter, in forward path of the voltage closed loop. A negative phase shift is introduced at lower frequency and limits the increase of voltage closed loop cross over frequency. A band pass filter-based SHC reduction method was proposed, which also had the same problem.

2.1. The Buck–Boost Converter

The buck converter combined with a boost is a type of DC–DC converter. A single inductor is used in the place of transformer as in fly back converter. Either incremental or decremental output voltage is observed in magnitude compared to input voltage.

2.2. The Inverting Topology

A similar circuit topology called inverting topology produces output voltage opposite to input polarity (Figure 23.1). Depending on the cycle of the switching transistor, output voltage is adjustable. But complication is caused in driving circuitry, due to absence of terminal at ground for the switch. Isolation of power supply from load circuit by using a battery, the diode polarity is reversed and the drawback is rectified. Buck and boost can be combined with the switch either on the ground or supply side during reversal.

A lower or higher output voltage with same polarity of the input voltage is achieved by using a single inductor for both inductor mode and switches instead of diodes. Four-switch buck–boost converter uses multiple inductors, with a single switch (Single-Ended Primary Inductor Converter (SEPIC) and Cuk topologies).

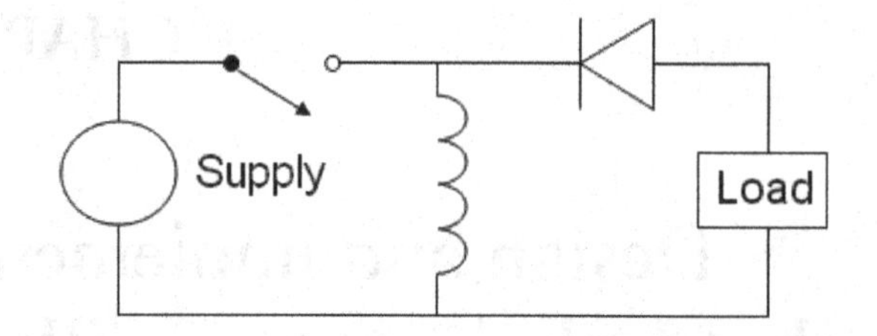

Figure 23.1: Inverting buck–boost converter

The operation involves directly, the connection of input voltage (Figure 23.2) to the inductor (L). At on-stage, the energy is supplied by the capacitor to output load and at off stage the output load and capacitor are connected by inductor and this results in polarity, which is inverse [2] of the input. At the switch on position, the inductor supplies the current via diode (D) toward the load. Two operating states are depicted in Figure 23.3.

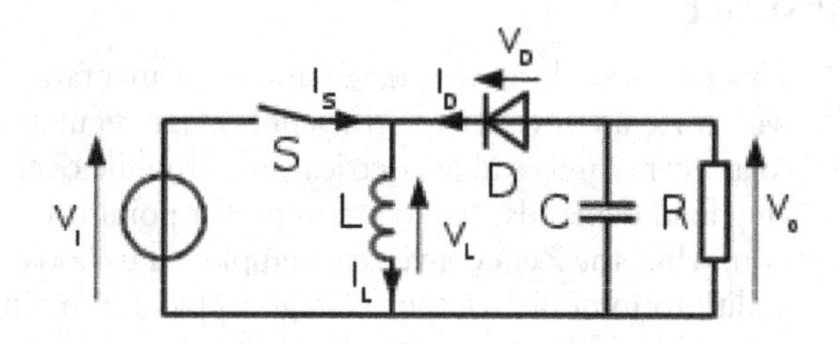

Figure 23.2: Schematic of a buck–boost converter

2.3. Conceptual Overview

The inductor's reluctance explains the operation in which the current changes rapidly. Initially on opening the switch, current flow is zero which flows through inductor and diode on closing and prevents flow of the current at right side of the circuit. There is no rapid current change in the inductor, which initially is low and slowly increases, where the stored energy is in magnetic field form.

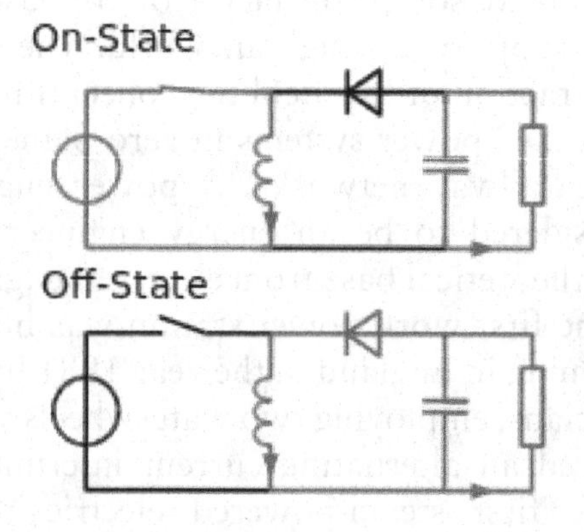

Figure 23.3: Operating states of a buck–boost converter

2.4. Continuous Mode

In this, the current will never be zero through the inductor L, during a commutation cycle. The fraction of commutation period T at switch on position, represents Duty cycle D, and the range of D is between 0 (S is never on) and 1 (S is always on). On opening the switch, inductor current flows through the load. At steady-state conditions, stored energy in its components remains the same throughout commutation cycle.

2.5. Discontinuous Mode

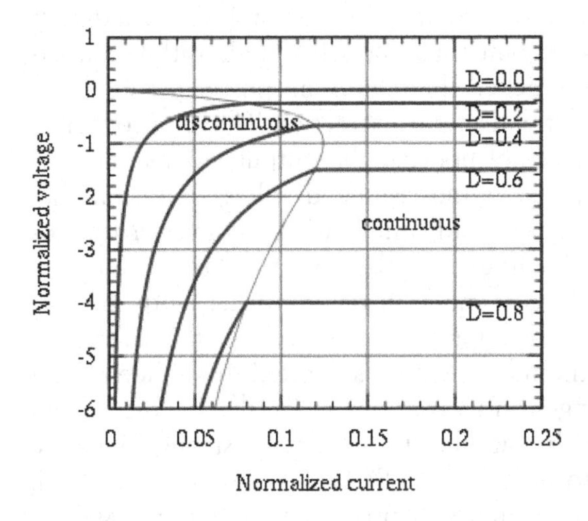

Figure 23.4: Evolution of normalized output voltage

Discontinuous mode involves, transfer of small required energy to the load within a shorter period of time than whole commutation period. Due to complete discharge of the inductor at final stage of the cycle, the current falls to zero. Output voltage equation shows strong effect, even if there is any small change that occurs. The current drawn by the load changes from discontinuous mode to continuous mode at low and high loads respectively (Figure 23.4). At the end of the cycle, the limit between them reaches zero-current fall at the inductor.

- Input filter is required for buck–boost converter.
- Output current is discontinuous (only output capacitors have to support entire load during portion of switching cycle).
- Buck–boost converter gives negative output voltage.
- Presence of right half zero in continuous conduction mode, makes the compensation of feedback loop difficult.

3. Proposed System

This work presents the solar [3] panel fed Z-Source converter and bidirectional [4–5] DC–DC converter which regulates intermediate DC bus voltage, in case of downstream single-phase inverter. The results are analyzed and a method is proposed to provide induced DC–AC inverter (Figure 23.5) by intermediate bus capacitor. Hysteresis control, relates the detection of voltage sap and swell. In remote areas power grid cannot be extended to photovoltaic power systems, wherein, standalone photovoltaic power system meets the need to small extent, though it works essentially in local loads. In this power system, as the instantaneous output power of DC–AC inverter pulsates, a doubled AC current component appears at input side.

3.1. Block Diagram

See Figure 23.5.

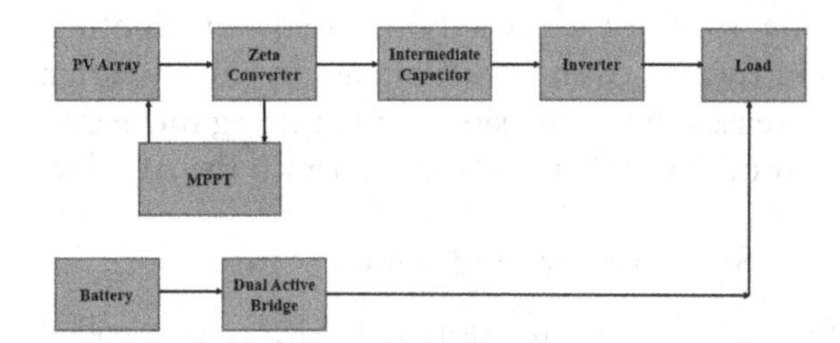

Figure 23.5: Block diagram of proposed system with zeta converter.

3.2. PV Array

Solar panels and several components are arranged which converts sunlight into electricity. The system's overall performance is improved by solar tracking system and integrated battery solution. A solar array with solar panels converts light into electricity [6–8].

3.3. MPPT

MPPT technique extracts maximum power every time [9]. Irrespective of the ultimate destination of the solar power, the best load is chosen by MPPT to present to the cells for a most usable power output. The efficiency of power transfer depends on the amount of incident radiations.

3.4. Zeta Converter

This consists of inductors and capacitors two each, capable to operate either in step-up or step-down mode, which is a 4th order DC–DC converter.

A positive output voltage below and above an input voltage is provided by zeta converter. It is advantageous over SEPIC converter in producing lower output-voltage ripple [10–11] with easy compensation. The drawback includes, larger flying capacitor, higher input voltage ripple and buck controller with high side P-channel Metal-Oxide-Semiconductor (PMOS). A coupled inductor is useful in minimizing the board space.

3.5. Inverter

DC input voltage is changed into a different AC output voltage by inverters with transformers. Inverters are used with transformer, which are not the same as the alternators that converts mechanical energy into AC.

DC generated from batteries and solar panels when connected to an inverter, only small household devices are met with electric power by a complex process, resulting in AC power. This form of electricity is used for lighting, microwave oven etc. Voltage by an inverter, increases the decreasing current using more current on DC side but a small part on the AC side.

3.6. Space Vector Modulation

Pulse width modulation (PWM), is controlled by an algorithm called space vector modulation (SVM) used for creating AC waveforms. Three phase AC motors class-D amplifiers are driven by SVM. The active area developed is, to reduce total harmonic distortion [12]. A series of switches helps in converting DC supply in three phase inverters to three output legs connected to motor. The switches are controlled in such a manner that both switches in the same leg is turned on, to avoid DC supply short. The lower order harmonics of stepped wave inverter is not eliminated [13, 14] by filters. Space vector pulse with modulation technique (SVPWM) provides 15% increment in maximum voltage than PWM.

4. Simulation Results

The simulation results demonstrate the effective performance of a standalone photovoltaic (PV) system integrated with a Zeta converter (Figure 23.6). The primary objective was to evaluate how well the system boosts and regulates output voltage under varying load and environmental conditions. The PV array output was analyzed under different irradiance levels, and the maximum power point tracking (MPPT) algorithm consistently tracked the maximum power point, ensuring optimal energy extraction. The Zeta converter successfully boosted the output voltage, delivering a stable and regulated voltage with minimal ripple, a key advantage of this converter topology. The simulation also confirmed the continuous conduction mode in the converter, with the current remaining stable throughout the operation. Furthermore, when subjected to sudden load changes and variations in irradiance, the system quickly stabilized, indicating strong dynamic performance. If a SVM inverter was included, the resulting AC output would exhibit low harmonic distortion, enhancing the power quality for variable-speed drives. Overall, the integration of the Zeta converter provides significant benefits in reducing voltage ripple and simplifying compensation, making it ideal for standalone PV systems in remote areas. The results confirm that this configuration offers an efficient solution for renewable energy applications with reliable voltage control and high energy conversion efficiency. Figure 23.7 shows output voltage coming out from the standalone [15] photovoltaic system.

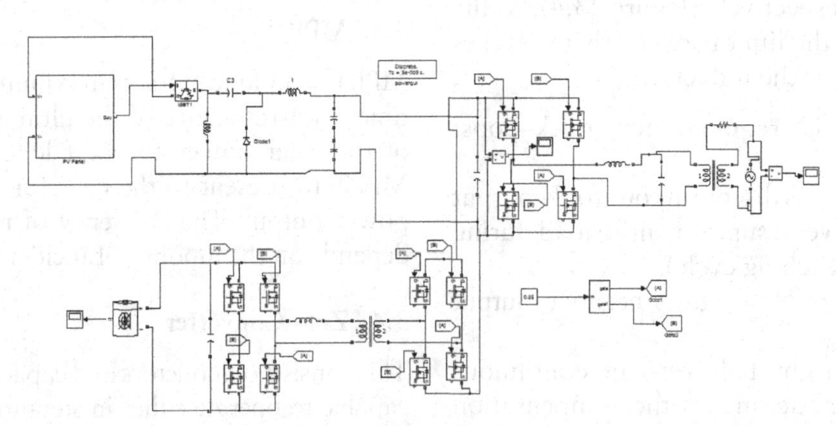

Figure 23.6: Simulation diagram

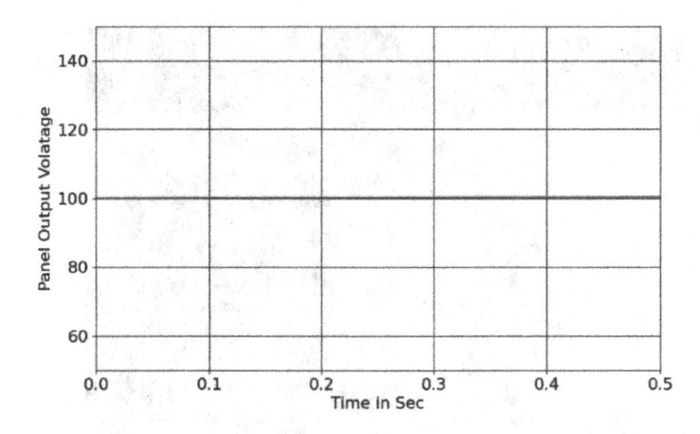

Figure 23.7: Panel Output Voltage

Figure 23.7 shows the output voltage of the solar panel under standard test conditions. The PV array efficiently converts sunlight into a DC voltage, which acts as the primary input for the Zeta converter. The output waveform indicates a stable DC voltage, reflecting the performance of the PV system in consistently delivering energy.

Figure 23.8 depicts, the boosted output voltage from the Zeta converter. The input DC voltage from the PV panel is stepped up to a higher level by the Zeta converter, which ensures minimal voltage ripple and stable output. The converter operates in continuous conduction mode, as seen from the smooth and consistent waveform, which is crucial for applications requiring reliable voltage regulation.

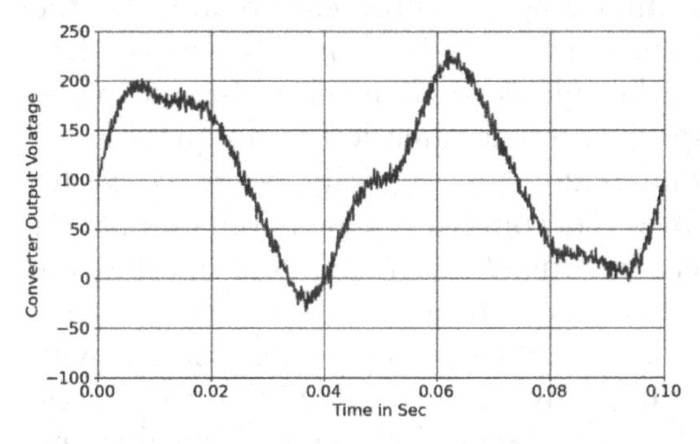

Figure 23.8: Output voltage coming from the converter after boosting the input voltage

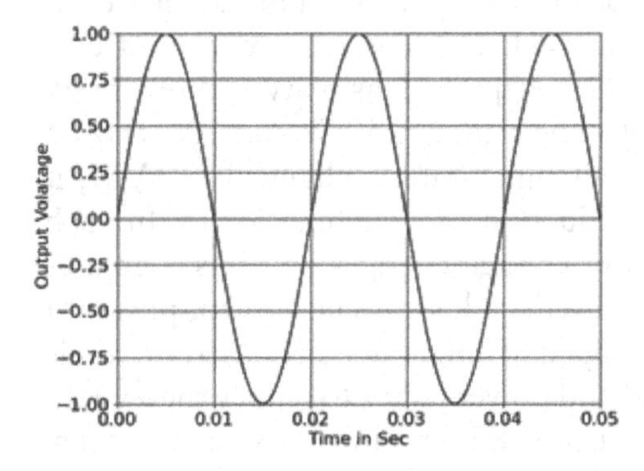

Figure 23.9: AC Output Voltage

The waveform depicted in Figure 23.9 represents the AC output voltage generated by the inverter after converting the boosted DC voltage. The sinusoidal AC waveform is clean and exhibits minimal harmonic distortion, which is vital for powering sensitive loads in remote areas. The inclusion of SVM enhances the output quality, providing a more efficient and stable AC voltage suitable for driving various loads.

4.1. Hardware Results

The hardware setup shown demonstrates a standalone photovoltaic (PV) system integrated with a Zeta converter. The arrangement includes a solar panel which converts sunlight into DC electricity. This generated power is then fed into the circuit. A digital multimeter, prominently displaying a reading of 12.7 V, is used to monitor the voltage in real-time during testing. The main hardware comprises several printed circuit boards (PCBs) that handle different functions, such as the Zeta converter, which steps up or steps down the input voltage to a stable level, and the MPPT circuit, which ensures maximum power extraction from the solar panel under varying light conditions. Additionally, the inverter circuit converts the DC output into a sinusoidal AC waveform, suitable for driving loads. The setup is interconnected through power lines and control signals, with indicator LEDs on the boards showing the operational status of various components. This setup is designed to experimentally validate the system's efficiency, focusing on voltage regulation, ripple minimization, and overall energy conversion performance for remote applications.

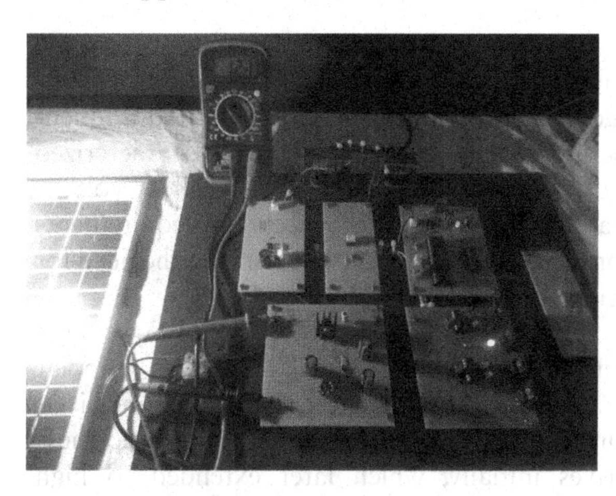

Figure 23.10: Solar input voltage

Figure 23.10 captures the input voltage obtained directly from the solar panel, which is around 12.7 V. The measured value reflects the initial DC voltage provided to the Zeta converter before boosting. This input voltage depends on factors like solar irradiance and panel efficiency, which directly influence the overall system performance.

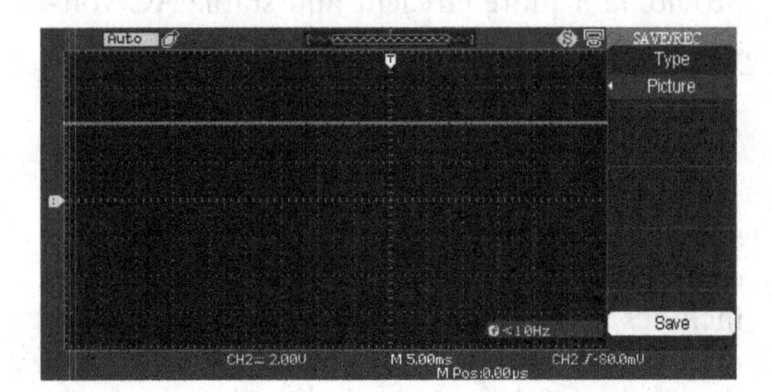

Figure 23.11: Solar input voltage

In Figure 23.11, the output DC voltage of the Zeta converter is shown. The converter successfully increases the input voltage from 12.7 V to a higher level (around 350 V), demonstrating the efficiency of the converter in boosting the voltage for subsequent conversion to AC by the inverter. The output is stable and suitable for grid-independent applications.

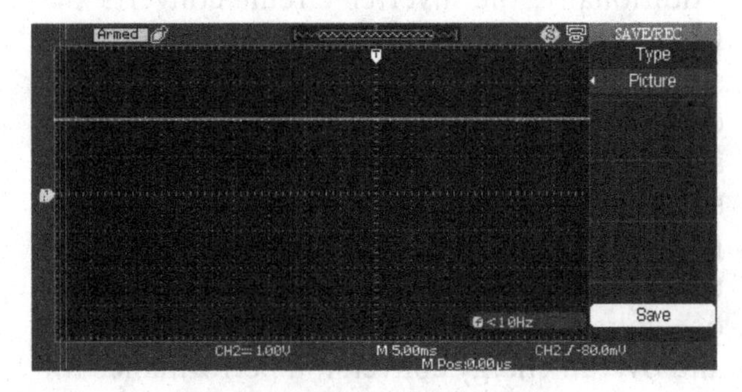

Figure 23.12: DC output voltage

Figure 23.12 represents DC-DC output from the converter. The inverter converts the high DC voltage into an AC output with the desired frequency and voltage level. The output is characterized by low harmonic content and excellent voltage stability, ensuring that it meets the requirements for remote area applications without the need for grid support (Figure 23.13).

Semiconductor technology employs high current-high voltage Insulated Gate Bipolar Transistor (IGBT) in invertors for traction drives. Only limited trams or regional trains used these drives initially, which later extended to high speed traction locomotives (Figure 23.14).

Figure 23.13: DC output voltage

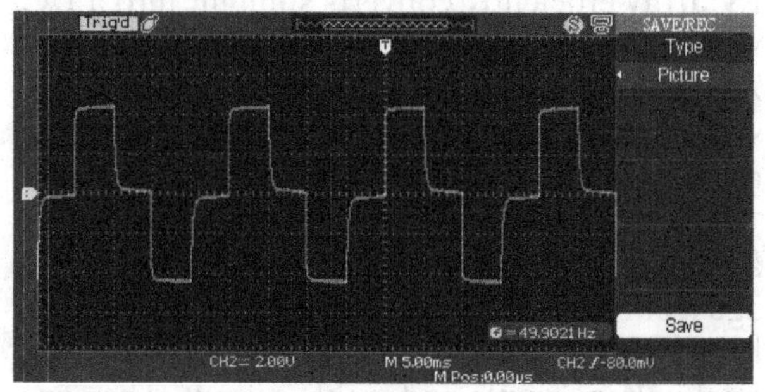

Figure 23.14: Output AC voltage

Neutral point clamped (NPC) inverters reduced, maximum drop voltage on the IGBTs. After analyzing the problems related to the development in this field, it was proposed to have 6-hexagons subdivision of 3-phase plane for space vector modulation technique.

Combination of quadratic-boost converter improves zeta converter. Sum of both output volt of zeta and quadratic boost accomplishes the output voltage boosting. The circuit is designed with same characteristics of the former with guaranteed advantages. In addition to a single switch that maintains low computational count, high current ripple [16–18] in quadratic-boost converter proposed circuit with an adequate filter design is used. This ensures low output current ripple [19–20]. The feasibility and reliability of lab prototype rates at 250 W and 100 kHz.

If any switching state is changed in SVM, it results in reduction of switching losses. In each single phase, voltage changes and has a greater flexibility to reduce losses. The control pulses of SVM is stopped at time intervals based on angle of the load and power factor to reduce the switching losses further. Elimination of extra switching, introduces different modulation indexes.

5. Conclusion

In conclusion, this work effectively demonstrates the integration of a Zeta converter in a standalone photovoltaic (PV) system for efficient energy conversion and voltage regulation. The simulation and hardware results validate the system's ability to maintain stable output voltage with minimal ripple, even under fluctuating irradiance levels and varying load conditions. The MPPT algorithm reliably extracts the maximum power from the PV array, thereby enhancing overall system efficiency. The Zeta converter's unique design facilitates both step-up and step-down operations while minimizing output voltage ripple, crucial for sensitive applications. The SVM technique, implemented in the inverter, ensures low harmonic distortion and superior power quality, making it suitable for driving variable-speed loads. In the hardware setup, the input DC voltage of 12.7 V was boosted to 350 V, and then converted to a stable AC output, demonstrating the system's practical viability for remote applications where grid access is limited. Future enhancements include the integration of advanced converter topologies, such as a hybrid quadratic-boost Zeta converter, which can achieve higher voltage gains with reduced component stress. Additionally, exploring more sophisticated control algorithms like adaptive MPPT and advanced SVM could further optimize performance, reduce switching losses, and enhance overall system reliability, making it scalable for large-scale renewable energy projects.

References

[1] Bojoi, R., Pica, C., Roiu, D., Tenconi, A.: New DC–DC converter with reduced low-frequency current ripple for fuel cell in single-phase distributed generation. In: Proceedings of the IEEE International Conference on Industrial Technology, pp. 1213–1218 (2010)

[2] Yao, W., Wang, X., Zhang, X., Tang, Y., Loh, P.C., Blaabjerg, F.: A unified Active damping control for single-phase differential mode buck inverter with LCL-filter. In: Proceedings of the IEEE 6th International Symposium on Power Electronics for Distributed Generation Systems, pp. 1–8 (2015)

[3] Carrasco, J.M., Franquelo, L.G., Bialasiewicz, J.T., Galvan, E., Guisado, R.C.P., Prats, M.A.M., Leon, J.I., Moreno-Alfonso, N.: Power-electronic systems for the grid integration of renewable energy sources: a survey. IEEE Trans. Ind. Electron. 53(4), 1002–1016 (2006)

[4] Dong, D., Cvetkovic, I., Boroyevich, D., Zhang, W., Wang, R., Mattavelli, P.: Grid interface bidirectional converter for residential DC distribution systems—part I: high-density two-stage topology. IEEE Trans. Power Electron. 28(4), 1655–1666 (2013)

[5] Agbossou, K., Kolhe, M., Hamelin, J., Bose, T.K.: Zet, 2004. IEEE Trans. Energy Convers. 19(3), 633–640

[6] Huang, Y., Peng, F.Z.: Survey of the power conditioning system for PV power generation. In: Proceedings of the IEEE Power Electronics Specialists Conference, June, pp. 1–6 (2006)

[7] Kwon, J.M., Nam, K.H., Kwon, B.H.: Grid-connected photovoltaic multi string PCS with PV current variation reduction control. IEEE Trans. Ind. Electron. 56(11), 4381–4388 (2009)

[8] Roman, E., Alonso, R., Ibanez, P., Elorduizapatarietxe, S., Goitia, D.: Intelligent PV module for grid-connected PV systems. IEEE Trans. Ind. Electron. 53(4), 1066–1073 (2006)

[9] Femia, N., Petrone, G., Spagnuolo, G., Vitelli, M.: A technique for improving P&O MPPT performance of dual-stage grid-connected photovoltaic systems. IEEE Trans. Ind. Electron. 56(11), 4473–4482 (2009)

[10] Blaabjerg, F., Teodorescu, R., Liserre, M., Timbus, A.: Overview of control and grid synchronization for distributed power generation systems. IEEE Trans. Ind. Electron. 53(5), 1398–1409 (2006)

[11] Itohand, J., Hayashi, F.: Ripple current reduction of a fuel cell for a single phase isolated converter using a DC Active filter with a center tap. IEEE Trans. Power Electron. 25(3), 550–556 (2010)

[12] Wang, W., Ruan, X., Wang, X.: A novel second harmonic current reduction method for dual-Active-bridge used in photovoltaic power system. In: Proceedings of the IEEE Energy Conversion Congress and Exposition, Denver, CO, USA, pp. 1635–1639 (2013)

[13] Zhang, L., Ruan, X., Ren, X.: Second-harmonic current reduction and dynamic performance improvement in the two-stage inverters: an output impedance perspective. IEEE Trans. Ind. Electron. 62(1), 394–404 (2015)

[14] Zhu, G., Ruan, X., Zhang, L., Wang, X.: On the reduction of second harmonic current and improvement of dynamic response for two stage single-phase inverter. IEEE Trans. Power Electron. 30(2), 1028–1041 (2015)

[15] Xiong, X., Tse, C., Ruan, X.: Bifurcation analysis of standalone photovoltaic-battery

hybrid power system. IEEE Trans. Circuits Syst. I Reg. Papers 60(5), 1354–1365 (2013)
[16] Kwon, J., Kim, E., Kwon, B., Nam, K.H.: High-efficiency fuel cell power conditioning system with input current ripple reduction. IEEE Trans. Ind. Electron. 56(3), 861–834 (2009)
[17] Wei, Z., Deng, X., Gong, C., Chen, J., Zhang, F.: A novel technique of low frequency input current ripple reduction in two-stage DC–AC inverter. In: Proceedings of the IEEE Industrial Electronics Society, pp. 139–143 (2012)
[18] Zhu, G., Tan, S., Chen, Y., Tse, C.K.: Mitigation of low-frequency current ripple in fuel-cell inverter systems through waveform control. IEEE Trans. Power Electron. 28(2), 779–792 (2013)
[19] Krein, P.T., Balog, R.S., Mirjafari, M.: Minimum energy and capacitance requirements for single-phase inverters and rectifiers using a ripple port. IEEE Trans. Power Electron. 27(11), 4690–4698 (2012)
[20] Liu, C., Lai, J.S.: Low frequency current ripple reduction technique with active control in a fuel cell power system with inverter load. IEEE Trans. Power Electron. 22(4), 1429–1436 (2007)

CHAPTER 24

In vitro assessment of a methanolic extract from *Coccinia indica* enhancing glucose uptake via the phosphoinositide 3-kinase dependent pathway

Anand Singaravelu,[1*,a] Kalaivani Arumugam,[1] Mettilda Manuel Swami Dorai[1]

[1]Department of Chemistry, Saveetha Engineering College, Thandalam, Chennai, Tamil Nadu, India

Email: [a]anands@saveetha.ac.in

Abstract

Type 2 diabetes is a chronic disorder of glucose and fat metabolism. Insulin resistance is a condition in which normal insulin levels fail to respond to metabolic disturbances. This study aims to elucidate the molecular mechanisms underlying glucose transport in L6 myotubes at the cellular level. *Coccinia indica* is known plant selected for its potential effects of glucose transport. Glucose uptake is performed on the different extracts obtained from *Coccinia indica*. Among these, the methanol extract is found to be the most active at a dose of 100 ng/ml, comparable to a positive control. The methanol extract of *Coccinia indica* increases the expression of essential molecules in glucose transport, such as IRS-1, PI3K, and GLUT4. The study suggested that methanol extract of *Coccinia indica* enhance glucose uptake in L6 myotubes by upregulating IRS-1, PI3K, and GLUT4 expression and utilizing a PI3K-dependent pathway. This could be valuable for future research on potential therapeutic agents for managing glucose metabolism and insulin resistance.

Keywords: *Coccinia indica*, L6 myotubes, GLUT4, PI3-Kinase, Type II DM

1. Introduction

Diabetes, a prevalent metabolic disorder affecting approximately 30 million people worldwide, is a complex syndrome characterized by abnormalities in glucose metabolism. About 85% of diabetes cases fall under the category of Non-Insulin Dependent Diabetes Mellitus (NIDDM), which predominantly impacts adults. NIDDM arises from poor glucose utilization in various tissues, including the pancreatic beta cells [1]. Insulin resistance occurs when the body's usual insulin levels are insufficient to trigger the expected response in adipose tissue, muscle, and liver cells. In adipocytes, insulin resistance diminishes insulin's effectiveness and promotes the breakdown of stored triglycerides without any interventions that would increase insulin. This results in an increased release of free fatty acids into the bloodstream. Moreover, insulin resistance in muscle cells leads to reduced glucose uptake and storage as glycogen, while insulin resistance in liver cells hinders glycogen storage and its release into the bloodstream when insulin levels drop. Consequently, both these factors contribute to elevated blood sugar levels.

The activation of insulin receptors initiates intracellular processes that directly influence glucose levels by controlling the quantity and function of protein molecules on the cell membrane responsible for glucose transportation into the cell. Insulin exerts its most significant impact on glucose uptake in two types of tissues: muscle cells and fat cells. Muscle cells are crucial for essential functions like movement, breathing, and blood circulation, while fat cells store surplus food energy for future requirements. Together, these two tissue types constitute approximately two-thirds of all the cells

DOI: 10.1201/9781003606611-24

in an average human body. Various therapeutic agents are currently under investigation for managing Type 2 diabetes. Given the challenge of maintaining blood glucose levels without affecting lipid metabolism using monotherapy, the concept of combination therapy appears increasingly appealing. Nevertheless, the search for novel anti-diabetic medications continues, as there have been failures with current treatments like Thiazolidinediones and Sulphonylureas, which can lead to adverse effects such as obesity, lactic acidosis, cardiovascular diseases, and significant damage to the liver and other organs instead of effectively reducing blood glucose levels [2].

Herbal medicines present a promising potential with minimal side effects, particularly in addressing metabolic syndromes, owing to the combined action of various active molecules. Despite their effectiveness and popularity, particularly in developing countries where approximately 80% of the population relies on traditional medicines, mainly herbal remedies, for primary healthcare, Ayurveda—an ancient Indian system of medicine—is classified as Class III by the USFDA (denoted as Therapy with concept of belief and no scientific validation). The limitations of herbal therapy lie in its complex nature, unclear mechanism of action, potential interactions with conventional drugs, and challenges in ensuring quality assurance.

A prior investigation established that *Coccinia indica* belongs to the Cucurbitaceae family. Various parts of this plant, including its roots, leaves, and fruits, have been used for a range of medicinal purposes, such as treating jaundice, diabetes, injuries, ulcers, abdominal pain, skin diseases, fever, and asthma. In a previous study, the lead of *Coccinia indica* and its constituents were found to exhibit multiple beneficial effects, including anthelmintic, antioxidant, anti-inflammatory, analgesic, antipyretic, antimicrobial, hypoglycemic, and hepatoprotective properties. Based on this background the current study focuses on *Coccinia indica* leaves due to their observed antidiabetic effects on L6 myotubes. However, despite these positive findings, the exact cellular and molecular mechanisms of *Coccinia indica* remain unknown, and further research is needed to better understand its therapeutic actions.

2. Materials and Methods

2.1. Preparation of Extracts

The dried pulverized plant powder (100 g) of *C. indica* was extracted sequentially using organic solvents in increasing polarity (hexane, ethyl acetate and methanol) at room temperature. Extracts were concentrated (Rotavap – Ika Instruments, Germany) under reduced pressure. Approximately, the yield of hexane extract is 1 g and ethyl acetate extract 2 g while methanol extract yielded 3 g. One milligram of all extracts were dissolved with DMSO and diluted 10 μg to 1 ng/ml as primary studies of GUA. The plant identification was confirmed by a taxonomist.

2.2. L6 Myoblasts and Myotubes

L6 myotubes were differentiated and cultured in DMEM under a 5% CO_2 atmosphere. Differentiation was initiated and transferred to DMEM after confluence of myotubes and treated with insulin for 15 min and rosiglitazone for 2 h.

2.3. Glucose Uptake Assay

Initially, differentiated myotubes was serum starvation for 5 h, followed by a 2-h incubation with *C. indica*. This was succeeded by insulin stimulation for 20 min added with 2-deoxy-D-1-[3H] glucose and washed three times by buffer and lysed in 0.1% SDS. The lysates were transferred into glass fiber paper in 96-well plate [3].

2.4. Cytotoxicity Assay

Myotube cells (0.2 million) were seeded in a 96-well plate of CIME. Maximal lysis was induced using 0.05% Triton X-100. The plate was then analyzed using a scanning multiwell spectrophotometer at 492 nm [4].

2.5. RT-PCR

The analysis was performed in myotubes, for the treatments of insulin for 15 min, 18 h for rosiglitazone and CIME), were homogenized and isolated total RNA. The RNA was transcribed into cDNA and amplified using specific primers for Glut4 and PI3K. The PCR were separated on a 1.5% agarose gel [5, 6].

2.6. Immunoprecipitation and Western Blot Analysis (IRS-1)

50 mg of protein were separated on a 10% SDS-polyacrylamide gel, transferred to a nitrocellulose membrane, blocked overnight at 4 °C with 5% nonfat milk, and incubated with primary and secondary antibodies. The blot was then visualized using chromogenic NBT/BCIP and photographed [3, 7].

2.7. Statistical Analysis

Statistical analysis was conducted using GraphPad Prism software with Dunnett's post hoc test. Results are expressed as means ± S.E.M., with statistical significance set at $P < 0.05$

3. Results

3.1. 2-Deoxy Glucose Uptake

Differentiation of myotubes was assessed based on glucose uptake in response to CIME. Figure 24.1 exhibited at 100 ng/ml, with no significant difference between 24 and 48 h. Consequently, a final concentration of 100 ng/ml was selected for subsequent studies, as depicted.

3.2. Effect of CIME on Cytotoxicity

To evaluate the potential toxic effects of CIME on cells, an assay was performed, and LDH levels were measured at 492 nm. As shown in Figure 24.2, CIME exhibited no cytotoxicity towards the cells. Cells treated with the extract displayed less than 25% cytotoxicity.

3.3. Effect of Wortmannin on Glucose Uptake

Investigate CIME in PI3K activation, glucose uptake in the presence of a PI3K inhibitor, Figure 24.3, insulin-treated cells showed complete inhibition of glucose uptake, indicating that this process occurs through a PI3K-dependent mechanism.

3.4. mRNA expression of CIME

The impact of CIME on GLUT and PI3K expression was assessed by RT-PCR. Compared to insulin and rosiglitazone, CIME led to an elevated intensity in GLUT4 and PI3K expression. This suggests that CIME operates through a PI3K-dependent pathway.

3.5. Effect of CIME on IRS-1 and PI3K Protein Levels

To evaluate the impact of CIME on the protein expression of IRS-1 and PI3K. Relative to untreated cells and insulin treatment, CIME demonstrated elevated protein levels of both IRS-1 and PI3K.

4. Discussion

Type 2 diabetes is a metabolic condition marked by inconsistent insulin resistance leading to elevated blood glucose levels are due to impaired glucose uptake by cells. This can be managed and sometimes reversed through various measures and drugs that improve insulin sensitivity or reduce the liver's production of glucose. As Type 2 diabetes progresses, the pancreas may become less effective at secreting insulin, further exacerbating the problem. This decline in insulin secretion can be due to the prolonged strain on the pancreas and the gradual deterioration of beta cells, which are responsible for producing insulin. In advanced stages, when insulin secretion is significantly impaired, therapeutic insulin replacement may be necessary to regulate blood sugar levels effectively. Type 2 diabetes caused by a known specific defect, such as certain genetic mutations, may be classified separately from the more common form of Type 2 diabetes [8].

Insulin, synthesized by the pancreas, is pivotal in maintaining glucose balance in the body. This translocation is facilitated by the phosphoinositide 3-kinase (PI3K) pathway. Insulin binding to its receptor on the surface activates PI3K, which in turn triggers a signaling cascade leading to GLUT translocation. It has been shown that AMPK activation can also lead to GLUT translocation, even in the absence of insulin. This pathway might be particularly relevant in situations where insulin sensitivity is compromised in NIDDM. The pathways for diabetes research, studying them in an in vitro model like L6 myotubes can provide valuable insights. Myotubes are skeletal muscle cells that can mimic some aspects of in vivo muscle tissue behavior. Researchers can manipulate various factors in this controlled environment to investigate how different pathways and molecules influence GLUT translocation and glucose uptake.

Different solvents were employed to extract compounds from *Coccinia indica* leaves, followed by assessing their glucose uptake potential in L6 myotubes. CIME increase glucose uptake at 100 ng/ml, when compared to a positive control.

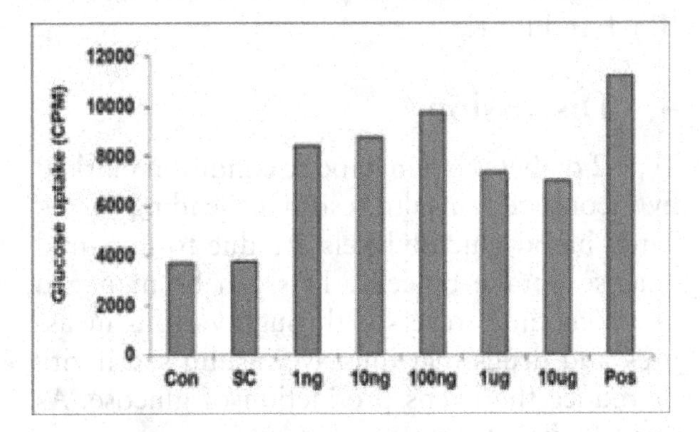

Figure 24.1 Effect of CIME on glucose uptake assay

To ensure cytotoxicity of CIME was evaluated using lactate dehydrogenase into the culture supernatant when cells undergo lysis. The release of LDH indicates cell membrane disruption, which can be a result of cytotoxic effects. The vehicle control likely refers to a sample that doesn't contain CIME, which serves as a baseline for comparison. If the percentage of cytotoxicity is low, it suggests that CIME did not cause significant cell damage or death. Triton-X is a detergent that is known to cause cell membrane disruption and release LDH, leading to cell death. Its use as a positive control helps validate the sensitivity and accuracy of the LDH measurement method.

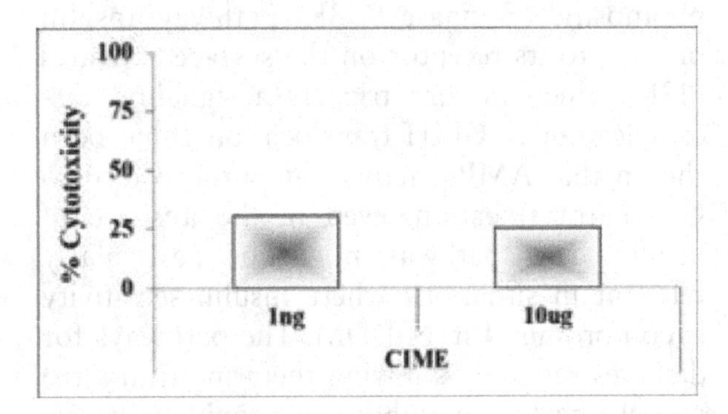

Figure 24.2 Effect of CIME on cell cytotoxicity assay

To elucidate the molecular mechanisms underlying CIME's antidiabetic properties. Glucose uptake assays were conducted using specific PI3K inhibitors. These inhibitors are designed to block the activity of PI3K, allowing researchers to study the role of PI3K in the process. The results of the experiment showed that the glucose uptake activity was completely blocked when specific PI3K inhibitors were present. This suggests that PI3K is essential for glucose uptake to occur. It indicates that CIME acts similarly to insulin (insulin-mimicking activity) by increasing glucose transport. This effect is achieved through a mechanism that is dependent on PI3K.

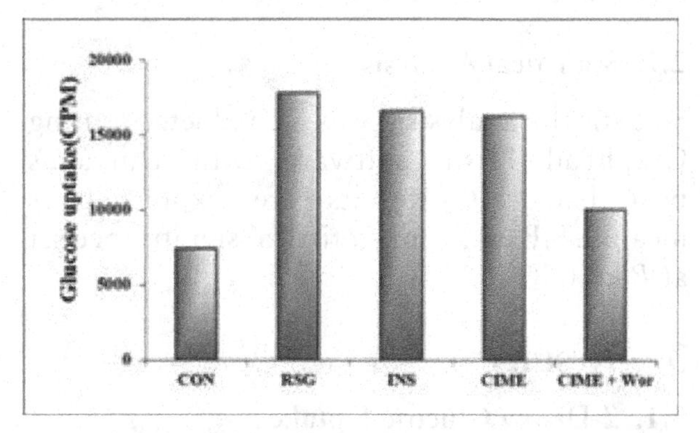

Figure 24.3 Effect of CIME on glucose uptake in PI3K inhibitor

Insulin signaling is a complex cascade of events that regulates glucose uptake and various other cellular processes. It begins with insulin binding to its receptor, leading to the activation of downstream signaling molecules, ultimately promoting glucose transport into target cells. The study suggests that CIME has antidiabetic effects, and one of the mechanisms through which it exerts these effects is by modulating the insulin signaling pathway. Specifically, it appears to enhance the activation of PI3K and increase the expression of GLUT4. This would lead to improved glucose transport into cells. Rosiglitazone is a thiazolidinedione (TZD) medication [10] (Figure 24.4).

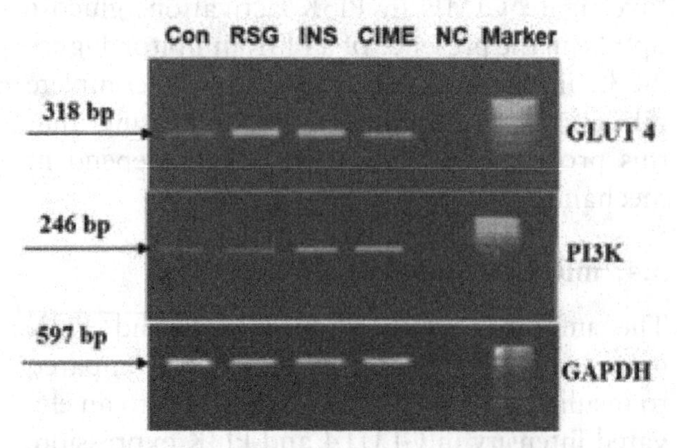

Figure 24.4 RT-PCR analysis of methanol extract of *Coccinia indica* by Glut-4 and PI3K

In contrast to CIME, the study suggests that rosiglitazone induces GLUT4 expression without affecting PI3K levels. This is consistent with the known mechanism of action of TZDs, which do not directly affect the insulin signaling pathway but improve insulin sensitivity through other mechanisms. By demonstrating that CIME enhances PI3K and GLUT4 expression. IRS-1 and PI3K was normalized to beta-actin. The result suggest that glucose uptake induction is mediated by insulin receptor activation. Furthermore, research indicates that CIME operates via an insulin-triggered, PI3K-dependent pathway (Figure 24.5) [9].

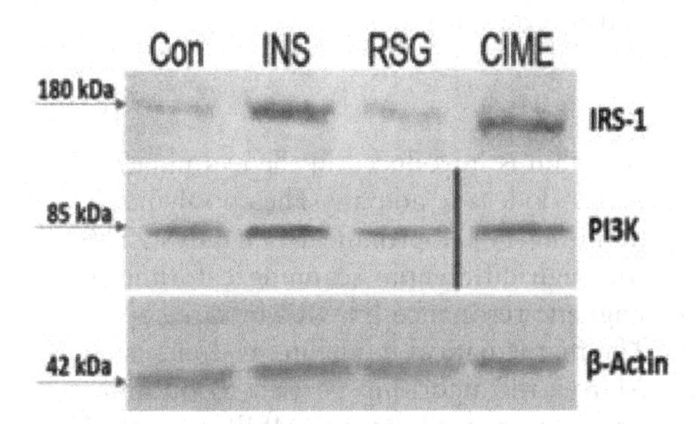

Figure 24.5 Western blot analysis of methanol extract of *Coccinia indica* by IRS-1 and PI3K

5. Conclusion

The results indicate that CIME enhances PI3K activity and increases GLUT4 expression, which could contribute to improved glucose uptake and potentially help mitigate insulin resistance. This research provides insights into a possible mechanism of action for CIME in the context of diabetes treatment. Further, is needed to purify the active compound responsible for the observed effects. Identifying this compound could pave the way for the development of more potent diabetes treatments and potentially anti-adipogenic agents.

References

[1] Haring HU (1999) Pathogenesis of type II diabetes: are there common causes for insulin resistance and secretion failure? Exp Clin Endocrinol Diabetes 107:S17–S23

[2] Klip A, Guma A, Ramlal T, Bilan PJ, Lam L, Leiter LA (1992) Stimulation of hexose transport by metformin in L6 muscle cells in culture. Endocrinology 130:2535–2544

[3] Anand S, Muthusamy VS, Sujatha S, Sangeetha KN, Bharthi Raja R, Sudhagar S, and Lakshmi, BS (2010) Aloe emodin glycosides stimulate glucose transport and glycogen storage through PI3K dependent mechanism in L6 myotubes and inhibit adipocyte differentiation in 3T3L1 adipocytes. FEBS Lett 584: 3170–3178

[4] Gayathri B, Manjula N, Vinay kumar KS, Lakshmi BS, Balakrishnan A (2007) Pure compound from *Boswellia serrata* extract exhibits anti-inflammatory properties in human PBMC's and mouse macrophages through inhibition of TNF alpha, IL-1 beta, NO and MAP kinases. Int Immunopharmacol 7:473–482

[5] Anand S, Raja RB, Lakshmi BS (2015) Enhancement of glucose uptake by *Ficus hispida* methanolic extract through phosphatidylinositol 3 kinase and inhibiting aldose reductase – an in vitro analysis. Eur J Integr Med 7:164–171

[6] Hall LR, Mehlotra RK, Higgins AW, Haxhiu MA, Pearlman E (1998) An essential role for interleukin-5 and eosinophils in helminth induced airway hyper-responsiveness. Infect Immun 66:4425–4430

[7] Muthusamy VS, Anand S, Sangeetha KN, Sujatha S, Balakrishnan Arun, Lakshmi BS (2008) Tannins present in *Cichorium intybus* enhance glucose uptake and inhibit adipogenesis in 3T3-L1 adipocytes through PTP1B inhibition. Chem Biol Interact 174:69–78

[8] Helping the pancreas produce insulin. HealthValue. Retrieved on 2007; 09-21

[9] Huang C, Somwar R, Patel N, Niu W, Török D, Klip A (2002) Sustained exposure of L6 myotubes to high glucose and insulin decreases insulin-stimulated GLUT4 translocation but upregulates GLUT4 activity. Diabetes 51:2090–2098

[10] Jung KH, Choi HS, Kim DH, Han MY, Chang UJ, Yim SV et al (2008) Epigallocatechin gallate stimulates glucose uptake through the phosphatidylinositol 3-kinase-mediated pathway in L6 rat skeletal muscle cells. J Med Food 11:429–434

CHAPTER 25

Mathematical modelling using Levenberg–Marquardt algorithm for polyethyleneglycoladipate–co–1, 12–dodecane adipate

B. Yamini[1, a] **and V. Anandan**[2, b]

[1]Department of Chemistry, Saveetha Engineering College, Chennai India

[2]Department of Mathematics, Saveetha Engineering College, Chennai India

Email: [a]chemistryyamini@gmail.com, [b]anandanworld@gmail.com

Abstract

A mathematical modelling using Levenberg-Marquardt algorithm is introduced in the examination of an aliphatic copolyester, polyethyleneglycoladipate–co–1,12–dodecane adipate. The copolymer is synthesised by a melt polycondensation technique using titanium (IV) isoproxide as a catalyst. The characteristics and structural properties are investigated through differential scanning calorimetry (DSC), gel permeation chromatography (GPC), nuclear magnetic resonance (NMR), infrared spectroscopy (FTIR) and thermogravimetric analysis (TGA). The mathematical modelling is done using a first-order reaction kinetics equation, and Levenberg-Marquardt algorithm is used to optimize the copolymers. The final output is investigated using root mean squared error (RMSE), statistical evaluation, and sensitivity analysis. The results determine a close association between the actual and predicted data, and the RMSE demonstrates a very good fit for the quantitative data. The observations provide a complete understanding of the polymerization, mathematical modelling. This study provides a better framework particularly in the domain of polymer chemistry to optimize and synthesise the copolymer polyethyleneglycoladipate–co–1,12–dodecane adipate for various applications.

Keywords: Levenberg-Marquardt algorithm, mathematical modelling, polyethylene glycol, polymers

1. Introduction

Polymers is a material which is found almost everywhere and plays a crucial role in the recent decades, providing a range of properties which are used for applications in various industries like packaging, aerospace, automotive, and biomedical domains [4]. The flexibility of polymer is based on its molecular structure, that are employed to attain appropriate mechanical, chemical, and thermal properties. The aliphatic copolyesters are widely used because of its flexibility, biodegradability, and biomedical properties among other polymers [5, 6].

The aliphatic polymers are extracted from plants and are used as an alternate for petroleum-based polymers, which is also economical comparing traditional polyolefins [1]. The aliphatic polymers are the main source for biobased fibers which have good mechanical properties. Several procedure routes such as olefin metathesis, hydrogenation, ADMET polymerization are used for polymerization. But, the synthesis at excess temperature leads to decomposition of catalyst which becomes difficult due to high viscosity [7, 9].

Mathematical modelling aids in understanding the kinetics of the synthesis procedure to enhance the reaction models, and to increase the production. Mathematical models shall be used to (i) simulate the molecular dynamics, (ii) forecast the distribution of molecular weight, and (iii) assess the influence of polymer properties.

The Levenberg–Marquardt algorithm is a self-corrective method, utilised to solve non-linear

DOI: 10.1201/9781003606611-25

least square equations. The algorithm is generally applied in parameter estimation, curve fitting and optimization fields. In polymerization, the Levenberg-Marquardt method is used to enhance the kinetic models by reducing the association among the actual and model predicted data. The robustness and simplicity of the method is found to be suitable for dealing complicated problems with multi parameters. The modelling permits the refinement of reaction environments, which leads to enhance and control the polymerisation and the resultant copolymer. The integration of Levenberg–Marquardt algorithm in polymerization improves the reliability and precision of the kinetic models. In the synthesis of polymers, the mathematical modelling contains the kinetic equations which defines the polymerization rate and molecular weight distribution depending on time. The optimization of the parameters in the kinetic equations indicates that the method is closely associated with the actual and predicted data.

Though the mathematical modelling and optimization algorithms are extensively used in polymer science, there is a gap in the application of these techniques. The present models concentrate on the understanding of the synthesis and enhancing copolyesters. This work intents to fill the gap by developing a comprehensive kinetic model for the polymerization of polyethyleneglycoladipate–co–1,12–dodecane adipate using the Levenberg–Marquardt algorithm.

2. Polyethylene Glycol

Polyethylene glycol (PEG) is a polymer which is used widely because of its flexibility and biocompatibility. The modelling of the polyethyleneglycoladipate–co–1,12–dodecane adipate (a PEG variant) handles the interaction of kinetic reactions among the various catalysts and monomers [13].

The polymerization is the foundation in polymer science, aiding the formation of newer materials with improved performances. The melt polycondensation technique controls the composition and its molecular weight precisely, which are significant in maintaining the composition of the resultant polymer [15]. The formation of the novel copolymer polyethyleneglycoladipate–co–1,12–dodecane adipate, contains complicated kinetic interactions and thermodynamics, forcing to understand the synthesis procedure comprehensively.

3. Mathematical Modelling

The mathematical modelling of the synthesis and the kinetics of polyethyleneglycoladipate–co–1,12–dodecane adipate are discussed in this section. Mathematical modelling is significant in polymerization for the following reasons: (i) the performance of the polymer reactions can be predicted, (ii) the synthesis environment can be optimized, and (iii) the characteristics and correlation among the available parameters of the product can be predicted. The integration of the mathematical modelling offers a robust model to improve and regulate the synthesis of copolymers.

3.1. Mathematical Modelling in Polymer Synthesis

Mathematical modelling is a useful tool in the synthesis of polymer, offering clear perception in the kinetics of polymer reactions. It helps to predict the distributions of reaction rates, molecular weight, and the effect of various parameters in the synthesis. This ability of predicting parameters is crucial in optimization to increase production and modify the polymer applications based on its properties.

The mathematical modelling is used to study the kinetics reaction and the optimization in the synthesis of polyethyleneglycoladipate–co–1,12–dodecane adipate. The copolymer formation rates and the influence of its concentration and reaction properties can be investigated based on the development of a proper kinetic model [2].

3.2. Developing the Kinetic Model

The copolymerization reaction is represented as follows:

$$\frac{d[P]}{dt} = k.[A].[B]$$

where:

- $[P]$ refers concentration of the copolymer formation,
- k refers rate constant,
- $[A]$ and $[B]$ refers concentrations of the reacting monomers (polyethyleneglycoladipate and 1, 12–dodecane adipate, respectively).

The rate constant k is defined using Arrhenius equation, which depends on temperature as follows:

$$k = A . e^{-\frac{E_\alpha}{RT}}$$

where:

- A refers pre-exponential factor,
- R refers gas constant,
- E_α refers activation energy,
- T refers temperature.

3.2. Copolymer Concentration Over Time

The concentration of the copolymer over time can be expressed as:

$$[P] = [P]_0 + \int_0^t k.[A].[B] dt$$

$[P]_0$ refers initial copolymer concentration.

3.3. Objective Function

The optimization and the model fitting are defined by the objective function using the copolymer concentration of predicted and actual data [8].

$$\text{ObjectiveFunction} = \sum_i \left(\frac{d[P]}{dt}_{\text{model}} - \frac{d[P]}{dt}_{\text{data}} \right)^2$$

It is also written as:

$$\text{Objective Function} = \sum_i \left(P_{\text{model}}(t_i) - P_{\text{data}}(t_i) \right)^2$$

where:

- $P_{\text{model}}(t_i)$ refers predicted copolymer concentration at any given time t_i,
- $P_{\text{data}}(t_i)$ refers actual copolymer concentration at any given time t_i.

3.4. Optimization Using Levenberg–Marquardt Algorithm [3]

The Levenberg–Marquardt Algorithm [3, 12] is optimized using:

$$\text{Min } F(x) = \frac{1}{2} \sum_i \left(P_{\text{model}}(t_i, x) - P_{\text{data}}(t_i) \right)^2$$

where:

- → $F(x)$ refers objective function,
- → x is a vector containing the parameters to be optimized (k, P_0).

3.5. Analysis and Interpretation Metrics

The root mean squared error (RMSE) is utilised to determine the performance of the model:

$$\text{RMSE} = \sqrt{\frac{1}{N} \sum_i \left(P_{\text{model}}(t_i, x) - P_{\text{data}}(t_i) \right)^2}$$

where:

- N refers total data points.

3.6. Sensitivity Analysis

Sensitivity analysis is done to study the changes of parameters and its impact in the objective function:

$$S_\theta = \frac{\partial F}{\partial \theta} \cdot \frac{\theta}{F}$$

where:

- F is the objective function,
- θ is a model parameter.

4. Results and Discussion

The optimization of reaction rate and the monomer concentrations are carried out using the mathematical modelling developed in the copolymerization method. Levenberg–Marquardt algorithm is used to optimize the nonlinear conditions exist in the problem which is found to be effective and accurate.

4.1. Model Validation

The optimal parameters offer a very good fit to the experimental conditions, which is confirmed by the lesser RMSE measure. The model exactly estimates the concentration of the copolymer with respect to over time, indicating less deviations from the responses.

Rate Constant (k) [10]: The optimal rate constant is determined as 0.085 min^{-1}, representing a reasonable reaction rate which is appropriate for any given experimental setup [11].

Initial Monomer Concentrations ($[M]_0$): The optimization of initial concentration of the monomers is observed to be 0.9 mol/L, which indicates high reactivity and suitable reaction atmosphere.

4.2. Visual Comparison

The visual comparison of the actual experimental data and the model predictions exhibits clear correlation, which confirms the accuracy of the model. The concentration period of generation of copolymers are relatively close, represents the reliability of the model in handling the reaction kinetics.

4.3. Statistical Evaluation

The optimized model obtains the RMSE value of 0.012, that demonstrates a better fit for the assumed data. The lesser RMSE measure indicates the precision of the model to forecast the concentration of the copolymer and its robust nature to deal experimental variability.

4.4. Sensitivity Analysis

Sensitivity analysis was conducted to determine the influence of the key parameters on the objective function. The sensitivity coefficient for (S_θ) each parameter was calculated as follows:

$$S_\theta = \frac{\partial F}{\partial \theta} \cdot \frac{\theta}{F}$$

The sensitivity analysis results are displayed in Table 25.1.

Table 25.1: Sensitivity analysis results

Parameter	Optimized Value	Sensitivity Coefficient $(S_è)$
Rate Constant	0.085 min^{-1}	0.45
Initial Concentration $([A]_0)$	0.9 mol/L	0.30
Initial Concentration $([B]_0)$	0.9 mol/L	0.25

4.5. Interpretation of Sensitivity Analysis

Rate Constant (k): The sensitivity coefficient for the rate constant is 0.45, indicating that a 1% change in k improves a 0.45% change in the objective function. This suggests that the rate constant has a significant impact on the reaction kinetics and the copolymer formation.

Initial Concentration ($[A]_0$): The sensitivity coefficient for the initial concentration of polyethyleneglycoladipate is 0.30. This indicates that the concentration of monomer A plays a crucial role in the copolymerization process.

Initial Concentration ($[B]_0$): The sensitivity coefficient for the initial concentration of 1,12-dodecane adipate is 0.25, showing its importance in the reaction but to a slightly lesser extent compared to monomer A [13].

4.6. Viscosity and Molecular Weight

The viscosity of the synthesised copolymer Polyethyleneglycoladipate–co–1,12–Dodecane Adipate is 0.3454 dl/g

The Molecular weight of the monomers

Polyethylene glycol	400 g/mol
Adipic acid	146.14 g/mol
1,12–Dodecane diol	202.33 g/mol

4.7. Advantages and Contributions

The integration of mathematical modelling with experimental data in this study provides several advantages:

- **Enhanced Understanding:** The kinetic model offers a deeper understanding of the copolymerization process, revealing the dynamics of monomer interactions and copolymer formation.
- **Predictive Capability:** The model allows for accurate prediction of copolymer concentrations under different conditions, facilitating optimization and control of the synthesis process.
- **Optimization Framework:** The use of the Levenberg-Marquardt algorithm provides an effective framework for parameter optimization, ensuring the best fit to experimental data.
- **Practical Applications:** The insights gained from the model can be applied to optimize industrial processes, improve material properties, and develop new copolymers with tailored characteristics.

This work contributes significantly to the field of polymer synthesis by combining experimental techniques with advanced mathematical modelling. The developed model not only improves the understanding of the copolymerization kinetics but also offer a useful tool in the optimization of the synthesis. The successful integration of the Levenberg–Marquardt algorithm demonstrates its utility in solving complex optimization problems in polymer chemistry [14, 16].

5. Conclusion

The study presents a comprehensive examination of the synthesis and characterization of polyethyleneglycoladipate–co–1,12–dodecane adipate, leveraging mathematical modelling to explain the copolymerization kinetics. The integration of experimental data with a first-order kinetic model optimized using the Levenberg-Marquardt algorithm provided accurate predictions of copolymer formation. The optimized rate constant (k) was determined to be 0.085 min^{-1}, and the initial monomer concentrations ($[A]_0$) and ($[B]_0$) were both 0.9 mol/L. The RMSE value of 0.012 represents a strong fit providing an high degree of accuracy.

The sensitivity analysis revealed that the rate constant (k) had the highest sensitivity coefficient (0.45), followed by the initial concentrations of polyethyleneglycoladipate (0.30) and 1,12-dodecane adipate (0.25). This highlights the significant impact of these parameters on the copolymerization process.

The advantages of this approach include enhanced understanding of the reaction dynamics, predictive capability, and an effective optimization framework. The contributions of this work to the field of polymer synthesis are significant, offering a robust methodology for optimizing polymerization processes and paving the way for future advancements in polymer chemistry and materials science.

References

[1] Azim H, Dekhterman A, Jiang Z, Gross RA (2006) Candida antarctica lipase B-catalyzed synthesis of poly(butylene succinate) shorter chain building blocks also work, Biomacromolecules 7:3093–3097

[2] Berry JS, Houstan SK (1995) Mathematical modelling. J. W. Arrowsmith Ltd., Bristol

[3] Broyden CG (2000) On the discovery of the "good Broyden" method. Math Program Ser B 87:209–213

[4] Burdick JA, Mauck RL (2010) Springer wein and Newyork. Coculture of human mesenchymal stem cells and articular chondrocytes reduces hypertrophy and enhances functional properties of engineered cartilage. In: Biomaterials for tissue engineering applications. A review of the past and future trends, vol 98. pp 243–277

[5] Dong AJ, Zhang JW, Jiang K, Deng LD (2008) Characterization and in vitro degradation of poly (octadecanoic anhydride). J Mater Sci 19:39–44

[6] Jayachandran K, Cingaram R, Viswanathan J, Sagadevan S, Jayaraman V (2016) Investigations on preparation and characterization of certain copolyesters. Mater Res 19:394–400

[7] Jiang Y, Woortman AJ, Van Ekenstein GORA, Loos K (2013) Enzyme-catalyzed synthesis of unsaturated aliphatic polyesters based on green monomers from renewable resources. Biomolecules 3:461–480

[8] Levenberg K (1944) A method for the solution of certain non-linear problems in least squares. Q Appl Math 2:164–168

[9] Liang Y, Xiao Y, Zhai L, Xie Y, Deng C, Dong L (2013) Preparation and characterization of biodegradable poly (sebacic anhydride) chain extended by glycol as drug carrier. J Appl Polym Sci 127:3948–3953

[10] Lourakis MIA (2005) A brief description of the Levenberg-Marquardt algorithm implemented by levmar. Technical report, Institute of Computer Science, Foundation for Research and Technology, Hellas

[11] Nielson HB (1999) Damping parameter in Marquardt's method. Technical report IMM-REP1999-05, Department of Mathematical Modeling, Technical University Denmark

[12] Tan L, Chen Y, Zhou W, Nie H, Li F, He X (2010) Novel poly (butylene succinate–co-lactic acid) copolyesters: synthesis, crystallization, and enzymatic degradation. Polym Degrad Stab 95:1920–1927

[13] Transtrum MK, Machta BB, Sethna JP (2010) Why are nonlinear fits to data so challenging? Phys Rev Lett 104:060201

[14] Transtrum MK, Sethna JP (2012) Improvements to the Levenberg-Marquardt algorithm for nonlinear least-squares minimization. Preprint submitted to Journal of Computational Physics, January 30

[15] Umare SS, Chandure AS, Pandey RA (2007) Synthesis, characterization and biodegradable studies of 1,3-propanediol based polyesters. Polym Degrad Stab 92:464–479

[16] Wang F, Li W, Zhang H (2012) A new extended homotopy perturbation method for nonlinear differential equations. Math Comput Model 55(3–4):1471–1477. https://doi.org/10.1016/j.mcm.2011.10.029, 2-s2.0-84855192570

CHAPTER 26

Quantitative analysis and optimization of aloe emodin glycosides using HPLC and HPTLC techniques

Anand Singaravelu,[1,a] Mettilda Manuel Swami Dorai,[1] Kalaivani Arumugam[1]

[1]Department of Chemistry, Saveetha Engineering College, Thandalam, Chennai, Tamil Nadu, India
Email: [a]anands@saveetha.ac.in

Abstract

In analytical chemistry, the development of methods for the separation, identification, and purification of compounds is crucial for both qualitative and quantitative analysis. These methods must be highly sensitive, accurate, specific, and fast to ensure reliable results. This is particularly important in the context of drug products, where the efficacy and safety of the product can be significantly impacted by the precision and accuracy of the analytical methods used. HPTLC is one of the modern, advanced techniques that has gained prominence in recent years. In the present study, various chromatographic methods were developed and optimized to estimate aloe emodin glycosides (AEG) from the crude fraction of methanolic extract (CFME). The stock solutions of both CFME and the AEG standard, each prepared at a concentration of 5 mg/ml, were used for the validation of these methods. The validation process involved a comprehensive assessment of key analytical parameters, including linearity, range, precision, and accuracy, ensuring the reliability of the results obtained. The analysis revealed that the AEG content in CFME was approximately 6.808%. This finding underscores the effectiveness of the developed methods in accurately quantifying AEG, which is essential for further applications in pharmaceutical and therapeutic contexts.

Keywords: Aloe emodin glycoside, HPTLC, HPLC, drug discovery, *Cassia fistula* methanolic extract (CFME)

1. Introduction

In the previous couple of decades seasonal medicine has experienced explosive growth and gained popularity in developed countries due to its natural origin and without side effects [1]. Plants play a vital role and are an exemplary source of medicine, and much of the medicine available today is directly or indirectly derived from them, and not even one percent of the approximately 2,50,000 higher plants have been used in the study of phytochemistry or pharmacological properties. An ethnomedicinal approach to developing herbal medicines is both cost-effective and logical.

Most of the active ingredients are derived from edible plants which will be undoubtedly be valuable due to their hypoglycemic potential. Some of the plant molecules such as flavonoids, alkaloids, glycosides, etc., have been obtained from many plant sources reported to anti-diabetic activity [2]. Many vital medicine equivalent to Taxol, camptothecin, and metformin have been isolated from plant sources [3].

Some of the plant compounds in therapeutic research continue to provide many active structures, which have developed the new drugs. Many medicinal plants have been used in Siddha medicines, but only 10–15% of the plant diversity has been used for medicinal purposes. Nearly 29 plant-based drugs have been approved during 2001–2008 which also include some of the novel molecule-based drugs like Nitisinone (Orfadin), Miglustat (Zavesca) and Galanthamine [4].

To investigate different extracts from *Cassia fistula* using various chromatography techniques and validated the compound by

DOI: 10.1201/9781003606611-26

HPTLC method [5]. *Cassia fistula* has astringent and laxative properties which is widely used in ancient remedies and Ayurvedic medicine for many years. This plant has been widely used to relieve and treat burns, cancer, constipation, convulsions, diarrhea and glandular tumors.

Cassia fistula is Caesalpinaceae family and cultivated throughout India and is commonly called Sarakonrai in Tamil. The whole part of the plant is used as medicinal properties for the remedy of inflammatory diseases [6, 7].

2. Materials and Methods

2.1. Chemicals

HPLC grade solvents, silica gel for column purification and thin layer chromatography plate were purchased from SISCO Laboratories, India. The mesh size 60–120 pre-coated TLC plates was bought from MERCK.

2.2. Extraction of *Cassia fistula* Using Different Organic Solvents

The flowers of *Cassia fistula* were selected for studying their anti-diabetic activity based on literature. The flower was shade dried and powdered. 100 g of dried, pulverized flower of *Cassia fistula* was extracted by sequential solvent extraction and repeated three times with the respective solvents and concentrated in rotary evaporator at the temperature of 45–50 °C in order to avoid the evaporation of plant products. All the extracts were carried out by thin layer chromatography to find active compounds for the different batches of the plant products. These extracts were again constituted with DMSO (1 mg/ml) and successively diluted by the ranges from 1 μg/ml to 1 pg/ml which leads to understand the screening method for selection of active extract [8].

2.3. Analysis of Thin Layer Chromatography

TLC is one the method used for *Cassia fistula* to perform solvents extraction. Various solvent system used, (a) 9% of chloroform 1% of methanol (b) 10% of ethyl acetate: 1% acetic acid: 1% formic acid and 2.5% of water to find a better solvent system to separate compounds [9].

2.4. Picturization of Molecules in TLC Plate

TLC plates were visualized under UV light to separate the extracts with two different spray reagents (a) Sulfuric acid methanol spray reagent (SAM)-10% H_2SO_4 in MeOH. (b) Ammonium molybdic acid reagent. Thus, the TLC plates were developed and documented.

2.5. Structural Elucidation by Bio-activity Guided Fractionation

Initially, 5 g of CFME was taken, fractionated between solvent and solvent, and dissolved in 60% methanol in water. CFME soluble compound in methanol/water was separated with increasing solvent such as hexane, chloroform and butanol to give in soluble water (0.7 g), soluble hexane (0.3 g), soluble chloroform (0.29 g), soluble butanol (1.72 g) and soluble hydroalcoholic fractions (3.05 g). For the above fractions was subjected to glucose uptake assay as primary selection. For the selection of bioactivity butanol fraction (2 g) was selected and subjected to methanol washing that resulted in a yellow colored amorphous powder (21.2 mg). All the fractions and the compound obtained were analyzed using thin layer chromatography and visualized by short and long UV and developed by spray reagents like sulfuric acid methanol and phosphomolybdic acid. The purity of the bioactive molecule was estimated using HPLC. The bioactive molecule was determined by 1H, 13C. Further, Mass spectra was confirmed, basic composition of the compound. The resultant data identified the active molecule as Aloe emodin glycosides (AEG).

2.6. Qualitative Analysis of CFME Using HPTLC Method

HPTLC was performed according to an earlier method [10]. 5 mg/ml of CFME as stock solution and 5 mg/ml of standard AEG was selected for TLC for quantification of AEG in CFME. The conditions were applied to chromatography using silica gel as the solid phase and a solvent such as 70% of butanol, 20% of acetic acid and 10% of water is used as mobile phase. Applicator scanner and 275 nm wavelength as scanning detection for LOD and LOQ were determined by purified AEG. The total amount of AEG from CFME were calculated, quantified and scanned by 275 nm.

3. Results

3.1. Extraction of Plant

Cassia fistula was successively extracted by increasing polarity of solvents. Extracts were reduced by rotary evaporator. Figure 26.1 shows that scheme of extraction followed by fractionation steps used in this study.

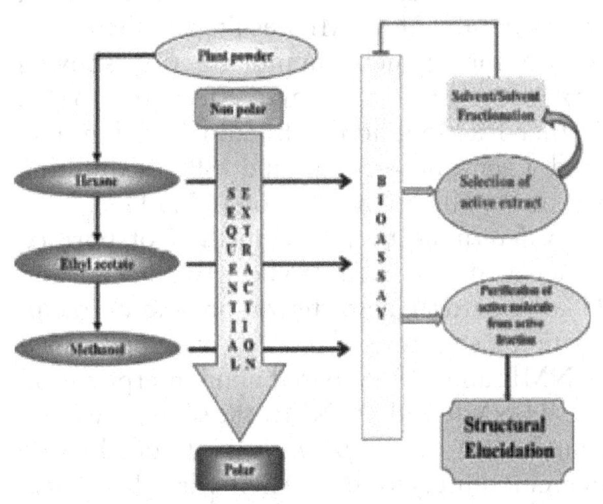

Figure 26.1: Sequential extraction of *Cassia fistula* from non-polar to polar. Each of these extracts was dried under reduced pressure in a rota-evaporator

3.2. TLC Finger Print

TLC was performed for all crude extracts of *Cassia fistula* to analyze the various compounds present in *Cassia fistula*. The solvent system was used for crude extracts such as 9% of Chloroform and 1% of methanol was found to be the suitable method. Separated compounds were viewed under short and long UV and after spraying of ammonium molybdate and sulfuric acid reagent. (Figure 26.2).

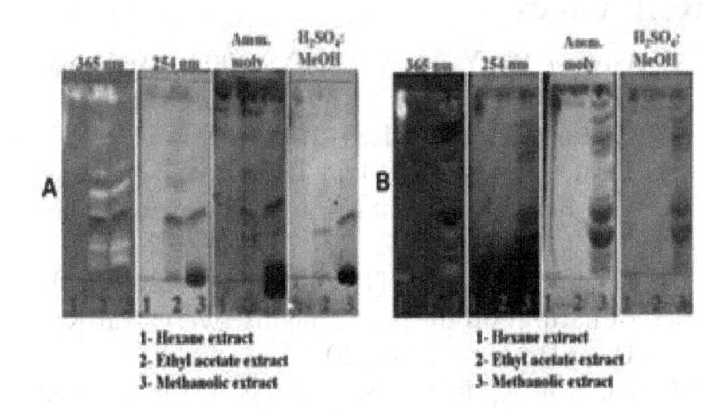

Figure 26.2: TLC with chloroform: methanol (90: 10)

3.3. Solvent-Solvent Fractionation

Fractionation of CFME was performed which yielded 5 fractions namely insoluble water-(F1), soluble-hexane-(F2), soluble-chloroform (F3), soluble-butanol (F4). and soluble in water (F5). The butanol soluble fraction exhibited maximum glucose uptake activity (data not shown). Hence, butanol soluble fraction was further fractionated to column chromatography for separation of bioactive molecule (Figure 26.3a–d).

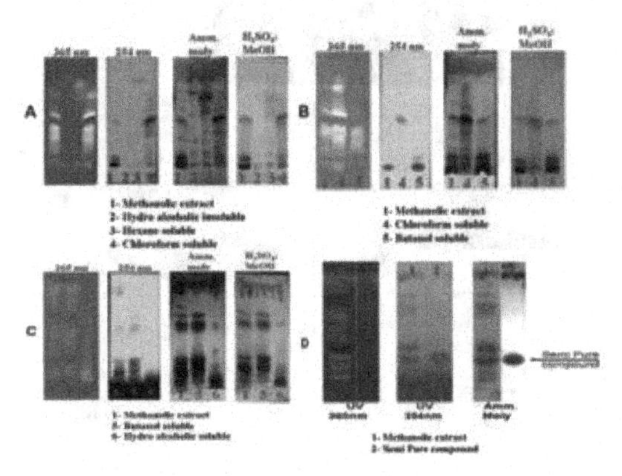

Figure 26.3: (a) TLC of CFME fractions with Chloroform: Methanol (90: 10)

3.4. Compound Characterisation Studies of a Semi Pure Compound

The semi pure molecule isolated from butanol fraction, was characterized using proton and carbon (NMR) and Mass spectroscopy. Based on the spectroscopic studies molecules identified as Aloe emodin glycosides ($C_{21}H_{22}O_{10}$) and molecular weight of 434 g/mol [10] (Figure 26.4).

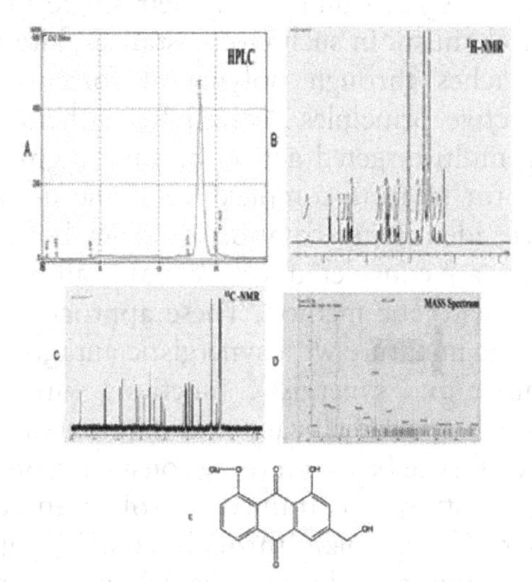

Figure 26.4: Characterisation of *Cassia fistula* by HPLC, NMR, MASS spectrum and active compound

3.5. Qualitative Analysis Using HPTLC

CFME and AEG TLC fingerprints in 10% of ethyl acetate, 1% of acetic acid, 1% of formic acid and 2.5% of water at long and short UV. Different concentration of CFME and AEG performed with UV spectrum followed by curve of linearity ranges from 200 to 1000 ng (Figure 26.5).

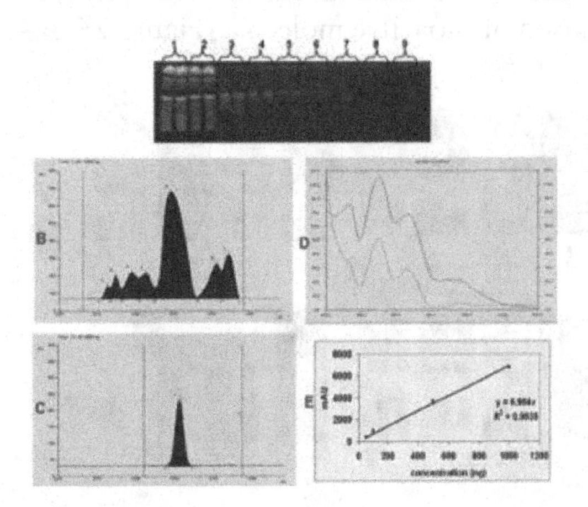

Figure 26.5: (a) TLC fingerprinting of CFME and AEG in ethyl acetate: acetic acid: formic acid: water (100: 11: 11: 25) with 100 μg and 50 μg of CFME and AEG with concentration of 50 ng to 10 ng

4. Discussion

Despite significant progress in drug development, diabetes is increasing worldwide. Although several drugs are available on the market for the treatment of hyperglycemic diseases, their usefulness has been limited by their single targeted mechanism. In such circumstances, alternate approaches through polyherbal formulations and active principles derived from them having a multi-targeted action become highly relevant for treating complex metabolic diseases. Hence, identifying compounds from botanical sources are considered to be a viable alternative to the synthetic method. These approaches are dynamic in nature with synergistic antagonistic, agonistic and synergistic functions with optimum therapeutic efficacy and minimal adverse effects. Hence based on these considerations the research study was framed to isolate an active principle from a plant through bioactivity guided screening and elucidate its role on a battery of targets involved in the insulin signaling pathway to delineate its multi-target of action.

Drugs isolated from plants are often considered less toxic and free of side effects. Based on the literature, *Cassia fistula* were selected and subjected to glucose uptake as a primary screening [11]. *Cassia fistula* was successively extracted by increasing polarity of solvents. Extracts were reduced by rotary evaporator.

CFME was subjected to bioactivity guided purification after solvent-solvent fractionation. Fractionation of CFME resulted in five fractions among which butanol fraction showed maximum glucose uptake effect of 153%. Further fractionation of butanol fraction, isolated a semi pure compound (SPC) exhibiting maximum glucose activity of 165% [12].

To determine by HPLC method of purified bioactive SPC showed it to be between 93% and 95%. Structural characterization studies to elucidate the structure of the SPC were performed by NMR and MASS. Based upon interpretation of a ^{1}H NMR and ^{13}C NMR the semi pure compound from *Cassia fistula* was decoded to be aloe emodin glycosides (AEG) with the molecular formula $C_{21}H_{22}O_{10}$ and molecular weight 434 g/mol [13]. A simple and robust quantitative HPTLC method developed to estimate the net content of AEG in the methanolic extract of *Cassia fistula* [14].

5. Conclusion

HPTLC is a widely used analytical method in various pharmaceutical industry, food and drug analysis. Aloe emodin glycosides were isolated from CFME using bioactivity based chromatographic separation and purification steps. The total content of AEG was found to be 6.808% w/w.

References

[1] Grover JK, Yadav S, Vats V (2002) Medicinal plants of India with anti-diabetic potential. J Ethnopharmacol 81:81–100

[2] Jarald E, Joshi SB, Jain DC (2008) Diabetes and herbal medicines. Iran J Pharmacol Ther 7:97–106

[3] Koehn FE, Carter GT (2005) The evolving role of natural products in drug discovery. Nat Rev Drug Discov 4:206–220

[4] Saklani A, Kutty SK (2008) Plant derived compounds in clinical trials. Drug Discov Today 13:161–171

[5] ICH Harmonized Tripartite Guideline (2008) Validation of analytical procedures. Text and methodology Q2 (R1). International Conference on Harmonization, Geneva

[6] Kirtikar KR, Basu BA (1991) Indian medicinal plants, vol II, 2nd edn. Periodical Experts Book Agency, New Delhi, pp 856–860

[7] Kuo YH, Lee PH, Wein YS (2002) Four new compounds from the seeds of *Cassia fistula*. J Nat Prod 65:1165–1167

[8] Anand S, Muthusamy VS, Sujatha S, Sangeetha KN, Bharthi Raja R, Sudhagar S, Lakshmi BS (2010) Aloe emodin glycosides stimulate glucose transport and glycogen storage through PI3K dependent mechanism in L6 myotubes and inhibit adipocyte differentiation in 3T3L1 adipocytes. FEBS Lett 584:3170–3178

[9] Prasad D, Singh AK, Solanki S (2020) Antioxidant determination and thin layer chromatography of extract *Withania somnifera*, *Terminalia arjuna*, *Bacopa monnieri*, *Ranunculus sceleratus* and *Acalypha indica*. Eur J Mol Clin Med 7:11

[10] Saravanababu C, Sunil AG, Vasanthi HR, Muthusamy VS, Ramanathan M (2007) Development and validation of an HPTLC method for simultaneous estimation of excitatory neurotransmitters in rat brain. J Liq Chromatogr Relat Technol 30(19):2891–2902

[11] Nile SH, Park SW (2014) HPTLC analysis antioxidant and antigout activity of Indian plants. Iran J Pharm Res 3(2):531–539

[12] Anand S, Raja RB, Lakshmi BS (2015) Enhancement of glucose uptake by *Ficus hispida* methanolic extract through phosphatidylinositol 3 kinase and inhibiting aldose reductase – an in vitro analysis. Eur J Integr Med 7:164–171

[13] Coopoosamy RM, Magwa ML (2006) Antibacterial activity of aloe emodin and aloin A isolated from *Aloe excels*. Afr J Biotechnol 5(11):1092–1094

[14] Jayachandran Nair CV, Ahamad S, Khan W, Anjum V, Mathur R (2017) Development and validation of high-performance thin-layer chromatography method for simultaneous determination of polyphenolic compounds in medicinal plants. Pharmacognosy Res 9(Suppl 1):S67–S73

CHAPTER 27

Design and analysis of vehicle suspension A-arm by generative design method

A. Thamarai Selvan[1,a], M. P. Saravanan[1], T. Dharshan Roy[1], V. Nithishnathan[1] and S. Shivaram[1]

[1]Department of Mechanical Engineering, Saveetha Engineering College, Chennai, Tamil Nadu, India

Email: [a]thamsmechian@gmail.com

Abstract

A design and analysis of mechanical properties such as yield strength, elongation and factor of safety of a vehicle suspension A-arm is carried out under static conditions. This research work presents a comparative static analysis of A-arm by using generative design and adaptive design. The A-arm is modelled by using Autodesk FUSION 360 tool for the structural analysis. The material used for analysis of the suspension A-arm are AISI 1018 106 HR, AISI 304 Steels and Aluminium 2014 T6. The simulated results of adaptive design and generative design are presented and compared. The results show that Aluminium 2014 T6 has more yield strength than AISI 1018. Stainless steel AISI 304 material performs better than other two materials in dynamic analysis by using both adaptive and generative methods

Keywords: Suspension A-arm, fusion 360, structural analysis, dynamic event simulation, generative design

1. Introduction

In any automobile, a well-designed suspension system is always responsible for its safety and comfort of driving. A suspension system is responsible for stability of the vehicle as it carries entire weight and all the forces transmitted by various moving parts, and traction on the road. It absorbs the vibrations produced due to rough terrains under dynamic conditions and provides stability irrespective of different conditions like acceleration, uneven surface, braking, loading, and unloading. A control arm, otherwise called as a wishbone plays an important role in a suspension system as it connects the steering knuckle to vehicle frame. In automobile manufacturing industry, various tools are used to optimize its weight and improve the performance of modern vehicles. However, static loads acting on automobile parts are so complex under transient conditions [1], especially on the lower arm of front suspension.

More researches were carried out on control arm of a suspension system in the recent years. A static structural analysis [2] was performed on a suspension arm using Finite Element Method [3–4] to find out the stress, deformations and safety factor of the component. The reason for failure of a lower wishbone on a light vehicle was investigated using finite element technique [5]. This optimization technique was carried out to enhance its structural strength. A modal and statical analyses of upper arm, lower arm and steering knuckle on a double wishbone was performed using Hyper mesh, Unigraphics, encode and Opti struct. The reliability and stress distribution under service loading were found by Finite element method. From metallographic study and hardness evaluation, it was found that the failure had occurred in high stress region of seam weldment.

A kriging interpolation technique [6] was adopted to optimize the shape and weight of an upper arm with static strength constraints. The experiments done with 1/4th weight of the vehicle ensures the correlation between its strength and durability. A stochastic approach was also performed in optimizing the strength

DOI: 10.1201/9781003606611-27

of a suspension A-arm A fatigue analysis [7] on upper arm of a wishbone under dynamic conditions estimated its life time. The cost of the arm was significantly reduced with improved product reliability. The progressive and linear rate of suspension was analysed by using Automated Dynamic Analysis of Mechanical Systems (ADAMS) tool where the results of multibody simulation were compared for statistical significance [8]. Online software vehicle suspension software (VSUSP) was used to design the suspension geometry [9] for the wishbone comparing the steel and aluminium materials. The aim was to reduce the weight of the component with optimized geometry for the same factor of safety. A lower arm suspension made of Carbon reinforced composites was analysed by using computer-aided engineering (CAE) software (ABAQUS) where 50% of weight was reduced with the proposed material [10].

All these above adaptive design [1–10, 12–13] produces better results in designing the wishbone, but the modelling and analysing times are more certainly for complex structures. The generative design uses artificial intelligence algorithms like evolutionary algorithm, neural networks [11] etc., and optimizes the design by reducing its production time. Generative design is an iterative method that Fusion 360-degree software adopts AI and cloud computing to enhance the optimized results.

2. Materials & Methods

In this research, three different designs were generated by using the materials Aluminium 2014 T6, Stainless Steel AISI 304 and AISI 1018. After performing simulation for all the three materials, the results were compared and interpreted. The most commonly used material for A-arm is AISI 1018 106 HR steel, which is cost efficient and good for fabrication. The material properties of AISI 1018 106 HR, AISI 304 Steels and Aluminium 2014 T6 are shown in Table 27.1. Reverse engineering was performed to design the lower suspension arm with the help of measuring tools like vernier gauge, radius gauge and divider etc. From measured dimensions, 3D model of lower suspension arm was created using FUSION 360 software. Generative design was carried out to analysis its performance with the nature of materials used, and manufacturing methods. After modelling the lower suspension arm, the 3D MODEL was imported to CAD software.

The load on lower suspension arm of one wheel is calculated with a total mass of 267 kg and constant speed of 80 km/h. The weight is found to be G = mass * acceleration due to gravity = 2620 N. The loads on front and rear axles are calculated as GFA = 1310 N and GRA = 1430.52 N. Loads on the front and rear wheels are found as G FAW = 655 N and G RAW = 715.26 N. From the above results of 655 N and 715.26 N, consider load on lower suspension arm 1000 N. For safety factor and dynamic load factor, a load of 1800 N acting on front lower suspension arm is considered.

2.1 Adaptive Design

Adaptive design is the process of redesigning the existing design in which the parameters of the existing design or machine or device is modified to improve quality, reduction in cost and increased safety. Based on the proposed dimensions, CAD model (Figure 27.1) was created using Autodesk Fusion 360 design workspace. While coming to the numerical simulation part,

Table 27.1 Material properties of AISI 1018 106 HR steel and aluminium 2014 T6

Properties	**AISI 1018 106 HR Stainless Steel**	**Aluminium 2014 T6**	**AISI 304 Stainless Steel**
Density	7870 Kg/m^3	2789 Kg/m^3	8000 Kg/m^3
Young's Modulus	207000 MPa	72400 MPa	195000 MPa
Poisson's ratio	0.33	0.33	0.29
Yield Strength	250 MPa	399.9 MPa	215 MPa
Ultimate Tensile Strength	354 MPa	455.1 MPa	505 MPa
Thermal Conductivity	51.9 W/m ^{0}C	154 W/m ^{0}C	16.2 W/m ^{0}C
Thermal Expansion Coefficient	0.0000115 / ^{0}C	2.3E-05 /^{0}C	1.73E-05 per ^{0}C
Specific Heat	486 J/Kg C	897 J/Kg ^{0}C	500 J/ Kg ^{0}C

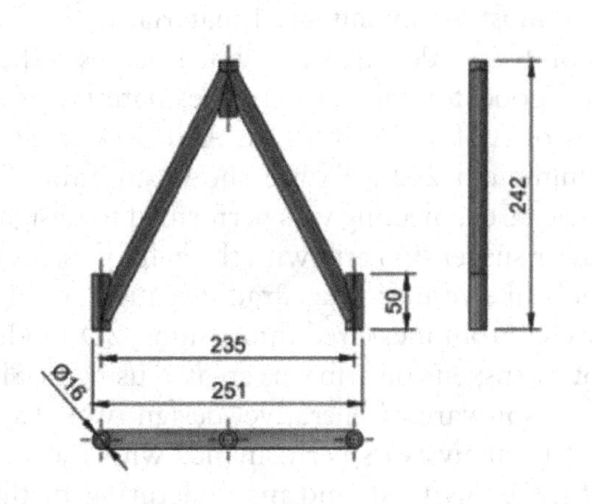

Figure 27.1 2D drawing of proposed A-arm

finite element method is quite familiar in structural analysis. In this study the adaptive design simulation is performed using Autodesk fusion 360 version 2.0.14337 simulation workspace.

This study analyses the deformation, factor of safety and stress in the model by strictly following the material properties, meshing criteria, structural loads and constraints which is given as input to simulation. In this simulation, physical material properties and engineering properties were applied to components and bodies in a Fusion 360 design. Since the structure is fabricated through welding process, there is no relative movement allowed throughout the analysis. Therefore, the contact between two bodies is considered as bonded joints which is more similar to welding in action.

2.2 Generative Design

In generative design, CAD model is imported into generative design workspace and input parameters are given step by step to generate new design. For generative design, manufacturing constrains need to be specified clearly. In this work additive manufacturing method is adopted for deigning the A-arm. The following inputs are required to initiate the design.

1. Orientation of design: include all six directions (+X, +Y, +Z, –X, –Y, –Z)
2. Overhang angle: 45°
3. Minimum thickness: 3 mm.

3. Results and Discussions

The typical vehicle parameters considered for static analysis for loads in front and rear axles are GFA = 1310 N and GRA = 1430.52 N. The loads in front suspension arm is 1800 N. The existing material used for suspension A-arm is AISI 1018 106 HR steel. In adaptive design, the factor of safety, von mises stress distribution and displacements under static analysis of Steel AISI 1018 106 HR are shown in Figure 27.2a–c. The minimum and maximum values of factor of safety, von mises and total stresses are found to be 10.71 & 15, 0.09522 MPa & 23.34 MPa and 0 & 0.007808 mm respectively where the static analysis of the same material by Analyzing Software for Structural and Fluid Dynamics Applications (ANSYS). [1] produced stress and displacement of 173.4 MPa and 3.423 mm and the stress generated for short glass fibres with polypropylene [2] under given loading condition is around 48.54 MPa

Table 27.2 shows the results of static analysis of AISI 1018 106 HR & AISI 304 stainless steel and Aluminium 2014 T6 which gives stress, displacement, reactions forces at the constraints, strain, contact forces and contact pressures.

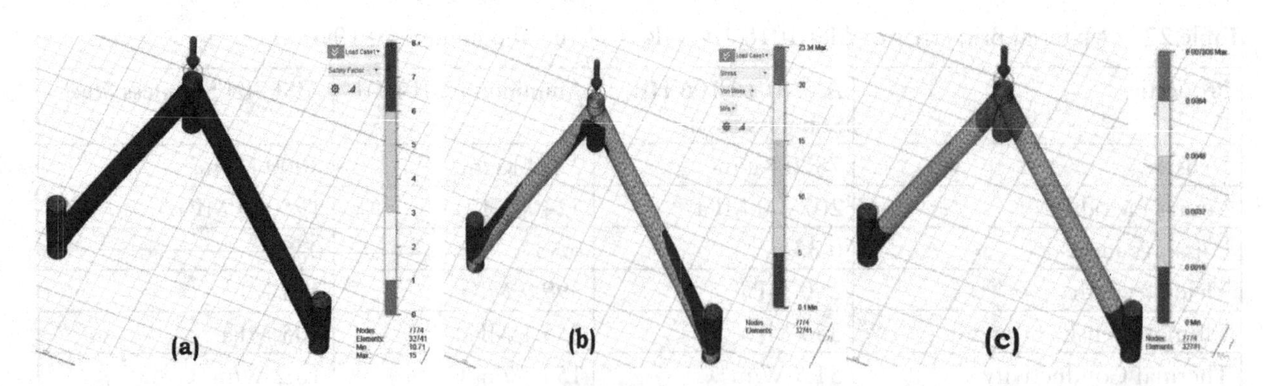

Figure 27.2: (a) Factor of safety after simulation (b) Von-misses stress distribution in A-arm and (c) displacement after simulation

Table 27. **2:** Static analysis results of stainless steel and aluminium alloys

S. No.	Parameters	Steel AISI 1018 106 HR	Aluminium 2014 T6	Stainless Steel AISI 304
1.	Max. Von Mises Stress (MPa)	23.34	10.36	9.445972
2.	Max. Displacement Global	0.0078	0.018	0.0080
3.	Min. 'Factor of Safety" (FOS)	10.71	27.489	22.76102
4.	Mass (Kg)	3.52	0.534	0.939666
5.	Stiffness (N/m)	2.3×10^8	5.56×10^8	2.25×10^8

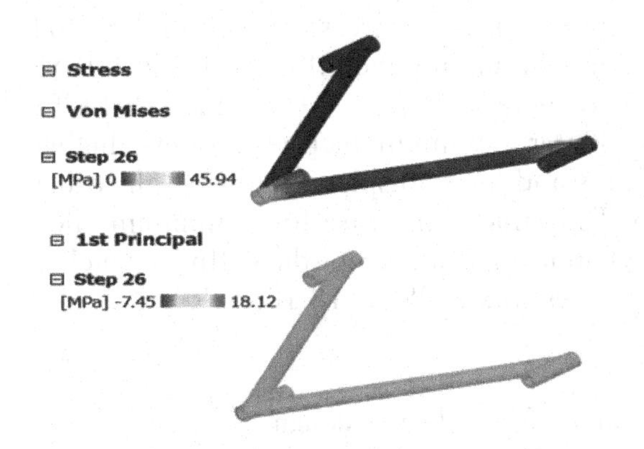

Figure 27. **3:** Von Mises and first principal stresses in the analysis

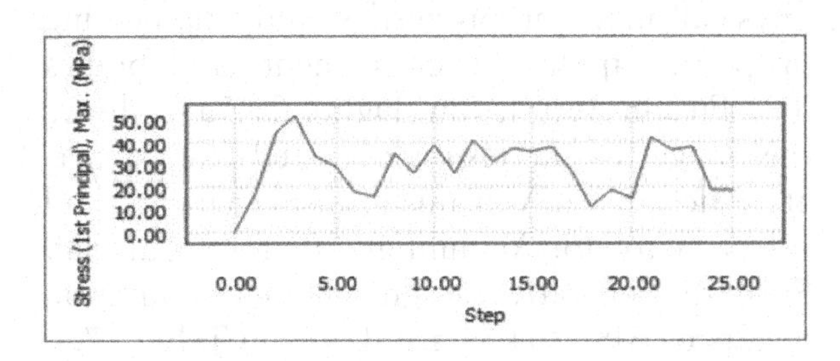

Figure 27. **4:** Stress versus step graph of von mises stress

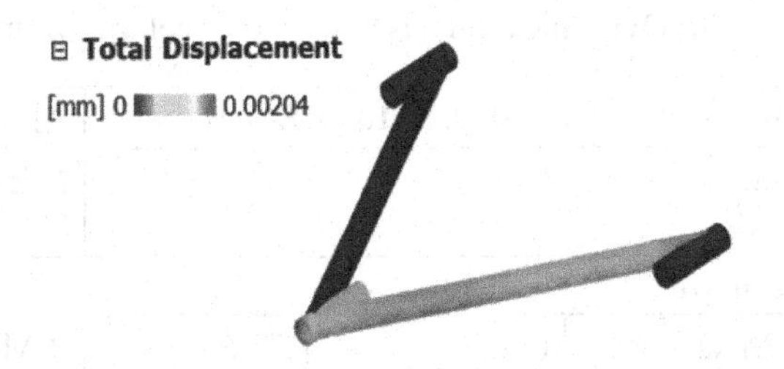

Figure 27. **5:** Image obtained for displacement after dynamic analysis

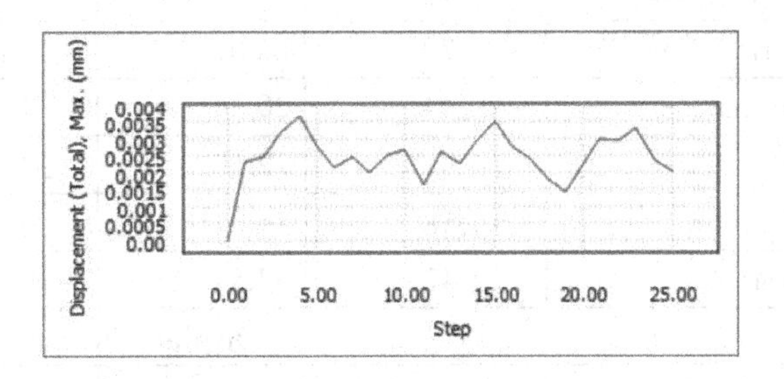

Figure 27. **6:** Displacement versus step graph of A-arm

3.1. Dynamic Simulation by Adaptive Design

In this dynamic simulation, the physical material properties and engineering properties were applied to components and bodies in a Fusion 360 design. Contact between two bodies are perfectly bonded together where they meet, since the structure is fabricated through welding process there is no relative movement allowed throughout the analysis. Therefore, the contact between two bodies is considered as bonded joints which is more similar to welding in action. In this study linear tetrahedral element with four triangular faces, six edges, and four nodes was preferred to perform simulation.

1) Curved mesh elements and 2) Aspect ratio

Figure 27.3 shows the images obtained after dynamic analysing for von mises and first principal stresses. The stress variations against number of steps is shown in Figure 27.4. The minimum and maximum von mises and first principle stress were found to be 0 and 45.94 MPa & -7.45 and 18.12 MPa respectively after 26 steps in the simulation. Figures 5 and 6 show the images obtained for displacement and its corresponding values on each steps after dynamic analysis.

The minimum and maximum displacements after 26 steps were obtained as 0 and 0.00204 mm. Table 27.3 shows dynamic analysis simulation results of A-arm made by AISI 1018 106 HR Steel, Aluminium 2014 T6 and AISI 304 Steel alloy under adaptive design.

3.2 Analysis of Aluminium 2014 T6 by Generative Design

Material selection is one of the important steps in generative design, since the target is to achieve maximum stiffness and mass reduction in A-arm so that new design is much more reliable and stronger than adaptive design. With the above considerations and type of manufacturing method, Aluminium 2014 T6 and AISI 304 Stainless steel were selected for analysing and simulation under generative design. In this simulation, physical material properties and engineering properties applied to components and bodies in a Fusion 360 design. Figure 27.7a–c shows the images of the results obtained from generative design for Von mises, displacements and safety factor for Aluminium 2014 T6 material. From the generative design, dynamic simulation results of AISI 304 SS are shown in Table 27.4.

The dynamic analysis shows the results for von mises stress and displacements values as 0 & 77.69 MPa and 0 & 0.003712 mm respectively.

The maximum velocity and acceleration are found to be 0.6787 m/s and 3.47E+06 m/s^2 respectively. For generative design, Aluminium 2014 T6 and AISI 304 Stainless Steel are used for static and dynamic analysis of suspension A-Arm. The comparative dynamic analysis results under generative design of stainless steel and aluminium alloys are shown in Table 27.5. From the above comparisons, it is clear that the maximum von mises stress is lower for AISI 304 SS than Aluminium and AISI 1081 106 HR SS in static analysis whereas Aluminium 2014 T6 has greater minimum factor of safety, higher stiffness and lower displacement than the other two. The same is the case for aluminium alloy in dynamic analysis except the stiffness which is slightly less than AISI 1018 106 HR SS.

Table 27.3: Dynamic analysis results of steel and aluminium alloys from adaptive design

Parameters	AISI 1018 106 HR steel		Aluminium 2014 T6		AISI 304 Steel alloy	
Minimum Maximum			Minimum	Maximum	Minimum	Maximum
Stress in MPa						
Von Mises	0	77.69	0 MPa	91.3	0	71.44
1st Principal	−15.23	51.94	−21.18	89.28	−14.5	73.92
3rd Principal	−73.31	8.306	−110.5	16.56	−80.85	11.45
Normal XX	−29.98	46.47	−63.11	56.23	−73.4	56.54
Normal YY	−20.79	26.02	−33.84	27.22	−21.69	18.21
Normal ZZ	−66.44	20.66	−72.87	55.96	−53.41	43.94
Shear XY	−16.9	17.58	−24.56	21.45	−15.54	14.26
Shear YZ	−14.58	14.22	−24.51	19.88	−15.75	18.17
Shear ZX	−32.42	33.59	−36.51	42.9	−35.3	32.5
Displacement in milli meter(mm)						
Total	0	0.003712	0	0.04006	0	0.01601
X	−5.787E-04	9.211E-04	−0.01645	0.01699	−0.005961	0.006066
Y	−4.88E-04	4.55E-04	−0.01707	0.001781	−0.009703	0.000368
Z	−0.003711	0.001293	−0.03642	0.007256	−0.01277	0.00265
Reaction Force in newton (N)						
Total	0	600	0	93.36	0	101.2
X	−331.5	378.4	−56.28	51.06	−67.79	65.73
Y	−424.3	356.7	−22.06	16.97	−14.75	15.61
Z	−24.38	200	−22.57	88.47	−26.57	79.62

Strain						
Equivalent	0	3.413E-04	0	0.001156	0	3.281E-04
1st Principal	−1.198E-05	2.871E-04	−6.94E-07	0.001065	−8.173E-07	3.336E-04
3rd Principal	−3.707E-04	4.843E-06	−0.001332	4.171E-10	−3.791E-07	5.079E-06
Normal XX	−1.309E-04	2.638E-04	−8.278E-04	5.39E-04	−3.731E-04	2.101E-04
Normal YY	−9.444E-05	1.116E-04	−3.055E-04	4.802E-04	−7.433E-05	1.348E-04
Normal ZZ	−3.042E-04	1.005E-04	−6.937E-04	4.589E-04	−2.038E-04	1.586E-04
Shear XY	−1.09E-04	1.134E-04	−4.469E-04	3.903E-04	−1.038E-04	9.523E-05
Shear YZ	−9.404E-05	9.171E-05	−4.461E-04	3.618E-04	−1.053E-04	1.214E-04
Shear ZX	−2.091E-04	2.167E-04	−6.645E-04	7.807E-04	−2.358E-04	2.171E-04
Plastic Strain	0	0	0	0	0	0
Acceleration in ms^{-2}						
Total	0	3.47E+06	0 m/s^2	1.58E+07	0	3.894E+06
X	−426807	332454	−1.408E+06	1.234E+06	−344753	204020
Y	−218462	354319	−1.111E+06	1.226E+06	−402261	211855
Z	−3.47E+06	446814	−1.582E+07	1.032E+06	−3.894E+06	505898
Velocity in ms^{-1}						
Total	0	0.6787	0	1.063	0	0.3679
X	−0.5094	0.3167	−0.6282	0.6095	−0.21	0.1999
Y	−0.1143	0.08946	−0.8128	0.6619	−0.2151	0.1948
Z	−0.6787	0.5659	−0.8597	0.9271	−0.301	0.3387

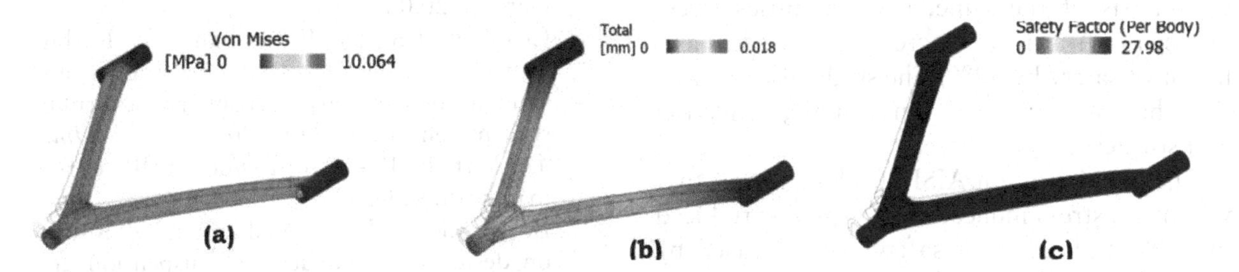

Figure 27.7: Images obtained from generative design of aluminium 2014 T6 A-arm for (**a**) Von mises stress (**b**) displacement (**c**) safety factor

Table 27.4: Properties of AISI 304 stainless steel obtained from generative design

Result No.	Recommend ation %	Manufacturing Method	Volume (mm^3)	Mass (kg)	Max Von Mises Stress (MPa)	F.O.S Limit	Min F.O.S	Max. Displacement Global (mm)
1	**95.7405866**	**Additive**	**117458.3**	0.939666	9.445972	2	22.76102	0.008064375
2	89.51533372	Additive	117738.6	0.941909	8.857858	2	24.27223	0.009590805
3	95.70383203	Additive	117251.6	0.938013	8.369659	2	25.68802	0.008129232
4	95.72614823	Additive	117421	0.939368	9.352289	2	22.98902	0.008081733
5	91.18082803	Additive	117646.2	0.941169	10.5073	2	20.46197	0.009367364
6	95.69354504	Additive	117265.1	0.938121	8.944642	2	24.03674	0.008133141

Table 27.5: Dynamic analysis results of stainless steels AISI 1080 & 304 and aluminium 2014 alloy

S. No.	Parameters	Steel AISI 1018 106 HR	Aluminium 2014 T6	Stainless Steel AISI 304
1.	Max. Von Mises Stress (MPa)	77.69	91.3	71.44
2.	Max. Displacement Global (mm)	0.003712	0.04006	0.01601
3.	Min. FOS	10.71	27.489	22.76102
4.	Mass (kg)	3.52	0.534	0.939666
5.	Stiffness (N/m)	4.84 x 108	4.49 x 108	1.12 x 108

4. Conclusion

From the results of dynamic analysis of AISI 1018 106 SS, Aluminium 2014 T6 and AISI 304 SS under adaptive design, the von mises stress, stiffness and safety factors were better for the latter two materials than the first one. Hence, those two materials were considered for dynamic analysis of suspension A-arm by generative design. From the static analysis of Aluminium 2014 T6, it was observed that the stress induced in A-arm was reduced by 44% and factor of safety got increased by 39%. In case of dynamic analysis of the same, the von mises stress induced in A-arm was increased by 17.5% and factor of safety by 39%. The study also showed that there is more variation in mass, deflection and stiffness.

The simulation on AISI 304 SS showed that von mises stress induced in A-arm was reduced by 40% and factor of safety got increased by 47% under adaptive design. The study also showed that there is no much variation in mass, deflection and stiffness. In generative design model, it was observed that the stress induced in A-arm was reduced only by 8% and factor of safety got increased by 47%. It also showed that there is no much variation in mass, deflection and stiffness. The structural advantage of A-arm was enhanced by adopting generative design.

Thus Aluminium 2014 T6 was considered as the best material in manufacturing vehicle suspension A-arm whose required mass was also less than 1 kg producing better results dominating the other two materials in generative design method.

References

[1] Pachapuri, M. Sadiq A., Ravi G. Lingannavar, Nagaraj K. Kelageri, Kritesh K. Phadate. 2021. Design and analysis of lower control arm of suspension system. *Materials Today: Proceedings* 47, Part 11: 2949–2956. https://doi.org/10.1016/j.matpr.2021.05.035.

[2] Anandakumar, P., Mallina Venkata Timmaraju, and R. Velmurugan. 2021. Development of efficient short/continuous fiber thermoplastic composite automobile suspension upper control arm. *Materials Today: Proceedings* 39, Part 4: 1187–1191. https://doi.org/10.1016/j.matpr.2020.03.543.

[3] Chauhan, Parnika, Katya Sah, and Rashmi Kaushal. 2021. Design, modeling and simulation of suspension geometry for formula student vehicles. *Materials Today: Proceedings* 43, Part 1: 17–27. https://doi.org/10.1016/j.matpr.2020.11.200.

[4] Darge, S.K., and S.S.S. Kulkarni. 2014. Review on design and evaluation of suspension arm for an automobile using FEM. International Journal of Advanced Engineering Research and *Studies*.

[5] Vivekanandan, N., Abhilash Gunaki, Chinmaya Acharya, Savio Gilbert, and Rushikesh Bodak. 2014. Design, analysis and simulation of double wishbone suspension system. *International Journal of Mechanical Engineering (IIJME)* 2 (6): 7.

[6] Khairolazar, Mohd Khairil Azirul Bin. 2009. *Robust design of lower arm suspension using stochastic optimization method.* University Malaysia Pahang. http://umpir.ump.edu.my/id/eprint/812

[7] Kim, J., S.-J. Kang, and B.-S. Kang. 2002. Computational approach to analysis and design

of hydroforming process for an automobile lower arm. *Computers & Structures* 80 (14–15): 1295–1304.

[8] Willy Prastiyo, Wiesław Fiebig, Multibody simulation and statistical comparison of the linear and progressive rate double wishbone suspension dynamical behavior, Simulation Modelling Practice and Theory, Volume 108, April 2021, 102273, https://doi.org/10.1016/j.simpat.2021.102273

[9] Sindhwani, Rahul, Ayan Bhatnagar, Abhi Soni, Ayushman Sisodia, Punj Lata Singh, Vipin Kaushik, Sumit Sharma, and Rakesh Kumar Phanden. 2021. Design and optimization of suspension for formula Society of Automotive Engineers (FSAE) vehicle. *Materials Today: Proceedings* 38, Part 1: 229–233. https://doi.org/10.1016/j.matpr.2020.07.077.

[10] Kim, Do-Hyoung, Dong-Hoon Choi, and Hak-Sung Kim. 2014. Design optimization of a carbon fiber reinforced composite automotive lower arm. Composites Part B: *Engineering* 58: 400–407. https://doi.org/10.1016/j.compositesb.2013.10.067.

[11] Thamarai Selvan, A., and S. Palani. 2023. Prediction of yield strength of AZ31 magnesium alloys by artificial neural network model. *Journal of Physics: Conference Series* 2484: 012015.

[12] Taksande, S.P., and A.V. Vanalkar. 2015. Design and modelling of car front suspension lower arm. *IJIRST –International Journal for Innovative Research in Science & Technology* 2 (02).

[13] Taksande, S.P., and A.V. Vanalkar. 2015. Methodology for failure analysis of car front suspension lower arm. *International Journal for Scientific Research & Development* 3 (05): 1158.

CHAPTER 28

Effect of infill parameters on mechanical properties of poly lactic acid material in 3D printing: A Review

B. Shenbagapandiyan[1, a*], E. Shankar[1], C. Devanathan[1], R. Prakash[2]

[1]Department of Mechanical Engineering, Rajalakshmi Engineering College, Chennai, India
[a]shenbagapandiyanb@gmail.com
[2]Department of Mechanical Engineering, Saveetha Engineering College, Chennai, India

Abstract

In a few decades, 3D printing is a budding technology that has metamorphosed in the field of additive manufacturing and unlocked endless possibilities for further research and development. Variations in material selection, infill parameters, techniques of printing and parameters of printing affect the mechanical properties of the final product. Of all these variations, infill parameters play a major role since they occupy more volume in the 3d printed object instead of other parameters like shell thickness. The biodegradable aliphatic polyester that is most extensively researched and utilized is Poly (lactic acid) (PLA). There are many applications for PLA, especially in the medicine field. The old petrochemical-based polymers are replaced by PLA. This article aims to provide an overview of how infill parameters affect the mechanical properties of PLA-based 3D printed products, namely their tensile strength, compressive strength, hardness, and flexural strength.

Keywords: PLA; 3d printer; mechanical properties; infill parameters

1. Introduction

One of the emerging processes in additive manufacturing is Three-Dimensional (3D) Printing. Here, the material is printed, or melted and then solidified and a 3D printed product is created by layer over the layer. The stepper motors with Arduino, which is an open-source program, enable the microcontroller to move the extruder in the axis of the Cartesian coordinate system [1]. When the material is manufactured using 3D printing the material waste reduction reduces between 70% and 90%. 3D printers are usually eco-friendly, economical and user-friendly when compared to other manufacturing technologies [2]. Acrylonitrile Butadiene Styrene (ABS) and Polylactic Acid (PLA) are the thermoplastics that are most commonly used in non-metallic 3D printing [3]. Because PLA is non-toxic, biodegradable, and uses less energy to 3D print than polymers derived from petroleum, it is considered an environmentally benign substance [4]. Moreover, when compared to ABS, the PLA's mechanical characteristics are higher [5, 6]. There is no review regarding the variation of tensile strength, compressive strength, hardness and flexural strength of PLA 3d printed material based on the variation of the parameters. This review mainly focuses on the variation of mechanical properties like tensile strength, compressive strength, hardness and flexural strength with respect to the infill parameters and has been provided in a detailed manner.

2. Properties of PLA

The parts that are built using 3d printers are anisotropic and brittle in nature which gives the importance of study of the mechanical properties of the 3D printed components [7, 8]. The measure of the polymer's crystalline region in relation to its amorphous content is called the rate of crystallinity which is a very important property of polymer. It is also proved that the

DOI: 10.1201/9781003606611-28

mechanical characteristics are affected by crystallinity [9,10]. Generally, it is difficult to obtain the mechanical characteristics of 3d printed objects since there are various combinations of possibilities available. In this review, we are going to focus only on the mechanical properties. There are many infill parameters in 3d printing. For instance, Dhinakaran Veeman & Sabarinathan Palaniyappan [11] say that infill density and printing orientation play a major role in affecting mechanical properties. Hence, all these parameters that affect the infill in 3d printing are reviewed.

3. Outlook and Summary

3.1 Tensile Strength

Most of the researchers used Filamento PLA premium BQ of 1.75 mm diameter as a PLA filament spool and Witbox 2 from Ultimaker company as fused deposition machine. One of the important parameters that control 3d printing is infill density. For example, the tensile strength is directly related to the infill density, according to Rismalia et al [12] research. The same is proved in many researches [13–15].

The trend indicates that the tensile characteristics grow in tandem with the infill density. However, the circular design with 75% infill density sees a decrease in ultimate strength, from 44.3 MPa to 42.2 MPa. The same is shown in Fig. 28.1. Generally, strength rises as density does for all patterns.

Table 28.1 shows the summary of infill patterns tested for tensile strengths. From the table, we conclude that infill patterns depend on the other infill parameters to yield the maximum tensile strength and it is difficult to select one infill pattern due to the wide range of possibilities. However, the honeycomb pattern makes the best choice of selection for yielding the maximum tensile strength since it is proven in many researches.

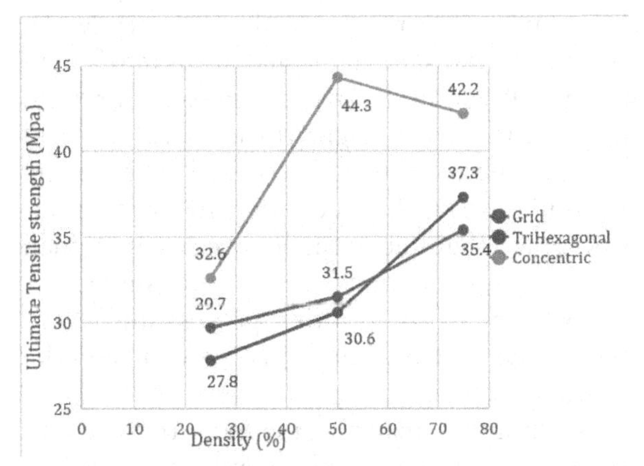

Figure 28.1 Changes in ultimate tensile strength for different infill patterns with infill density

Tensile strength drops as print layer thickness rises. In comparison to the specimen with a 0.4 mm layer thickness, the specimen with a 0.15 mm layer thickness exhibits a higher tensile strength [21–23]. It is discovered that when the filaments align with the direction of loading and align longitudinally and when the raster angle is 45°, the best tensile properties are obtained [24–26]. After reviewing the results above obtained from various researches, we conclude that the maximum tensile strength is obtained when the extrusion temperature is 220 °C, layer height is 0.3 mm, infill density is 100%, print speed is 50 mm/s and infill pattern is the hexagonal pattern.

Table 28.1 Summary of Infill patterns tested for tensile strengths

Sl. No.	Tested infill patterns	Pattern with the highest tensile strength	Reference
1	Hexagonal, rectilinear	Hexagonal	[16]
2	Zig zag, grid, concentric	Zig zag	[17]
3	Solid, rectangular, diamond, hexagonal	Hexagonal	[18]
4	Rectilinear, concentric, honeycomb and hilbert curve	Rectilinear	[19]
5	Line, grid, concentric	Line	[20]

3.2 Compressive Strength

Generally, PLA has more resistance in compression than tension [27]. Variation of compressive strength based on infill density, layer height and print speed is analyzed [28]. It was observed that in almost all cases, compressive strength increases at first up to 0.2 mm layer height and then starts to decrease. On the contrary, when it comes to the sample printed at 50 mm/s having an 80% infill density, compressive strength drops as the layer height rises to 0.2 mm and starts to increase. It is discovered that raising the infill ratio can boost compressive strength. It is observed that due to the higher density which means a higher amount of PLA is consumed per unit volume in the printing process and eventually each unit has small free space inside the printed specimen.

The more the material, the more the strength of the product. In general, the infill ratio directly relates to the compressive strength [29–31]. Table 28.2 shows the summary of infill patterns tested for compressive strength. After reviewing the results obtained from various researches, we conclude that the highest compressive strength is obtained when the material is heated and held for a certain period, layer height is 0.2 mm, infill density is 100%, print speed is 40 mm/min and infill pattern is grid pattern.

3.3 Hardness Test

The experiment conducted by Muammel M. Hanon et al. [37] yielded results about the hardness of 3D printed PLA. Three locations at the gauge, curvature, and gripping portions were used to measure the hardness. In flat, on-edge & upright orientations, the highest hardness value is obtained in on-edge orientation. In [0/45°], [45/135°] & [45/90°] raster direction angle, the highest hardness value is obtained in [0/45°]. On the whole, the maximum hardness value is obtained in a flat orientation, [45/135°] raster angle and 0.1 mm layer thickness. Cheow Keat Yeoh et al. [17] performed the hardness test and found that grid pattern has higher hardness than zig-zag and concentric infill patterns.

After reviewing the results above obtained from various researches, we conclude that the highest hardness is obtained when the layer height is 0.1 mm, the orientation is flat and the infill pattern is a grid pattern.

3.4 Flexural Strength

Of rectilinear, honeycomb, Hilbert curve, concentric, octagram, spiral and Archimedean chord infill patterns, concentric infill pattern corresponds to the maximum strength [38]. The maximum flexural strength of specimens is obtained when the printed specimen is printed at 0° orientation, 38 mm/s printing speed, and 0.2 mm layer thickness, according to an experiment conducted by K. G. Jaya Christiyan et al. [39]. When the nozzle diameter is increased, the strength initially decreases up to a certain level and starts to increase [40].

After reviewing the results obtained from various researches, we conclude that the highest flexural strength is obtained when the material is heated and holding it for a certain period of time, layer height is 0.2 mm, print speed is 38 mm/s and infill pattern is concentric pattern.

Table 28.2 Summary of Infill patterns tested for compressive strengths

Sl. no.	Infill patterns	Pattern with the highest compressive strength	Reference
1	Tri hexagon, gyroid, linear	Linear	[32]
2	Hilbert, honeycomb, line, rectilinear, archimedean, octagram spiral	Hilbert	[33]
3	Triangle, gyroid and line	Triangle	[34]
4	Triangle, line, rectilinear, honeycomb	Honeycomb	[35]
5	Honeycomb, zigzag, square, triangle	Square	[36]

4. Conclusion

The study of the mechanical properties is essential for further development in this field. When it comes to 3d printing, there are various input parameters required to produce a 3d printed product. These factors influence the nature and properties of the product. If engineering is considered, mechanical properties play a significant part in it. These input parameters may alter the mechanical properties of the product. Hence it is necessary to review and study the mechanical behavior of the product with respect to the input parameters. When the extrusion temperature is 220 °C, the layer height is 0.3 mm, the infill density is 100%, the print speed is 50 mm/s, and the infill pattern is hexagonal, the maximum tensile strength is achieved. When the material is heated and kept for a specific amount of time, layer height is 0.2 mm, infill density is 100%, print speed is 40 mm/min, and the infill pattern is grid pattern, the highest compressive strength is achieved. When the orientation is flat, the layer height is 0.1 mm, and the infill pattern is grid-like, the maximum hardness is achieved. Finally, layer height of 0.2 mm, print speed of 38 mm/s, concentric infill pattern and heating the material for a specific amount of time yield the maximum flexural strength.

This review is concentrated on the main mechanical properties of the material i.e., tensile strength, compressive strength, hardness and flexural strength. The influence of input parameters in 3d printing on these properties is analyzed. There are a wide range of variations observed in the properties due to these parameters. A summary of those variations is also provided. Moreover, the best input parameters to obtain the maximum strength of each property based on the referenced research papers are also provided. In future, the properties like impact test and surface characteristics variation with respect to the change in printing parameters can also be reviewed.

Reference

[1] Dong B, Qi G, Gu X, Wei X (2008) Web service-oriented manufacturing resource applications for networked product development. Adv Eng Inform 22:282–295

[2] Weller C, Kleer R, Piller FT (2015) Economic implications of 3D printing: Market structure models in light of additive manufacturing revisited. Int J Prod Econ 164:43–56

[3] Tian X, Liu T, Yang C, Wang Q, Li D (2016) Interface and performance of 3D printed continuous carbon fiber reinforced PLA composites. Compos Part Appl Sci Manuf 88:198–205

[4] Rodríguez-Reyna SL, Díaz-Aguilera JH, Acevedo-Parra HR, García CJ, Gutierrez-Castañeda EJ, Tapia F (2023) Design and optimization methodology for different 3D processed materials (PLA, ABS and carbon fiber reinforced nylon PA12) subjected to static and dynamic loads. J Mech Behav Biomed Mater 150:106257

[5] Tymrak BM, Kreiger M, Pearce JM (2014) Mechanical properties of components fabricated with open-source 3-D printers under realistic environmental conditions. Mater Des 58:242–246

[6] Ebel E, Sinnemann T (2014) Fabrication of FDM 3D objects with ABS and PLA and determination of their mechanical properties. https://www.semanticscholar.org/paper/Fabrication-of-FDM-3D-objects-with-ABS-and-PLA-and-Ebel-Sinnemann/cc96693074847df4b312ea47017bf2ff3e49f40a. Accessed 4 Jan 2024

[7] Sood AK, Ohdar RK, Mahapatra SS (2012) Experimental investigation and empirical modelling of FDM process for compressive strength improvement. J Adv Res 3:81–90

[8] P. Jawahar, Viresh Payak, J. Chandradass, P. Prabhu (2021) Optimization of mechanical properties of CNT-rubber nanocomposites. Materials Today Proceedings 45, 7183–7189

[9] Perego G, Cella GD, Bastioli C (1996) Effect of molecular weight and crystallinity on poly (lactic acid) mechanical properties. J Appl Polym Sci 59:37–43

[10] Balaji Venkatesan, Selvam Ramasamy, Jayabalakrishnan Duraivelu, Joji Thomas, Karthikeyan Thangappan, Subburaj Venkatesan, Prabhu Paulraj (2022) Mechanical, Thermal Conductivity and Water Absorption of Hybrid Nano-Silica Coir Fiber Mat Reinforced Epoxy Resin Composites. Materiale Plastice 59, Issue 2, 194–203

[11] Veeman D, Palaniyappan S (2023) Process optimisation on the compressive strength property for the 3D printing of PLA/almond shell composite. J Thermoplast Compos Mater 36:2435–2458

[12] Rismalia M, Hidajat SC, Permana IGR, Hadisujoto B, Muslimin M, Triawan F (2019) Infill pattern and density effects on the tensile

properties of 3D printed PLA material. J Phys Conf Ser 1402:044041

[13] Żur P (2019) Influence of 3D-printing Parameters on Mechanical Properties of PLA defined in the Static Bending Test. Eur J Eng Sci Technol. https://doi.org/10.33422/EJEST.2019.01.52

[14] P. Prabhu, M. Sivakumar, S. Mohamed Iqbal, A. Balaji and B. Karthikeyan (2022) Research and Review of clay and glass fiber reinforced polyester nanocomposite materials using optimization techniques, Surface Review and Letters 29 (01), 2250014, 2022.

[15] Farazin A, Mohammadimehr M (2021) Effect of different parameters on the tensile properties of printed Polylactic acid samples by FDM: Experimental design tested with MDs simulation. https://doi.org/10.21203/rs.3.rs-273321/v1

[16] Yeoh CK, Cheah CS, Pushpanathan R, Song CC, Tan MA, Teh PL (2020) Effect of infill pattern on mechanical properties of 3D printed PLA and cPLA. IOP Conf Ser Mater Sci Eng 957:012064

[17] Farbman D, McCoy C (2016) Materials Testing of 3D Printed ABS and PLA Samples to Guide Mechanical Design. In: Vol. 2 Mater. Biomanufacturing Prop. Appl. Syst. Sustain. Manuf. American Society of Mechanical Engineers, Blacksburg, Virginia, USA, p V002T01A015

[18] Harpool TD, Alarifi IM, Alshammari BA, Aabid A, Baig M, Malik RA, Mohamed Sayed A, Asmatulu R, EL-Bagory TMAA (2021) Evaluation of the Infill Design on the Tensile Response of 3D Printed Polylactic Acid Polymer. Materials 14:2195

[19] Khan SF, Zakaria H, Chong YL, Saad MAM, Basaruddin K (2018) Effect of infill on tensile and flexural strength of 3D printed PLA parts. IOP Conf Ser Mater Sci Eng 429:012101

[20] Patel DM (2017) Effects of Infill Patterns on Time, Surface Roughness and Tensile Strength in 3D Printing. 5:

[21] Marşavina L, Vălean C, Mărghitaş M, Linul E, Razavi N, Berto F, Brighenti R (2022) Effect of the manufacturing parameters on the tensile and fracture properties of FDM 3D-printed PLA specimens. Eng Fract Mech 274:108766

[22] V Balaji, S Ravi, P Naveen Chandran (2018) Optimization on cryogenic Co2 machining parameters of AISI D2 steel using Taguchi based grey relational approach and TOPSIS, International Journal of Engineering and Technology (UAE), 2018, 7(3.12 Special Issue 12), pp. 885–893

[23] Sukindar NA, Mohd ariffin M khairolanuar, Baharudin BT, Ismail MIS, Jaafar CNA (2017) Analysis on the impact process parameters on tensile strength using 3d printer repetier-host software. J Eng Appl Sci 12:3341–3346

[24] Dizon JRC, Espera AH, Chen Q, Advincula RC (2018) Mechanical characterization of 3D-printed polymers. Addit Manuf 20:44–67

[25] Subburaj Venkatesan, Ramesh Babu Chokkalingam, P. Prabhu, Balaji V (2022) Preparation and Characterization of Biocement Mortar Containing Silane-treated Nano-Si3N4 Prepared from Rice Husk Ash *Silicon* **15**, 1669–1677

[26] Letcher T, Waytashek M (2014) Material Property Testing of 3D-Printed Specimen in PLA on an Entry-Level 3D Printer. In: Vol. 2A Adv. Manuf. American Society of Mechanical Engineers, Montreal, Quebec, Canada, p V02AT02A014

[27] Hanon MM, Dobos J, Zsidai L (2021) The influence of 3D printing process parameters on the mechanical performance of PLA polymer and its correlation with hardness. Procedia Manuf 54:244–249

[28] Brischetto S, Torre R (2020) Tensile and Compressive Behavior in the Experimental Tests for PLA Specimens Produced via Fused Deposition Modelling Technique. J Compos Sci 4:140

[29] Al Khawaja H, Alabdouli H, Alqaydi H, Mansour A, Ahmed W, Al Jassmi H (2020) Investigating the Mechanical Properties of 3D Printed Components. In: 2020 Adv. Sci. Eng. Technol. Int. Conf. ASET. IEEE, Dubai, United Arab Emirates, pp 1–7

[30] Asmatulu E, Alonayni A, Subeshan B, Rahman MM (2018) Investigating compression strengths of 3D printed polymeric infill specimens of various geometries. In: Varadan VK (ed) Nano- Bio- Info-Tech Sens. 3D Syst. II. SPIE, Denver, United States, p 21

[31] V Balaji, S Ravi, P Naveen Chandran (2018) FEM method structural analysis of pressure hull by using hyper mesh *International Journal of Engineering and Technology(UAE)*, 2018, 7(1.5 Special Issue 5), pp. 258–263

[32] Abdulridha H, Abbas T (2023) Analyzing the Impact of FDM Parameters on Compression Strength and Dimensional Accuracy in 3D printed PLA Parts. Eng Technol J 41:1611–1626

[33] Wu W, Geng P, Li G, Zhao D, Zhang H, Zhao J (2015) Influence of Layer Thickness and Raster Angle on the Mechanical Properties of 3D-Printed PEEK and a Comparative

Mechanical Study between PEEK and ABS. Materials 8:5834–5846

[34] Aloyaydi B, Sivasankaran S, Mustafa A (2020) Investigation of infill-patterns on mechanical response of 3D printed poly-lactic-acid. Polym Test 87:106557

[35] Yadav P, Sahai A, Sharma RS (2021) Strength and Surface Characteristics of FDM-Based 3D Printed PLA Parts for Multiple Infill Design Patterns. J Inst Eng India Ser C 102:197–207

[36] Ma Q, Rejab M, Kumar AP, Fu H, Kumar NM, Tang J (2021) Effect of infill pattern, density and material type of 3D printed cubic structure under quasi-static loading. Proc Inst Mech Eng Part C J Mech Eng Sci 235:4254–4272

[37] Do TD, Le MC, Nguyen TA, Le TH (2022) Effect of Infill Density and Printing Patterns on Compressive Strength of ABS, PLA, PLA-CF Materials for FDM 3D Printing. Mater Sci Forum 1068:19–27

[38] Saniman MNF, Bidin MF, Nasir RM, Shariff JM, Harimon MA (2020) Flexural Properties Evaluation of Additively Manufactured Components with Various Infill Patterns. Int J Adv Sci Technol 29:4646–4657

[39] Jaya Christiyan KG, Chandrasekhar U, Venkateswarlu K (2018) Flexural Properties of PLA Components Under Various Test Conditions Manufactured by 3D Printer. J Inst Eng India Ser C 99:363–367

[40] Khatwani J, Srivastava V (2019) Effect of Process Parameters on Mechanical Properties of Solidified PLA Parts Fabricated by 3D Printing Process. In: Kumar LJ, Pandey PM, Wimpenny DI (eds) 3D Print. Addit. Manuf. Technol. Springer Singapore, Singapore, pp 95–104

5. Acknowledgment

We thank Rajalakshmi Engineering College for their expertise and assistance throughout all aspects of our review and for their help in completing the manuscript.

CHAPTER 29

Enhancing mechanical properties of coir fibre composites through $CaCO_3$ reinforcement

R. Selvam,[1*, a] Arulmurugan Seetharaman,[2] and Joji Thomas[3]

[1]Department of Mechanical Engineering, Saveetha Engineering College, Chennai, India.
[2]Department of Mechanical Engineering, Rajalakshmi Engineering College, Chennai, India
[3]Department of Mechanical Engineering, Chouksey Engineering College, Bilaspur, India
Email: [a]selvamr@saveetha.ac.in, selvam_r7@yahoo.com, 0000-0002-6829-0396

Abstract

Driven by environmental concerns, the scientific community is actively pursuing the substitution of synthetic fibres with natural alternatives within the polymer composite fabrication paradigm. To address this challenge, researchers have proposed the development of hybrid composites, aiming to synergistically enhance the mechanical and thermal response of natural fibre-reinforced composites. Through the strategic hybridization of natural fibres with each other or with synthetic counterparts within a uniform matrix, a novel class of composites (hybrid composites) emerges. These hybrid composites exhibit demonstrably superior mechanical and thermal properties when compared to their singular fibre-reinforced polymer composite counterparts. Notably, the development of hybrid composites presents a twofold benefit: enhanced reliability for diverse applications and a more ecologically sustainable material solution.

Keywords: Coir/epoxy, hand lay-up method, $CaCO_3$, mechanical properties

1. Introduction

Natural fibres are increasingly being considered in engineering and technology due to their advantageous properties. Natural fibres are biodegradable, reducing their environmental impact compared to conventional materials. They exhibit good mechanical strength and can be resistant to corrosion. Environmental factors can lead to inconsistencies in cross-sectional area and shape along the fibre length. These inconsistencies can affect the ultimate load-bearing capacity in structural applications.

This section delves into specific research on improving natural fibre composites, but it can be separated from the main points about coir and natural fibres. Studies explored the use of additives like nano-calcium carbonate (nano-$CaCO_3$) and chemical treatments (NaOH) to improve properties like compressive strength, thermal stability, and mechanical performance [1–7]. Research also investigated the impact of these treatments on factors like the filler-polymer interface and dynamic mechanical behaviour [7, 8]. One study examined the effect of using eggshell calcium to improve the flexural strength of coir-vinyl ester composites [9].

2. Experimental Work

The experimental methodology for creating coir fibre laminates reinforced with calcium carbonate ($CaCO_3$).

Here's a breakdown of the key steps: The initial weight percentage of the coir fibres was measured. The fibres were treated with a 5% sodium hydroxide (NaOH) solution for 48 h to remove unwanted materials. The treated fibres were thoroughly washed with water and dried in ambient air conditions. Fiber-to-Epoxy Ratio: A constant weight ratio of 10:1 was maintained between the coir fibres and the

DOI: 10.1201/9781003606611-29

Table 29.1: ASTM standards

S.No	Type of Test	ASTM Standard	Dimension	
			Length (mm)	Width (mm)
1	Compressive	ASTM D695	50	50
2	Flexural	ASTM D790	120	15
3	Impact	ASTM D256	65	15

Table 29.2: Arrangement of fibre in laminates

Laminate 1	Coir fibre/epoxy with 5% of $CaCo_3$
Laminate 2	Coir fibre/epoxy with 10% of $CaCo_3$
Laminate 3	Coir fibre/epoxy with 15% of $CaCo_3$
Laminate 4	Coir fibre/epoxy with 20% of $CaCo_3$
Laminate 5	Coir fibre/epoxy with 25% of $CaCo_3$

epoxy matrix material (including hardener). $CaCO_3$ Reinforcement: Varying weight percentages (5%, 10%, 15%, 20%, and 25%) of $CaCO_3$ were incorporated into the epoxy using a stirring method.

Treated coir fibres were placed between two matching metal moulds measuring 300 x 300 mm. The $CaCO_3$-reinforced epoxy was poured over the fibres. The entire assembly was subjected to compression in a compression moulding machine. The fibres were cured at elevated temperature and pressure to bind them with the matrix material. Post-Curing: After curing, the laminates were cut using water jet cutting technology. The arrangement of fibres within the laminates likely follows a specific pattern as referenced in Table 29.1. The passage mentions adherence to ASTM standards (American Society for Testing and Materials), possibly referring to specific testing procedures detailed in Table 29.2.

3. Result and Discussion

Mechanical Property Evaluation: The laminates were tested to determine their compressive strength, flexural strength, and impact resistance. Data Analysis: The results obtained from these tests were likely compiled into tables for analysis. Property Determination: Based on the test data, the researchers aimed to understand the overall mechanical properties of the composite materials.

3.1. Compression Test

Specimen Preparation: The coir fibre laminate specimens were prepared following the ASTM D695 standard. This standard likely specifies dimensions of 50 mm length, 50 mm width, and 3 mm thickness for the test specimens. The compression test aimed to determine the mechanical properties of the laminates under compressive load. Specifically, compression load and compression strength were measured. The results for compression load and compression stress for all tested specimens were likely tabulated, Figures visually represent the comparison of compression load and stress across different specimens, respectively. Based on Figure 29.1, the passage suggests a direct proportional relationship between the increasing percentage of $CaCO_3$ and the compression load and stress. The test aimed to understand how much force the laminates could withstand before breaking under compression. Initial observations suggest that adding more $CaCO_3$ increases the compressive strength of the laminates.

The study suggests further analysis of the results is likely presented later. It would be interesting to see if the trend holds true for all mechanical properties tested and how the addition of $CaCO_3$ affects other aspects like flexural strength or impact resistance.

The study delves deeper into the connection between the amount of $CaCO_3$ and the material's ability to resist compression. For instance, samples with 25% $CaCO_3$ withstood a compressive force of 5.74 KN and a maximum stress of 2.296 N/mm^2, compared to just 2.36 KN and 0.944 N/mm^2 for those containing only 5% $CaCO_3$. Table 29.3 likely provides detailed data on compressive force and stress for all tested $CaCO_3$ percentages. Access to this table or a summary of the complete results would solidify the observed trend of increasing compressive strength with higher $CaCO_3$ content.

3.2. Flexural Test

Three-point bending, illustrated in Figure 29.4, was the method used for the flexural test. Specimens were prepared according to ASTM D 790 specifications, measuring 120 mm in length, 15 mm in width, and 3 mm in thickness. As evidenced by Figures 2, a clear trend emerges: increasing the $CaCO_3$ content enhances the specimen's performance under flexural load and strength. Specifically, raising the $CaCO_3$ weight percentage from 5% to 25% can significantly boost the flexural load value to 4.70 KN and the flexural strength value to a maximum of 6266.6 N/mm^2.

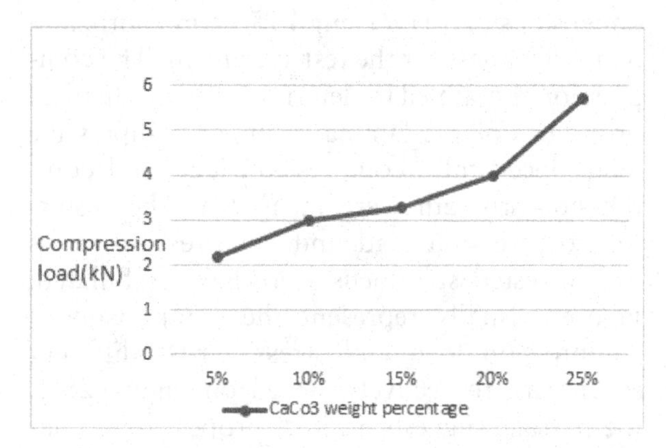

Figure 29.1: Compression load for different $CaCO_3$ weight percentage

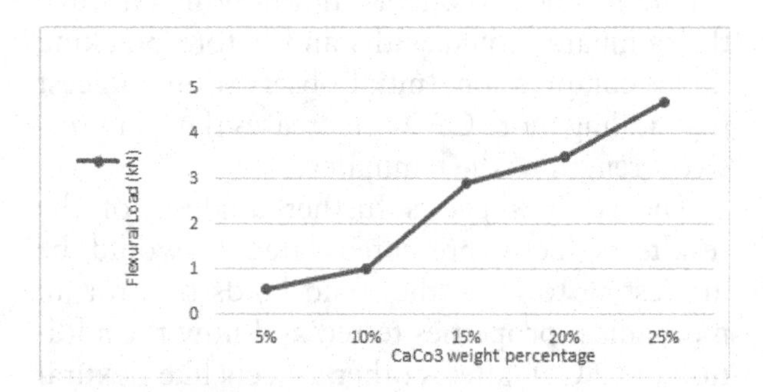

Figure 29.2: Flexural load for different CaCO3 weight percentage

Table 29.3: Compression load and stress for different $CaCO_3$ weight percentage

S.No	Specimen (%)	Compression Load (kN)	Compression Stress (N/mm^2)
1	5	2.36	0.944
2	10	3.08	1.232
3	15	3.24	1.296
4	20	4.09	1.636
5	25	5.74	2.296

The specimen has prepared according to ASTM D 256 where specimen size is 65 mm length, 15 mm width and 3 mm thickness. The property such as impact load was determined and tabulated. Table 29.4 which shows the results of charpy impact test for different percentage of $CaCO_3$ and Figure 29.8 shows a comparison of the impact load of various specimens tested under impact test. By Figure 29.8 it becomes clear that, increasing the percentage of $CaCO_3$ is directly proportional to the impact load.

Table 29.4: Impact load for different $CaCO_3$ weight percentage

S.No	Specimen (%)	Flexural Load (kN)
1	5	0.56
2	10	1.00
3	15	2.90
4	20	3.50
5	25	4.70

4. Conclusion

Materials and Fabrication: Coir fibres were treated with sodium hydroxide (NaOH) for improved bonding. Epoxy resin served as the matrix material. $CaCO_3$ was incorporated into the epoxy at varying weight percentages (5%, 10%, 15%, 20%, and 25%). Laminates were fabricated using a compression moulding process. Mechanical Testing: The laminates were subjected to three types of mechanical tests: Flexural test: Evaluates the material's response to bending forces. Compression test: Measures the material's ability to withstand compressive loads. Impact test: Assesses the material's resistance to sudden impacts. Results and Observations: The study observed a general trend of increasing mechanical behaviour (strength) in all three tests with a rise in the $CaCO_3$ content. This suggests that adding $CaCO_3$ strengthens the coir fibre laminates. Additionally, the absence of delamination (separation between layers) throughout the laminate thickness indicates good bonding between the coir fibres and the epoxy matrix. Overall, the study suggests that $CaCO_3$ reinforcement can be a viable approach to improve the mechanical performance of coir fibre composites. It would be beneficial to see the complete results for a more comprehensive understanding of the impact of $CaCO_3$ content. Further research could explore the optimal $CaCO_3$ content for maximizing mechanical properties without compromising other factors like weight or cost. Investigating the effect of $CaCO_3$ on other properties like thermal stability or moisture resistance could also be valuable.

References

[1] He, Hongwei, Zheng Zhang, Jianlong Wang, and Kaixi Li. 2013. Compressive properties of nano-calcium carbonate/epoxy and its fibre composites. *Composites: Part B* 45: 919–924.

[2] Narendar, R., K. Priya Dasan, and Muraleedharan Nair. 2014. Development of coir pith/nylon fabric/epoxy hybrid composites: Mechanical and ageing studies. *Materials and Design* 54: 644–651.

[3] Ramadhas, A.S., S. Jayaraj, and C. Muraleedharan. Dual fuel mode operation in diesel engines using renewable fuels: Rubber seed oil and coir-pith producer gas. Received 25 Nov 2006; accepted 30 Nov 2007. Available online 28 Jan 2008.

[4] Mustata, Fanica, Nita Tudorachi, and Dan Rosu. 2012. Thermal behavior of some organic/inorganic composites based on epoxy resin and calcium carbonate obtained from conch shell of *Rapana thomasiana. Composites: Part B* 43: 702–710.

[5] Fang, Qinghong, Bo Song, Tiam-Ting Tee, Lee Tin Sin, David Hui, and Soo-Tueen Bee. 2014. Investigation of dynamic characteristics of nano-size calcium carbonate added in natural rubber vulcanizate. *Composites: Part B* 60: 561–567.

[6] Gu, Huang. 2009. Tensile behaviours of the coir fibre and related composites after NaOH treatment. *Materials and Design* 30: 3931–3934.

[7] He, Hongwei, Kaixi Li, Jian Wang, Guohua Sun, Yanqiu Li, and Jianlong Wang. 2011. Study on thermal and mechanical properties of nano-calcium carbonate/epoxy composites. *Materials and Design* 32: 4521–4527.

[8] Geethamma, V.G. G. Kalaprasad, Gabrie Groeninckx, and Sabu Thomas. Dynamic mechanical behavior of short coir fiber reinforced natural rubber composites. Received 17 May 2004; revised 25 Jan 2005; accepted 3 March 2005.

[9] Lakshmanan, P., S. Sathiyamurthy, K. Chidambaram, and S. Jayabal. 2010–2011. Flexural properties of natural fiber calcium (boiled egg shell) impregnated coir-vinyl ester composites. *IOSR Journal of Mechanical and Civil Engineering (IOSR-JMCE).*

CHAPTER 30

Finite element analysis and functional performance of mecanum wheel

R. Selvam[1*], S. Raghavan[1], N. Gokul[3], K.V. Kiran Raaj[1], and S. Damoodaran[1]

[1]Department of Mechanical Engineering, Saveetha Engineering College, Chennai India
Email: [a]selvamr@saveetha.ac.in, selvam_r7@yahoo.com, ORCID: 0000-0002-6829-0396

Abstract

The mecanum wheel is a unique and innovative technology that has transformed the mobility for various sectors. This type of wheel features a series of small rollers that are arranged in a specific pattern, allowing the mobility platforms to move in any direction, including sideways and diagonally. The benefits of the mecanum wheel have been recognized by wheelchair manufacturers, warehouse and logistics robots, etc. Who are incorporating this technology into their designs to meet the growing demand for more versatile mobility platforms. In addition to its practical benefits, the mecanum wheel also has a sleek and modern aesthetic, making it a popular choice among automobile users who value style and design. Overall, the mecanum wheel represents a significant advancement in mobility technology, providing users with greater independence and a higher quality of life. With its unique capabilities and practical benefits, the mecanum wheel is poised to become a standard feature in modern mobility designs.

Keywords: Dynamic analysis, concept generation, manoeuvrability, DoF-degree of freedom, mobility platforms

1. Introduction

The potential of wheeled omnidirectional running systems has been the subject of extensive investigation over the last few years. Among these, mecanum omnidirectional vehicles have emerged as a highly versatile solution with widespread applications in domains ranging from military to storage and transportation, as well as social services. Typically designed to operate on flat or smooth surfaces with three degrees of freedom on the ground, omnidirectional vehicles offer a range of steering motions, including longitudinal, lateral, and centre-point, or any combination thereof.

2. Literature Review

Omnidirectional mobile robots (OMRs) move in three degrees of freedom on a plane, allowing them to move in any direction without changing the direction of their wheels. There has been study on a number of omnidirectional wheels, including the continuous alternate wheel (CAW) with passive rollers for OMRs [1]. Trilok Mistry's [2] paper presents an overview of the initial design phase involved in creating omni-directional mobile robots utilizing mecanum wheels and omni wheels. An omni-directional platform with mechatronics systems, omni-directional wheels, and intelligent movements made possible by microcontroller interfacing is being designed as part of the development process [3].

3. Rollers Design

The maintenance of continuous motion is one of the most important considerations when building a new drive system. In essence, this implies ensuring that the contact line and quantity of rollers are sufficient to cover the wheel's

DOI: 10.1201/9781003606611-30

curvature, which is effectively a circumference. Finding the ideal balance between a few large rollers per wheel and a lot of little rollers per wheel is also essential. The rollers in a Mecanum wheel are typically made of rubber or polyurethane, and they are arranged around the circumference of the wheel at a 45° angle to the wheel axis.

3.1. CAD Model—Mecanum Wheel

The design created in Fusion 360 was to find the best silhouette for the roller (Figure 30.1). According to the vehicle load, we did this by running 1000 iterations with a roller count between 8 and 12, while randomly choosing the wheel width and moving continuity coefficient. The range for breadth was 3–10 cm, while the range for movement. The curvature of the roller's surface was determined by assigning pre-defined values to parameters α, θ, τ, and R. It's important to note that the desired figures for α and θ were approximately 45° and 11°, respectively, whereas τ was about 70°. The value of R was established based on the project's objective, which will be elaborated on in the subsequent section.

3.1 Finite Element Analysis (FEA) Model—Mecanum Wheel

Meshing is one of the key components to obtaining accurate results from an FEA model. The elements in the mesh must take many aspects into account to be able to discretize stress gradients accurately. The design which is created in the fusion 360 was imported into Ansys software for the FEA analysis. Then discretize the CAD model into fine mesh of finite elements. The mesh density should be fine enough to accurately capture the stress and deformation of the wheel under load. Then mention the material properties of the wheel which includes the Young's modulus, Poisson's ratio, and yield strength of the materials used in the wheel. Then We applied appropriate loads and boundary conditions to the model. This will calculate the stress and deformation of the wheel under load. Which gives the analytic result of the FEA Model which may involves the stress distribution, deformation, and safety factors of the wheel (Figure 30.2).

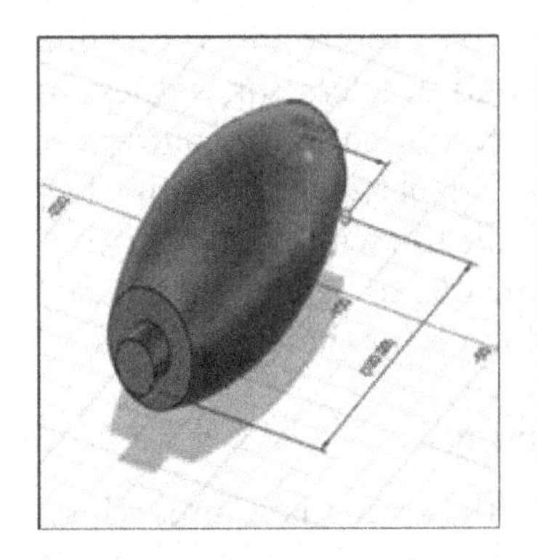

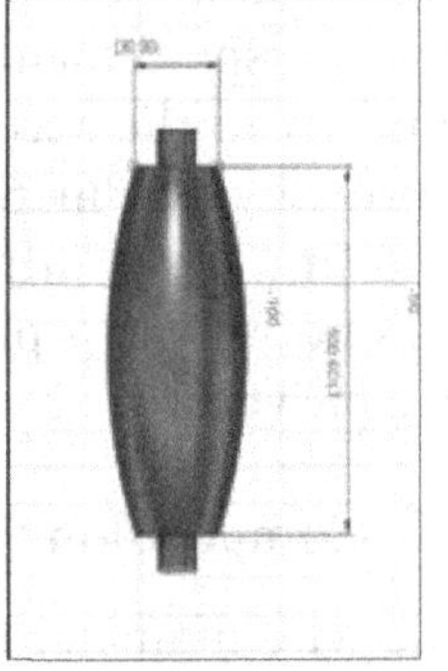

Figure 30.1 Top View and Side View of Roller

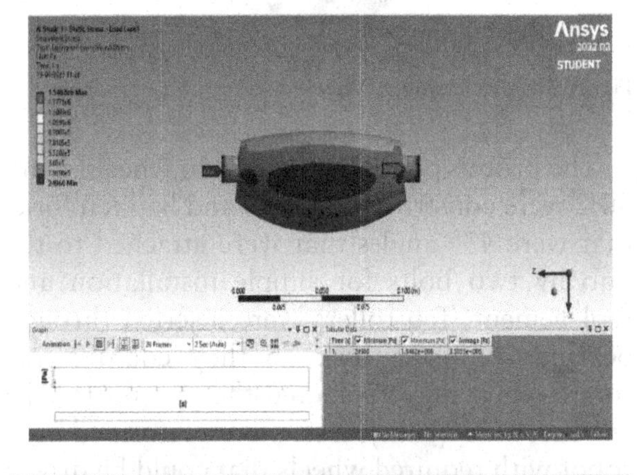

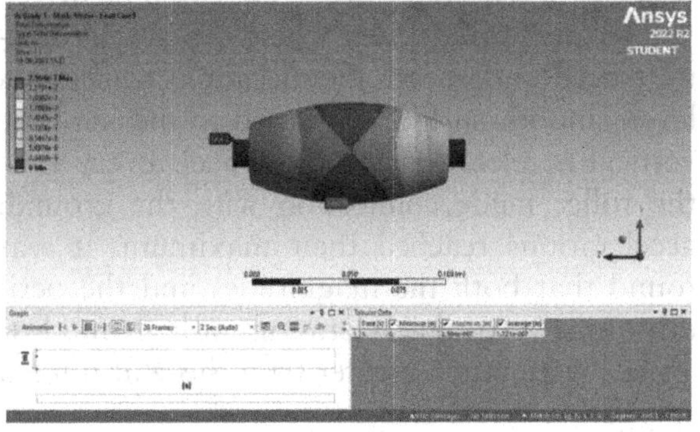

Figure 30.2: Stress distribution of the Roller

4. Model for a Proposed Mecanum Wheel's Dynamic Simulation

Ansys software was used to run dynamic simulations in order to study the asymmetric acceleration patterns of a vehicle with integrated mechanical wheels [4]. To mimic real-world driving situations, the roller slip was considered (Figure 30.3). After analysing the vehicle's displacements and accelerations, the best range of curvature to reduce vertical vibration was suggested.

Figure 30.3: Mecanum Wheel

The prototypes of the mecanum wheel Figure 30.2 were constructed of steel and had ten forks that were 45° angles that were attached to the hub by two bolts for simple installation and replacement. Ten rollers were inserted through bearings on each fork so they could freely rotate when they contacted the ground [5]. To enhance ride quality, tests were performed on a robot with required wheels that could be driven and suspension on each wheel. An accelerometer placed underneath the mecanum wheel on a level, smooth surface was used to measure the vertical accelerations of the wheel. Every time the roller made connection with the ground; accelerations reached their maximum. It was found that both the magnitudes and the periods of the vertical accelerations at high and low speeds were much greater than those at limited speeds.

4.1. Kinematic Model and Simulation

A kinematic model is a mathematical representation of the motion of a system, in this case, a mecanum wheel [6]. Mecanum wheels are omnidirectional wheels that allow a robot to move in any direction without turning. They are commonly used in mobile robotics and automation applications. The kinematic model of a mecanum wheel involves the analysis of the movement of the wheel in the x-y plane. The mecanum wheel has ten rollers that are oriented at a 45° angle with respect to the motion's axis. Since the rollers are not producing any lateral motion, the robot moves in a straight line while the wheels rotate in a straight line. The rollers, on the other hand, cause the wheel to produce a lateral motion while the wheel is moving at an angle. The robot may move in any direction without turning thanks to this lateral motion. The mecanum wheel's kinematic model considers the robot's location in the x-y plane as well as the speed and direction of the wheel.

5. Wheel Movement and Vehicle Direction

The coefficient mentioned previously plays a crucial role in determining the ability of mecanum wheels to maintain movement continuity. It should be greater than 1, but as close to 1 as possible [7]. To evaluate this, simulations were conducted using varying numbers of rollers ranging from 8 to 12. It was determined that the optimal performance was achieved with 10 rollers due to the undesired vibrations caused by roller contact when using 10 rollers during displacement. The vehicle body and four mechanised wheels make up the vehicle model used in the simulations. In order to prevent any asymmetric effects brought on by the body design, the vehicle body was chosen to have a cuboid shape. The vehicle body's dimensions were set by according to the vehicle load. The driving torque was imparted to each of the four revolute joints that joined the four mecanum wheels to the body [7]. So some additional values have been chosen as the static and dynamic friction coefficient between the rollers and the ground based upon the vehicle loads.

5.1 A Spring and a Revolute Joint Are Used in the Original and Proposed Designs

According to the literature reviews, the roller's edges that are too sharp or too rounded provide larger displacement and acceleration values. In contrast to the acceleration data, this pattern is more obvious in the displacement results [8]. Since it may be difficult to switch between rollers smoothly, it is plausible to believe that excessive curvature is to blame for these increases.

5.2 Compression Springs

The wire diameter of the spring is 2 mm, which refers to the thickness of the wire used to make the spring. The outer diameter of the spring is 20 mm, which is the diameter of the spring at its widest point. The nominal length of the spring is 30 mm, which is the length of the spring when it is not compressed [9]. The spring rate is a measure of the spring's stiffness and is determined by the wire diameter, number of active coils, and the material properties of the spring. The maximum load capacity is the maximum amount of force the spring can withstand before it becomes permanently deformed or fails. The compression distance is the distance the spring compresses under a given load and is determined by the spring rate and the amount of force applied [10].

5.3 Wire Material: Spring Steel and Closed and Ground

A mechanical wheel's ideal curvature is essential for reducing displacements and accelerations during the transition between rollers. While a suitable curvature ensures a smooth transition between rollers and lowers the displacement and acceleration magnitudes, a badly chosen curvature might cause a rough motion of the vehicle during the changeover of rollers [11]. Local minima were seen in both the Root mean square (RMS) of displacements and the acceleration data, suggesting that the ideal curvature range lies between the third and seventh curvatures. In addition to curvature, the fork's design was changed to include a spring to lessen impact forces and stop the mecanum wheel's vertical oscillations. In order to allow for spring deformation during dynamic simulation, the original bridge was replaced with a spring and a frictionless revolute joint. Lower peak-to-peak values were obtained as a result of the spring's shock absorption greatly reducing vertical accelerations in comparison to the case when there was no spring.

6. Conclusion

The roller geometry for a power autonomous vehicle with mecanum wheels was the primary focus of this project. The ideal roller curvature for increased vehicle manoeuvrability was established by the researchers using FUSION 360 and ANSYS simulations. However, there was little active research done on the vibration brought on by the roller's constant contact [12]. The researchers developed a prototype mecanum wheel and quantitatively analysed the vertical vibrations to address this. In order to analyse the vertical vibration characteristic of the vehicle body, they ran dynamic simulations and noticed asymmetric acceleration in the experimental results. The researchers suggested an ideal roller curvature to reduce vertical vibration based on the peak to peak and RMS values of the displacements and accelerations. To further reduce vibration, they also recommended changing the fork's design and increasing vertical accelerations. In conclusion, this study emphasises how vital it is to take vibration into account when designing mechanical wheels and offers a methodology for maximising roller curvature and other aspects of vibration reduction in power autonomous wheelchairs.

References

[1] Park, Y.K., P. Lee, J.K. Choi, and K.-S. Byun, Analysis of factors related to vertical vibration of continuous alternate wheels for omnidirectional mobile robots, *Intelligent Service Robotics*, vol. iss. 2016.

[2] Soni, Sanket, Trilok Mistry, and Jayesh Hanath. "Experimental Analysis of Mecanum wheel and Omni wheel". May, 2014.

[3] Bae, Jong-Jin, and Namcheol Kang. "Design Optimization of a Mecanum Wheel to Reduce Vertical Vibrations by the Consideration of Equivalent Stiffness". Feb, 2016

[4] Li, Yunwang, Sumei Dai, Yuwei Zheng, Feng Tian, and Xucong Yan. "Modeling and Kinematics Simulation of a Mecanum Wheel Platform in RecurDyn". Jan, 2018.

[5] Gorash, Y., T. Comlekci, and R. Hamilton, CAE-based application for identification and verification of hyperelastic parameters, *Proceedings of the Institution of Mechanical Engineers, Part L: Journal of Materials: Design and Applications*, 2015.

[6] Guo, S., Y. Jin, S. Bao, et al. 2016. Accuracy analysis of omnidirectional mobile manipulator with mecanum wheels. *Adv. Manuf.* 4, 363–370 (2016).

[7] Ramirez-Serrano, A., and R. Kuzyk. 2017. "Modified Mecanum Wheels for Traversing Rough Terrains", The 6thIntl Conf on Autonomic and Autonomous Systems (ICAS) March 7-13, 2010, Cancun, Mexico.

[8] Ransom, S., O. Krömer, and M. Lückemeier, "Planetary rovers with mecanum wheels," 16thISTVS Intl Conf, 25-28 Nov, 2008, Torino, Italy.

[9] Gfrerrer, A. "Geometry and kinematics of the Mecanum wheel". *Computer Aided Geometric Design 25* (2008) 784-791.
[10] Kiran Kumar, N.A., Vijay M. Patil, and Santosh Kunnur. "Design and Fabrication of Improved Mecanum Wheel by Drag Reduction Method". 2019.
[11] Doroftei, Ioan, Victor Grosu, and Veaceslav Spinu. "Omnidirectional Mobile Robot – Design and Implementation". Bio inspiration and Robotics: *Walking and Climbing Robots*, Book edited by: Maki K. HabibISBN 978-3-902613-15-8, pp. 544, I-Tech, Vienna, Austria, EU.
[12] Sumbatov, A.S, Lagrange's equations in nonholonomic mechanics (in Russian), *Russian Journal of Nonlinear Dynamics* 2013, 9, 39.
[13] Kudra, G., and J. Awrejcewicz Appication and experemental validation of new computational models of friction forces and rolling resistance, Acta Mech. 2015, 226, 2831.
[14] Tatar, M.O., C. Popovici, D. Mandru, I. Ardelean, and A. Plesa, "Design and development of an autonomous omni-directional mobile robot with Mecanum wheels," in *Proceedings of the 2014 19th IEEE International Conference on Automation, Quality and Testing, Robotics, AQTR 2014*, May 2014.

CHAPTER 31

Synthesis and photo-physical investigation of nano SiC particulate reinforced material for binder jet3D printing

Selvam R.,[1] Prakash R.[1*, a], Ilakkiya T.[2], Joji Thomas[3], and Devanathan C.[4]

[1]Department of Mechanical Engineering, Saveetha Engineering College, Chennai, Tamil Nadu, India
[2]Department of Management Studies, Sri Sairam institute of Technology, Chennai, Tamil Nadu, India
[3]Department of Mechanical Engineering, Chouksey Engineering College, Bilaspur, India
[4]Department of Mechanical Engineering, Rajalakshmi Engineering College, Chennai, Tamil Nadu, India
Email: [a]prakash8781@gmail.com, prakash@saveetha.ac.in

Abstract

In the view of insinuating an enhanced material for binder jetting 3D printing, nano silicon carbide (SiC) particulate is synthesized and studied for its superior properties which is then bonded with polyester for its application in binder jet 3D printing. In order to predict the physical characteristics of the developed SiC particulate for application in binder jet products, physical investigations are carried out that includes tensile, compression, hardness, flexural and impact tests to superscribe the preponderance of this developed material. In the perspective of letting homogeneity of this product and to understand the morphology, Scanning electron microscopy (SEM) analysis, Transmission electron microscopy (TEM) analysis, FTIR, Energy dispersive X-ray analysis (EDAX) and XRD analysis are conducted. This experimental analysis proves that the developed product has good thermal stability by conducting Thermo gravimetric analysis (TGA) test and results proves the thermal degradation occurs in the beginning itself for initial mass change of 9.56% at 390 °C. Also, greater resistance to water absorption of 0.001%. This nano silicon carbide particulate is used to form polyester matrix composite nPMC, which is heat-resistant, lightweight and noteworthy compressive strength.

Keywords: Binder jetting, nano silicon carbide particulate, 3D printing, nano polymer composite, physical testing, heat resistant

1. Introduction

The geometrical representation of an object can be easily converted and created as physical entity using this evolving technique Binder jet 3D printing by adding material successively [1]. There are several 3D printing technologies developed in recent years that produces product doing different functions. It is classified in to seven different groups viz., binding jetting, directed energy deposition, material extrusion, material jetting, powder bed fusion, sheet lamination and photopolymerization. These technologies have their own uniqueness and meet their novel applications [2]. Nowadays, product visualization and testing by 3D printing technologies at all the stages of prototyping such as the alpha stage and beta stage also are used for visualizing the product while making new product in a large scale [3].

High quality materials are needed to for binder jetting 3D printing process resulting in development of high-quality devices [4]. The procedure of making, material requirements and control are to be seriously practiced between the suppliers, purchasers and end-users, this can produce several functional components for machines and also precision equipment's using different materials and its combinations (ceramic, metallic, polymers) [5]. One of the evolving technologies in 3D printing process is

DOI: 10.1201/9781003606611-31

binder jetting, this system has a feeder head that can move in all three axes (x, y and z direction). The complex shaped products are built by the head in a thin layer over layer with nano SiC particulate. This SiC nano particulate is built with a binding agent namely polyester, which is sprayed through a nozzle to interconnect the particulate cluster by cluster [6, 7]. The desired property of the binding agent is habituation in less viscosity. Whatever the complex shapes can be manufactured by the binder jet. This printer jet head is controlled by a servo-controlled programmable integral derivative (PID) system. The controller makes the head to translate in x and y-direction. A roller is also provided to roll on the mixer of nano particulate and binder to make it compact product at the end. Polyester is used as a binding agent being spread over the first layer, then the nano silicon carbide particulate powder is being bound together. The powder bed moves progressively down after completing each layer of material deposition. When the working table moves down, the nano silicon carbide particulate paved over the next layer by the roller and then polyester resin is sprayed again [8–10]. This manufacturing technique is classified as a bottom-up method. Another major advantage in binder jetting 3D printing is the "filling material", which gives support to the printed product during printing itself with no need of any additional support. Hence, it saves time and cost by eliminating time required to remove supporting materials from the finished object. It is easy to clean the product from the extra powder left on the platform. Different entities in the product can be denoted by different color using the coloring agent [9, 11]. It is easy to clean the product from the extra powder left on the platform. Coloring agent can be added to the reinforcement or binder. To enhance the strength of the product thermal treatment can be suggested for improvement. Binder jet 3D printers can be used for mass production in industry. These 3D binder jet process also have demerit like, the material properties are not homogenized throughout the inner structure. To ensure the desired strength of the product additional thermal treatment is required. In addition, macro sized SiC particulate is used in 3D jet binding processes which is less in structural properties than nano size SiC particulate.

2. Method

There is complication in additive manufacturing of metals and polymers especially, ceramics have high melting point which cannot be cast or machined easily, but the evolution of 3D printing techniques will overcome these problem. The 3D printing binder jet technology which is one of AM technique uses the nano silicon carbide particulate reinforcement with polymer matrix structures that can be produce complex shape and cellular architecture of greater geometrical flexibility [8, 12]. This paper concentrates and exposes the synthesis and testing of nano silicon carbide polymer matrix composite (nPMC). Photo investigation like SEM and TEM analysis is carried out to validate under spectroscopic properties. Further, physical properties like tensile, compressive, hardness and structural properties are also being evaluated. In addition to that thermo-gravimetric analysis and water absorption test are also conducted to validate the investigations. This developed material has its application in additive manufacturing of protection systems, porous burners, micro electromechanical systems and electronic device packaging [9, 11, 13].

3. Experimental

3.1 Preparation of nPMC Used in Binder Jet 3D Printing

The binder jetting process uses nano silicon carbide (SiC) particulate as reinforcement powder and polyester matrix as a binding agent. This binder acts as an adhesive between nano silicon carbide particulate layers and it is mostly in the liquid form which builds the desired shape by this particulate [14, 15].

3.1.1 SiC Nano Particle Synthesis

The natural raw material is quartz which is converted into SiC powder by thermo-chemical reduction. The source of the ore is collected from the bank of Kallar river – Tamil Nadu. The nano-sized particulate SiC is mill downed by a top-down approach by a high energy planetary ball at appropriate Ball to powder ratio (BPR) [4]. Amalgamation is avoided during this process, after a specified duration of time micro sized SiC are converted into nanoparticulate as cluster by cluster. Figure 31.1 shows the physical appearance of synthesized nano SiC particulates [10, 16].

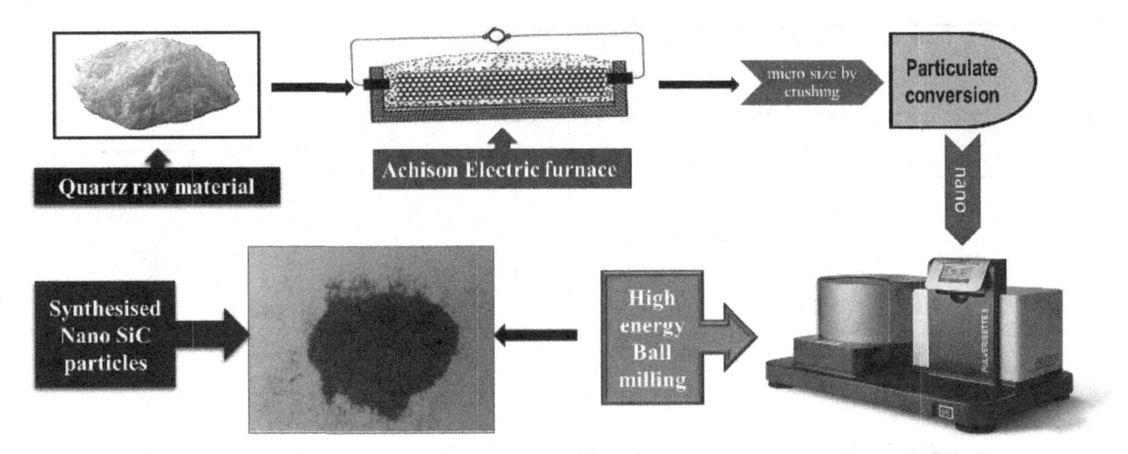

Figure 31.1: Nano SiC particle synthesis

The usage of high determinative and attractive focal point ascertains the proper structure with size of the nano structure's carbide particles. The morphological investigation is the microscopic analysis to assess particle structure. Figure 31.2a shows the SEM image whereas Figure 31.2b depicts TEM.

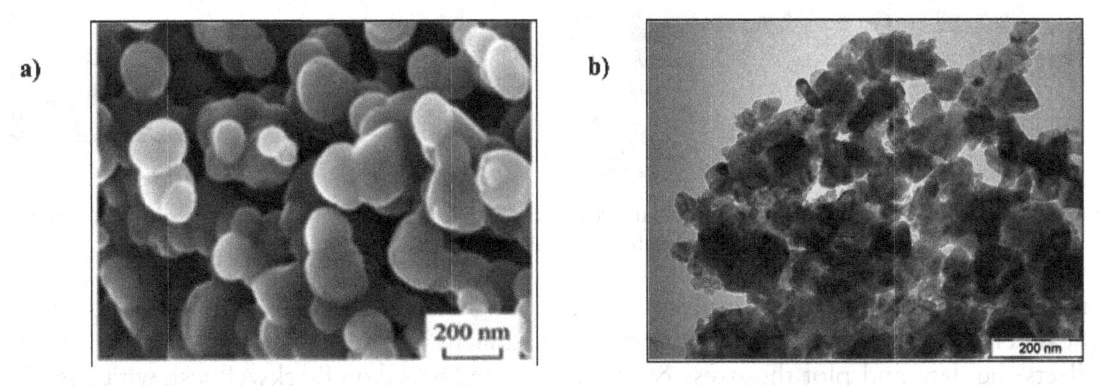

Figure 31.2: (**a**) Nano silicon carbide (α type) particulate: SEM image, (**b**) TEM image

From the SEM picture, the nano silicon carbide particulate shows it has 200 nano meter (nm) size and from the appearance of TEM picture it is clear that it has irregular shapes and sharp edges as shown in Figure 31.2b.

3.2 Analyse the Spread of Size of Particle Using PSA (Particle Size Analyser)

Using particle size analysis, the exact size or mean size of the nano particulate silicon carbide is estimated. The milled SiC is found to have nano size and the particle size analysis confirms the same. From Figure 31.3 it is evident that, an average particulate size of 117 d. nm is attained at 100% intensity. Also, the Figure 31.3 shows the quality report of particle size analysis.

4. Testing and Investigation

The developed nano SiC particulate (α type) is subjected to the following testing viz., tensile test, compressive test, test of flexural, test for hardness, test of wear and test of water absorption [17, 18]. It is used to characterize the product for its physical behavior.

4.1 Structural Test

As per ASTM D368, the test specimen is prepared and is shown in Figure 31.4a. It is the specimen made for characterization of nano SiC particulate reinforced polyester matrix binder jet 3D printed component as per the particulate size and percentage of mixture. Prepared specimen is placed in the universal testing machine (UTM) to carry a tensile load [19]. The

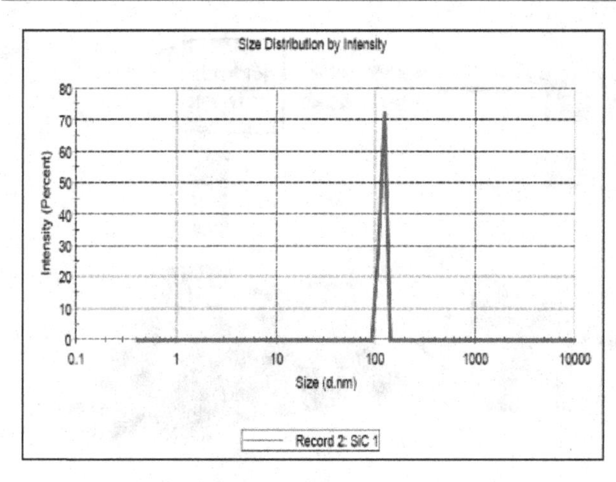

Results

			Size (d.nm...	% Intensity:	St Dev (d.n...
Z-Average (d.nm):	8785	Peak 1:	117.8	100.0	7.450
PdI:	0.774	Peak 2:	0.000	0.0	0.000
Intercept:	0.717	Peak 3:	0.000	0.0	0.000
Result quality	Refer to quality report				

Figure 31.3: PSA chart and table

extensometer in the UTM automatically measures the changes in gauge length of the specimen. The test specimen is placed in UTM and it get elongated with continuous load [20, 21]. The specimen after tensile test is depicted in Figure 31.4b.

a)
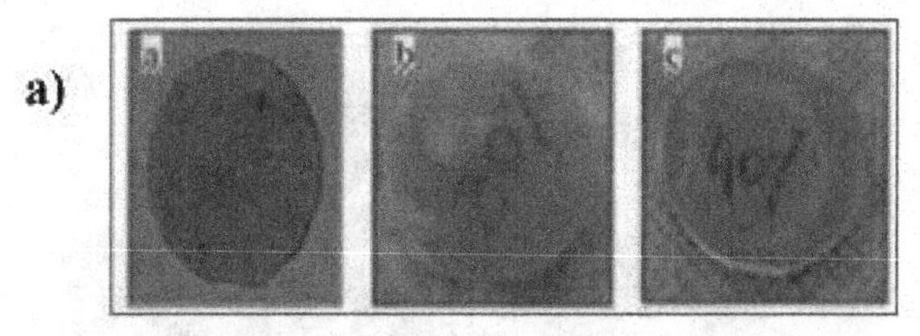
b)

Figure 31.4 (**a**) Nano silicon carbide particulate reinforced polyester matrix specimen in different proportion, (**b**) specimen (30% nano SiC polyester) after tensile test

During the test the control system observes the significant changes in the specimen. The computer collects the data and plot the stress & strain graph which shows the ultimate tensile strength of the specimen as 106 N/mm^{2}for 30% of SiC nano composite. The compressive strength of 515.5 N/mm^2 and flexural strength is 1.62 N/mm^2 which is investigated by utilizing 3-point-bowing technique as shown in Table 31.1.

Table 31.1: Mechanical properties of nPMC

Sl. No.	Properties	Obtained results
1	Tensile strength [N/mm^2]	106
2	Compressive Strength [N/mm^2]	515.5
3	Flexural Strength [N/mm^2]	12.3
4	Strength of impact [J/m]	29
5	Hardness [HRB]	75
6	Water absorption [%]	0.001

The impact test is accomplished by using a Charpy test and it is found that 29J. Hardness is tested based on Rockwell test, which is 75 HRR. To assess the internal structure of the specimen aggregate water absorption is also carried out and is found to be 0.001%.

4.2 Thermogravimetric Analysis (TGA)

The Nano Sic polymer matrix composite sample of mass ~10 mg is taken in this research and heated in gravimetric balance at the rate of 20 °C/min. Upon careful observation it is noticed that thermal degradation occurs in the beginning itself which is 9.56% of initial mass change, whereas the losing of mass and its changes get intensified and deep at 3900 °C which is estimated as 36.45%. At 2400 °C some low-level volatiles are being emitted from the taken sample at isothermal conditions. Around 24.55% mass residue is observed with at time lapse of 11 min at 1008.80 °C [4].

4.3 Analysis Using Scanning Electron Microscopic

Figure 31.5a shows the morphological changes of synthesized SiC nano-material evaluated under scanning electron microscopy at specific stages. The dark zone indicates the richness of SiC particles and patches of white are resin rich locations and also it is evident that 75% of particles are uniformly distributed [22–24]. The experiment is done by taking a sample size 5 mm and the magnification factor of 15×10^3 is obtained to visualize the distribution of SiC nano-particulates in polyester matrix. Overall, it indicates the developed material has greater chemical compatibility between matrix and reinforcement as shown in Figure 31.5b (Energy dispersive X-ray analysis (EDAX) chart).

a)

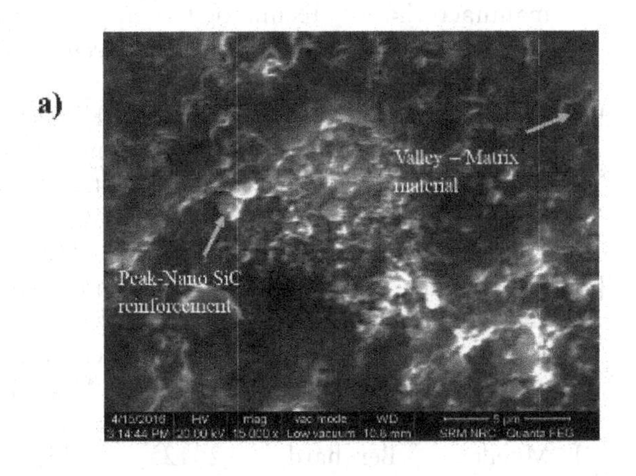

b)

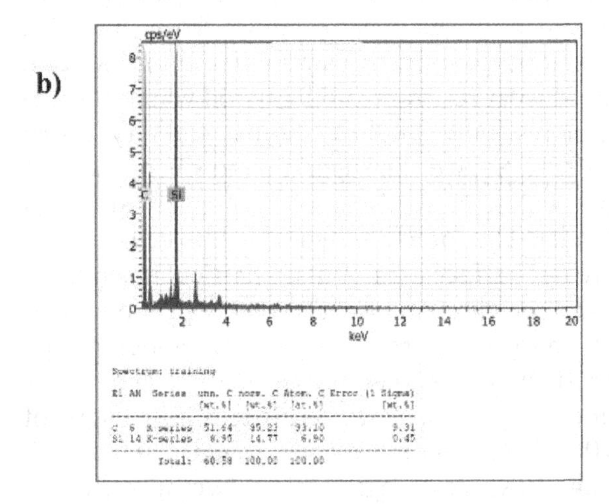

Figure 31.5: (**a**) SEM image of nPMC, (**b**) EDAX chart (30% nano SiC polyester)

5. Result and Discussion

Structural test involving tension, compression, flexibility and impact shows the surface area of contact is more in the developed SiC nano composite material. The capability of uniform dispersion of particle with no agglomerations is a basic factor that signifies the effectiveness of the synthesized material. Adding the material Tulane at the synthesis stage itself will overcome the problem of dispersion in nano particulate. It is confirmed by analysing the various results obtained from the use of particle size analysis, FE-SEM, HR-TEM, FT-IR, Thermal conductivity test. From PSA test, the average size of SiC particulates is found to be 117 d.nm (14 h milling time). TEM results also confirms that nano particulate SiC has a crystalline spherical structure with sharp edges. From the SEM result, it has been understood that nano particles have a spherical shape and has slight changes in its size and shape due to its agglomeration. Further, the particulate size and method of handling decides and enhance the composite material properties. The major portion of SEM image is dark and it indicates that it is associated the capability of reinforcement. Formation of chemically homogenous composite (polyester and SiC) is evidently confirmed with the enlarged SEM image [25, 26].

Polymer composite having the particulate as in nano scale with more surface energy provokes tensile, flexural and compressive strength. nPMC is shown in FE-SEM, the occurrence of dark region spectrum confirms proper reinforcement. The magnified picture of polyester and SiC evidently indicate, the formation of composite and thus the nPMC is chemically homogeneous, transparent and well dispersed.

Most critical processing parameter is to have the capability for uniform distribution and agglomerations reduce the strength of nPMC. It has good wear resistance capability also resistance to absorb the water. From the TGA analysis, SiC and polyester nano composite undergoes thermal degradation which begun at 390 °C. The mass loss is gradually taken due to the decomposition; break down reactions, gas reaction, chemisorption. The residual mass observed after 11 min was 24.55% at 1008.8 °C. According to the test and investigation, the nPMC (prepared with 30% of nano SiC particulate) is better for light weight, high strength applications. This nPMC material property can be improved by particulate size and the processing method. The nPMC properties further can be improved by particulate size and the processing method.

6. Conclusions

Thus, the nano silicon carbide particulate reinforced polyester matrix binder jetting product has good compressive strength and low weight, good wear resistant and has good water absorption resistance. It has thermal stability up to 390 °C. So, we can suggest nano silicon carbide particulate as reinforcement and polyester as a binder. The interface region is boosted and the greater surface area of nano particulate enhances the product quality. The rheological behavior of nano SiC particulates in 3D binder jetting products yielded good performance. The results under structural test reveals tensile strength of 106 N/mm^2, compressive strength of 515 N/mm^2, flexural strength of 12.3 N/mm^2, hardness of 75 HRB. The water absorption and thermal behavior of the suspensions essentially determines the influence of developed material in 3D printing processes. The nano silicon carbide particulate reinforced polyester matrix binder jetting 3D product is important for the realizable resolution and accuracy of manufactured components [21, 22].

Acknowledgements

I would like to say that "I have not received any financial support from any organization for my research work.

References

[1] Scheithauer, U., T. Slawik, E. Schwarzer, H.J. Richter, T. Moritz, and A. Michaelis. 2015. Additive manufacturing of metal-ceramic-composites by thermoplastic 3D-printing (3DTP). *Journal of Ceramic Science and Technology* 6 (2): 125–132.

[2] Fayazfar, Haniyeh, Mehrnaz Salarian, Allan Rogalsky, Dyuti Sarker, Paola Russo, Vlad Paserin, and Ehsan Toyserkani. 2018. A critical review of powder-based additive manufacturing of ferrous alloys: Process parameters, microstructure and mechanical properties. *Materials & Design* 144: 98–128.

[3] Kruth, J.-P., Ming-Chuan Leu, and Terunaga Nakagawa. 1998. Progress in additive manufacturing and rapid prototyping. *CIRP Annals* 47 (2): 525–540.

[4] Selvam, R., S. Ravi, and R. Raja. 2018. Wear resistance and water absorption study of SiC reinforced polyester composite. *Materials Today: Proceedings* 5 (6): 14567–14572.

[5] Sharke, Paul. 2001. Rapid transit to manufacturing. *Mechanical Engineering* 123 (03): 63–65.

[6] Guo, Nannan, and Ming C. Leu. 2013. Additive manufacturing: Technology, applications and research needs. *Frontiers of Mechanical Engineering* 8: 215–243.

[7] Alcisto, J., A. Enriquez, H. Garcia, S. Hinkson, T. Steelman, E. Silverman, P. Valdovino, et al. 2011. Tensile properties and microstructures of laser-formed Ti-6Al-4V. *Journal of Materials Engineering and Performance* 20: 203–212.

[8] Do, Truong, Patrick Kwon, and Chang Seop Shin. 2017. Process development toward full-density stainless steel parts with binder jetting printing. *International Journal of Machine Tools and Manufacture* 121: 50–60.

[9] Mueller, Bernhard. 2012. Additive manufacturing technologies—Rapid prototyping to direct digital manufacturing. *Assembly Automation* 32 (2).

[10] Konda Gokuldoss, Prashanth, Sri Kolla, and Jürgen Eckert. 2017. Additive manufacturing processes: Selective laser melting, electron beam melting and binder jetting—Selection guidelines. *Materials* 10 (6): 672.

[11] Selvam, R., S. Ravi, and R. Raja. 2017. Study of thermal behavior of silicon carbide reinforced polyester matrix nanocomposite. *Journal of Surface Science and Technology* 33 (1–2): 29–33.

[12] Selvam, R., Ravi, S., & Balasubramanian, K. (2018), Mechanical Testing of Plastoceramic (nPMC Sheet- SiC Reinforced Polyester Nano Composite). International Journal of Engineering & Technology, 7(3.12), 1195-1198. https://doi.org/10.14419/ijet.v7i3.12.17785.

[13] Conner, Brett P., Guha P. Manogharan, Ashley N. Martof, Lauren M. Rodomsky, Caitlyn M. Rodomsky, Dakesha C. Jordan, and James W. Limperos. 2014. Making sense of 3-D printing: Creating a map of additive manufacturing products and services. *Additive Manufacturing* 1: 64–76.

[14] DebRoy, Tarasankar, H.L. Wei, J.S. Zuback, Tuhin Mukherjee, J.W. Elmer, J.O. Milewski, Allison Michelle Beese, A. de Wilson-Heid, Amitava De, and Wei Zhang. 2018. Additive manufacturing of metallic components–

process, structure and properties. *Progress in Materials Science* 92: 112–224.

[15] Gao, Wei, Yunbo Zhang, Devarajan Ramanujan, Karthik Ramani, Yong Chen, Christopher B. Williams, Charlie C.L. Wang, Yung C. Shin, Song Zhang, and Pablo D. Zavattieri. 2015. The status, challenges, and future of additive manufacturing in engineering. *Computer-Aided Design* 69: 65–89.

[16] Konda Gokuldoss, Prashanth, Sri Kolla, and Jürgen Eckert. 2017. Additive manufacturing processes: Selective laser melting, electron beam melting and binder jetting—Selection guidelines. *Materials* 10 (6): 672.

[17] Wang, Yujia, and Yaoyao Fiona Zhao. 2017. Investigation of sintering shrinkage in binder jetting additive manufacturing process. *Procedia Manufacturing* 10: 779–790.

[18] Chen, Han, and Yaoyao Fiona Zhao. 2016. Process parameters optimization for improving surface quality and manufacturing accuracy of binder jetting additive manufacturing process. *Rapid Prototyping Journal* 22 (3): 527–538.

[19] Frykholm, Robert, Yoshinobu Takeda, Bo-Göran Andersson, and Ralf Carlström. 2016. Solid state sintered 3-D printing component by using inkjet (binder) method. *Journal of the Japan Society of Powder and Powder Metallurgy* 63 (7): 421–426.

[20] Scheithauer, Uwe, Eric Schwarzer, Hans-Jürgen Richter, and Tassilo Moritz. 2015. Thermoplastic 3D printing—An additive manufacturing method for producing dense ceramics. *International Journal of Applied Ceramic Technology* 12 (1): 26–31.

[21] Herzog, Dirk, Vanessa Seyda, Eric Wycisk, and Claus Emmelmann. 2016. Additive manufacturing of metals. *Acta Materialia* 117: 371–392.

[22] Prabhu, P., G. Manohar, T. Karthikeyan, et al. 2024. Characterization of vinyl silane-treated biomass cellulose and banana fiber on reinforced unsaturated polyester composite. *Biomass Conversion and Biorefinery*. https://doi.org/10.1007/s13399-024-06029-8.

[23] Prabhu, P., G. Gokilakrishnan, S. Hanish Anand, et al. 2024. Comparative analysis of tamarind shell biomass powder and roasted chickpeas powder in kenaf fiber reinforced vinyl ester composite. *Biomass Conversion and Biorefinery*, 1–10

[24] Venkatesan, B., S. Ramasamy, J. Duraivelu, J. Thomas, K. Thangappan, S. Venkatesan, and P. Paulraj. 2022. Mechanical, thermal conductivity and water absorption of hybrid nano-silica coir fiber mat reinforced epoxy resin composites. *Materiale Plastice* 59 (2): 194–203. https://doi.org/10.37358/MP.22.2.5598.

[25] Prabhu. P., M. Siva Kumar, S. Mohamed Iqbal, A. Balaji, and B. Karthikeyan. 2022. Research and review of clay and glass fiber reinforced polyester nanocomposite materials using optimization techniques. *Surface Review and Letters* 29 (01): 2250014.

[26] Subburaj, V., R.B. Chokkalingam, P. Prabhu, et al. 2023. Preparation and characterization of biocement mortar containing silane-treated Nano-Si_3N_4 prepared from rice husk ash. *Silicon* 15: 1669–1677. https://doi.org/10.1007/s12633-022-02135-2.

CHAPTER 32

An innovative method for integrating solar photovoltaic systems with battery storage using multiport converters compared to direct coupling methods with an emphasis on enhancing battery storage capacity

K. Aarthi[1] and M. Lavanya[2,*]

[1]Research Scholar, [1]Department of Electrical and Electronic Engineering, Saveetha School of Engineering, Saveetha Institute of Medical and Technical Sciences, Chennai, Tamilnadu, India.

[2]Project Guide, Department of Electrical and Electronic Engineering, Saveetha School of Engineering, Saveetha Institute of Medical and Technical Sciences, Chennai, Tamilnadu, India.

[*]lavanyam@saveetha.com

Abstract

The primary aim of this research is to develop an innovative method for integrating solar photovoltaic (PV) systems with battery storage using novel multiport converters, focusing on enhancing battery storage capacity and comparing the efficiency and effectiveness of this integration against traditional direct coupling methods. Two distinct groups are formed to facilitate a comparative analysis of the integration methods for solar PV systems with battery storage. Group 1 utilizes traditional direct coupling methods, while Group 2 employs the novel Multiport DC Converters. Each group consists of 20 samples, making a total sample size of 40. The implementation of the systems and the collection of output data are conducted through simulations using MATLAB software. The performance comparison focuses on battery storage capacity under identical operational conditions. Statistical analysis is applied to evaluate the effectiveness and efficiency differences between the two groups, aiming to provide a robust comparison of the two technologies. The sample size for the study is calculated to ensure adequate power to detect significant differences between the two groups. Using clinCalc.com, with a statistical power (1-β) set at 80% and an alpha (α) of 0.05, the necessary sample size is determined to provide reliable results while maintaining a 95% confidence interval. The multiport DC converter exhibited a battery storage capacity of 647.00 kWh, while those using direct coupling methods showed a capacity of 506.00 kWh. A two-tailed significance test was conducted to analyze these results, yielding a p-value of 0.029, which is below the threshold of 0.05. The study conclusively demonstrates that novel multiport DC converters offer a significant improvement in battery storage capacity over traditional direct coupling methods in solar PV systems.

Keywords: Solar PV systems, battery storage, direct coupling, novel multiport DC converter, photovoltaic, renewable energy, solar panel

1. Introduction

The integration of solar PV systems with battery storage is increasingly critical in today's energy landscape, where the demand for renewable sources is driven by the urgent need to mitigate climate change and enhance energy security (Khezri, Mahmoudi, and Aki 2022). Solar PV systems, which convert sunlight directly into electricity, are among the most promising renewable technologies due to their ability to harness a clean and abundant energy source.

DOI: 10.1201/9781003606611-32

However, the intermittent nature of solar energy necessitates effective storage solutions to ensure a stable and reliable power supply (Mustafa et al. 2022). Traditionally, solar PV systems have been integrated with battery storage using direct coupling methods. This approach, while straightforward, involves several drawbacks, including inefficient energy transfer, potential overloading of batteries, and faster degradation due to the strain of continuous charge and discharge cycles. These limitations can significantly impede the practical utility and efficiency of solar energy systems. In response to these challenges, the proposed study introduces an innovative solution: the novel Multiport DC Converter (Abomazid, El-Taweel, and Farag 2022). This technology promises to revolutionize the integration of solar PV systems with battery storage by optimizing energy management. The multiport converter allows for more efficient and flexible control of power flows between the solar panels, the battery, and the load, potentially enhancing the battery's storage capacity and extending its lifespan. This research aims to thoroughly analyze and compare the battery storage capacities of systems using the novel multiport converter against those employing traditional direct coupling methods (Salameh et al. 2022). The applications of this study extend to various sectors, enhancing efficiency and sustainability. Key areas include residential and commercial buildings for improved energy solutions, industrial operations with scalable storage, and agriculture through reliable renewable energy sources.

Over the last five years, around 97 articles related to this subject have appeared on Google Scholar, and 141 articles have been featured on ScienceDirect (Chaurasia, Gairola, and Pal 2022; Gul et al. 2022; A. Hassan et al. 2022; Ud-Din Khan et al. 2022). Babatunde et al. (Babatunde, Munda, and Hamam 2022) explored the potential of energy routers in optimizing the flow of energy from various sources directly to loads or storage units, thereby minimizing conversion losses and enhancing efficiency. Hassan et al. (Q. Hassan et al. 2022) demonstrated how hybrid inverters could streamline the integration of wind and solar systems by saving space and simplifying installation while improving power output quality. Rana et al. (Rana et al. 2022) highlighted the advantages of IoT-based real-time monitoring systems in reducing downtime and enhancing system reliability through proactive maintenance. Majji et al. (Majji, Mishra, and Dongre 2022) employed phase-locked loop systems to synchronize outputs from diverse renewable sources, facilitating stable integration into the power grid. Rekioua et al. (Rekioua 2023) developed a dynamic load balancing technique that adjusts the load between wind and solar components, addressing the issues of overloading and underutilization effectively. Abualigah et al. (Abualigah et al. 2022) investigated blockchain technology for decentralized energy trading, which not only increased economic benefits for system owners but also improved grid stability. Among these, the study by Abualigah et al. (Abualigah et al. 2022) stands out as particularly transformative. Their research into using blockchain for decentralized energy trading within hybrid systems represents a groundbreaking approach to energy distribution. By allowing households to efficiently buy and sell surplus energy, Norton et al.'s work not only enhances the economic benefits for individual system owners but also contributes significantly to overall grid stability by distributing energy production more evenly.

Despite the significant advancements in hybrid energy systems, there remains a notable research gap in optimizing the integration of various energy sources with storage solutions to minimize energy losses and enhance overall system efficiency. While studies like those of Norton et al. (2023) and others have introduced innovative solutions, the integration techniques often still suffer from conversion losses and limited capacity to manage fluctuating energy inputs effectively. There is a pressing need for technologies that can seamlessly integrate diverse energy sources into the grid without compromising on efficiency or storage capabilities. The proposed novel Multiport DC Converter aims to address this gap by providing a versatile solution that enables direct integration of multiple energy sources with enhanced battery storage capacities. This converter is designed to reduce energy wastage and improve system reliability and sustainability, leveraging expertise in electrical engineering and renewable energy systems to develop a more efficient and robust framework for managing hybrid energy setups.

2. Materials and Methods

The study was conducted in the Department of Electrical and Electronics Engineering at Saveetha University. The materials required for this research include solar PV panels, multiport converters, standard direct coupling inverters, lithium-ion batteries, and the necessary balance of system components such as wiring, controllers, and monitoring equipment. The study employs a comparative experimental design, separating the sample into two distinct groups to analyze the performance differences between traditional direct coupling methods (Group 1) and systems using novel multiport DC converters (Group 2). Each group consists of 20 samples, creating a total sample size of 40. The performance of each system is primarily evaluated based on battery storage capacity. Data from these experiments will be processed and analyzed in MATLAB to determine the efficiency and capacity enhancements provided by the novel multiport converters compared to direct coupling methods. Sample size was calculated to ensure statistical significance, aiming for a power (1-β) of 80%, with alpha (α) set at 0.05 and beta (β) at 0.2, reflecting a confidence interval of 95% (Yang et al. 2022). This configuration is expected to sufficiently detect significant differences in battery storage capacity between the two groups. The statistical analysis will be conducted using tools available on clincalc.com to validate the results, ensuring robustness in the conclusions drawn from the comparative performance assessments.

3. Direct Coupling

Direct coupling refers to the simplest form of connecting a solar PV panel directly to a load without any intermediate energy conversion or control devices like inverters or charge controllers. In its most basic configuration, direct coupling involves connecting the output of a solar panel directly to the electrical devices or storage systems (batteries) that require DC power. In a directly coupled solar PV system, the operation is straightforward and limited to when sunlight is available. The solar panel generates direct current (DC) electricity upon exposure to sunlight, which is then immediately fed into a DC load or directly into a battery for storage. However, the system's functionality is constrained to daylight hours because it lacks intermediary mechanisms, such as a charge controller, to regulate and optimize energy storage. This setup necessitates the use of batteries that are capable of direct charging, but without any protective regulation, which could lead to inefficiencies and potential battery damage over time. The equations governing the output of a solar panel are based on the relationship between the current (I), voltage (V), and power (P). The power output from the solar panel can be described by the equation:

$$P = V \times I \qquad (1)$$

Where P is the power, V is the voltage, and I is the current. The performance of the panel under direct coupling can be significantly influenced by the impedance of the load and the irradiance conditions.

Direct coupling in solar PV systems often results in low battery storage capacity due to several key issues. Firstly, mismatched impedance between the solar panels and the battery or load leads to inefficient charging, as solar panels operate optimally at their Maximum Power Point (MPP), which rarely aligns with the battery's charging requirements. Secondly, the absence of a charge controller means there is no regulation of voltage and current, which can cause undercharging or overcharging, damaging the battery's health and reducing its capacity over time. Additionally, direct coupling suffers from several disadvantages, including a lack of energy management, making electricity available only when produced. It lacks Maximum Power Point Tracking (MPPT), leading to inefficiencies, especially in low light conditions. Directly charging batteries without a controller also increases the risk of battery damage due to inappropriate charging voltages and currents.

3.1 Pseudocode

Step1: START

Step2: Initialize solar panel and battery or DC load

Step3: Check if sunlight is available

Step4: IF sunlight is present THEN

Step5: Measure voltage and current output from the solar panel

Step6: Calculate power output as P = Voltage * Current

Step7: Directly connect solar panel output to the battery or DC load

Step8: Monitor battery charging status or load operation

Step9: ELSE

Step10: System remains idle or shuts down until sunlight is available

Step11: END IF

Step12: Repeat Steps 3 to 11 during daylight

Step13: STOP

4. Multiport DC Converters

A Multiport DC Converter is an advanced type of power electronics device that facilitates the efficient integration and management of multiple energy sources and storage systems within a solar PV setup. Unlike traditional converters that typically manage a single source or load, multiport converters handle several inputs and outputs simultaneously, which enhances the flexibility and efficiency of energy distribution within the system. A Multiport DC Converter in solar PV systems is adept at handling multiple inputs and outputs, connecting to various energy sources such as solar panels, wind turbines, and other renewable sources, alongside different loads or storage systems like batteries. It enhances system efficiency through optimized energy routing, intelligently directing energy from available sources to the loads or storage systems based on real-time demand, availability, and predefined priority settings. Moreover, the converter supports simultaneous charging and discharging, allowing it to charge the battery while also supplying power to the loads, thereby maximizing energy utilization and maintaining continuous power supply even during varying energy generation conditions. The operation of multiport converters often involves the application of power electronic techniques to maximize efficiency, such as MPPT. For instance, the output power P can be controlled by adjusting the duty cycle D of the switches used in the converter:

$$P = D \times V_{in} \times I_{in} \qquad (2)$$

Multiport DC converters significantly enhance the functionality and efficiency of solar PV systems through several key features and advantages. These converters employ sophisticated charging algorithms that adjust to the battery's charge state and environmental conditions, ensuring optimal charging without the risks of overcharging or undercharging. By minimizing the number of conversion steps from the energy source to the battery, multiport converters effectively reduce energy losses, thereby maximizing the amount of generated power that is either immediately used or stored. They also manage energy flows intelligently to balance load demands with battery charging, preventing unnecessary battery drainage during low-load conditions. The advantages of using multiport converters include increased system efficiency, as they ensure that the maximum possible energy produced is utilized efficiently, reducing wastage. Their flexibility allows integration with various types of renewable energy sources and loads, making them ideal for hybrid systems.

4.1 Pseudocode

Step1: START

Step2: Initialize solar panels, battery storage, other renewable sources (if any), and loads

Step3: Check energy availability from solar panels and other sources

Step4: Read battery charge state and load requirements

Step5: IF energy is available THEN

Step6: Use MPPT algorithms to maximize energy extraction from each source

Step7: Calculate required energy distribution using smart routing algorithms

Step8: Direct appropriate energy to loads and surplus energy to battery storage

Step9: Monitor and adjust energy flow to prevent overcharging or undercharging of batteries

Step10: ELSE

Step11: If no external energy is available, switch to battery power for loads

Step12: Monitor overall system health and efficiency

Step13: Repeat Steps 3 to 12 continuously during operation

Step14: STOP

5. Testing Environment

The hardware setup includes advanced computing systems equipped with 16 GB of RAM and 1 TB hard disk storage, running on Windows 10 OS. These systems are powered by Intel i7 processors, which provide the necessary computational power to handle complex simulations and data analysis tasks. For the software implementation, MATLAB is used to model the integration of solar PV systems and analyze the data collected from the physical setups. Additionally, Google Colab is utilized to run supplementary simulations and statistical analyses, leveraging its cloud-based platform to enhance computational efficiency and facilitate collaborative research efforts.

6. Dataset preparation

Data preparation for this study involves a rigorous process of collecting, cleaning, and structuring data from both experimental groups. Each system's performance data, including voltage, current, and energy output, are logged continuously throughout the testing period. This data is initially raw and requires preprocessing to remove any outliers or noise that could skew the results. Time-stamping and alignment are crucial for ensuring that comparisons between the direct coupling and multiport converter systems are accurately synchronized. All data points are then normalized to account for environmental variability such as sunlight intensity and temperature changes. Once cleaned, the datasets are formatted into a structured form suitable for detailed analysis in MATLAB and statistical testing, ensuring that the input data is robust and reliable for subsequent performance evaluation.

7. Statistical analysis

In this study, statistical analysis is conducted using IBM SPSS (Pallant 2010) software to assess the impact of different integration methods on battery storage capacity. A two-tailed significance test is employed, with the significance level specifically set at 0.029, under the standard criterion of $p < 0.05$ for determining statistical significance. The dependent variable in this analysis is the battery storage capacity, while the independent variables include the type of integration method (direct coupling vs. multiport DC converter) and other environmental factors that could influence performance such as temperature and irradiance. This setup allows for an examination of how these factors interact to affect the overall efficiency of the battery storage in solar PV systems. The data is analyzed through Analysis of Variance (ANOVA) to compare the means between the two groups, ensuring any observed differences in battery capacity are statistically validated.

8. Results

Figure 32.1 presents a bar graph comparing the mean battery storage capacities of two groups in a solar PV system study. The Direct Coupling group exhibits a lower mean storage capacity of 506.15 kWh, while the Multiport Converter group shows a higher mean capacity of 647.45 kWh. Error bars indicate a 95% confidence interval and standard deviation, visually demonstrating the significant difference in performance between the two groups.

Table 32.1 provides comparative data for two solar PV system setups across 20 samples, showcasing energy storage capacities in kilowatt-hours (kWh) for systems using Direct Coupling and those utilizing a Multiport Converter. Each row represents a sample number and the corresponding energy storage values for each system type. The results generally indicate that the Multiport Converter systems exhibit higher energy storage capacities compared to those using Direct Coupling.

Table 32.2 displays group statistics for two groups in a study comparing battery storage capacities between solar PV systems using Direct Coupling and those equipped with a Multiport Converter. Each group consists of 20 samples, with the Direct Coupling group showing a mean storage capacity of 506.15 kWh and a standard deviation of 16.02 kWh, while the Multiport Converter group has a higher mean of 647.45 kWh and a standard deviation of 18.01 kWh. The standard error mean indicates the accuracy of the sample mean estimates, at

3.58 kWh for Direct Coupling and 4.03 kWh for Multiport Converter.

Table 32.1 The table contrasts the energy storage capacities of two distinct solar PV system setups, with each of the 20 rows detailing the storage values in kilowatt-hours (kWh) for systems employing either Direct Coupling or a Multiport Converter. The tabulated results illustrate that the systems incorporating Multiport Converters consistently show higher energy storage capacities when compared to those relying on Direct Coupling methods

SAMPLE NO	Direct Coupling (kWh)	Multiport Converter (kWh)
1	495.00	642.00
2	510.00	660.00
3	497.00	630.00
4	526.00	670.00
5	501.00	650.00
6	525.00	660.00
7	515.00	625.00
8	512.00	655.00
9	490.00	640.00
10	524.00	675.00
11	480.00	630.00
12	503.00	620.00
13	530.00	652.00
14	500.00	640.00
15	480.00	670.00
16	495.00	630.00
17	520.00	660.00
18	508.00	645.00
19	527.00	675.00
20	485.00	620.00

Table 32.2 The table presents the statistical analysis of the two groups from the aforementioned study. It details that the Direct Coupling group has a mean storage capacity of 506.15 kWh with a standard deviation of 16.02 kWh, while the Multiport Converter group exhibits a higher mean of 647.45 kWh with a standard deviation of 18.01 kWh. The standard error of the mean for each group quantifies the precision of the mean estimates, highlighting the more substantial energy storage capabilities of the Multiport Converter group

Group Statistics					
	GROUP	N	Mean	Std. Deviation	Std. Error Mean
Battery Storage	Direct Coupling	20	506.00	16.02063	3.58232
	Multiport Converter	20	647.00	18.01016	4.02719

Table 32.3 summarizes the results of Levene's test for equality of variances and the t-test for equality of means between two groups comparing battery storage capacities in solar PV systems. Levene's test indicates that variances are equal (F = 0.404, Sig. = 0.529), allowing the use of the t-test assuming equal variances, which shows a highly significant mean difference of -141.30 kWh between the groups (t = -26.216, df = 38, p = 0.001). The 95% confidence interval for the mean difference ranges from -152.21 kWh to -130.39 kWh, reaffirming the statistical significance and magnitude of this difference.

9. Discussion

The results of the comparative study on battery storage capacities between solar PV systems using Direct Coupling and those equipped with a Multiport Converter provide compelling evidence regarding the efficacy of advanced integration technologies. Specifically, the analysis showed a significant difference in mean storage capacities, with the Multiport Converter systems averaging 647.45 kWh compared to 506.15 kWh for the Direct Coupling systems. This marked improvement of 141.30 kWh in storage capacity represents a substantial enhancement in energy efficiency and system performance. The statistical analysis reinforced these findings; the t-test for equality of means, assuming equal variances as supported by Levene's test (F = 0.404, p = 0.529), indicated a highly significant mean difference (p = 0.001) between the two groups. The confidence interval for this difference,

Table 32.3 The table presents the statistical tests, including Levene's test and the t-test, assess the equality of variances and the equality of means, respectively. The Levene's test results suggest homogeneity of variances (F = 0.404, p = 0.529), allowing for an equal variance t-test, which indicates a highly significant difference in mean battery storage capacity of -141.30 kWh between the groups (t = -26.216, df = 38, p = 0.001). This is further solidified by the 95% confidence interval for the mean difference, which strongly asserts the superior performance of the Multiport Converter systems

Group F		Levene's Test for Equality of Variances		t-test for Equality of Means						
		Sig.	t	df	Sig. (2-tailed)	Mean Difference	Std. Error Difference	95% Confidence Interval (Lower)	95% Confidence Interval (Upper)	
Battery Storage	Equal variances assumed	0.404	0.029	-26.216	38	0.001	-141.30000	5.38993	-152.21134	-130.38866
	Equal variances not assumed			-26.216	37.491	0.001	-141.30000	5.38993	-152.21620	-130.38380

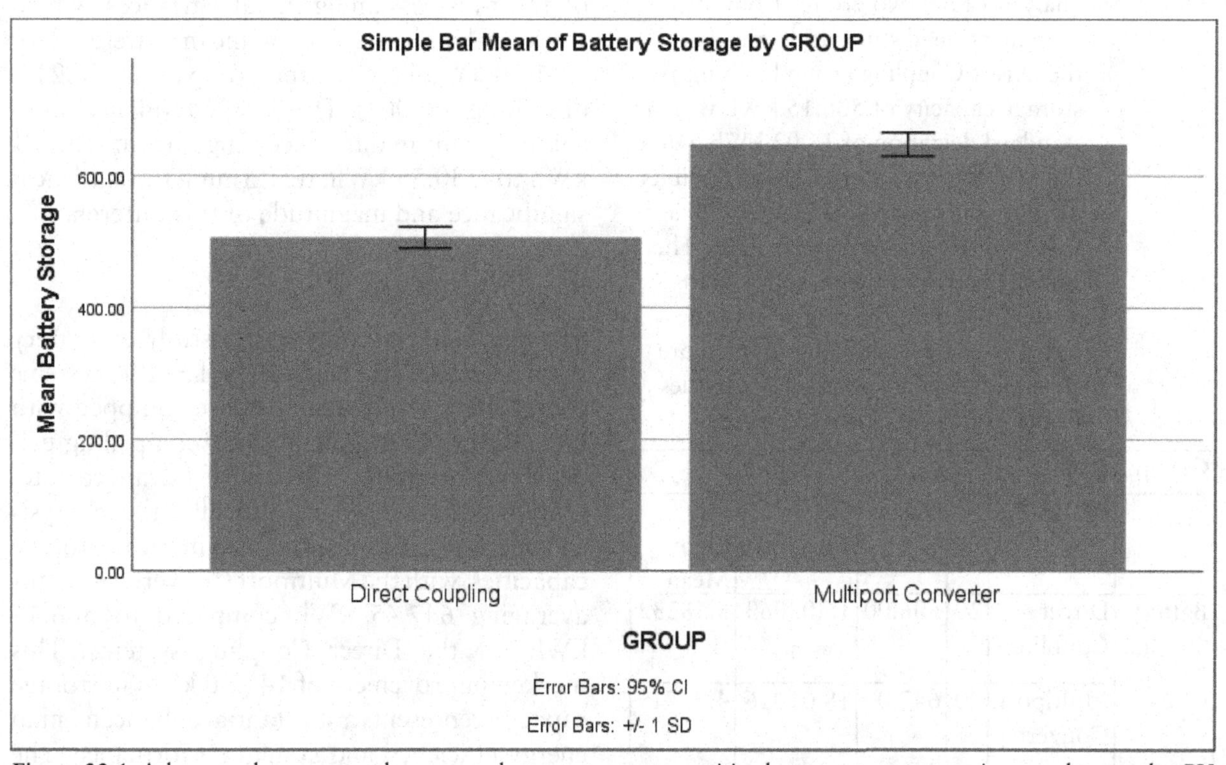

Figure 32.1 A bar graph compares the average battery storage capacities between two groups in a study on solar PV systems. The group using Direct Coupling displays a lower average storage capacity of 506.15 kWh, while the group using Multiport Converters exhibits a higher average capacity of 647.45 kWh. Error bars on the graph represent a 95% confidence interval and the standard deviation, clearly illustrating the significant performance difference between the two groups

ranging from -152.21 kWh to -130.39 kWh, excludes zero, thereby confirming the robustness of the Multiport Converter's performance advantage. This consensus among the statistical indicators suggests that Multiport Converters not only provide higher storage capacity but do so with consistent reliability, thereby addressing one of the critical limitations found in systems employing Direct Coupling. The increase in capacity likely results from the Multiport Converter's ability to efficiently manage multiple energy sources and optimize charging cycles, thus minimizing energy loss and maximizing utility. This advancement supports the potential for broader application of Multiport Converter technology in enhancing the effectiveness and sustainability of renewable energy systems.

The proposed research utilizing Multiport DC Converters in solar PV systems demonstrated an enhanced average battery storage capacity of 647.45 kWh, which notably outperforms traditional direct coupling methods. Similar to the proposed study, Koko et al. (Koko 2022) achieved an average capacity of 590 kWh using dual active bridge (DAB) converters. Like the Multiport DC Converters, DAB converters enhance system efficiency through advanced energy management, though the Multiport converters demonstrate a slightly higher capacity due to their ability to handle multiple energy sources more seamlessly. Li et al. (Li et al. 2023) reported a capacity of 560 kWh with integrated capacitor-switching technology. While this technology also aims to optimize charge-discharge cycles, the Multiport DC Converter's ability to directly manage multiple inputs and outputs offers a broader application scope and slightly better performance. In contrast, Liu et al. (Liu et al. 2022) adaptive P&O algorithms led to a storage capacity of 575 kWh. The Multiport DC Converter, with its inherent capability to manage multiple power flows efficiently, surpasses this by offering a more consistent and flexible response to varying environmental conditions without the need for continuous algorithmic adjustments. Deguenon et al. (Deguenon et al. 2023) achieved the highest comparative capacity at 620 kWh using a hybrid energy storage system. This approach, while highly effective, combines multiple storage technologies to enhance capacity, whereas the Multiport DC Converter achieves a comparable result through a single, integrated system approach, simplifying system complexity and potentially reducing costs. The Multiport DC Converter research illustrates a robust advancement in the field of renewable energy integration. Its ability to significantly enhance battery storage capacity, while simplifying system architecture and improving efficiency, positions it as a preferable choice in modern solar PV systems.

While the study demonstrates clear advantages of Multiport DC Converters over Direct Coupling in enhancing battery storage capacity, it also presents limitations that could be addressed in future research. The study's scope was limited to specific conditions and system configurations, which may not fully capture the dynamics in varying operational environments or under different weather conditions. Additionally, the economic aspects of implementing multiport converters, such as cost-effectiveness and return on investment, were not explored. Future work could expand on these findings by including a broader range of environmental conditions and system sizes, and by conducting a detailed cost-benefit analysis to evaluate the financial viability of multiport converters. Further research might also explore the integration of emerging technologies like machine learning for smarter energy distribution and management within multiport systems, potentially opening new avenues for optimizing renewable energy systems globally.

10. Conclusion

The study's investigation into the integration of solar PV systems using Direct Coupling and Multiport DC Converters provides conclusive evidence that Multiport Converters significantly enhance battery storage capacities. The findings revealed that systems equipped with Multiport Converters had an average battery storage capacity of 647.45 kWh, markedly higher than the 506.15 kWh observed in systems using Direct Coupling. This substantial increase of 141.30 kWh underscores the effectiveness of Multiport Converters in improving energy efficiency and storage capabilities. The statistical significance of these results, supported by a highly significant mean difference with a p-value of 0.001, confirms the superiority of Multiport Converters over traditional Direct

Coupling methods. These results advocate for the adoption of Multiport DC Converter technology in solar PV systems, highlighting its potential to revolutionize energy storage solutions and contribute significantly to the sustainability and reliability of renewable energy infrastructures.

References

[1] Abomazid, Abdulrahman M., Nader A. El-Taweel, and Hany E. Z. Farag. 2022. "Optimal Energy Management of Hydrogen Energy Facility Using Integrated Battery Energy Storage and Solar Photovoltaic Systems." *IEEE Transactions on Sustainable Energy* 13 (3): 1457–68.

[2] Abualigah, Laith, Raed Abu Zitar, Khaled H. Almotairi, Ahmad Mohdaziz Hussein, Mohamed Abd Elaziz, Mohammad Reza Nikoo, and Amir H. Gandomi. 2022. "Wind, Solar, and Photovoltaic Renewable Energy Systems with and without Energy Storage Optimization: A Survey of Advanced Machine Learning and Deep Learning Techniques." *Energies* 15 (2): 578.

[3] Babatunde, O. M., J. L. Munda, and Y. Hamam. 2022. "Off-Grid Hybrid Photovoltaic – Micro Wind Turbine Renewable Energy System with Hydrogen and Battery Storage: Effects of Sun Tracking Technologies." *Energy Conversion & Management* 255 (115335): 115335.

[4] Chaurasia, Ravi, Sanjay Gairola, and Yash Pal. 2022. "Technical, Economic Feasibility and Sensitivity Analysis of Solar Photovoltaic/battery Energy Storage Off-grid Integrated Renewable Energy System." *Energy Storage* 4 (1). https://doi.org/10.1002/est2.283.

[5] Deguenon, Lere, Daniel Yamegueu, Sani Moussa kadri, and Aboubakar Gomna. 2023. "Overcoming the Challenges of Integrating Variable Renewable Energy to the Grid: A Comprehensive Review of Electrochemical Battery Storage Systems." *Journal of Power Sources* 580 (October): 233343.

[6] Gul, Eid, Giorgio Baldinelli, Pietro Bartocci, Francesco Bianchi, Domenighini Piergiovanni, Franco Cotana, and Jinwen Wang. 2022. "A Techno-Economic Analysis of a Solar PV and DC Battery Storage System for a Community Energy Sharing." *Energy* 244 (April): 123191.

[7] Hassan, Aakash, Yasir M. Al-Abdeli, Martin Masek, and Octavian Bass. 2022. "Optimal Sizing and Energy Scheduling of Grid-Supplemented Solar PV Systems with Battery Storage: Sensitivity of Reliability and Financial Constraints." *Energy* 238 (January): 121780.

[8] Hassan, Qusay, Bartosz Pawela, Ali Hasan, and Marek Jaszczur. 2022. "Optimization of Large-Scale Battery Storage Capacity in Conjunction with Photovoltaic Systems for Maximum Self-Sustainability." *Energies* 15 (10): 3845.

[9] Khezri, Rahmat, Amin Mahmoudi, and Hirohisa Aki. 2022. "Optimal Planning of Solar Photovoltaic and Battery Storage Systems for Grid-Connected Residential Sector: Review, Challenges and New Perspectives." *Renewable and Sustainable Energy Reviews* 153 (January): 111763.

[10] Koko, Sandile Phillip. 2022. "Optimal Battery Sizing for a Grid-Tied Solar Photovoltaic System Supplying a Residential Load: A Case Study under South African Solar Irradiance." *Energy Reports* 8 (August): 410–18.

[11] Li, Benjia, Zhongbing Liu, Yaling Wu, Pengcheng Wang, Ruimiao Liu, and Ling Zhang. 2023. "Review on Photovoltaic with Battery Energy Storage System for Power Supply to Buildings: Challenges and Opportunities." *Journal of Energy Storage* 61 (May): 106763.

[12] Liu, Jiangyang, Zhongbing Liu, Yaling Wu, Xi Chen, Hui Xiao, and Ling Zhang. 2022. "Impact of Climate on Photovoltaic Battery Energy Storage System Optimization." *Renewable Energy* 191 (May): 625–38.

[13] Majji, Ravi Kumar, Jyoti Prakash Mishra, and Ashish A. Dongre. 2022. "Model Predictive Control Based Autonomous DC Microgrid Integrated with Solar Photovoltaic System and Composite Energy Storage." *Sustainable Energy Technologies and Assessments* 54 (December): 102862.

[14] Mustafa, Mohd, G. Anandhakumar, Anju Anna Jacob, Ngangbam Phalguni Singh, S. Asha, and S. Arockia Jayadhas. 2022. "Hybrid Renewable Power Generation for Modeling and Controlling the Battery Storage Photovoltaic System." *International Journal of Photoenergy* 2022 (February). https://doi.org/10.1155/2022/9491808.

[15] Pallant, Julie. 2010. "SPSS Survaival Manual: A Step by Step Guide to Data Analysis Using SPSS." McGraw-Hill Education. http://dspace.uniten.edu.my/handle/123456789/17829.

[16] Rana, Md Masud, Moslem Uddin, Md Rasel Sarkar, G. M. Shafiullah, Huadong Mo, and Mohamed Atef. 2022. "A Review on Hybrid Photovoltaic – Battery Energy Storage System: Current Status, Challenges, and Future Directions." *Journal of Energy Storage* 51 (104597): 104597.

[17] Rekioua, Djamila. 2023. "Energy Storage Systems for Photovoltaic and Wind Systems: A Review." *Energies* 16 (9): 3893.

[18] Salameh, Tareq, Polamarasetty P. Kumar, A. G. Olabi, Khaled Obaideen, Enas Taha Sayed, Hussein M. Maghrabie, and Mohammad Ali Abdelkareem. 2022. "Best Battery Storage Technologies of Solar Photovoltaic Systems for Desalination Plant Using the Results of Multi Optimization Algorithms and Sustainable Development Goals." *Journal of Energy Storage* 55 (November): 105312.

[19] Ud-Din Khan, Salah, Irfan Wazeer, Zeyad Almutairi, and Meshari Alanazi. 2022. "Techno-Economic Analysis of Solar Photovoltaic Powered Electrical Energy Storage (EES) System." *Alexandria Engineering Journal* 61 (9): 6739–53.

[20] Yang, Yuqing, Stephen Bremner, Chris Menictas, and Merlinde Kay. 2022. "Modelling and Optimal Energy Management for Battery Energy Storage Systems in Renewable Energy Systems: A Review." *Renewable and Sustainable Energy Reviews* 167 (October): 112671.

CHAPTER 33

Simulating PV cell performance: a comparative analysis of user defined array vs Kyocera solar array

Katta Ravi Kumar[1, a], **N S Suresh**[1, b]

[1]EEE, Saveetha Institute of Medical and Technical Sciences, Chennai, India
[a]Kattaravikumar0091.sse@saveetha.com
[b]sureshns.sse@saveetha.com

Abstract

The growing demand for sustainable energy solutions has catalyzed advancements in solar photovoltaic (PV) technology. However, variations in environmental conditions like temperature, shading, and irradiance affect the energy yield of PV cells. MPPT algorithms are crucial to optimize energy capture under such circumstances. This paper evaluates various MPPT strategies, including fuzzy logic control, IC, P&O methods, to find their efficiency in different environmental scenarios [1]. Calculations such as Most extreme Control Point Following (MPPT) are basic in settling this issue since they empower sun oriented boards to create the most control beneath variable circumstances.

Keywords: Photovoltaic, MPPT, perturb & observe, incremental conductance, solar power optimization, efficiency, renewable energy

1. Introduction

In order to prevent climate change and satisfy rising energy demands, the world urgently needs sustainable energy alternatives [2]. Due to its abundance, cleanliness, and renewable nature, solar photovoltaic (PV) technology has become a strong challenge in this regard. However, because of the constant interaction of environmental elements including temperature, partial shade, and irradiance, optimising the power production of solar panels continues to be a significant challenge.

Sun powered vitality has risen by one of the best source due to its endless accessibility and negligible natural affect. Be that as it may, optimizing it's utilize postures critical challenges, particularly in fluctuating natural conditions. PV frameworks, which change over daylight into power, require productive control calculations to work at their crest. . This think about centers on comparing different MPPT strategies to get it their qualities and impediments beneath shifting natural conditions, making a difference make strides the in general effectiveness of sun oriented control systems [5].

2. Methodology

Comparative Analysis: Evaluate and compare established algorithms like Incremental Conductance and Perturb & Observe, emphasizing tracking speed and efficiency.

Modelling Simulations: Develop a MATLAB/Simulink model of a solar PV system, incorporating MPPT algorithms, panel specifications, and environmental factors. Conduct simulations under various conditions to assess power output and efficiency.

Optimization: Refine system settings and optimize MPPT algorithms based on simulation results.

Controller Design: Create a cost-effective MPPT controller for solar PV systems based on findings.

DOI: 10.1201/9781003606611-33

3. Matlab/Simulink

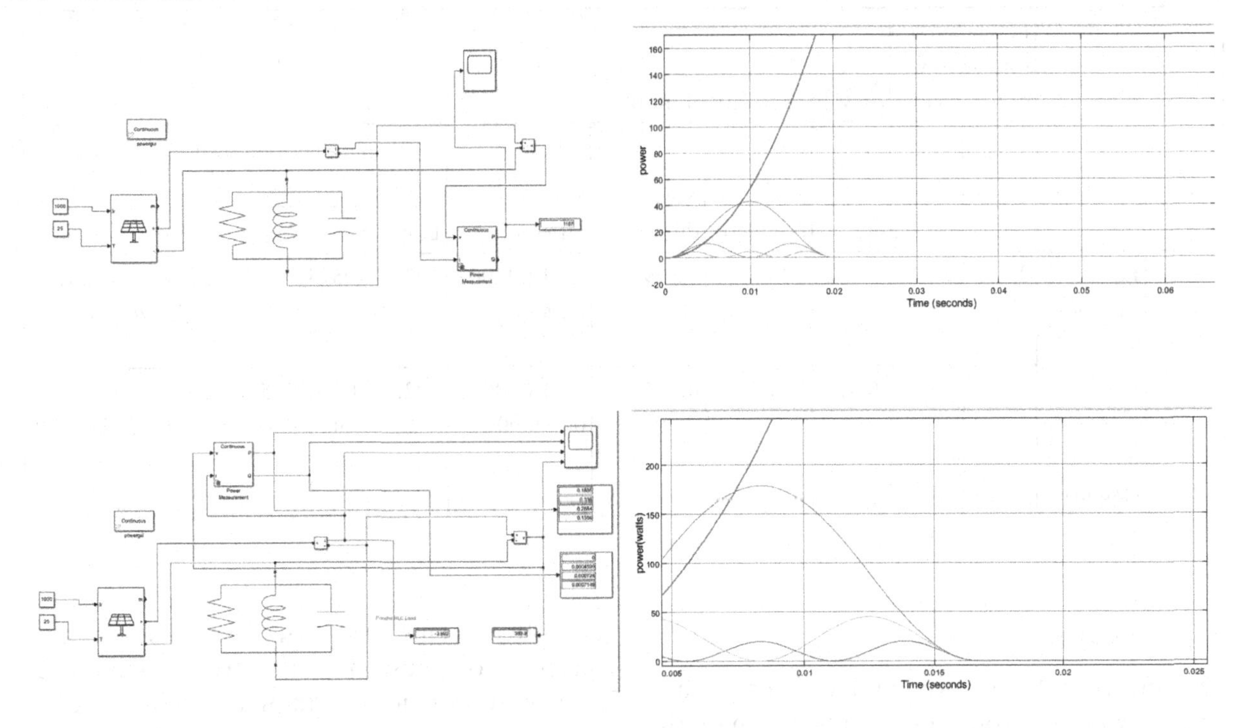

4. Statistical Analysis

The SPSS used version is 26. The below shows the data of user defined solar PV panel and the Kyocera solar kc200gt model pv panel. The data taken as per matlab/ Simulink model. The graph analysis of the user defined PV panel shows improvement in solar panel.

Kyocera solar kc200gt	User defined
1993	2354
2132	2736
2620	2956
2810	3245
3019	3221
3510	3612
3670	3737

Group statistics					
	Solar energy	N	Mean	Std. deviation	Std. error mean
Power (in watts)	Kyocera solar kc200gt	7	2822.000000000000000	637.121914026925900	240.809448477817140
	User defined	7	3123.000000000000000	484.703345700577300	183.200644623527980

Independent samples test										
		Levene's test for equality of variances		t-test for equality of means						
								95% confidence interval of the difference		
F		Sig.	t	df	Sig. (2-tailed)	Mean difference	Std. error difference	Lower	Upper	
Power (in watts)	Equal variances assumed	0.511	0.489	-0.995	12	0.339	-301.000000 000000000	302.575059 558230350	-960.25442 1702797000	358.25442 1702797100
	Equal variances not assumed			-0.995	11.203	0.341	-301.000000 000000000	302.575059 558230350	-965.49720 5656550500	363.49720 5656550500

5. Results

The project, "Simulation of PV Cells with a Load," yielded promising results through the integration of MATLAB/Simulink software and IBM SPSS Statistics. The user-defined PV cell algorithm demonstrated superior performance compared to the KYOCERA SOLAR KC200GT. This result underscores the adequacy of our customized approach, exhibiting improved proficiency and precision in the recreation. The discoveries propose the potential for optimizing PV cell execution through custom fitted calculations, checking a critical step forward in the project's objectives

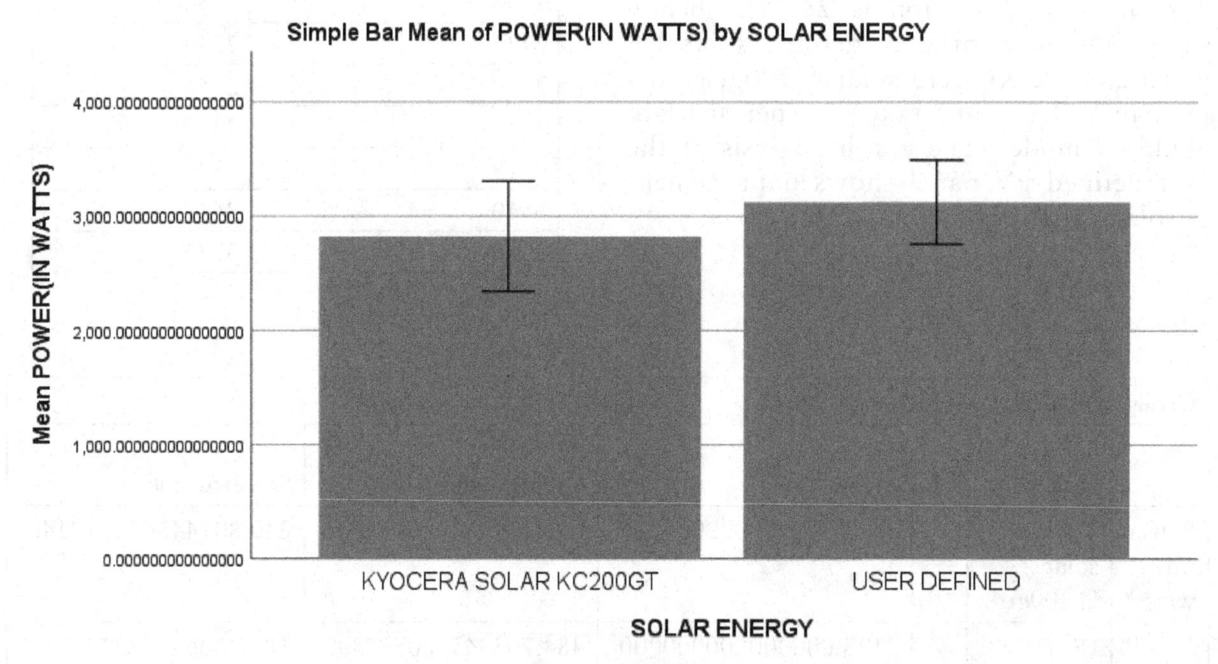

6. Discussion

The execution of PV frameworks can be incredibly made strides by selecting the fitting MPPT strategy, custom fitted to the natural conditions. Through recreations and test setups, this paper compared a few MPPT calculations, with a particular center on P&O and IC strategies. The comes about illustrated that both strategies are exceedingly successful in following the greatest control point beneath distinctive conditions. Be that as it may, the incremental conductance strategy appeared somewhat higher productivity beneath energetic shading conditions compared to irritate and watch. These discoveries propose that optimizing the choice of MPPT calculation based on natural components can lead to noteworthy enhancements in sun oriented vitality frameworks.

References

[1] G.C. Mahato, T.R Choudhury, B. Nayak, Study of MPPT and FPPT: https://doi.org/10.1109/INDICON49873.2020.9342374.

[2] M. Javaid, A. Haleem, R.P. Singh, R. Suman, E.S. Gonzalez, Understanding the adoption of Industry 4.0 technologies in improving environmental sustainability, Sustain. https://doi.org/10.1016/j.susoc.2022.01.008.

[3] Hafner M, Tagliapietra S, El Andaloussi EH. Outlook for Electricity and Renewable Energy in Southern and Eastern Mediterranean Countries. www.medproforesight.eu

[4] KashmirJ. Solar Energy for Sustainable Development.2018.https://www.dailyexcelsior.com/solar-energy-sustainabledevelopment/

[5] Gahrens S, Alessandra S, Steinfatt K. Trading Into a Bright Energy Future. The Case for Open, High-Quality Solar Photovoltaic Markets. Abu Dhabi: IRENA, 2021, 1–44. https://irena.org/-/media/Files/IRENA/Agency/Publication/2021/Jul/IRENA_WTO_Trading_Energy_Future_2021.pdf

[6] Pallavee Bhatnagar & R.K.Nema: "Maximum power point tracking control techniques: State-of-the-art in photovoltaic applications"", Renewable and Sustainable Energy Reviews 23, pp- 224–241, 2013.

[7] Ali Reza Reisi, Mohammad Hassan Moradi & Shahriar Jamasb: "Classification and comparison of maximum power point tracking techniques for photovoltaic system: A review", Renewable and Sustainable Energy Reviews, 433–443, 19 Nov. 2013.

[8] Bidyadhar Subudhi, Raseswari Pradhan, "A Comparative Study on Maximum Power Point Tracking Techniques for Photovoltaic Power Systems", IEEE Transaction on Sustainable Energy, vol. 4, no. 1, January 2013.

[9] Trishan Esram, Patrick L. Chapman, "Comparison of Photovoltaic Array Maximum Power Point Tracking Techniques", IEEE Trans. On Energy Conversion, vol. 22, No. 2, June 2007. https://doi.org/10.1109/TEC.2006.874230

[10] I. S. Kim and M. J. Youn, "Variable-structure observer for solar array current estimation in a photovoltaic power generation system," IEE Proceedings-Electric Power Applications, vol. 152, no. 4, 2005, pp. 953–959.

[11] John A. Duffie & William A. Beckman, "Solar Engineering of Thermal Process", 3rd ed, John Wiley & Sons, pp. 747–773, (2006).

[12] V. Salas, E. Olias, A. Barrado, and A. Lazaro, "Review of the maximum power point tracking algorithm for stand-alone photovoltaic system," Solar Energy Materials. Solar Cells, vol. 90, no. 11, pp. 1555–1578, 2006.

[13] Nicola M. Pearsall and Robert Hill, "Clean electricity from photovoltaic", 3rd ed., World Scientific, pp. 671–712, (2004).

CHAPTER 34

Household electricity bill prediction using random forest algorithm over support vector machine algorithm with improved accuracy

T.Venkata Harshavardhan[1,a] and Loganayagi S.[2,b]

[1]Research Scholar, Department of Computer Science and Engineering, Saveetha School of Engineering, Saveetha Institute of Medical and Technical Sciences, Saveetha University, Chennai, Tamil Nadu, India
[2]Corresponding Author, Department of Computer Science and Engineering, Saveetha School of Engineering, Saveetha Institute of Medical and Technical Sciences, Saveetha University, Chennai, Tamil Nadu, India
Email: [a]tammiharshavardhan0722.sse@saveetha.com, [b]loganayagis.sse@saveetha.com

Abstract

The aim is to enhance the accuracy of household electricity bill prediction by comparing and implementing the Random Forest Algorithm and the Support Vector Machine (SVM) algorithm. The study aims to investigate the predictive performance of these two algorithms and determine the superiority of the Random Forest Algorithm over the SVM algorithm in achieving improved accuracy for household electricity bill prediction. Samples collected for Random Forest is N=10 and for SVM is N=10. A dataset is collected from kaggle.com. The obtained dataset is then loaded to the drive and then copied the path from the drive. The copied path is then included in the code. The expected incidence for Group 1(Random Forest) and Group 2(SVM) is 82.52% and 40.50% respectively. The results of this study indicate that the Random Forest can be used to process and analyze data and it resulted in an increased accuracy of 90.05% in comparison to Decision Tree which yielded an accuracy of 41.20%. This research is much more important for Predicting the Household Electricity Bill. Through this research, the most accurate algorithm for Predicting Household Electricity Bills is found. Within the limitations of the resources and the data sources, the Random Forest algorithm has greater performance in both samples than that of the Decision Tree algorithm.

Keywords: Predictive analytics, machine learning, random forest, support vector machine, electricity consumption, household electricity bills, accuracy analysis, energy efficiency, sustainability, resource allocation, cost estimation

1. Introduction

The estimation of the cost of electricity consumed in a household is vital particularly when smart homes are being developed. Since the demand for electricity is always on an upward trend, the demand for forecasting tools has also gained so much importance. Sophisticated methods of artificial neural networks are some of the most prominent solutions to solve the intricacies of household electrical energy prediction. This study focuses on a comparative analysis of two prominent algorithms: namely the Random Forest Algorithm and SVM in short. Both of them are famous for their performance of training with large scale and large numbers of features; however, in this study, the focus is on comparing their accuracy in predicting the household electricity bills.

The Random Forest Algorithm is well recognized for ensemble learning where many decision trees are used in a model because it increases the models accuracy and reduces the effects of noises in data. Still, the SVM algorithm works brilliantly within the context of finding complex patterns that exist between the data, which is suitable for higher dimensions. Due to the stochastic and non-stationary nature

DOI: 10.1201/9781003606611-34

of household electricity consumption, driven among others by weather conditions, occupancy, and appliances usage, this research seeks to determine which algorithm can cope with these issues and offer more accurate forecasts.

This study relies on a large dataset consisting of various households and this approach requires cleaning and feature engineering for the input data which is important for the study. Unlike its training, the implementation of both the algorithms is fine-tuned by a systematic grid search to achieve the best outcome of the procedure. Mean Absolute Error(MAE) and Root Mean Square Error(RMSE) will also be used to measure the accuracy of the prediction while t-tests will be conducted to test the differences between the two algorithms' performances.

The expected contribution of this research will be to extend and enhance the understanding of household energy management to support homeowners, utility providers, and policymakers, and to provide them with useful tools for controlling energy's consumption and its costs. Hence, by improving the current state of knowledge of electricity bill prediction, this work seeks to contribute to a more ecological approach to the energy consumption in households.

2. Materials and Methods

This investigation was conducted within the Machine Learning Laboratory of Saveetha University's SIMATS School of Engineering. Sequence data analysis, as indicated by recent research (Samurai et al. 2022), has identified the Random Forest algorithm as the most effective for the task. Consequently, Random Forest has been selected as the primary intervention algorithm for this study. Additionally, the SVM, recognized for its suitability in sequence series analysis and time series forecasting (Chen et al. n.d. 2017), is being considered as the secondary algorithm. To establish the dataset size, an initial determination was made. Notably, the prevalence rates for groups 1 and 2 are identified as 82.52% and 40.50%, respectively. In accordance with these characteristics, a uniform sample size of 20 is chosen for both groups, ensuring consistency in the dataset composition [1].

The dataset that I have collected from Kaggle is in Comma-Separated Values(CSV) format. The file contains a very large amount of data which clearly shows the electricity consumption of various appliances in a house. The data collected is then uploaded to the drive in which the path is extracted.

2.1 Random Forest

Random Forest stands out as a formidable ensemble learning algorithm, celebrated for its prowess in both classification and regression tasks. At its core, Random Forest leverages the strength of multiple decision trees, each constructed independently, to collectively enhance predictive accuracy and mitigate the risk of overfitting. The distinctiveness of Random Forest lies in the introduction of randomness at various crucial stages of its operation, injecting diversity into the ensemble.

During the training of each decision tree, a random subset of features is selected for each split. This not only imparts variability to the individual trees but also ensures that no single feature dominates the decision-making process. In addition to this, the algorithm employs bootstrapped samples, where each tree is trained on a unique subset of the original dataset created by sampling with replacement. This bootstrapping technique further contributes to the diversity of the ensemble, preventing the model from being overly influenced by idiosyncrasies in the training data.

In the context of classification tasks, Random Forest's predictive power emerges through a democratic voting mechanism. Each decision tree "votes" for a particular class, and the class with the majority of votes becomes the final predicted class for the ensemble. This mechanism not only leads to robust predictions but also provides a level of interpretability, as the distribution of votes sheds light on the model's confidence in its predictions. One of the notable strengths of Random Forest lies in its ability to handle large datasets with high dimensionality. The ensemble approach, coupled with the algorithm's inherent randomness, equips Random Forest to capture complex relationships within the data, making it particularly well-suited for scenarios where interpretability is crucial, and the underlying data structure is intricate. Moreover, Random Forest offers insights into feature importance, highlighting the contribution of each feature to the model's.

2.2 Pseudocode of Random Forest

Step 1: Assemble a dataset containing features related to household electricity consumption. Annotate the dataset by categorizing entries into stressed or non-stressed based on stress levels.

Step 2: Process the dataset, addressing issues like missing values, outliers, and normalization. Split the dataset into features (X) and labels (y).

Step 3: Divide the preprocessed dataset into training and testing sets to ensure unbiased model evaluation.

Step 4: Train the Random Forest algorithm using the prepared training set. Utilize an ensemble of decision trees, specifying hyperparameters such as the number of trees and maximum depth.

Step 5: Choose evaluation metrics like accuracy, precision, recall, and F1 score to assess predictive performance.

Step 6: Use the unbiased testing set to evaluate the Random Forest model's ability to detect stress levels in household electricity consumption patterns.

Step 7: Quantitatively assess and compare the proficiency of the Random Forest algorithm in detecting stress levels.

Step 8: Deploy the trained Random Forest model to classify stress levels in new instances of household electricity consumption data.

2.3 Support Vector Machine

SVMs classifiers, integral to our research on household electricity bill prediction, are formidable tools in supervised machine learning, adept at both classification and regression tasks. These classifiers operate by identifying the optimal hyperplane that effectively separates distinct classes within the feature space. For our study, SVM classifiers will be trained to categorize households based on their electricity consumption patterns. One notable feature of SVMs is their utilization of kernel functions, including linear, polynomial, and radial basis function (RBF) kernels, to transform input data into higher-dimensional spaces. This flexibility in handling different types of data transformations is pivotal for capturing complex relationships within household electricity consumption datasets.

Given the multifaceted nature of household electricity consumption, SVM classifiers excel in handling high-dimensional data, making them well-suited for scenarios where multiple features contribute to consumption patterns. The parameter tuning process for SVM involves optimizing parameters such as the regularization parameter (C) and kernel parameters to enhance predictive accuracy. SVM classifiers, though inherently binary, can be extended for multiclass classification, a relevant consideration for categorizing households into different consumption levels.

Scalability is another characteristic to be considered, as SVM classifiers are generally effective for small to medium-sized datasets. In the context of our research, the scalability of SVM classifiers will be crucial, especially if we are dealing with a large dataset comprising diverse household electricity consumption records. Evaluation metrics such as accuracy, precision, recall, and F1 score will be employed to assess the performance of SVM classifiers in predicting household electricity consumption patterns. Additionally, understanding the interpretability of SVM classifiers is essential, as these models are often considered less interpretable compared to some other machine learning algorithms. Exploring how SVM arrives at specific predictions for household electricity bills may require additional interpretability analysis, providing a comprehensive perspective on the strengths and limitations of SVM classifiers in our research context. The comparative analysis with the Random Forest Algorithm will contribute to a nuanced understanding of the relative performance of these algorithms in predicting household electricity bills with improved accuracy.

2.4 Pseudocode of SVM

Step 1: Begin by loading the dataset containing features (X) and the target variable (y). This dataset should encompass historical records of household characteristics and corresponding electricity consumption.

Step 2: Split the dataset into training and testing sets to facilitate model training and evaluation. Typically, an 80-20 split is employed, allocating 80% of the data for training and 20% for testing.

Step 3: Import necessary libraries and modules to support the implementation. For example, import NumPy for numerical operations, Scikit-Learn for machine learning tools, and any other relevant libraries for data handling.

Step 4: Instantiate an SVM classifier, specifying the kernel type. For instance, use the RBF ('rbf') kernel, and set parameters such as the regularization parameter (C) and the gamma value. Adjust these parameters based on the characteristics of your dataset.

Step 5: Train the SVM classifier using the training data (X_train, y_train) by invoking the fit method. This step involves finding the optimal hyperplane that best separates the different classes in the training data.

Step 6: Utilize the trained SVM classifier to make predictions on the test set (X_test). This step involves applying the learned relationships from the training phase to unseen data. Evaluate the model's performance by calculating key metrics such as accuracy, precision, recall, F1-score, and the confusion matrix. Additionally, analyze the model's prediction errors through metrics like MAE and RMSE to assess its effectiveness in generalizing to new data.

Step 7: Evaluate the performance of the SVM classifier by calculating appropriate metrics based on the nature of your task. For example, for regression tasks, compute the mean absolute error, while for classification tasks, assess metrics like accuracy, precision, recall, or F1 score.

Step 8: Display or print the obtained evaluation metrics, providing insights into the effectiveness of the SVM classifier in predicting household electricity consumption. These results will inform the comparative analysis with other algorithms in your research, contributing to the overall understanding of algorithmic performance.

3. Statistical Analysis

The parametric values in our study were statistically represented using International Business Machines Corporation(IBM) Statistical Package for the Social Sciences(SPSS) Statistics, an online application downloaded and installed on our system. Utilizing the accuracy values obtained from the Random Forest and SVM, we conducted a comprehensive statistical analysis for our project. IBM SPSS Statistics is a versatile statistical package enabling the analysis and interpretation of datasets through various tools like descriptive statistics, t-tests, regression analysis, mean comparisons, and more [2]. The group statistics, encompassing the number of accuracies (N), mean, statistical standard error, and standard mean error, were computed under an independent sample t-test. This analysis also included tables for Levene's test for variance equality and the t-test for mean equality. The subsequent graph illustrates accuracy bars, with the X-axis representing group or algorithm parameters and the Y-axis depicting mean accuracy values for the machine learning algorithms.

4. Results

The parametric values in our study were statistically represented using IBM SPSS Statistics, an online application downloaded and installed on our system. Utilizing the accuracy values obtained from the Random Forest and SVM we conducted a comprehensive statistical analysis for our project. IBM SPSS Statistics is a versatile statistical package enabling the analysis and interpretation of datasets through various tools like descriptive statistics, t-tests, regression analysis, mean comparisons, and more [3].

The group statistics, encompassing the number of accuracies (N), mean, statistical standard error, and standard mean error, were computed under an independent sample t-test. This analysis also included tables for Levene's test for variance equality and the t-test for mean equality. The subsequent graph illustrates accuracy bars, with the X-axis representing group or algorithm [4] parameters and the Y-axis depicting mean accuracy values for the machine learning algorithms. The Random Forest (N=10) has gained the maximum accuracy of 90.05% compared to the accuracy gained by the SVM algorithm (N=10), which gained 41.20% as the mean accuracy.

In Table 34.1, a comparison is made between the Random Forest and SVM algorithms using N = 10 samples of the dataset. The best performance achieved is 95%.

Table 34.2 presents the mean values of accuracies for the Random Forest and SVM algorithms, along with their standard deviations and standard errors.

Table 34.1: Comparative among the random forest and SVM algorithms using N=10 samples of the dataset, with the best performance of 95% in the sample (when N=1) using the dataset size of 200 and the proportions of training & testing data being 70% & 30%, respectively

Samples	Accuracy of Random Forest	Accuracy of SVM
1	89.50	39.50
2	86.00	37.50
3	89.00	42.00
4	91.50	41.00
5	91.50	43.00
6	88.50	39.50
7	87.50	44.00
8	90.50	42.00
9	92.50	39.50
10	94.00	44.00
Average Accuracy (%)	90.05	41.20

Table 34.2: The mean value of accuracies of random forest and SVM algorithms standard deviation and standard error

	Algorithm	N	Mean	Std. Deviation	Std. Deviation
Accuracy	Random Forest	10	90.0500	2.42040	0.76540
	SVM	10	41.2000	2.17562	0.68799

Table 34.3: The independent sample T-test results carried out for the two algorithms, and represented as Levene's test for equality of variances and t-test for equality of means

		Levene's test for equality of variances		T-test for Equality of Means						
		F	Sig.	t	df	sig. (2-tailed)	Mean difference	Std Error difference		95% confidence interval of the Difference
									lower	upper
ACCU	Equal variances assumed	0.081	0.779	47.466	18	<0.01	48.85000	1.02916	46.68792	51.01218
RACY	Equal variances not assumed			47.466	17.799	<0.001	48.85000	1.02916	46.68607	51.01393

Table 34.3 displays the results of the independent sample T-test conducted for the two algorithms, represented through Levene's test for equality of variances and the T-test for equality of means.

The mean accuracies of the proposed and current algorithms are contrasted in a bar graph, with accuracy represented on the Y-axis and the proposed and existing algorithms plotted on the X-axis. The Random Forest algorithm's accuracy is 90.05%, compared to 41.20% for the SVM algorithm. The typical detection precision is ±2SD (see Figure 34.1).

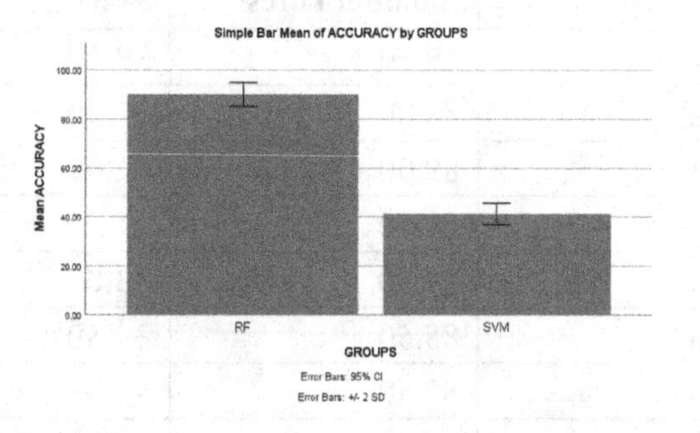

Figure 34.1: The mean accuracies of the proposed and current algorithms are contrasted in a bar graph. Accuracy is represented on the Y-axis, while the proposed and existing algorithms are plotted on the X-axis. The Random Forest algorithm's accuracy is 90.05%, compared to 41.20% in the SVM Algorithm. The typical detection precision is +/-2SD

5. Discussion

The research presented herein delves into the critical domain of household electricity bill prediction, employing a comparative analysis of two prominent machine learning algorithms: the Random Forest Algorithm and the SVM Algorithm("Energy Consumption Prediction by Using Machine Learning for Smart Building: Case Study in Malaysia" 2021). The empirical results substantiate the superior performance of the Random Forest Algorithm, showcasing its commendable accuracy rate of 90.05% in contrast to the SVM's 41.20%. This stark difference in accuracy underscores the Random Forest Algorithm's efficacy in providing more precise predictions for household electricity consumption patterns.

The observed success of the Random Forest Algorithm can be attributed to its ensemble learning approach, which amalgamates the predictive power of multiple decision trees. This allows the algorithm to capture complex relationships within the dataset, making it adept at handling the nuanced nature of household electricity consumption. Additionally, the algorithm's resilience to overfitting and robustness to noisy data further contribute to its superior performance [5]. On the other hand, the SVM, while a powerful algorithm in its own right, exhibits a comparatively lower accuracy rate in this specific context. SVMs are known for their proficiency in handling high-dimensional data, making them suitable for complex tasks. However, the intricacies of household electricity consumption patterns may require a more nuanced approach, a characteristic where the Random Forest Algorithm excels.

The implications of these findings extend beyond the realm of academia, resonating strongly with stakeholders in the energy sector. The Random Forest Algorithm emerges as a robust and reliable tool for estimating and managing household electricity expense [6]. Its high accuracy rate translates into tangible benefits for decision-makers in utilities and energy management, offering a more precise understanding of consumer behavior and aiding in the development of efficient energy management strategies [7]. Throughout the research process, a meticulous approach was adopted, ensuring the integrity and reliability of the results. The dataset, sourced from Kaggle.com, underwent careful curation, and both algorithms were trained and tested on unbiased datasets. This methodological rigor enhances the credibility of the study's findings and supports their applicability to real-world scenarios. However, it is crucial to acknowledge the study's limitations. The conclusions drawn are bounded by the available resources, dataset characteristics, and algorithm parameters chosen for evaluation.

Future research endeavors could explore additional refinements, consider the incorporation of more features, or expand the dataset to gain a more comprehensive understanding of predictive modeling for household electricity bills. In conclusion, this research not only contributes to the academic discourse in machine learning but also provides valuable insights for industry practitioners [8]. The practical significance of choosing the Random Forest Algorithm over the SVM is evident in the potential for more accurate, efficient, and reliable models. As the energy sector navigates the challenges of modern energy consumption, this study stands as a testament to the transformative power of advanced machine learning algorithms in optimizing household electricity management and enhancing the precision of predictive modeling applications.

6. Conclusion

The empirical results obtained from our study reveal a noteworthy performance difference between the Random Forest algorithm and the SVM. The Random Forest algorithm demonstrated superior accuracy, achieving an impressive rate of 90.05%, while the SVM achieved a comparatively lower accuracy of 41.20%. This discrepancy in accuracy rates highlights the effectiveness of the Random Forest algorithm over the SVM in predicting household electricity consumption patterns [5]. These findings carry significant implications for stakeholders within the energy sector and related industries. The Random Forest algorithm emerges as a robust and reliable tool for estimating and managing household electricity expenses. Its high accuracy rate signifies its potential to make more precise predictions, providing valuable insights for decision-making processes.

Throughout the study, a meticulous and systematic approach was employed in various phases. The data collection process involved obtaining a carefully curated dataset from Kaggle.com, ensuring its relevance and representativeness for the task at hand. Preprocessing steps, including cleaning and normalization, were conducted to prepare the data for algorithmic training and testing [9]. The algorithms, both Random Forest and SVM, underwent training and testing on unbiased datasets, ensuring a fair assessment of their performance. The research methodology was designed to guarantee a rigorous evaluation, instilling confidence in the validity and generalizability of the results. While celebrating the success of the Random Forest algorithm, it is essential to acknowledge the study's limitations.

The conclusions drawn are constrained by the available resources and data sources. Future research endeavors may explore further refinements, consider additional features, or expand the dataset to achieve a more comprehensive understanding of predictive modeling in the context of household electricity bills. This research contributes not only to the academic discourse in machine learning but also provides tangible benefits to industries involved in electricity management [10]. The practical implications of choosing the Random Forest algorithm over the SVM are evident in the potential for more accurate, efficient, and reliable models. This enhancement in precision has direct implications for improving the accuracy of household electricity bill predictions. In essence, the project's findings signify a notable advancement in predictive modeling applications within the energy sector. The practical significance of leveraging sophisticated algorithms, such as the Random Forest, becomes evident in their ability to facilitate more effective decision-making. As we navigate the complexities of modern energy consumption, this study stands as a testament to the power of machine learning in optimizing household electricity management and contributing to the overall efficiency of energy-related processes.

Refrences

[1] Al-Khazraji, H., Nasser, A. R., Hasan, A. M., Al Mhdawi, A. K., Al-Raweshidy, H., & Humaidi, A. J. (n.d.). Aircraft engines remaining useful life prediction based on a hybrid model of autoencoder and deep belief network. *IEEE Xplore*. Accessed January 7, 2024. https://ieeexplore.ieee.org/abstract/document/9815246.

[2] Anantha Prabha, P., Arjun, N., Gogul, J., & Divya Prasanth, S. (2022). Two-way economical smart device control and power consumption prediction system. In *Proceedings of International Conference on Recent Trends in Computing* (pp. 415–429).

[3] An innovative random forest-based nonlinear ensemble paradigm of improved feature extraction and deep learning for carbon price forecasting. (2021). *The Science of the Total Environment*, 762, 143099. https://doi.org/10.1016/j.scitotenv.2020.143099.

[4] Cai, H., Shen, S., Lin, Q., Li, X., & Xiao, H. (n.d.). Predicting the energy consumption of residential buildings for regional electricity supply-side and demand-side management. *IEEE Xplore*. Accessed January 13, 2024. https://ieeexplore.ieee.org/abstract/document/8651280.

[5] Chen, X., Xiang, S., Liu, C.-L., & Pan, C.-H. (n.d.). Aircraft detection by deep belief nets. *IEEE Xplore*. Accessed January 7, 2024. https://ieeexplore.ieee.org/abstract/document/6778281.

[6] Cody, C., Ford, V., & Siraj, A. (n.d.). Decision tree learning for fraud detection in consumer energy consumption. *IEEE Xplore*. Accessed January 13, 2024. https://ieeexplore.ieee.org/abstract/document/7424479.

[7] Energy consumption prediction by using machine learning for smart building: Case study in Malaysia. (2021). *Developments in the Built Environment*, 5, 100037. https://doi.org/10.1016/j.dibe.2020.100037.

[8] González-Briones, A., Hernández, G., Corchado, J. M., Omatu, S., & Mohamad, M. S. (n.d.). Machine learning models for electricity consumption forecasting: A review. *IEEE Xplore*. Accessed January 13, 2024. https://ieeexplore.ieee.org/abstract/document/8769508.

[9] Kaya, B., Oysu, C., Ertunc, H. M., & Ocak, H. (2012). A support vector machine-based online tool condition monitoring for milling using sensor fusion and a genetic algorithm. *Proceedings of the Institution of Mechanical Engineers, Part B: Journal of Engineering Manufacture*. https://doi.org/10.1177/0954405412458047.

[10] Li, S., Han, Y., Yao, X., Yingchen, S., Wang, J., & Zhao, Q. (2019). Electricity theft detection in power grids with deep learning and random forests. *Journal of Electrical and Computer Engineering*, 2019, 4136874. https://doi.org/10.1155/2019/4136874.

[11] Sumari, A. D. W., Millatina Nugraheni, A., & Yunhasnawa, Y. (2022). A novel approach for recognition and identification of lowlevel flight military aircraft using Naive Bayes classifier and information fusion. International Journal of Artificial Intelligence Research, 6(2). https://doi.org/10.29099/ijair.v6i1.248.

[12] Naghibi, S. A., Ahmadi, K., & Daneshi, A. (2017). References of support vector machine, random forest, and genetic algorithm optimized random forest models in groundwater potential mapping. *Water Resources Management*, 31(9), 2761–2775. https://doi.org/10.1007/s11269-017-1660-3.

[13] Short-term wind speed forecasting using wavelet transform and support vector machines optimized by genetic algorithm. (2014). *Renewable Energy*, 62, 592–597. https://doi.org/10.1016/j.renene.2013.08.002.

[14] Sumari, A. D. W., Millatina Nugraheni, A., & Yunhasnawa, Y. (2022). A novel approach for recognition and identification of low-level flight military aircraft using Naive Bayes classifier and information fusion. *International Journal of Artificial Intelligence Research, 6*(2). https://doi.org/10.29099/ijair.v6i1.248.

[15] Xia, X., Luo, Z., & Lyu, Q. (n.d.). Analysis of aircraft delay with deep belief network and causal graph methodology. *IEEE Xplore*. Accessed January 7, 2024. https://ieeexplore.ieee.org/abstract/document/9712696.

[16] Yan, S., Li, K., Wang, F., Ge, X., Lu, X., Mi, Z., Chen, H., & Chang, S. (n.d.). Time–frequency feature combination based household characteristic identification approach using smart meter data. *IEEE Xplore*. Accessed January 13, 2024. https://ieeexplore.ieee.org/abstract/document/9042298.

[17] Zhao, D., Liu, H., Zheng, Y., He, Y., Lu, D., & Lyu, C. (2018). A reliable method for colorectal cancer prediction based on feature selection and support vector machine. *Medical & Biological Engineering & Computing, 57*(4), 901–912. https://doi.org/10.1007/s11517-018-1935-3.

CHAPTER 35

"Perception of stakeholders on digital learning" a case of digital India initiative

S. Shankarkumar[1,a], Nellore Pratika Reddy[2,b] and K.A. Harish[3]

[1]Assistant Professor, Department of Business Administration, Faculty of Management, SRM Institute of Science and Technology, Vadapalani Campus, Chennai, Tamil Nadu, India

[2]Assistant Professor, Department of Psychology, Faculty of Science and Humanities, SRM Institute of Science and Technology, Vadapalani Campus, Chennai, Tamil Nadu, India

[3]Associate Professor, Vel Tech Multi Tech Dr Rangarajan Dr Sagunthala Engineering College, Tamil Nadu, India

[a]shankarkumar698@gmail.com

[b]pratikanellore@gmail.com

Abstract

The Digital India initiative is an integrated programme for the Indian citizens to practice and use the government services in a digital platform. The Digital Learning is lined up with the Digital India vision and pillars which would aid in changing the lives of stakeholders. The aim of this research is to understand the outcome of digital learning in the light of stakeholder's perception among digital India initiative. This study adopted descriptive research design and quota sampling technique with a sample size of 297 stakeholders. The research data were analyzed through IBM SPSS by factor analysis using principle component analysis (PCA) to reduce multiple dimension, SPSS AMOS was used for confirmatory factor analysis (CFA) and Structural Equation Model (SEM). From the results 6 prominent factors were identified which are Interpersonal Barrier, Access and Usage, Opportunity to Learn, Digital Empowerment, Flexible Education, Feasibility barrier were identified. The CFA showed that the statements of the different factors showcased a perfect fit with a chi-square degrees of freedom of 2.55. The SEM confirmed the hypothesized model with a chi-square degrees of freedom of 2.554 and Goodness of fit Index of 0.992. The modern style of digital learning aids the stakeholder to ease the process of knowledge attainment by making learning more flexible and cost-effective. Digital learning is very significant as it helps the learners attain continuous and constant learning without any interruptions

Keywords: Digital India, digital learning, perception, stakeholder, knowledge, flexible

1. Introduction

Digital India initiative is an integrated programme for the Indian citizens to practice and use the government services in a digital platform (Nayyar and Singh, 2017). The Digital Learning is lined up with the Digital India vision and pillars which would aid in changing the lives of stakeholders like student community, parents and teaching fraternity with the support of Information and Communication Technology, Internet of Things, Artificial Intelligence and so on (Aiswarya, 2019). The digital India vision includes areas that are stated below:

- Digital Infrastructure
- Digital Governance and services on demand
- Digital Empowerment to Citizen

Digital Learning and Digital India as an interrelated initiative have been proactively taken by the Government of India with specific goals on Education 4.0. Education 4.0 (student centric learning) is a learning connected with learners, focused towards learners, demonstrated by the learners and led by the learners. The digital learning would make education both affordable and accessible for all. Digital

DOI: 10.1201/9781003606611-35

India is a common platform for digital learning platforms like Swayam, Swayam Prabha, National Digital Library, National Academic Depository, E-Shodu Sindhu, Spoken Tutorial and Virtual Lab (Jeyakumari and Balu, 2020). The government is bringing a digital revolution in the higher education landscape and is taking various initiatives in promoting digital learning and digital education (Ambadkar, 2020).

E-learning has become an influential tool, used by instructors and learners to transform the education pattern from conventional to digital learning in our nation (Arkorful and Abaidoo, 2014). With various digital platforms, learning has become simple and at ease for learners to enhance themselves in a fast paced manner, to obtain updated material and to spread information and Knowledge to several people around them (Gil-Jaurena and Domínguez, 2018). The faculty members and students are highly benefitted from the digital learning platforms, which provide a collection of various courses from different streams (Zhou et al., 2020).

Bhattacharjee and Deb (2016) indicate that both learners and instructors are having positive outlook towards making use of digital learning as an aided tool for teaching where Internet plays the foremost part. People are ready to update themselves through digital learning platforms which would cause a dramatic shift in the education system in our nation. At the same time, it becomes easy for the learners and instructors to use these platforms to share resources, materials and knowledge. The student commitment towards the blended MOOC course is designed based on four leading elements, in this they can enhance the student community to engage on: "interactive learning, higher-order thinking skills, learning strategies and collaborative learning" (Almutairi and White, 2018). The MOOCs play a major part in changing the pattern of nation by offering digital experience of learning and qualification based employment opportunities (Devgun, 2013). At present, students and teachers are using both digital learning and traditional learning practices to update on recent trends (Dhir et al., 2017).

2. Literature review

2.1 Interpersonal Barrier

Mojtahedzadeh et al. (2024) indicated that educators should use innovative pedagogical skills to enhance interactions with the learners. Providing constructive feedback and the use of various communication tools and social networks are important to enhance interpersonal relationships. According to Peltier et al. (2024), interaction is a crucial aspect in the learning process, which also sets traditional learning apart from modern digital learning. Communication influences the interaction between the teacher and the learner, the comfort in which the students are able to raise questions, and to put forward their ideas and how they are able to attend to the instructor and respond to the teaching. Singh et al. (2022) demonstrated that in order to overcome the distance that persists in online learning it is important that the facilitators or educators have the ability to help the learners to engage in meaningful interactions. Ong and Quek (2023) determined that insufficient interaction between the student and the teacher is a major contributing in the increasing in ineffective learning.

2.2 Access and Usage

Dhawan (2020) emphasised that online teaching and learning are the new norms for development in the education sector which is accompanied by varying challenges caused by the changes in the distance, teaching personalisation and learning through online mode. The solutions put forward in the present may not address all the hindrances thereby failing to compensate the effectiveness of conventional learning practices. Afzal et al. (2023) in their study pointed out the differences caused due to the digital divide among the student population which reduces the opportunity to access and engage with digital resources and online platforms which has become a vital part to modern education. Their study demonstrated the need to educate and promote digital literacy and skill building programs to empower students with the necessary aptitude to use digital technology and take advantage of the existing valuable resources to uplift their skills and education. UNESCO in 2020 mentioned issues such as unequal access to ICT infrastructure, lack of sufficient internet facilities, skill and power gaps in effectively utilising distance learning platforms. The digital divide was further highlighted by Devara (2020) where it was acknowledged that India is the second largest country with around 49.8% of the population being Internet users, but more than half of the nation's populace do not have adequate Internet Facilities. The underprivileged communities are still lagging behind in this transition into the digital race. Reddy et al. (2020) mentioned the digital divide that exists between the urban and rural places in India. Majority of the Indians accounting to over 66% live in villages in which only 15% rural households have internet access as against 42% of the urban households as determined by the NSSO data in 2017–18. Only 17% of the rural students as against 44% of the urban students have internet access in the country. Students from better socio economic backgrounds have better chances to access good digital infrastructures as opposed to students from the marginalised and poorest income groups. Nikore and Uppadhayay (2021) elucidated that the technical advancements don't bridge the gender gap in the ownership and usage of digital services which is a serious concern in South Asia. Among the Indian women, about nearly 15% of the women are less likely to have a mobile and 33% less likely to use internet services than men. Mohammadi and Fadaiyan (2014) stated that although e-learning has made it possible for more people to access valuable resources and the betterment of their education and their skills, the success of the programs offered digitally is highly dependent on the users rather than the complete reliance on technology.

2.3 Feasibility Barrier

According to Babinčakova and Bernard (2020), exorbitant and weak internet connections are a major barrier for the students as well as the teaching faculty towards pursuing online mode of education. It has been highly prevalent among the non-urban areas of developing nations leading to increased waiting time thereby increasing student's frustration and confusion. Rani et al. (2022) stated that though students are able to learn in a conducive environment through digital learning it becomes difficult for the students when it comes to doubt clearance and the lack of proximity between the students and the teachers. The major problem in implementing online learning is the lack of awareness among the stakeholders. Sumallika et al. (2022) found that along with significant potential in development of online learning in India there are many future hurdles as well. According to their report, hurdles such as inexperienced professors, lack of impact with practical oriented classes, lack of sufficient internet access, lack of infrastructure, and lack of awareness.

2.4 Opportunity to Learn

Rosemarie (2022) stated that learners will be able to easily use digital education as it is more accessible, convenient and can be personalised providing a flexible learning material. It is not only beneficial to the learners but it also helps the educators to connect and propel their knowledge and engage the learners. Patil et al. (2023) determined that digital education provides the opportunity to use the educational resources at a place and time that is convenient to both the educator and the learner. This can aid in supporting lifelong learning thereby expanding the horizons of education. Online and blended mode of education provides opportunities for the students to study their areas of interest even if they are from distant places, or are working individuals. From classroom interaction in tradition learning, learners and educators are able to connect through various mediums such as through email, messages, video chat, online forums, social media, learning materials etc. with convenience of time and place. Budhia and Behera (2023) stated that students will meet the industry expectations by engaging in online education which helps to bridge the gap between the candidates and respective employers. Naqvi and Sahu (2020) pointed out that the great evolutionary advancement of digitalization will change the entire education system. Technology is accelerating an educational trajectory while removing barriers related to learning, teaching, and social functionalities. Balaji (2019), studied the implementation of e-learning in higher education in India. He discussed about the great potential that e-learning brings to the table by improving accessibility and flexibility while at the same time presenting substantial hurdles such as infrastructure limitations, faculty readiness and digital divide. He also focused on how the online courses offered by Indian Institutes should focus on industrial requirements thereby enhancing the chances of the employability of graduates.

2.5 Digital Empowerment

According to Mäkinen (2006), people tend to obtain new opportunities and room for growth to engage and show their opinions in a networked society through Information Technology. Digital empowerment is not a direct cause towards the need to use technical facilities, but it is a way that helps to enhance communication, networking and corporation opportunities. It enhances the competence of individuals and communities to gather together to act as influential people of the information society. Gambo Danmuchikwali and Muhammad Suleiman (2020) determined a progressive change in the way students interact and generate opinions with the use of various internet technologies and advancements. The study stipulated the presence of ICT resources in classrooms and teachers have been putting a lot of efforts to involve digital technology thereby enhancing access to collaborative activities and information for the learners. Sadiku et al. (2017) enumerated that teachers are put under pressure to use digital technologies to educate students thereby preparing them to face the globalized digital economy. Digital education benefits the students by helping them in acquiring skills for growing in the digital world. Irwansyah and Hardiah (2020) highlighted that there is a need to know the projections as well as the context in which the ability of the teachers and students rely on and to see if there is self-access to relevant technology elements on information and communication, using which required measures can be formulated in response to the digitalisation of the teaching learning processes.

2.6 Flexible Education

Bergamin et al. (2012) mentioned the need to individualize learning process by covering all activities from beginning to the end as it is important that these courses are flexible beyond the feasibility of place and time. Sahin and Shelley (2008) stated that there was a positive relationship between the student's ability to partake and use online tools and student's satisfaction. A study conducted in Vietnam with 1232 university students as participants showed a positive relationship between quality of e-learning services and student loyalty (Pham et al., 2019). Muller et al. (2023) investigated blended learning design with flexible study programmes showcasing the need to pay more attention to structuring the course, activation of learning tasks, thought-provoking initiation and social presence of teachers, and on time feedback regarding the process and outcomes of learning. Soffer et al. (2019) determined that factors related to learners such as place, time, learning resources, pace of learning, interaction are the dimensions of flexibility in e-learning. Bergamin et al. (2010) formulated a model indicating dimensions of flexible learning which are Flexibility of time management, flexibility of teacher contact, and flexibility of content. Soffer et al. (2019) mentioned that the student's achievement was associated to flexibility patterns which are centred on learning time and access to learning resources in online learning environment. Various existing studies showcase that the academic performance of students can be improved with high level flexibility in e-learning environments in terms of technical tools, education, resources and activities (Bergamin et al., 2010). Eom and Ashill (2016) found education system through e-learning gives a platform for the learners with learner specific styles of learning, by maintaining proper interaction, giving timely and significant feedback to them; it can be a greater mode to modify the education system.

The existing literature demonstrates that there are a lot of benefits when it comes to incorporating digital learning among the learners but it also comes with it's own barriers. The benefits that the stakeholders are obtaining from these digital learning platforms thereby substantiate the need to establish the digital India initiative. With sufficient evidence for both the benefits as well as the costs that comes with promoting digital learning among the stakeholders it is important that we understand the perception of people regarding the relevance the Digital India initiative holds among the beneficiaries. The current study understands the perception of people and determines whether Digital India initiative is beneficial or is it a space that requires further development.

3. Research Gap & Research Objective

Previous literature has demonstrated that online education comes with a lot of barriers as well as potential for enhancing education among the stakeholders. This thereby brings in the question as to how different stakeholders such as the students, faculty members and parents perceive the participation of learners in digital learning platform. With the growing need for Online education and limited research present to validate the need to promote and grow e-learning it is important that we understand the perception of stakeholders on digital learning in the case of digital India Initiative. This study addresses various outcomes of digital learning. The current study will support the government as this study examines the relationship among socio economic profiles of stakeholders like students, faculty members and parents and their perception towards digital learning and evaluating the opinion of stakeholders on digital learning.

3.1 Research Objective

- To analyse the socio economic profile of stakeholders using digital learning platform.
- To examine the sources of information about digital learning platform and access of digital learning platform using digital devices.
- To identify the different outcomes of digital learning in the light of stakeholder's perception among digital India initiative.
- To establish a theoretical model of the determinants of digital learning outcome.

4. Research Methodology

4.1 Research Design and Paradigm

A descriptive method was adopted for the study. The data was obtained from the respondents to determine the perception of stakeholders on digital learning.

4.2 Target Population and Unit of Analysis

Population in research study refers to the entire group of people from whom we need to draw inferences. In some studies, the population may be finite and in some studies, it would be infinite. The target population and unit of analysis for the current study are the student community, parents and teaching fraternity, as they are considered as a source of knowledge hub and more aware of digital learning

4.3 Sample Size and Sampling Technique

Sample refers to a specific group from which the data is collected. The sample size consists of 297 respondents who are 99 students, 99 parents and 99 Faculty members. The sampling design used in this study is non-probability quota sampling where the participants were selected based on the type of the stakeholder (Student, Parent and Faculty categories).

4.4 Sources of Data

- Primary Data: Structured Questionnaire and survey method
- Secondary Data: Internet sources, publications, research articles and journals

4.5 Research Instrument Development

The data for the study was collected using dichotomous questions, close-ended questions, and a structured questionnaire with a Likert five point rating scale where questions were rated from 1 to 5 (Strongly Disagree to Strongly Agree). This questionnaire contains 37 questions divided into 3 segments as stated below:

- First part includes questions related to demographic details (Gender; Age; Location; Stakeholder; Digital Learning Platform).
- Second part represents the fundamental questions related to digital learning (Sources of Information about Digital Learning; Access to Digital Learning using Digital Devices).
- Third part focused on 30 statements where digital learning outcomes are explored using statements obtained from focus group discussion conducted among stakeholders and literature reviewed.

Table 35.1 Digital learning outcome

Statement	Question
S1	Digital learning provides the opportunity, flexibility and convenience to study at your own time to enhance your academic credentials.
S2	Digital learning helps you to complete an entire course while working, and enhances the academic degree that will help to raise standard of living.
S3	Digital learning offers various courses to choose and explore in the digital platform.
S4	Feedback mechanism is a part of e-learning to express our views and opinions
S5	Listening to lectures and submitting assignments online is beneficial to save time and other resources
S6	Self-discipline is essential for any individual to accept and to successfully complete any Digital learning course.
S7	Digital learning aids in easy access of course lectures, course videos and downloading materials.
S8	Digital learning provides the chance to build a network with others across different countries through discussion forums.
S9	Digital learning will enhance research skills and helps in submitting creative and innovative assignments to the course administrator.
S10	Digital learning connects the rural villages thereby empowering the students.
S11	Digital learning discussion forum helps to construct a reasonable response that will develop communication skills and logical thinking.
S12	Digital learning is cost-effective that helps to eliminate both transportation cost and accommodation cost
S13	Our government intends to promote quality education through digital learning methods
S14	Everyone is having adequate knowledge on digital learning methods.
S15	The stakeholders of digital learning accept the transformation from conventional to digital method.
S16	Digital learning platform makes it difficult to understand and read course modules, study materials and manage deadlines for submission of assignments online.
S17	There is a high chance of disappointment in digital learning because most of the time it is a one way learning system.
S18	There is huge pressure and academic workload in digital learning oriented education
S19	Digital learning methods are applicable mostly for technical students.
S20	There is lack of awareness regarding various digital learning platforms available.
S21	Due to insufficient time and personal commitments, it becomes difficult to concentrate on online course modules and materials.
S22	Digital learning does not create any chance to interact with others in digital platforms, which is essential for student exposure.
S23	Technological glitches and power cut interrupts effective digital learning thereby making it a burden.
S24	In digital learning, the learning environment is not favourable for concentrating on the lectures.
S25	It is not easy for everyone to learn the subject concepts using digital learning
S26	The lack of technological resources and technical knowledge causes concerns handling courses and studying in digital platform.

S27	There is no face to face interaction and clearing doubts with the course administrator in digital learning thereby leading to poor performance in test
S28	Digital resources with low quality instructional materials and recorded videos reduce quality of learning.
S29	Digital learning will reduce the role of a teacher in the classroom.
S30	Digital learning usage helps me to clarify my doubts instantly and precisely.

(Source: Author's compilation)

4.6 Methods of Data Analysis

The data was analysed using IBM SPSS Version 25 and SPSS AMOS 4.0 by factor analysis methods such as PCA to reduce multiple dimensions and CFA to measure validity of the constructs. Structured Equation Model was used to determine if the hypothesized model is a good fit.

5. Research Findings and Discussion

5.1 Demographic Profile

Table 35.2 Sampling frame (N = 297)

Parameter	Target Audience	Frequency	Percentage
Gender	Male	178	59.9%
	Female	119	40.1%
Age	Below 30	106	35.7%
	31-50	126	42.4%
	Above 50	65	21.9%
Location	Urban	168	56.6%
	Rural	129	43.4%
Stakeholder	Students	99	33.4%
	Faculty Members	99	33.3%
	Parents	99	33.3%
Digital Learning Platform	Swayam	71	24%
	Swayam Prabha	46	16%
	National Digital Library	42	14%
	National Academic Depository	39	13%
	E-Shodu Sindhu	36	12%
	Spoken Tutorial	33	11%
	Virtual Lab	30	10%

(Source: Primary Data)

The above table determines that most of the respondents are Male (59.9%) and majority of the respondents are between 31 and 50 (42.4%) years of age. The highest responses are collected from the urban area (56.6%). The data was equally collected from stakeholders like Students (33.4%), Faculty Members (33.3%) and Parents (33.3%) and majority of the respondents use digital learning platforms such as Swayam (24%), Swayam Prabha (16%) and so on.

5.2 Ranking of Dichotomous Questions

Table 35.3 Ranking of dichotomous questions

Items	**Yes (%)**	**No (%)**	**Rank**
Source of Information about Digital Learning			
1.Friends	71	29	II
2.Relatives	40	60	IV
3.Faculty Members	93	07	I
4.Advertisement	42	58	V
5.Social Media	51	49	III
Access to Digital Learning using Digital Devices			
1.Smartphone	64	36	I
2.Personal Computer	52	48	III
3.Tablet	58	42	II

(Source: Primary Data)

Table 35.3 specifies that around 64% of the participant's access digital learning through their own

Smartphones, 58% of the participants use their own Tablet, 52% use their Personal Computer. This shows that respondents show high inclination towards smart phones and the least inclination to Personal Computer. The source of information about digital learning is from Faculty Members (93%), Friends (71%), Social Media (51%), Relatives (44%) and Advertisement (42%).

5.3 *Factor analysis*

Dimension reduction technique (Factor analysis) was applied to reduce the number of variables measuring the outcomes of digital learning. The reliability tests were conducted for the 30 statements using Cronbach alpha's test and the result was 0.922, this shows good internal consistency and high reliability needed for further analysis.

Table 35.4 Reliability statistics

Cronbach's Alpha	N of Items
.922	30

(Source: Primary Data)

KMO sampling adequacy and Bartlett test of sphericity is tested to confirm the data suitability. It is assumed that variance proposition is high and the data is appropriate for factor analysis (Kaiser, 1974). Since Bartlett's test of sphericity is significant ($p < .001$) and the value of KMO statistics is greater than 0.8 the data is suitable for factor analysis. Dimension reduction is applied based on the Rotated Component results.

Table 35.5 KMO and Bartlett's test for digital learning outcome

Kaiser-Meyer-Olkin Measure of Sampling Adequacy		.892
Bartlett's Test of Sphericity	Approx. Chi-Square 3459.099	
	Df	435
	Sig.	.001

Table 35.6 Factor loading based on digital learning outcomes

Statement	Factor Loading						Eigen Value	% of Variance	Cumulative %
	1	2	3	4	5	6			
S16	.500						3.818	12.728	12.728
S17	.514								
S18	.519								
S24	.616								
S25	.642								
S27	.603								
S28	.585								
S29	.608								
S3		.603					3.111	10.369	23.097
S4		.638							
S5		.655							
S6		.631							
S7		.591							
S20			.696				3.013	10.043	33.140
S21			.778						
S22			.747						
S23			.524						
S26			.526						

S11				.495					
S12				.798					
S13				.673			2.876	9.586	42.726
S14				.630					
S15				.502					
S9					.564		2.053	6.842	49.569
S10					.721				
S1						.725			
S2						.646	1.727	5.755	55.324
S30						.463			

Rotation Method: Varimax with Kaiser Normalization.
Note: Rotation converged in 9 iterations.

(Source: Primary Data)

The analysis of Table 35.6 shows the Rotated Component Matrix of digital learning outcomes, out of the 30 statements 28 statements were reduced to 6 factors with 9 iterations. The first iteration has 8 factor loadings, the second iteration has 5 factor loadings, third iteration has 5 factor loadings, fourth iteration has 5 factor loadings, fifth iteration has 2 factor loadings and the sixth iteration has 3 factor loadings. In this analysis there are 6 factors with a total of 55.324 cumulative percentage variance in the complete dataset. The Eigen values of the 6 factors were 3.818, 3.111, 3.013, 2.876, 2.053 and 1.727 respectively. The statements 8 and 19 were removed has the factor loadings were below 0.40 and did not best represent any of the factors.

After the Rotated Component Analysis is done, the factors were clubbed and suitable names were given based on the statements and the literature review. The factor Interpersonal barrier has 8 statements, Access and usage has 5 statements, the third factor is feasibility barrier with 5 statements, the fourth factor is opportunity to learn with 5 statements, the fifth factor is Digital Empowerment with 2 statements and the sixth factor is flexible education with 3 statements.

5.4 Convergent Validity and Construct Reliability of the digital learning outcomes

From the below table it was found that all the factors have an average variance of extracted above 0.5 thereby indicating that the factors are consistent in measuring the latent variable. The CR value for all the six factors are greater than 0.7 ranging between 0.7 and 0.95 thereby indicating that the constructs relate to the latent variable. These findings thereby determine that the model has a good convergent validity and construct reliability.

Table 35.7 Convergent validity and construct reliability of the digital learning outcomes

Variables	Indicator	Validity Test		Reliability Test		
		Factor Loading	Conclusion	AVE>0.5	CR>0.7	Conclusion
Interpersonal Barrier	S16	0.613	Valid	0.615	0.833	Reliable
	S17	0.669	Valid			
	S18	0.594	Valid			
	S24	0.634	Valid			
	S25	0.59	Valid			
	S27	0.642	Valid			
	S28	0.598	Valid			
	S29	0.617	Valid			

Access And Usage	S7	0.6	Valid	0.606	0.740	Reliable
	S6	0.561	Valid			
	S5	0.651	Valid			
	S4	0.628	Valid			
	S3	0.571	Valid			
Feasibility Barrier	S26	0.782	Valid	0.531	0.935	Reliable
	S23	0.919	Valid			
	S22	0.826	Valid			
	S21	0.831	Valid			
	S20	0.944	Valid			
Opportunity To Learn	S15	0.664	Valid	0.532	0.848	Reliable
	S14	0.629	Valid			
	S13	0.671	Valid			
	S12	0.877	Valid			
	S11	0.777	Valid			
Digital Empowerment	S9	0.691	Valid	0.627	0.768	Reliable
	S10	0.881	Valid			
Flexible Education	S1	0.654	Valid	0.711	0.755	Reliable
	S2	0.754	Valid			
	S30	0.726	Valid			

(Source: Primary Data)

Table 35.8 Confirmatory factor analysis (CFA) and structural equation model of the digital learning outcomes

Indices	CFA Value	Remarks	SEM Value	Remarks	Suggested Value
Chi Square Value	854.17		7.663		-
DF	335		3		-
Chi-Square Value/DF	2.550	Accepted	2.554	Accepted	< 5.00 (Hair et al., 1998)
P Value	0.01	Not Accepted	0.054	Accepted	> 0.05 (Hair et al., 1998)
GFI	0.840	Not Accepted	0.992	Accepted	>0.90 (Hu and Bentler, 1999)
AGFI	0.806	Not Accepted	0.941	Accepted	> 0.90 (Hair et al., 1998)
NFI	0.743	Not Accepted	0.989	Accepted	> 0.90 (Hu and Bentler, 1999)
CFI	0.823	Not Accepted	0.993	Accepted	>0.90 (Hooper et al., 2008)
RMR	0.048	Accepted	0.144	Not Accepted	< 0.08 (Hair et al. 2006)
RMSEA	0.072	Accepted	0.072	Accepted	< 0.08 (Hair et al. 2006)

(Source: Primary Data)

From Table 35.8 it was found that the CFA model has a chi square vale of 854.175 with chi-square degrees of freedom of 2.55 which is lesser than 5 thereby indicating that the model fits perfectly. The values of Root Mean square Residuals (RMR) and Root Mean Square Error of Approximation (RMSEA) are 0.048 and 0.072 which is less than 0.8 thereby representing a good fit of the model.

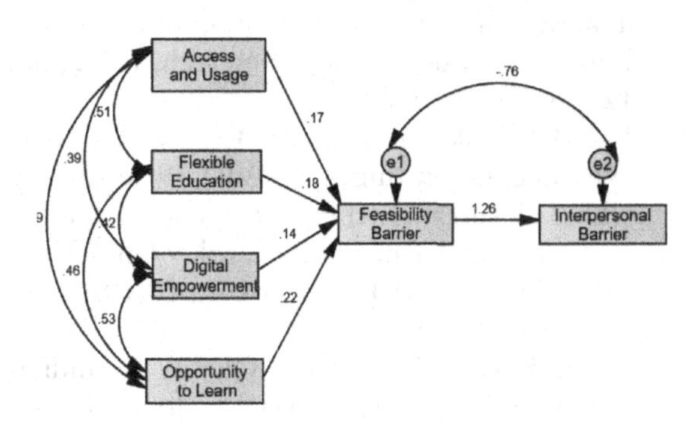

Figure 35.1 Structural equation model of the digital learning outcomes

Table 35.8 also shows the values of the SEM Model. It has a P Value of 0.054 which is greater than 0.05 thereby indicting that the model fits perfectly. The model has a chi square vale of 7.663 with chi-square degrees of freedom of 2.554 which is lesser than 5 thereby indicating that the model fits perfectly. With a goodness of fit Index of 0.992 and Adjusted Goodness of fit Index of 0.941 both of which are greater than the suggested value of 0.90 thereby indicating that there is a good fit between the model that was hypothesized and the observed covariance matrix. The values of Normed fit Index (0.989), Comparative Fit Index (0.993), and RMSEA (0.072) further demonstrate that model is perfectly fit.

6. Implication of Study

The current study showcased perception of stakeholders on digital learning, the findings revealed that the stakeholders who are the students, faculty members, and parents have a positive view towards various digital learning platforms. These digital learning services can be segmented to be free, easy to use and available at need for school, college, and research education. This segmentation thereby helps and benefits people of varying age groups who have the intent to up skill themselves thus providing holistic educational service. The various digital learning platforms offers certification as well as domain expert as a sign of achievement and credit transfers improve significance to the individual's profile. The various determinants outlined in this research paper which aligns with the Digital India initiative given by the students, faculty members, and parents will provide the government with the opportunity to modify and further improve the services provided.

Based on the findings, some recommendations can be stated in this regard. First, there are several methods to construct a better digital learning ecosystem, but; it is purely determined by the country, students, faculty members, parents, government, and policymakers to re-examine the need and importance. Second, the Government strives to create a digital ecosystem that would highly benefit the stakeholders of the digital learning Initiative and it is important that they are included when any important change or decision is made and make consuming digital learning a more live experience. The Appropriate Government and Education Ministry of a country should allot necessary funds for increasing standards of digital learning. The government is responsible for proper implementation of the different stages of digital learning platform for all the age groups

7. Conclusion

Digital India and Digital India initiative has the potential to make a significant impact on both academic as well as the corporate world and also has the potential to overcome barriers such as digital divide, economic divide and information divide at various levels. The modern style of digital learning aids the stakeholder to ease the process of knowledge attainment by making learning more flexible and cost-effective. Digital learning is very significant as it helps the learners attain continuous and constant learning without any interruptions. The current study does not limit to showcasing the perception of the different stakeholders but also helps us to understand various ways in which online mode of learning can be improvised. The current study is however not without its own limitation. The usage of quota sampling method with a sample size of 297 may not be sufficient to represent the entire population to accomplish the research objectives. Thus, future research can look towards delving deep into the perceptions of each of the stakeholders with a greater sample size. However, the current study gives us the platform to understand that digital learning is not only introduced for benefiting the student community but it is to bridge the gap among various other stakeholders with a standard segmentation. Digital learning fully turns into an instrument for global dissemination of expertise and empowerment. India's educational environment has the potential to become more inclusive and brighter as long as technology keeps evolving. The determinants of digital learning and digital India initiative are possible to change the face of the nation like "Participation of All, Growth for All".

8. Future Research Directions

The digital learning platform has wide applications and opportunities for knowledge sources. The current study focused on the overall perception of various stakeholders thereby narrowing down the perception to various determinants that can be further explored upon. Future research thereby can look into exploring on these various determinants and doing an in-depth analysis of each of these factors. Future research can also delve into analysing the secondary data of different digital learning platforms to compare on the number of beneficiaries and the amount of impact different platforms have created. A retrospective study can be conducted to analyse if the digital divide that exists between the different regions of the country as seen any kind of improvement throughout the course of years. Future research can explore on predetermined variables like awareness, usage, adoption, satisfaction and empowerment in response to digital learning initiatives.

References

[1] A. Afzal, S. Khan, S. Daud, Z. Ahmad, and A. Butt, "Addressing the digital divide: Access and use of technology in education," Spring 2023, vol. 3, no. 2, pp. 883–895, 2023. https://doi.org/10.54183/jssr.v3i2.326.

[2] V. Aiswarya, "Digital India – A Roadmap to Sustainability," International Journal of Innovative Technology and Exploring Engineering. vol. 8, no. 5, pp. 571-576, 2019.

[3] F. Almutairi, and S. White, "How to measure student engagement in the context of blended-MOOC," Interactive Technology and Smart Education, vol. 15, no. 3, pp. 262-278, 2018.

[4] R. S. Ambadkar, "E-Learning through Swayam MOOCs – Awareness and Motivation among Commerce Students," International Journal of Scientific & Technology Research, vol. 9, no. 2, pp. 3529-3538, 2020.

[5] V. Arkorful, and N. Abaidoo, "The role of e-learning, the advantages and disadvantages of its adoption in Higher Education," International Journal of Education and Research, vol. 2, no. 12, pp. 397–410, 2014.

[6] M. Babinčáková and P. Bernard, "Online experimentation during COVID-19 secondary school closures: Teaching methods and student perceptions," J. Chem. Educ., vol. 97, no. 9, pp. 3295–3300, 2020. https://doi.org/10.1021/acs.jchemed.0c00748.

[7] M. S. Balaji, "E-Learning Quality in Indian Universities," International Journal of Management, vol. 10, no. 5, pp. 216-221, 2019.

[8] P. B. Bergamin, S. Ziska, E. Werlen, and E. Siegenthaler, "The relationship between flexible and self-regulated learning in open and distance universities," Int. Rev. Res. Open Distrib. Learn., vol. 13, no. 2, p. 101, 2012. https://doi.org/10.19173/irrodl.v13i2.1124.

[9] P. Bergamin, S. Ziska, and R. Groner, "Structural equation modeling of factors affecting success in student's performance in ODL-programs: Extending quality management concepts," Open Praxis, vol. 4, no. 1, pp. 18-25, 2010.

[10] B. Bhattacharjee, and K. Deb, "Role of ICT in 21st Century's Teacher Education," International Journal of Education and Information Studies, vol. 6, no. 1, pp. 1-6, 2016.

[11] N. Budhia and S. Behera, "Challenges and oppourtunities of digital education in India," Asian Journal of Education and Social Studies, vol. 45, no. 3, pp. 1–7, 2023. https://doi.org/10.9734/ajess/2023/v45i3982.

[12] R. Devara, "Broadband internet access: a luxury or a right?," Econ Pol Wkly, vol. 55, no. 48, pp. 21–25, 2020.

[13] P. Devgun, "Prospects for success of MOOC in higher education in India," International Journal of Information and Computation Technology, vol. 3, no. 7, pp. 641-646, 2013.

[14] S. Dhawan, "Online learning: A panacea in the time of COVID-19 crisis," J. Educ. Technol. Syst., vol. 49, no. 1, pp. 5–22, 2020. doi: 10.1177/0047239520934018.

[15] S.K. Dhir, D. Verma, M. Batta, and D. Mishra, "E-learning in medical education in India," Indian Pediatrics, vol. 54, no. 10, pp. 871–877, 2017.

[16] S. B. Eom and N. Ashill, "The Determinants of Students Perceived Learning Outcomes and Satisfaction in University Online Education: An Update. Decision Sciences," Journal of Innovative Education, vol. 14, no. 2, pp. 185–215, 2016. doi: https://doi.org/10.1111/dsji.12097.

[17] B. Gambo Danmuchikwali and M. Muhammad Suleiman, "Digital education: Opportunities, threats, and challenges," J. Eval. Pendidik., vol. 11, no. 2, pp. 78–83, 2020. https://doi.org/10.21009/10.21009/JEP.0126.

[18] I. Gil-Jaurena, and D. Domínguez, "Teachers' roles in light of massive open online courses (MOOCs): Evolution and challenges in higher distance education," International Review of Education, 64(1):197–219, 2018.

[19] J. F. Hair, R. E. Anderson, R. L. Tatham, and W. C. Black, "Multivariate Data Analysis," Prentice- Hall, Upper Saddle River, New Jersey, 1998. In: Marcin Pont and Lisa McQuilken (2002). "Testing the Fit of the BANKSERV Model to BANKPERF Data," ANZMAG conference proceedings.865.

[20] J. F. Hair, R. E. Anderson, and R. L. Tatham, Multivariate Data Analysis.10th edn., Prentice Hall: New Jersey, 2006. In: Malek, A. L. Majali, Nik Kamariah Nik Mat "Modeling the antecedents of internet banking service adoption (IBSA) in Jordan: A Structural Equation Modeling (SEM) approach," Journal of Internet Banking and Commerce, Vol. 16, No. 1, pp.8-13, 2011.

[21] J. F., Hair, R. E. Anderson, R. L. Tatham, W. C. Black, and B. J. Babin, Multivariate Data Analysis. 6th edn., Pearson Education, New Delhi. 734-735, 2006.

[22] D, Hooper, J. Coughlan, and M. R. Mullen, "Structural equation modelling: guidelines for determining model fit," Electronic Journal of Business Research Methods, vol. 6, no. 1, pp. 53-60, 2008.

[23] L.-T. Hu and P. M. Bentler, "Cutoff criteria for fit indexes in covariance structure analysis: Conventional criteria versus new alternatives," Struct. Equ. Modeling, vol. 6, no. 1, pp. 1–55, 1999. https://doi.org/10.1080/10705519909540118.

[24] I. Irwansyah and S. Hardiah, "Digital collaboration in teaching and learning activities: The reflexivity study on educational digital empowerment," Int. J. Learn. Teach. Educ. Res., vol. 19, no. 10, pp. 355–370, 2020. https://doi.org/10.26803/ijlter.19.10.20.

[25] K. Jeyakumari, and A. Balu, "Awareness of Swayam courses among the prospective teachers in sivaganga district," International Journal of Advanced Science and Technology, vol. 29, no. 6, pp. 5665-5671, 2020.

[26] M. O. Kaiser, Kaiser-Meyer-Olkin measure for identity correlation matrix. Journal of the Royal Statistical Society. 1974.

[27] M. Mäkinen, "Digital Empowerment as a Process for Enhancing Citizens Participation," E-Learning and Digital Media, vol. 3, no. 3, pp. 381–395,2006 doi:10.2304/elea.2006.3.3.381.

[28] Z. Mohammadi and B. Fadaiyan, "Surveying the six effective dimensions on E-learners' satisfaction (case study: Bushehr legal medicine organization)," Procedia Soc. Behav. Sci., vol. 143, pp. 432–436, 2014. https://doi.org/10.1016/j.sbspro.2014.07.509.

[29] R. Mojtahedzadeh, S. Hasanvand, A. Mohammadi, S. Malmir, and M. Vatankhah, "Students' experience of interpersonal interactions quality in e-Learning: A qualitative research," PLoS One, vol. 19, no. 3, p. e0298079, 2024. https://doi.org/10.1371/journal.pone.0298079.

[30] C. Müller, T. Mildenberger, and D. Steingruber, "Learning effectiveness of a flexible learning study programme in a blended learning design: why are some courses more effective than others?," Int. J. Educ. Technol. High. Educ., vol. 20, no. 1, 2023. https://doi.org/10.1186/s41239-022-00379-x.

[31] W. M. Naqvi and A. Sahu, "Paradigmatic shift in the education system in a time of COVID 19," J. Evol. Med. Dent. Sci., vol. 9, no. 27, pp. 1974–1976, 2020. https://doi.org/10.14260/jemds/2020/430.

[32] S. Nayyar, and Y. Singh, "A Study on the Awareness of Digital India Programme in College Students (Special Reference to Raipur City)," International Journal of Research In Science & Engineering, vol. 2, no. 3, pp. 54–59, 2017.

[33] M. Nikore, and I. Uppadhayay. "India's Gendered Digital Divide: How the Absence of Digital Access is Leaving Women Behind," Observer Research Foundation, 22, 2021. August. https://www.orfonline.org/expert-speak/indias-gendered-digital-divide/.

[34] S. G. T. Ong and G. C. L. Quek, "Enhancing teacher–student interactions and student online engagement in an online learning environment," Learn. Environ. Res., vol. 26, no. 3, pp. 681–707, 2023. https://doi.org/10.1007/s10984-022-09447-5.

[35] D. W. Patil, S. S. Markad, and B. B. Ghute, "Digital education challenges and opportunities in India," In Prof. Virag.S.Gawande, Dr. Vilas Aghav, Dr. Sachin L. Patki, & Dr. Prashantkumar P. Joshi (Eds.), B.Aadhar Peer-Reviewed & Refereed Indexed Multidisciplinary International Research Journal, (p. 9), 2023. https://www.researchgate.net/publication/372195240.

[36] J.W. Peltier, J. A. Schibrowsky, and W. Drago, "The interdependence of the factors influencing the perceived quality of the online learning experience: A causal model," Journal of Marketing Education, vol. 29, no. 2, pp. 140–153, 2007.

[37] L. Pham, Y. B. Limbu, T. K. Bui, H. T. Nguyen, and H. T. Pham, "Does e-learning service quality influence e-learning student satisfaction and loyalty? Evidence from Vietnam," Int. J. Educ. Technol. High. Educ., vol. 16, no. 1, 2019. https://doi.org/10.1186/s41239-019-0136-3

[38] G. Rani, P. Kaur, and T. Sharma, "Digital Education Challenges and Opportunities," J. Eng. Educ. Transform., vol. 35, no. 4, pp. 121–128, 2022. DOI: 10.16920/jeet/2022/v35i4/22111

[39] B. Reddy, A.S. Jose, and R. Vaidehi, Of access and inclusivity digital divide in online education. Econ Pol Wkl. vol. 55, no. 36, pp. 23–26, 2020.

[40] M. Rosemarie,"The Watt works quick guide 14- Introduction to digital education," Heriot-Watt University, UK, 2022 https://lta.hw.ac.uk/wp-content/uploads/GuideNo14_Introduction-to-digitaleducation.pdf

[41] M.N. Sadiku, A.E. Shadare, and S.M. Musa, "Digital Education," Journal of Educational Research and Policies, vol. 4, no. 12, pp. 1-10, 2017 DOI:10.53469/jerp.2022.04(12).16

[42] I. Sahin, and M. Shelley, "Considering students' perceptions: The distance education student satisfaction model," Journal of Educational Technology & Society, vol. 11, no. 3, pp. 216–223, 2008.

[43] J. Singh, L. Singh, and B. Matthees, "Establishing social, cognitive, and teaching presence in online learning—A panacea in COVID-19 pandemic, post vaccine and post pandemic times," J. Educ. Technol. Syst., vol. 51, no. 1, pp. 28–45, 2022. https://doi.org/10.1177/00472395221095169

[44] T. Soffer, T. Kahan, and R. Nachmias, "Patterns of students' utilization of flexibility in online academic courses and their relation to course achievement," Int. Rev. Res. Open Distrib. Learn., vol. 20, no. 3, 2019. https://doi.org/10.19173/irrodl.v20i4.3949

[45] T. Sumallika, P. V. M. Raju, D. N. V. S. L. S. Indira, and P. Rajya Lakshmi, "An Overview of E-Learning and its Challenges in India," International Journal of Mechanical Engineering, vol. 7, no. 2, pp. 314–316, 2022

[46] UNESCO, UNITWIN/UNESCO Chair holders institutional responses to COVID-19: preliminary results of a survey conducted in April, 2020a

[47] T. Zhou, S. Huang, J. Cheng, and Y. Xiao, "The distance teaching practice of combined mode of massive open online course Micro-Video for interns in emergency department during the COVID-19 epidemic period," Telemedicine and e-Health, vol. 26, no. 5, pp. 584–588, 2020.

CHAPTER 36

Using comparative molecular docking, the lead chemical design of the Phosphatidylinositol 3,4,5-trisphosphate 3-phosphatase and dual-specificity protein phosphatase is intended for use in oesophageal cancer

Reddy Sekhar[1,a] **& M. Kannan**[1,b]

[1]Department of Bioinformatics, Saveetha School of Engineering, Saveetha Institute of Medical and Technical Sciences, Saveetha University, Chennai, Tamil Nadu, India

[a]192020015.sse@saveetha.com

[b]kannanm.sse@saveetha.com

Abstract

This work explores the potential of Phosphatidylinositol 3,4,5-trisphosphate 3-phosphatase and dual-specificity protein phosphatase as a target for cancer therapy, as well as the capacity of proteolysis targeting chimaera (PROTAC) to degrade cancer cells.

The Phosphatidylinositol 3,4,5-trisphosphate 3-phosphatase and dual-specificity protein phosphatase's amino acid sequence was obtained from the UniProt database, and the MODELLER application was utilised to create the protein's three-dimensional structure. We conducted protein-ligand interactions and molecular docking using free software and databases.

The Ramanchandran plot and ERRAT agreed well with the expected structure. By generating one hydrogen bond and eleven hydrophobic contacts, the newly discovered PROTAC molecules exhibit good interaction with the Phosphatidylinositol 3,4,5-trisphosphate 3-phosphatase and dual-specificity protein phosphatase protein.

Our research indicates that the novel PROTAC chemicals have the ability to target and inhibit the phosphatidylinositol 3,4,5-trisphosphate 3-phosphatase and dual-specificity protein phosphatase protein, hence halting the progression of malignant illnesses.

1. Introduction

Various factors, including genetics, lifestyle, and environment, can impact oesophageal cancer, also known as oesophageal carcinoma (Han et al. 2024). Variables and processes related to inflammation, cell proliferation, and protein expression may have a role in the development of oesophageal cancer, even if individual proteins may not be the main cause of the disease (Papadakos et al. 2024; Hofstetter 2013). The earliest sign of oesophageal cancer is usually seen in the muscular tube known as the oesophagus, which transports food and drinks from the mouth to the stomach. This type of cancer arises from abnormal alterations in the oesophageal lining cells, leading to excessively high cell proliferation and the development of tumours (Papadakos et al. 2024). Prolonged exposure to irritants and inflammation caused by an oesophageal stent can raise the risk of cancer. Among the proteins associated with the inflammatory pathway are chemokines and cytokines, which may intensify oesophageal chronic inflammation and foster an environment favourable to the development of cancer (Xin, Song, and Jiang 2024). Uncontrolled cell division can result from alterations in specific genes and

DOI: 10.1201/9781003606611-36

proteins, and this can ultimately cause cancer. The survival, proliferation, and resistance to cell death of oesophageal cancer cells are caused by the abnormal expression of certain proteins. Dysregulated proteins in signalling pathways may have an impact on the ability of oesophageal cancer cells to proliferate and survive (Hirose et al. 2024). Gene expression patterns can be influenced by proteins linked to epigenetic modifications such as DNA methylation and histone modification. The majority of proteins involved in several cellular signalling pathways, including receptor tyrosine kinases (RTKs) and their downstream effectors, govern cell growth and survival. Aberrant epigenetic modifications have the potential to either mute tumour suppressor genes or activate oncogenes. Constant irritation from disorders including gastroesophageal reflux disease (Papadakos et al. 2024), alcohol and tobacco use, and dietary choices are risk factors for oesophageal (Finze et al. 2024) cancer. A person's experience with cancer will also vary according to the way certain biological events and many circumstances combine. Further investigation is required to fully understand the molecular processes behind oesophageal cancer. Once this information is obtained, more advanced preventive and early detection techniques as well as specially designed medications may be created. If you have any specific risk factors or concerns regarding oesophageal cancer, it is recommended that you consult a healthcare professional for tailored advice and screening. Oesophageal cancer is a global health concern due to its varying incidence rates and risk factors among various populations and regions. If you have any specific risk factors or concerns regarding oesophageal cancer, it is recommended that you consult a healthcare professional for tailored advice and screening (Saba and El-Rayes 2019). A thorough review of the patient's medical history, a physical examination, and a number of diagnostic techniques are required to diagnose oesophageal cancer. Common techniques include biopsies obtained for microscopic examination and endoscopy, which visualises the oesophagus via a flexible tube equipped with a camera. Imaging tests like PET and CT scans are useful in determining the extent of a tumour, while X-rays are used in barium swallows to identify anomalies. Endoscopic ultrasound provides a comprehensive tissue examination by combining endoscopy and ultrasound. If there is a suspicion of airway involvement, bronchoscopy may be used. Additional information is obtained via blood tests, which include liver function tests and tumour markers. Cancer spread is determined by staging, and genetic testing may reveal particular characteristics affecting therapy choices (Schlottmann, Molena, and Patti 2018). When these techniques are combined, a precise diagnosis and a customised treatment plan can be created. For those who are at risk or exhibiting symptoms of oesophageal cancer, early diagnosis through routine screenings is critical to better outcomes. Changing to a healthier way of living is essential to avoiding oesophageal cancer. Giving up smoking and abstaining from all tobacco products is crucial because they significantly raise the chance of contracting the illness. Additionally, since alcohol consumption is strongly associated with an increased risk of oesophageal cancer, it is imperative to limit or refrain from excessive alcohol consumption. Eating a well-balanced diet rich in fruits and vegetables may help reduce the incidence of oesophageal cancer by providing important antioxidants and other health benefits. In addition, regular exercise that helps one reach and maintain a healthy weight improves general health and lowers the risk of obesity-related complications. Controlling acid reflux is another crucial preventive step (Finze et al. 2024), as it is frequently linked to the development of Barrett's oesophagus. In addition to drugs or surgery, lifestyle modifications like sleeping with your head propped up and avoiding late-night meals can help manage acid reflux and reduce its dangers. Preventing infection with the human papillomavirus (HPV) is one way to avoid developing some forms of oesophageal cancer. Preventing HPV infection and its linked cancer risks can be achieved in part by practising (Papadakos et al. 2024) safe sexual behaviour and, for those who qualify, receiving an HPV vaccination. The temperature of the beverage should also be taken into account. Reducing the amount of really hot beverages consumed could help reduce the risk of oesophageal cancer. Lastly, frequent exercise has been linked to a lower risk of several malignancies, including oesophageal cancer, in addition to its benefits for managing weight. One way to support general cancer prevention efforts is to participate in at least 150 minutes of moderate-intensity exercise every week. Even while these lifestyle changes can dramatically reduce the chance of oesophageal cancer, (Schlottmann, Molena, and Patti 2018) it is still advisable for anyone with particular concerns or risk factors to speak with medical professionals. Preventive measures can be made even more effective with the use of tests, early detection techniques, and personalised counsel.

1.1 PROTAC

Within the field of medication research and discovery, the PROTAC molecules represent a distinct class of medicinal substances. Unlike traditional small-molecule inhibitors (Suo et al. 2024), PROTACs are designed to specifically break down specific proteins inside of cells, offering a fresh approach to controlling protein levels. In contrast to conventional small-molecule inhibitors, PROTACs are engineered to selectively degrade particular proteins within cells, providing a novel method of regulating protein levels. This combination promotes ubiquitination of the target protein, designating it for proteasome degradation. Targeted protein degradation using the PROTAC approach can be helpful, particularly for proteins that are challenging to block with traditional small-molecule medications (Shah et al. 2024). Interest has been piqued by the possibility that PROTAC technology could treat conditions involving dysregulated protein expression, such as neurological illnesses and various forms of cancer. This method gives precise and flexible control of protein levels by utilising the biological mechanisms that lead to protein breakdown. It is significant since PROTAC research and development is an ever-evolving field and because there may have been advancements since my last report (Xu et al. 2024). To obtain up-to-date knowledge on PROTAC molecules and their applications in drug development, I recommend perusing current scientific papers and sources. PROTAC research is expanding rapidly because of ongoing efforts to enhance selectivity, increase the range of targetable proteins, and improve molecular design. As long as research and development into PROTAC technology are sustained, it appears to have great potential as an adaptable tool for developing more accurate and effective treatments for a variety of ailments.

1.2 E3 LIGASE

Enzymes known as E3 ligases are necessary for the post-translational modification known as ubiquitination(Huang et al. 2024), which attaches tiny proteins known as ubiquitins to target proteins. This ubiquitination pathway adds to cellular signalling networks and regulates not only the function but also the destruction of proteins. E1 (ubiquitin-activating enzyme), E2 (ubiquitin-conjugating enzyme), and E3 (ubiquitin ligase) are the three primary kinds of enzymes that are involved in the ubiquitination process. The main function of E3 ligases is to transfer ubiquitin molecules from E2 enzymes to specific target proteins. The target protein is marked for breakdown by proteasomes or other regulatory processes (Kim et al. 2024) through this transfer of ubiquitin. Based on how they function, E3 ligases are divided into two primary groups: HECT ligases, which are homologous to the E6-AP carboxyl terminus, and RING ligases (Wang et al. 2024), a fascinating new class of gene. HECT E3 Ligases Before adding ubiquitin to the target It is not necessary to form an intermediate complex when ubiquitin is transferred directly from the E2 enzyme to the target protein thanks to the action of ring E3 ligases. There exist two subtypes of RING ligases that have been identified: single-subunit and multi-subunit RING ligases. MDM2 (mouse double minute 2) and c-Cbl (Casitas B-lineage lymphoma) are two instances of RING E3 ligases. Because it controls protein turnover and degradation, the ubiquitin-proteasome system—which is regulated by E3 ligases—is crucial for preserving cellular homeostasis. Numerous illnesses, including as autoimmune diseases (Müller et al. 2024), cancer, and neurological problems, have been linked to the disturbance of this system. To improve our understanding of how cells function and to create tailored treatments for illnesses, it is essential to comprehend the functions and regulatory processes of E3 ligases.

Protein, these ligases establish an intermediate complex with it. The ubiquitin is transported to the target protein once it has been conjugated to the E3 ligase.

2. Materials and Methods

2.1 Protein Modelling

Initially, the UniProt database (http://www.uniprot.org) is used to retrieve the Phosphatidylinositol 3,4,5-trisphosphate 3-phosphatase and dual-specificity protein phosphatase protein(Li et al. 2022).To find similarities between homologous templates, the retrieved sequence is run via the BLAST search engine. The PDB ID for the isoform is obtained from www.rcsb.org. The obtained templates are later used to build isoforms based on the Modeller 9v8 program (Topno, Kannan, and Krishna 2016). On the basis of discrete optimised protein energy (DOPE) the precise models are chosen based on its score. The chosen similarities and cross checks through a verification server known as SAVES (Marshall Freeman 2022) and the structural analysis for the same are performed. The server utilises various tools such as ERRAT (Gaidhani et al. 2024) and PROCHECK. Finally, docking and simulation studies are performed on the chosen isoforms (Panneerselvam, Muthu, and Ramadas 2015).

2.2 Molecular Docking

A computational method called "molecular docking" is used to forecast how tiny compounds, such potential drugs, would bond to target proteins. Molecular docking studies can offer important insights into the binding mechanisms and affinities of ligands to PTEN and Dual-Specificity Protein Phosphatase DUSP. Tumour suppressor PTEN is important in controlling cell division and proliferation, while DUSP proteins(Bansal and Srinivasan 2013) are involved in phosphorylation serine/threonine and tyrosine residues in different cellular signalling pathways. The following procedures are usually followed by researchers in a molecular docking investigation involving the PTEN (Fang et al. 2024) and DUSP proteins: PTEN and DUSP protein three-dimensional (3D) structures are derived from databases or experimental techniques like NMR spectroscopy and X-ray crystallography. Subsequently, the protein's shape is optimised, hydrogen atoms are added, and water molecules are eliminated to create these structures. Mall compounds or ligands that have the potential to interact with DUSP or PTEN are chosen. These ligands may be interesting molecules or possible therapeutic candidates. The target proteins' binding pocket or active site is surrounded by a three-dimensional grid. During the docking simulation (Fang et al. 2024; Saquib et al. 2024), the search space for the ligands is defined by this grid. The binding of ligands to target proteins is simulated using molecular docking software, such as AutoDock or Vina. In order to forecast the most advantageous binding modes, the software investigates various ligand conformations and orientations within the binding site. Based on variables like binding energy or affinity, docking software (Gaidhani et al. 2024) rates the anticipated binding conformations (Rosenzweig et al. 2018). More advantageous binding is indicated by lower energy scores. To determine probable binding mechanisms, important interacting residues, and the degree of ligand-protein interactions, researchers examine the data. To evaluate the dependability of the docking results, the anticipated binding modes and interactions are frequently verified against experimental data, if available. PTEN and DUSP molecular docking studies can help guide the development of novel drug candidates or treatment approaches that target these proteins by providing insight into putative ligand-protein interactions (Gaidhani et al. 2024). It is significant to remember that although molecular docking offers useful information, experimental validation is required to verify the real binding and biological consequences of ligands that have been identified (Gaidhani et al. 2024).

3. Results

The protein structure, which was built with a modeller programme, has alpha helices and beta sheets that extend from the N-to C-terminal portions of the protein. The protein's single domain from its structure is also visible (Bansal and Srinivasan 2013). The anticipated structure was uploaded to the SAVES server in order to assess the model's quality. According to the Ramchandran plot, 80.1% of the anticipated structure favours the region, while 14.8% of the expected structure is in the additionally allowed region. The structure's overall quality factor (94.58) on the ERRAT indicates that it was in good shape. Table 36.1 displays the results of screening and docking 30 warhead PROTAC compounds based on their interaction residues and energy score. Consideration was given to the binding energy or score of the docking poses (Thakur and Gauba 2024).

Figure 36.1 The three dimensional structure of phosphatidylinositol 3,4,5-triphosphate 3-phosphate and dual specificity protein phosphatase protein

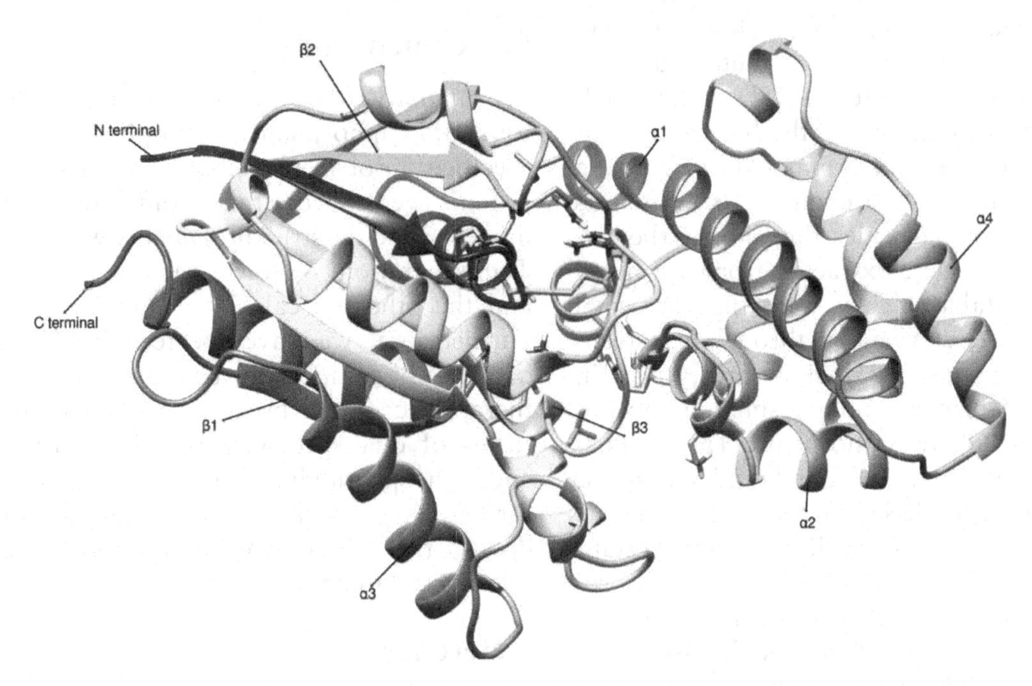

Figure 36.2 Three dimensional structures docked phosphatidylinositol 3,4,5-triphosphate 3-phosphate and dual specificity protein phosphatase protein to the warhead compounds

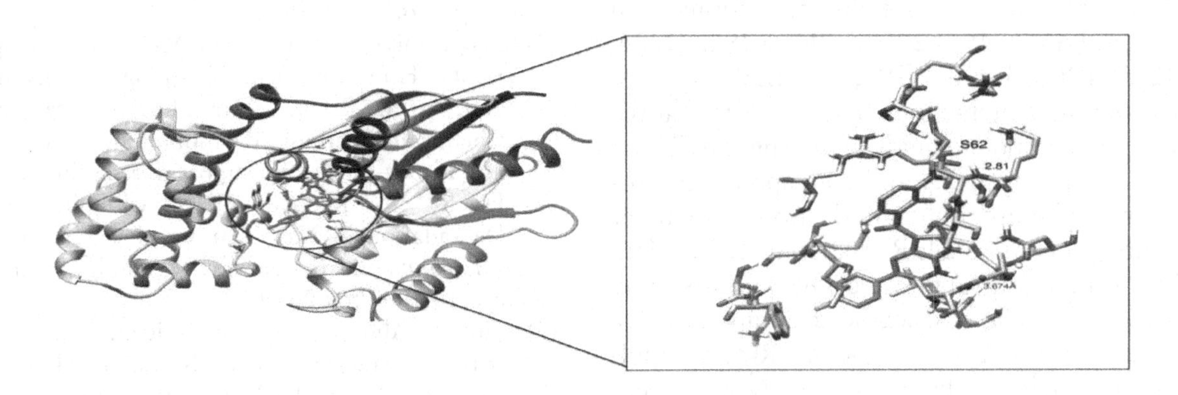

Table 36.1 Describes the interacting residues and its distance of warhead and phosphatidylinositol 3,4,5-triphosphate 3-phosphate and dual specificity protein phosphatase protein

DONOR			ACCEPTOR			DISTANCE
SER	62	N	UNL	1	03	2.81

4. Discussion

In addition to 80% of the residues lying in the permitted and generously allowed region of the Ramachandran plot, which further validates the stability of the structure, the modelled structure exhibits good ERRAT values (Gaidhani et al. 2024), which amply demonstrate the quality of the model. While alpha helices and beta strands work together to contribute (Gaidhani et al. 2024) to the protein's CTD (C-terminal) domain, the arrangement of the N-terminal helices adds to its NTD (N terminal) domain. Using Pyrex software, the 30 compounds known as "warheads," or molecules that inhibit the targeted proteins, were extracted from the novel PROTAC database. They were then screened using an auto dock vina high throughput screening process against the dual specificity protein phosphatase protein and phosphatidylinositol 3,4,5-trisphosphate 3-phosphate. The optimal energy molecules were found and examined for the interaction based on the docking results (Table 36.1). Here, the warhead complex establishes three hydrogen bonds with the dual specificity protein phosphatase protein and the S62 residue of phosphatidylinositol 3,4,5-triphosphate 3-phosphate, with a contact distance of 2.81Å for each (Figure 36.2) (Gaidhani et al. 2024; Radha et al. 2024). Additionally, the warhead compound forms tight interactions with the residues T349, L295, A294, K292, Q226, G225, T203, R200, A199, and L198 through hydrophobic interactions. This suggests that the warhead compound binds to phosphatidylinositol 3,4,5-triphosphate and dual specificity protein phosphatase protein in the optimal manner (Gaidhani et al. 2024; Radha et al. 2024; Ham Sembiring, Nursanti, and AisyahRahmania 2023). By using the warhead compound, we are able to extract the unique PROTAC molecules, which will be utilised in the treatment of malignant disorders and in the drug design process against PTEN and DUSP. Higher interaction energy and better binding energy are demonstrated by the identified PROTAC molecule against PTEN and DUSP, indicating that this molecule may be the lead candidate for inhibiting PTEN and DUSP. Consequently, this aids in breaking the bond with the protein's progenitor. After the E3 ligand is fully attached to the protein and PROTAC, selective protein degradation pathways are successfully started by increasing the activity of the proteasome complex. This will eradicate the phosphatidylinositol 3,4,5-triphosphate 3-phosphate and dual specificity protein phosphatase protein in the cancer which can halt the progression of cancerous cells.

5. Conclusion

In summary, the mutant phosphatidylinositol 3,4,5-triphosphate 3-phosphate and dual specificity protein phosphatase protein domain protein were successfully stimulated using a modeller, and the structural quality was assessed. The separation between each active residue and the others is calculated and summarised. Furthermore, we have located the PROTAC database of compounds that can be used in the therapeutic design process to combat mutant forms of dual-specificity protein phosphatase and phosphatidylinositol 3,4,5-triphosphate 3-phosphate. As the protein target is broken down, the ubiquitination process proceeds effectively, killing the cancer cells.

References

[1] Bansal, Manju, and N. Srinivasan. 2013. *Biomolecular Forms and Functions: A Celebration of 50 Years of the Ramachandran Map*. World Scientific.

[2] Fang, Shiyi, Lanzhu Peng, Mengmin Zhang, Rentao Hou, Xing Deng, Xiaoning Li, Jianyang Xin, et al. 2024. "MiR-2110 Induced by Chemically Synthesized Cinobufagin Functions as a Tumor-Metastatic Suppressor via Targeting FGFR1 to Reduce PTEN Ubiquitination Degradation in Nasopharyngeal Carcinoma." *Environmental Toxicology*, March. https://doi.org/10.1002/tox.24197.

[3] Finze, Alida, Guy HejVijgen, Johanna Betzler, Vanessa Orth, Svetlana Hetjens, ChristophReissfelder, Mirko Otto, and Susanne Blank. 2024. "Malnutrition and Vitamin Deficiencies after Surgery for Esophageal and Gastric Cancer: A Metanalysis." *Clinical Nutrition ESPEN* 60 (April): 348–55.

[4] Gaidhani, Prerna Mahesh, SwastikChakraborty, Kheerthana Ramesh, PadmanabanVelayudhaperumalChellam, and Eric D. van Hullebusch. 2024. "Molecular Interactions of Paraben Family of Pollutants with Embryonic Neuronal Proteins of DanioRerio: A Step Ahead in Computational

Toxicity towards Adverse Outcome Pathway." *Chemosphere* 351 (March): 141155.

[5] Ham Sembiring, Mikael, OktaNursanti, and TesiaAisyahRahmania. 2023. "Molecular Docking and Toxicity Studies of Nerve Agents against Acetylcholinesterase (AChE)." *Journal of Receptor and Signal Transduction Research* 43 (5): 115–22.

[6] Han, Batter, Ying Ma, Pengjie Yang, Fangchao Zhao, Haiyong Zhu, Shujun Li, Rong Yu, and Subudao Bao. 2024. "Novel Histone Acetylation-Related lncRNA Signature for Predicting Prognosis and Tumor Microenvironment in Esophageal Carcinoma." *Aging* 16 (March). https://doi.org/10.18632/aging.205636.

[7] Hirose, Toshiharu, Shun Yamamoto, Yoshitaka Honma, Kazuki Yokoyama, Hidekazu Hirano, Natsuko Okita, Hirokazu Shoji, et al. 2024. "Preoperative Docetaxel, Cisplatin, and 5-Fluorouracil for Resectable Locally Advanced Esophageal and Esophagogastric Junctional Adenocarcinoma." *Esophagus: Official Journal of the Japan Esophageal Society*, March. https://doi.org/10.1007/s10388-024-01050-2.

[8] Hofstetter, Wayne. 2013. *Evolving Therapies in Esophageal Carcinoma, An Issue of Thoracic Surgery Clinics,*.Elsevier Health Sciences.

[9] Huang, Tao, Qi You, Dengjun Huang, Yan Zhang, Zhijie He, XuguangShen, Fei Li, et al. 2024. "A Positive Feedback between PDIA3P1 and OCT4 Promotes the Cancer Stem Cell Properties of Esophageal Squamous Cell Carcinoma." *Cell Communication and Signaling: CCS* 22 (1): 60.

[10] Kim, Daewon, Su Jeong Jeon, Jeum Kyu Hong, Min Gab Kim, Sang Hee Kim, Ulhas S. Kadam, Woe-Yeon Kim, Woo Sik Chung, Gary Stacey, and Jong Chan Hong. 2024. "The Auto-Regulation of ATL2 E3 Ubiquitin Ligase Plays an Important Role in the Immune Response against in." *International Journal of Molecular Sciences* 25 (4).https://doi.org/10.3390/ijms25042388.

[11] Li, Fengjiao, Long Liu, Xiao Yu, Christopher Rensing, and Dun Wang. 2022. "The PI3K/AKT Pathway and Gene Are Involved in 'Tree-Top Disease' of." *Genes* 13 (2).https://doi.org/10.3390/genes13020247.

[12] Marshall Freeman, J. 2022. *Barnabas Bopwright Saves the City*. Bold Strokes Books Inc.

[13] Müller, Fabian, JuleTaubmann, Laura Bucci, Artur Wilhelm, Christina Bergmann, Simon Völkl, Michael Aigner, et al. 2024. "CD19 CAR T-Cell Therapy in Autoimmune Disease - A Case Series with Follow-Up." *The New England Journal of Medicine* 390 (8): 687–700.

[14] Papadakos, Stavros P., Alexandra Argyrou, Vasileios Lekakis, Konstantinos Arvanitakis, Polyxeni Kalisperati, Ioanna E. Stergiou, Ippokratis Konstantinidis, et al. 2024. "Metformin in Esophageal Carcinoma: Exploring Molecular Mechanisms and Therapeutic Insights." *International Journal of Molecular Sciences* 25 (5).https://doi.org/10.3390/ijms25052978.

[15] Radha, Gudapureddy, PratyushPragyandipta, Pradeep Kumar Naik, and Manu Lopus. 2024. "Biochemical and Analysis of the Binding Mode of Erastin with Tubulin." *Journal of Biomolecular Structure & Dynamics*, February, 1–8.

[16] Rosenzweig, Cynthia, William D. Solecki, Patricia Romero-Lankao, ShagunMehrotra, ShobhakarDhakal, and Somayya Ali Ibrahim. 2018. *Climate Change and Cities: Second Assessment Report of the Urban Climate Change Research Network*.

[17] Saba, Nabil F., and Bassel F. El-Rayes. 2019. *Esophageal Cancer: Prevention, Diagnosis and Therapy*. Springer Nature.

[18] Saquib, Quaiser, Ahmed H. Bakheit, Sarfaraz Ahmed, Sabiha M. Ansari, Abdullah M. Al-Salem, and Abdulaziz A. Al-Khedhairy. 2024. "Identification of Phytochemicals from Arabian Peninsula Medicinal Plants as Strong Binders to SARS-CoV-2 Proteases (3CL and PL) by Molecular Docking and Dynamic Simulation Studies." *Molecules* 29 (5).https://doi.org/10.3390/molecules29050998.

[19] Schlottmann, Francisco, Daniela Molena, and Marco G. Patti. 2018. *Esophageal Cancer: Diagnosis and Treatment*. Springer.

[20] Shah, Rishi R., Elena De Vita, Preethi S. Sathyamurthi, Daniel Conole, Xinyue Zhang, Elliot Fellows, Eleanor R. Dickinson, et al. 2024. "Structure-Guided Design and Optimization of Covalent VHL-Targeted Sulfonyl Fluoride PROTACs." *Journal of Medicinal Chemistry*, March. https://doi.org/10.1021/acs.jmedchem.3c02123.

[21] Suo, Yuying, Daohai Du, Chao Chen, Hongwen Zhu, Xiongjun Wang, Nixue Song, Dayun Lu, et al. 2024. "Uncovering PROTAC Sensitivity and Efficacy by Multidimensional Proteome Profiling: A Case for STAT3." *Journal of Medicinal Chemistry*, March. https://doi.org/10.1021/acs.jmedchem.3c02371.

[22] Thakur, Preeti, and Pammi Gauba. 2024. "Expression Analysis of Nitrogen Metabolism Genes in Lelliottia Amnigena PTJIIT1005,

Comparison with Escherichia Coli K12 and Validation of Nitrogen Metabolism Genes." *Biochemical Genetics*, February. https://doi.org/10.1007/s10528-024-10677-w.

[23] Wang, Linsheng, Xi Yang, Kaiqiang Zhao, Shengshuo Huang, Yiming Qin, Zixin Chen, Xiaobin Hu, Guoxiang Jin, and Zhongjun Zhou. 2024. "MOF-Mediated Acetylation of UHRF1 Enhances UHRF1 E3 Ligase Activity to Facilitate DNA Methylation Maintenance." *Cell Reports* 43 (3): 113908.

[24] Xin, Gaojie, Naicheng Song, and Ke Jiang. 2024. "Esophageal Squamous Cell Carcinoma Transformed into Neuroendocrine Carcinoma after NeoadjuvantImmunochemotherapy: A Case Report." *Oncology Letters* 27 (4): 184.

[25] Xu, Mengxia, Yuyang Yun, Changjun Li, Yiling Ruan, Osamu Muraoka, Weijia Xie, and Xiaolian Sun. 2024. "Radiation Responsive PROTAC Nanoparticles for Tumour-Specific Proteolysis Enhanced Radiotherapy." *Journal of Materials Chemistry. B, Materials for Biology and Medicine*, March. https://doi.org/10.1039/d3tb03046f.

CHAPTER 37

Enhancing user excellence evaluating satisfaction via quantitative survey in electrical maintenance for a comparative study between iOS and web platforms

Divya B.[1] and Malathi K.[2,a]

[1]Research Scholar, Department of Embedded systems, Saveetha School of Engineering, Saveetha Institute of Medical and Technical Sciences, Saveetha University, Chennai, Tamil Nadu, India

[2]Project Guide, Corresponding Author, Department of Embedded systems, Saveetha School of Engineering, Saveetha Institute of Medical and Technical Sciences, Saveetha University, Chennai, Tamil Nadu, India

Email: [a]malathi.sse@saveetha.com

Abstract

The purpose of this work is to compare the analysis between iOS and Web Applications. To optimize the evaluation of rating accuracy in hardware system failure rate assessments, both a web application (N=14) and an iOS application (N=14) are employed for text summarization. The sample size for the study will be determined using G power analysis, aiming for an 80% power level to ensure statistical reliability. The web application outperforms the iOS counterpart in terms of rating accuracy for the evaluation of offers and requests. However, the statistical analysis indicates a non-significant difference between the two platforms, as demonstrated by a p-value of 4.344 ($p > 0.047$). The rating accuracy for the iOS application is recorded at 74.14 percentage, while the web application boasts a rating accuracy of 92.79 percentage.

Keywords: Web, iOS, maintenance, rating, equipment, electrical

1. Introduction

This research is focused on advancing user excellence in Electrical Equipment Maintenance through a quantitative survey designed to assess user satisfaction (Starr 1997). The study compares user experiences between a web application and an iOS application tailored for Electrical Equipment Maintenance (Qaid et al. 2022). Utilizing a quantitative survey methodology, the research aims to systematically collect insights into user satisfaction metrics, facilitating a comprehensive comparison of user experiences across both platforms (López-Campos, Márquez, and Fernández 2018). The outcomes will guide strategies for optimizing user engagement and satisfaction in Electrical Equipment Maintenance applications on both web and iOS platforms. This study is committed to enhancing user excellence in Electrical Equipment Maintenance by conducting a thorough evaluation of user satisfaction (Trushkin et al. 2022). Employing a quantitative survey method, the research systematically examines user experiences with web and iOS applications dedicated to Electrical Equipment Maintenance. The primary goal is to quantify and compare user satisfaction metrics, offering valuable insights into the strengths and areas for improvement for each platform (Bustamante et al. 2023). The findings will contribute to strategic enhancements, ensuring a user-centric approach for optimizing user engagement and satisfaction in Electrical Equipment Maintenance applications.

In the field of academic research platforms, our proposal is a genuinely unique and original undertaking. Unlike other apps on the market, it not only collects a staggering amount of scholarly papers—89,54,000 from IEEE, 33,40,000 from Google Scholar,

DOI: 10.1201/9781003606611-37

99,80,000 from ScienceDirect, and an enormous 4,80,000 from Springer—but it also offers novel features and functionalities that completely change the way users interact with the application (Zhang, Qi, and Zhang 2023). Our project is unique because it integrates cutting-edge technology including natural language processing, machine learning, and artificial intelligence. Together, these components give users an extremely sophisticated and personalized search experience that enables them to find and examine academic literature in ways that are not achievable with conventional platforms (Zeng, Zhuang, and Su 2016). Furthermore, our project's dynamic adaptability and response to user needs are what really make it stand out. In contrast to static databases, our platform is updated often to keep up with new research trends, giving users access to a large number of articles as well as the most up-to-date and pertinent material in their particular sectors (Alseiari and Farrell 2021). Our dedication to user interaction is further demonstrated via interactive elements that promote community building, discussion, and cooperation (Gonçalves et al. 2023). Because of this collaborative spirit, our initiative stands out as a live ecosystem rather than just a repository where academics, professionals, and researchers can interact, exchange ideas, and progress knowledge together (Crook 1935). To put it briefly, our project is a pioneer because it provides a novel and revolutionary method for accessing, examining, and interacting with scholarly literature.

By integrating the concept of user excellence, this research is dedicated to delving into user perceptions and ratings, aiming to provide valuable insights into user satisfaction and preferences. The literature review comprehensively explores current studies on various aspects, including cross-platform development, optimization strategies, mobile and web application performance, and response times specifically in maintenance apps. Through a synthesized examination of these critical components, the study aspires to offer in-depth analyses and strategic recommendations. The overarching goal is not only to enhance the functionality of the Electrical Equipment Maintenance App but also to actively promote user excellence. This involves optimizing services and workflow across both web and iOS platforms, aligning the application with the highest standards of user satisfaction and usability.

2. Materials and Methods

This study adopts a dual-group design under the guidance of the robotics lab within the department of computer science and engineering at the Saveetha School of Engineering, an integral part of the broader Saveetha Institute of Medical and Technical Sciences. Group 1 delves into the intricacies of iOS application development, while Group 2 explores the dynamic realm of online applications. The determination of an ideal iteration sample size (N=14) for each group is based on meticulous analysis conducted using the Clincalc website, ensuring a robust 80% statistical power specified in Gpower. This precise calculation necessitates 14 iterations for each group, resulting in a balanced total sample size of 28. Following platform-specific best practices, the research design ensures a comprehensive exploration of both web and iOS application development.

The meticulous creation of database tables in the XAMPP environment aligns seamlessly with the provided instructions, tailoring these tables to the specific needs of our application. This tailored approach ensures that the tables provide a fundamental structure for efficiently organizing and safeguarding crucial data. The effectiveness and integrity of the database are guaranteed by a meticulously crafted schema, rooted in a thorough examination of data relationships (Li, Lei, and Qu 2024). Each table is designed with a distinct function, and the selection of fields is carefully curated to incorporate relevant data, resulting in a robust database structure. This thoughtful design not only ensures the system's organization and responsiveness but also facilitates the application's seamless interaction with and retrieval of data from this well-structured database. The XAMPP tables represent a strategic consideration of essential requirements and serve as the technological cornerstone of the project. Three dedicated databases have been created specifically for our collection, with detailed designs outlined without initial rows or columns. Despite their current empty state, these databases serve as official repositories for our custom dataset, deliberately sized to accommodate anticipated data (Zhu and

Shi 2022). This intentional design underscores a methodical approach to database architecture, emphasizing the personalized nature of the collection. As the project progresses, these databases stand ready to efficiently store and organize distinct data, enabling smooth interactions and comprehensive analysis. The initial setup in XAMPP establishes the framework for subsequent stages, streamlining the integration of our dataset into these databases and laying the groundwork for insightful discoveries and practical applications.

The iOS application written in Swift communicates with the XAMPP server using the Hypertext Preprocessor (PHP) Application Programming Interface (API). PHP scripts on the server act as Application Programming Interface handlers when they receive Hypertext Transfer Protocol (HTTP) GET requests to particular API endpoints. These scripts get JavaScript Object Notation (JSON)-formatted data from the MyStructured Query Language (SQL) database hosted on XAMPP. In a similar vein, the web application connects to the XAMPP server via PHP and utilizes scripts to do SQL queries on the MySQL database. SQL queries and PHP scripts provide a smooth interface between the web application and the XAMPP database, ensuring effective data processing and interchange.

2.1 Electrical Maintenance Web Development

To enhance the overall user experience, our primary focus is on refining the system architecture, specifically within Preparation Group 2. This phase involves a meticulous analysis of the web application's performance, particularly during Course Offer and Assignment Submission within the field of Electrical Maintenance. Our goal is to address challenges such as response time and navigation latency. Our strategy encompasses a comprehensive optimization approach, involving the fine-tuning of data, server-side modifications, and the implementation of streamlined algorithms. The key objective is to minimize latency, enhance real-time updates, and ensure a smooth and engaging user interaction. The web application's adaptability to dynamic demands is guaranteed through a responsive feedback loop, integrating user input and iterative testing—integral components of our continuous development process. For a detailed insight into the web application's implementation, refer to Table 37.1, accompanied by the provided pseudocode.

2.2 Electrical Maintenance Ios Development

The primary objective of this project is to enhance the efficiency of Electrical Maintenance, fostering a smoother and more responsive user experience. In the preliminary phase, designated as Preparation Group 1, our focus centers on an in-depth exploration of the iOS application as a case study. Specifically, we aim to address and rectify issues related to navigation delays and response times, particularly during critical phases such as course offering and donation requests.

Our approach involves the implementation of innovative strategies and streamlined procedures designed to facilitate prompt communication. The ultimate goal is to achieve zero delays, significantly improving the overall effectiveness of Electrical Maintenance, especially in the context of the dynamic nature of transactions associated with course offers. Table 37.2 provides a comprehensive overview of the iOS application, including details, pseudocode, and encapsulation of the Central Processing Unit(CPU). These components collectively contribute to the anticipation of increased output and an enhanced user experience.

2.3 Statistical Analysis

International Business Machines (IBM) SPSS stands as an indispensable tool for statistical data classification across various industrial sectors, proving to be a valuable asset for research endeavors. Its robust capabilities extend to standard computations, effortlessly handling tasks such as computing averages, determining F values, calculating degrees of freedom (df), and establishing standard errors—critical metrics for evaluating significance in research. SPSS simplifies the computation of reaction times by extracting resource allocation data from datasets. Addressing program execution delays is crucial, and the layout and reaction time of the application play pivotal roles in this regard (Almeida-Filho, Ferreira, and Almeida 2013). The latest version, IBM SPSS Statistics 27.0, underscores the program's commitment to staying current and adaptive. Widely utilized in diverse fields including market research, social sciences, and healthcare, SPSS remains the standard platform for comprehensive statistical analysis.

3. Result

Table 37.1 Outlines a systematic approach for monitoring the data accuracy of an iOS application by employing the APIHandler class to timestamp the initiation and conclusion of API queries. The process involves creating a Uniform Resource Locator (URL), scheduling the API call's execution with URLSession, and executing the request. After receiving the API time is recorded, and error handling is implemented for potential issues. Following data ingestion, JSONDecoder is utilized for decoding, and the data accuracy is calculated using the formula [accuracy = calculateAccuracy]. This comprehensive method provides an effective way to measure and monitor the accuracy of data retrieved through API requests in iOS applications.

Table 37.2 Describes a procedure for evaluating the adaptability of a web development project, which involves seamlessly integrating database interaction with session management for user authentication. The process begins by initiating a PHP session, recording the start time, and establishing a secure database connection. User input is then processed using "POST" or "GET" methods, and a SQL query is executed to retrieve user data. Successful verification triggers the authentication process, updating visit and database variables while comparing the entered username against stored data. The assessment of system responsiveness includes recording the conclusion time, and

Table 37.1: Web application pseudocode

Input: Rating Accuracy
Output: Calculated Time
Step 1: Set start time let startTime = Date()
Step 2: Make an API request guard let url = URL(string: apiUrl) else { return} var request = URLRequest(url: url)
Step 3: Execute the API call let task = URLSession.shared.dataTask(with: request) { (data, response, error) in if let error = error {onCompletion(.failure(error)) return }
Step 4: Implement error handling if let error = error {onCompletion(.failure(error)) return }
Step 5: Decode the received data do { let decodeData = try JSONDecoder().decode(type, from: data)
Step 6: Rating accuracy let accuracy = calculateAccuracy()

Step 7: Print accuracy
print("Accuracy: \(accuracy)")
onCompletion(.success((decodeData, accuracy)))
} catch {onCompletion(.failure(error)) }
task.resume()
funccalculateAccuracy() -> Double {
return 0.8
}

Table 37.2: iOS pseudocode for response time

Input: Response Time
Output: Accuracy
Step 1: Launch a PHP session. session_start()
Step 2: first note the Start Time $start_time = microtime(true)
Step 3: Apply the Database Connection $conn = new mysqli($dbname, $password, $servername, $username)
Step 4: Obtain user input using the "POST" or "GET" methods. File_get_contents('php://input') = $request_data $_POST['username'] = $username $_POST['password'] = $password
Step 5: Execute SQL query to pull user information from the database $stmt = $conn->prepare("SELECT user_id, username, password FROM user_info WHERE username = ?"); $stmt->bind_param("s", $username); $stmt->execute(); $stmt->bind_result($user_id, $db_username, $db_password); $stmt->fetch(); $stmt->close();
Step 6: Check if the entered username is in use if ($password === $db_password) {
Step 7: Verification process is successful $_SESSION['user_id'] = $user_id; $_SESSION['username'] = $db_username;
Step 8: Note the conclusion time and determine the reaction time $end_time = microtime(true); $response_time = (($end_time - $start_time) * 1000);

Step 9: Calculate accuracy and print it in the error log
$accuracy = calculateAccuracy();
error_log("Login Accuracy: " . $accuracy);
} else {error_log("Login failed");
}
function calculateAccuracy() {
$accuracy = 0.8;
return $accuracy;
}
?>

this valuable information is logged in error log, offering insights into the overall effectiveness of the authentication process.

Table 37.3 presents rating accuracy for iOS and web applications, calculated based on a sample size of N = 14 for text data entry. The findings indicate that the web version exhibits quicker response times compared to the iOS application.

Table 37.4 presents a comparison between iOS and web applications utilizing statistics on independent samples. The average rating accuracy for the web application is 92.79, while for the iOS application, it is 74.14. The standard deviation of the iOS application is 16.219,

Table 37.3: Outlines the experimental setup with a sample size of N=14, where text data is entered. Every ten iterations, the rating accuracy for both Web applications and iOS is computed. Notably, the iOS response time is observed to be faster than that of the Web application

S. No	iOS Application Rating Accuracy (%)	WEB Application Rating Accuracy (%)
1	67	97
2	80	100
3	89	95
4	30	92
5	85	88
6	66	94
7	82	98
8	81	90
9	50	70
10	84	90
11	82	99
12	80	98
13	80	99
14	82	89

Table 37.4: Information for independent samples comparing iOS-based web applications. For Web apps, it is 19.43, whereas in iOS, it is usually 24.36. An iOS application has a standard deviation of 3.177, while a web application has a standard deviation of 4.146. The standard error means for Web Application 2.071 and iOS are 4.335

	Algorithm	N	Mean	Std. Deviation	Std. Error Mean
Rating Accuracy	iOS	14	74.14	16.219	4.335
	Web	14	92.79	7.748	2.071

whereas the web application has a standard deviation of 7.748. Both the web application (2.071) and iOS App (4.335) support standard error approaches.

Figure 37.1 shows the database of web and iOS applications.

The start page of the web-based application seamlessly guides users to the login page, facilitated by its visually appealing user interface and prominently placed "Login" button (Figure 37.2).

The page shows the 360 deg view of the selected equipment called Air Conditioner, it shows the history of services and repairs, service charge and work done by invoice and vendors (Figure 37.3).

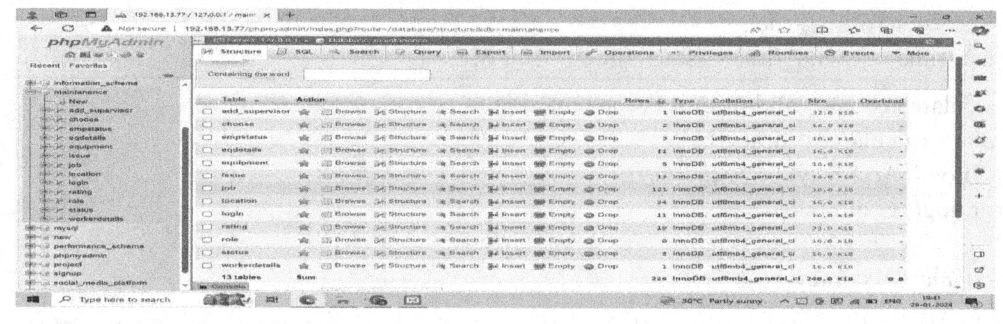

Figure 37.1: Database

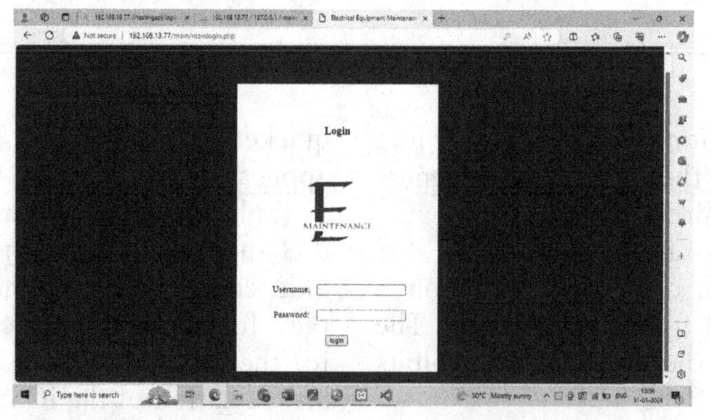

Figure 37.2: Users have the option to access their accounts using their unique usernames and passwords

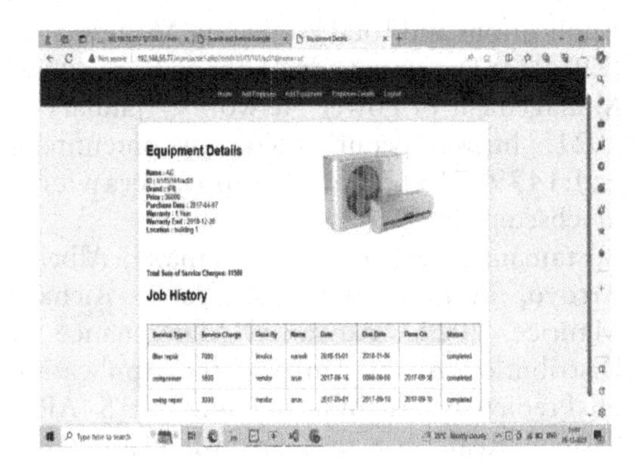

Figure 37.3: 360 degree of selected equipment that shows the history of services and repairs

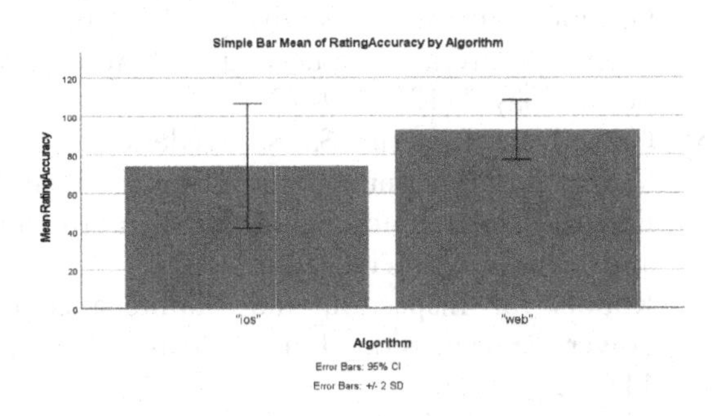

Figure 37.4: Rating accuracy comparison bar chart between iOS and web platforms

Figure 6.4 Represents the rating accuracy comparison bar chart between IOS and web platforms. Which shows the error bars of 95% confidence index and + or - error bars.

4. Discussion

The determination of the significance value stems from the analysis of an independent sample T-test. In summary, the Web application exhibits a faster performance, outpacing the iOS application by 23.15 ms, with an overall advantage for iOS at 29.31 ms. This discrepancy proves to be statistically significant at a level of 0.0, slightly below 0.023. The evaluation of rating accuracy during real-time network modifications is a pivotal consideration for both web and iOS applications (Syazwan et al. 2023). Elevating these elements across all platforms is imperative for ensuring a seamless and engaging user experience. In the case of iOS, the primary focus revolves around leveraging the platform's inherent features to facilitate swift and user-friendly interactions, particularly in the realms of user interface and rating accuracy.

Venturing into the domain of Electrical Maintenance, this study, centered on enhancing user excellence and evaluating satisfaction through a quantitative survey, particularly in the context of web and iOS applications, presents a compelling exploration into the integration of technology with the intricacies of maintaining electrical systems (Xu et al. 2013). The research intricately examines user experiences, responsiveness, and overall satisfaction, providing a valuable perspective on the efficacy of digital tools tailored for electrical maintenance. In the evolving landscape where technology intertwines with our daily lives, this study captures the attention not only of tech enthusiasts and developers but also resonates with professionals seeking modern and user-centric solutions in electrical system management (Kittson 1990). The use of quantitative surveys introduces a systematic approach to understanding user preferences and perceptions, shedding light on the nuances that contribute to an enriched experience in electrical maintenance (Zuccolotto et al. 2015). The findings of this study are poised to contribute not only to the development of effective electrical maintenance applications but also to the broader discourse on the intersection of technology and infrastructure management in our modern lifestyles.

As we embark on the journey to improve user excellence and assess satisfaction in Electrical Maintenance through a quantitative survey across web and iOS applications, it's imperative to acknowledge and address potential limitations that may impact the study's outcomes. The specific functionalities embedded in the chosen applications could introduce bias, potentially limiting the generalizability of the results to a broader context. Variability in user preferences and geographical factors may contribute to differences, affecting the study's applicability in diverse settings. Furthermore, survey responses are subject to individual interpretations and biases, introducing a potential influence on the overall accuracy of the findings (Wilson 2008). A thoughtful recognition and proactive management of these limitations are essential for ensuring a nuanced interpretation of the study's outcomes. Beyond the immediate comparison of user satisfaction in Electrical Maintenance applications, this study sets the stage for broader investigations into user preferences and experiences at the crossroads of technology and maintenance in the electrical domain. Future research endeavors could explore specific features significantly impacting user satisfaction, offering actionable insights for developers aiming to enhance application functionalities (Wang et al. 2016). The findings from this study might play a pivotal role in shaping the design of more tailored and user-friendly applications for Electrical Maintenance, aligning with evolving user expectations. Additionally, the structured framework of employing a quantitative survey methodology opens avenues for replication and expansion in similar studies across different domains. This contributes to the accumulation of knowledge in the ever-evolving field of user experience and satisfaction, fostering a deeper understanding of user dynamics in the digital age.

5. Conclusion

In conclusion, the accuracy of user ratings emerges as a crucial factor intricately tied to the effective execution of electrical maintenance tasks. The diligent efforts of supervisors and workers in maintaining equipment play a pivotal role in shaping user perceptions and ratings. Users evaluate the quality and efficiency of electrical maintenance services, with their ratings serving as a direct reflection of the overall performance of the maintenance team. As users engage with and assess the functionality of electrical equipment, the meticulousness and responsiveness demonstrated by maintenance personnel significantly influence their satisfaction levels and, consequently, the accuracy of their ratings. This underscores the interconnectedness of user feedback, the proficiency of maintenance practices, and the collaborative efforts of supervisors and workers in ensuring a high standard of electrical maintenance service.

References

[1] Almeida-Filho, Adiel, Rodrigo J. P. Ferreira, and Adiel Almeida. 2013. "A DSS Based on Multiple Criteria Decision Making for Maintenance Planning in an Electrical Power Distributor." Evolutionary Multi-Criterion Optimization, 787–95.

[2] Alseiari, Abdulla, and Peter Farrell. 2021. "Optimising Maintenance Strategies through Integrated Artificial Intelligence (AI) Applications and Total Productive Maintenance (TPM) for Developing and Enhancing the Asset Management of Power Networks." January 1, 2021. https://openurl.ebsco.com/contentitem/gcd:147947151?sid=ebsco:plink:crawler&id=ebsco:gcd:147947151.

[3] Bustamante, Sergio, Mario Manana, Alberto Arroyo, Antonio González, and Richard Maurice. 2023. "Advanced Maintenance of Distribution Assets Through the Application of Predictive Techniques Using GE'S APM System: Real Case in a Spanish DSO." 16th WCEAM Proceedings, 195–204.

[4] Crook, W. E. 1935. "Maintenance of Electrical Equipment: A Complete Guide for Prospective Ground Engineers Studying for an 'X' Licence." Aircraft Engineering and Aerospace Technology 7 (12): 309–13.

[5] Gonçalves, Rogério Sales, Frederico Costa Souza, Rafael Zimmermann Homma, Daniel Edgardo TioSudbrack, Paulo Victor Trautmann, and Bruno CordeioClasen. 2023. "Review: Robots for Inspection and Maintenance of Power Transmission Lines." Robot Design, 119–42.

[6] Kittson, John E. G. 1990. "Department of National Defence's Use of Thermography for Facilities Maintenance." In Thermosense XII: An International Conference on Thermal Sensing and Imaging Diagnostic Applications, 1313:2–5. SPIE.

[7] Li, Wei, Zhenchao Lei, and Mingfei Qu. 2024."Research on the Integration of Multimodal Online Teaching Resources for Aircraft Electromechanical Equipment Maintenance Specialty." E-Learning, E-Education, and Online Training, 383–97.

[8] López-Campos, Mónica Alejandra, Adolfo CrespoMárquez, and Juan Francisco Gómez Fernández. 2018. "The Integration of Open Reliability, Maintenance, and Condition Monitoring Management Systems." Advanced Maintenance Modelling for Asset Management, 43–78.

[9] Qaid, Mohammed Saleh Ahmed, AnasMohd Noor, Ahmad NasrulNorali, ZulkarnayZakaria, A. Z. Ahmad Firdaus, AsyrafHakimi Abu Bakar, and Chong Yen Fook. 2022. "Remote Monitoring and Predictive Maintenance of Medical Devices." Proceedings of the International E-Conference on Intelligent Systems and Signal Processing, 727–37.

[10] Starr, A. G. 1997. "A Structured Approach to the Selection of Condition Based Maintenance," January, 131–38.

[11] Syazwan, M. M. Syafiq, F. Yusop, M. S. Saad, K. A. Mohd Sari, A. A. M. Damanhuri, N. N. Mat Hassan, and A. R. Fahmi. 2023. "Identification of Competency Supervisor and Technician Heating Ventilating and Air Conditioning (HVAC) Maintenance in Oil and Gas Industry Malaysia." AIP Conference Proceedings 2544 (1): 040029.

[12] Trushkin, V. A., M. A. Levin, O. N. Churlyaeva, and A. S. Guzachev. 2022. "Conditions for Constructing a Schedule of Maintenance When Using a Risk-Based Strategy of Technical Operation." In Computer Applications for Management and Sustainable Development of Production and Industry (CMSD2021), 12251:149–53.SPIE.

[13] Wang, Youyuan, Hang Liu, Jiangang Bi, Feng Wang, Chunyu Yan, and Taiyun Zhu. 2016. "An Approach for Condition Based Maintenance Strategy Optimization Oriented to Multi-Source Data." Cluster Computing 19 (4): 1951–62.

[14] Wilson, R. M. 2008. "Design for Maintenance," January, 363–85.

[15] Xu, Hongke, Yong Pan, Qi Wang, Yanyan Qin, and Peichao Dong. 2013. "Expressway Electro-Mechanical Equipment Maintenance Management System Functional Research Based on B/S," November, 410–15.

[16] Zeng, Yonghua, Jiandong Zhuang, and Zhenglian Su. 2016. "Construction of Domain Ontology for Engineering Equipment Maintenance Support." Knowledge Graph and Semantic Computing: Semantic, Knowledge, and Linked Big Data, 33–38.

[17] Zhang, Q., J. Qi, and X. Zhang. 2023. "Analysis of the Application of RCM Maintenance Ideas in the Operation and Maintenance of Air Traffic Control Equipment in High Altitude Airports," January, 482–89.

[18] Zhu, Dongshan, and Dongdong Shi. 2022. "Research on Intelligent Operation and Maintenance Scheme of Railway Signal System Based on Big Data." In International Conference on Electronic Information Technology (EIT 2022), 12254:432–39.SPIE.

[19] Zuccolotto, Marcos, Luca Fasanotti, Sergio Cavalieri, and Carlos Eduardo Pereira. 2015. "A Distributed Intelligent Maintenance Approach Based on Artificial Immune Systems." Engineering Asset Management - Systems, Professional Practices and Certification, 969–81.

CHAPTER 38

Analyzing the user satisfaction levels of the preoperative risk assessment device on the iOS and Android platforms

M.S. Kiruba[1] and T. Suresh Balakrishnan[2,a]

[1]Research Scholar, Department of Networking, Saveetha School of Engineering, Saveetha Institute of Medical and Technical Sciences

[2]Project Guide, Corresponding Author, Department of Networking, Saveetha School of Engineering, Saveetha Institute of Medical and Technical Sciences

[a]sureshbalakrishnant.sse@saveetha.com

Abstract

The main focus of this investigation is to assess the impact of preoperative risk assessment applications on user satisfaction across iOS and Android platforms. This research aims to uncover the user-perceived benefits and drawbacks in terms of satisfaction, as well as the responsiveness of the platforms to user input. Through evaluating user satisfaction during key interactions, identifying factors influencing satisfaction levels, and proposing enhancement strategies, this study endeavors to offer valuable insights for improving both user experience and platform efficiency in satisfying user needs. **Materials and Methods:** The objective of this study is to compare user satisfaction levels between iOS and Android applications, utilizing a sample size of ten (N=10) and achieving a G power of 80%. **Result:** Based on the data collected, it appears that the average user satisfaction rating for Android applications is 66.0000 percentage, slightly lesser than the average rating of 73.8889 percentage for iOS applications. **Conclusion:** User satisfaction scores for iOS and Android applications, respectively, stand at 73.8889 percentage and 66.0000 percentage.

Keywords: User satisfaction, features, iOS, percentage, medical, data, scores, database

1. Introduction

The preoperative risk assessment application's user-friendly interface and API, which automate total score computations based on chosen symptoms, facilitate users' selection of symptoms (Deighan, Ayobi, and O'Kane 2023). By effectively categorizing dangers into pre-operative, intraoperative, and post-operative stages, it guarantees a well-defined and organized strategy (Hamilton et al. 2014). Real-time risk assessment capabilities provide healthcare staff with rapid insights and enable them to make well-informed decisions during critical patient care moments (Muessig et al. 2013). Ensuring patient confidentiality is of utmost importance, and data security is largely dependent on safe storage features. Smooth integration with electronic health record systems enhances workflow efficiency in clinical settings, and flexible scoring criteria provide greater adaptation to a variety of medical treatments. The program is multi-device compatible and simplifies clinical settings by making it easy to export, print, or store evaluation results for comprehensive documentation requirements (Morita et al. 2019). Its intuitive interface makes it easier to connect with and navigate, and its advanced algorithms ensure accuracy and reliability while doing risk assessments. The program is an essential resource for modern healthcare, enabling educated and efficient decision-making that improves patient outcomes. Rapid reaction to evolving patient circumstances is made possible

DOI: 10.1201/9781003606611-38

by real-time updates, and interoperability minimizes needless data entry.

Following assessment of the study, a substantial amount of publications were discovered in Google Scholar (about 44,289 articles), IEEE Xplore (21,318 research papers), and Springer (5,812 research articles). The article (Meirte et al. 2020)This study aims to assess the advantages, obstacles, and drawbacks of digitally collecting qualitative electronic patient-reported outcome measures (ePROMs). Through a systematic review of 32 articles retrieved from PubMED and Web of Science, we provide a comprehensive overview following PRISMA guidelines. The article (Lewis and Wyatt 2014) The widespread adoption of medical apps among clinicians introduces the potential for both enhanced care delivery and patient safety risks. This article outlines various risks associated with medical apps and introduces a generic risk framework to aid stakeholders in assessing and mitigating these risks, ultimately aiming to improve patient safety. The article (Kalz et al. 2014) which has 102 citations. This study evaluates CPR mobile apps for both medical content and usability, involving medical experts and end-users. Out of 61 apps assessed, 46 underwent medical evaluation, with 13 progressing to usability testing by non-medical end-users. The study highlights 5 recommendable apps based on both medical and usability criteria. The article (Al Ayubi et al. 2014)This project focuses on developing a persuasive mHealth app for promoting physical activity, integrating social networking systems. Through a structured analytical study and technical development, the app was deployed for a 4-week feasibility study. With 14 participants, high usability scores were reported, with an average of 7.57 hours of app usage per day per participant over the study period. The article (Al Ayubi et al. 2014; Goyal et al. 2017)A randomized controlled trial evaluated the impact of the diabetes self-management app "bant" on adolescents with type 1 diabetes. Although no significant changes were observed in primary and secondary clinical outcomes overall, increased self-monitoring of blood glucose (SMBG) was associated with improved HbA1c in the intervention group, particularly among users with higher SMBG frequency.

There are a number of issues with preoperative risk assessment applications that could make them less useful in practical settings (Ulmer et al. 2004). First of all, the assessment procedure may become inaccurate and inconsistent due to their reliance on subjective patient reporting. Due to its subjective nature, the assessment's overall accuracy may be compromised by misinterpretation or underestimation of risk factors. Furthermore, these applications might not be as predictive as they could be, particularly when it comes to foreseeing uncommon or unanticipated surgical problems (Ishihara et al. 2021). This constraint makes it more difficult for them to offer comprehensive risk assessments, which could result in insufficient planning for unfavorable occurrences. Furthermore, disparities in how medical professionals interpret risk scores might lead to uneven decision-making procedures. Accurate risk assessment is further complicated by the inherent danger of overestimation or underestimating depending on unique conditions. Moreover, for integration into clinical workflows to be successful, practitioners must have the necessary training and experience to make efficient use of the data these applications provide (Semik et al. 2001). These problems are made worse by technical problems or system failures, which interfere with the evaluation procedure and jeopardize the accuracy of the findings. To fully exploit the promise of preoperative risk assessment tools in assisting clinical decision-making and enhancing patient outcomes, it is imperative to solve these issues.

2. Materials and Methods

Several faculty members provided me with significant support throughout my research stint at the Robotics Lab in the Saveetha School of Engineering. Their direction and assistance were crucial to the successful completion of my study. My work has progressed greatly because of the lab's accommodating environment. The modern technology and facilities in the lab improved the efficacy and efficiency of my research projects. My learning experience was enhanced by the collaborative environment, which encouraged meaningful interactions and knowledge exchange between students and professors. In addition, the lab's accommodations were excellent, providing a wealth of tools and room for creativity and exploration. In summary, my stay at the Robotics Lab was marked by innovative teamwork, state-of-the-art equipment, and a nurturing atmosphere that substantially enabled the execution of my studies.

XAMPP was chosen to handle the database during the development of the iOS application, which resulted in a much more streamlined backend infrastructure. Using XAMPP ensured effective data storage and retrieval procedures by offering a stable and adaptable framework for database operations. The frontend and backend components of the iOS application were able to communicate with each other more easily thanks to the integration of XAMPP, which improved the overall functionality and user experience. The scalability and ease of deployment made possible by XAMPP's flexibility enabled the application to grow and expand as needed. Furthermore, XAMPP's extensive feature set made database management jobs simpler, which cut down on development time and effort. Overall, the successful and effective development of iOS applications was aided by the wise choice to utilize XAMPP server for database management.

PHP code was used in the iOS application development process to connect to the XAMPP server, which acted as the backend infrastructure. This methodology enabled smooth connection between the database and the front-end interface, guaranteeing effective operations for data administration and retrieval. Using PHP for backend functions offered a stable and adaptable approach that allowed for dynamic database interaction. To run queries and manage data transactions safely and efficiently, PHP scripts were used. Furthermore, the application's flexibility and scalability were improved by PHP's interoperability with a number of database management systems, including MySQL, which is supported by XAMPP. All things considered, the XAMPP server backend's integration of PHP was essential to the development of a unified and effective iOS application with strong backend capabilities.

2.1 Preoperative Risk Assessment Ios Application

The Preoperative Risk Assessment for iOS application was created in large part by utilizing the Xcode Integrated Development Environment (IDE), which was designed especially for creating iOS applications. Because of Xcode's extensive feature set, jobs like code management, version control, and teamwork were significantly improved from the beginning. The integrated debugging features of Xcode expedited the identification and resolution of any possible issues, further ensuring the stability and reliability of the program. Furthermore, the native support of the Swift programming language by Xcode made it easier to develop complex algorithms required for accurate preoperative risk assessment computations. By utilizing Xcode's comprehensive documentation and user-friendly interface, a state-of-the-art application was constructed quickly, enabling medical professionals to do precise preoperative risk assessments on iOS devices.

2.2 Preoperative Risk Assessment Android Application

With the help of Android Studio, a cutting-edge IDE designed especially for Android app development, the Preoperative Risk Assessment Android application was completed. Our development team exceeded the preoperative risk assessment standards and created a dependable and easy-to-use application by utilizing the rich toolkit and capabilities of Android Studio. From the beginning to the end of the development process, Android Studio offered a smooth and user-friendly platform for creating, evaluating, and refining each aspect of the application. Its strong debugging features guaranteed the reliability of the application and allowed for quick detection and fixing of any problems that arose. Furthermore, the ability of Android Studio to handle a wide range of programming languages including Java made it easier to integrate sophisticated features that are essential for precise risk assessment.

2.3 Statistical Analysis

A robust program for statistical analysis and data management is called SPSS (Statistical Package for the Social Sciences). It provides a full range of tools for doing intricate statistical studies, such as regression analysis, hypothesis testing, and descriptive statistics. SPSS makes data management and interpretation easier with its comprehensive documentation and user-friendly interface. Because SPSS can handle enormous datasets effectively and generate readable, informative graphical representations, it is a trusted tool for researchers and analysts. Because of its adaptable qualities, it can be used in a wide range of sectors, such as market research, healthcare, and the social sciences. Through the analysis of data patterns and relationships, SPSS makes evidence-based research and well-informed decision-making possible. Because of its many capabilities, versatility, and user-friendliness, SPSS is an invaluable resource for researchers and analysts who want to extract significant insights from their data.

3. Result

Table 38.1 indicates that, through scrutinizing user satisfaction on the Android platform, developers gain invaluable insights into the preferences and experiences of Android users. This understanding empowers developers to customize app features and functionalities to align with user expectations, thereby enhancing the overall user experience. By refining the user interface and fostering increased user engagement, developers can significantly elevate satisfaction levels with Android applications.

Table 38.2 analyzing user satisfaction for iOS apps offers crucial insights for project management, steering endeavors towards enriching user experience and fulfilling user expectations. By closely monitoring satisfaction levels, developers can pinpoint areas necessitating enhancement, fine-tune features, and prioritize updates to refine the app's performance and usability, ensuring its alignment with users' needs and preferences.

Table 38.3 indicates, the systematic collection of user satisfaction ratings, conducted every 10 iterations for both iOS and Android platforms with a sample size of N = 10, underscores the consistent superiority of iOS user satisfaction over Android. This insight informs ongoing development efforts to address user preferences and optimize the user experience across both platforms.

Table 38.4 indicates, when comparing independent samples, the average user satisfaction rating for iOS users is 73.8889 percentage, compared to 66.0000 percentage for Android users. Notably, both platforms display variability in user ratings, with standard deviations of 2.58414 for Android and 1.77639 for iOS. Additionally, the standard error means for iOS and Android user satisfaction remains consistent at 1.77639.

Furthermore, the diagram offers a holistic view of each doctor's professional profile, facilitating efficient communication and coordination within medical teams (Figure 38.1(a)). By including essential contact details alongside

Table 38.1: By analyzing user satisfaction on the Android platform, developers can understand the preferences and experiences of Android users, enabling them to tailor app features and functionalities to meet their expectations. This insight aids in optimizing the user interface, enhancing user engagement, and ultimately improving overall satisfaction with Android applications

Input: User satisfaction
Output: Percentage
Step 1: Distribute the Android application to users for comprehensive usage, instructing them to utilize all available features.
Step 2: Request users to provide ratings based on their experience with the application, emphasizing the importance of honest feedback.
Step 3: Allow users sufficient time to explore and interact with the app, ensuring they have ample opportunity to form opinions on its functionality and usability.
Step 4: Encourage users to rate various aspects of the application, including user interface, performance, features, and overall satisfaction.
Step 5: Use the collected ratings as valuable feedback for future iterations of the Android application, guiding improvements and enhancements to better meet user expectations.

Table 38.2: Tracking user satisfaction for iOS apps provides valuable insights for project management, guiding efforts to enhance user experience and meet user expectations. By monitoring satisfaction levels, developers can identify areas for improvement, refine features, and prioritize updates to optimize the app's performance and usability

Input: User satisfaction
Output: Percentage
Step 1: Deploy the iOS application to users and encourage them to engage with all available features extensively.
Step 2: Prompt users to rate the application based on their overall experience, emphasizing the importance of providing detailed feedback.
Step 3: Encourage users to rate specific aspects of the iOS application, such as usability, design, functionality, and performance.
Step 4: Offer users the option to provide additional comments or suggestions to elaborate on their ratings and offer insights into areas for improvement.
Step 5: Analyze the collected user ratings and feedback systematically to identify trends, patterns, and common themes in user satisfaction or dissatisfaction.
Step 6: Utilize the feedback gathered from users to inform future iterations of the iOS application, prioritizing enhancements that address user concerns and align with their preferences.

Table 38.3: With a sample size of N = 10, user satisfaction ratings for both iOS and Android platforms are collected every 10 iterations. Findings reveal higher satisfaction ratings for iOS users compared to Android users

S. no	Android Development time (hrs)	iOS Development time (hrs)
1	64	75
2	69	76
3	62	71
4	60	70
5	63	71
6	66	73
7	68	72
8	65	75
9	67	74
10	61	76

Table 38.4: When comparing independent samples, the mean user satisfaction rating for iOS users is 73.8889 percentage, whereas for Android users, it is 66.0000 percentage. With standard deviations of 2.58414 for Android and 1.77639 for iOS user satisfaction, both platforms exhibit variability in user ratings. The standard error means for iOS and Android user satisfaction are 1.77639 and 1.77639, respectively.

	Algorithm	N	Mean	Std.Deviation	Std.Error Mean
User_satisfaction	"Android"	10	66.0000	4.08248	1.29099
	"iOS"	10	73.8889	3.14024	1.04675

Figure 38.1: *(a)* The diagram provides comprehensive information about doctors, including unique identifiers such as doctor ID, as well as essential contact details like email and phone number. Additionally, it encompasses personal details like the doctor's name and address. Moreover, it specifies the doctor's designation, offering insight into their professional role or specialization within the medical field

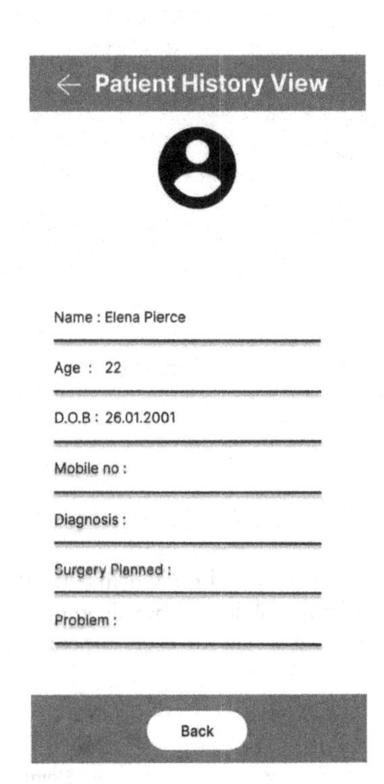

Figure 38.1: *(b)* The diagram illustrates the Patient History view, containing crucial details such as the patient's name, age, date of birth, mobile number, diagnosis, planned surgeries, and associated medical issues. This comprehensive overview offers a detailed snapshot of the patient's medical background, facilitating efficient healthcare management

personal information such as name and address, the diagram ensures seamless access to pertinent information for effective collaboration and patient care. Additionally, specifying the doctor's designation provides clarity on their expertise and responsibilities, aiding in task delegation and patient referrals.

Additionally, the Patient History view in the diagram provides a comprehensive overview of the patient's medical background, enabling healthcare professionals to access crucial information efficiently (Figure 38.1(b)). By including essential details such as the patient's name, age, and date of birth, along with diagnosis, planned surgeries, and associated medical issues, this comprehensive snapshot aids in effective healthcare management. This detailed overview supports healthcare providers in making informed decisions and delivering personalized care to patients. Furthermore, the table offers a comprehensive compilation of Patient details, encompassing vital information such as patient ID, name, age, and date of birth, as well as contact details like mobile number (Figure 38.2). Additionally, it includes medical specifics such as diagnosis, planned surgeries, and existing medical issues. This comprehensive dataset provides a holistic insight into each patient's medical profile, facilitating efficient healthcare management and informed treatment planning by healthcare professionals.

The depicted bar chart illustrates the mean user satisfaction scores for the Preoperative Risk Assessment application across both iOS and Android platforms (Figure 38.3). Android users convey an average satisfaction rating of 66.0000 percent, while iOS users exhibit a notably higher satisfaction level, averaging at 73.8889 percent. The x-axis categorizes the data based on the platform utilized, distinguishing between Android and iOS, while the y-axis denotes the corresponding satisfaction levels. This visualization provides valuable insights into user perception and platform preferences, aiding in informed decision-making for app development and optimization strategies.

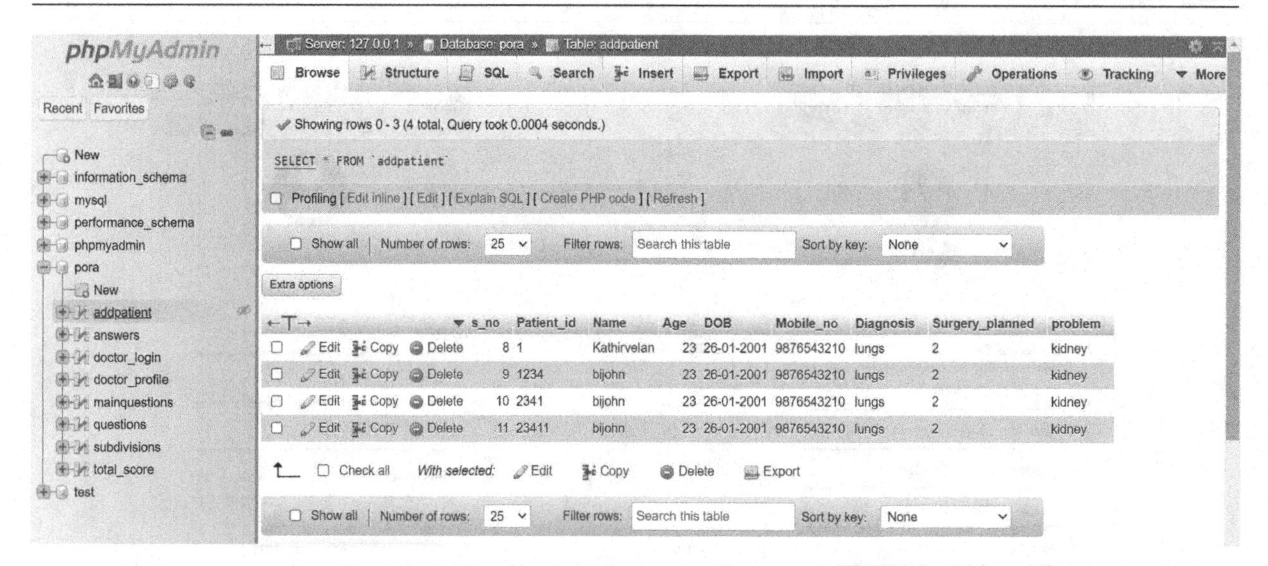

Figure 38.2: The provided table delineates Patient details, comprising essential information including patient ID, name, age, date of birth, mobile number, diagnosis, planned surgeries, and existing medical issues. This comprehensive dataset offers a holistic overview of each patient's medical profile, aiding in efficient healthcare management and treatment planning

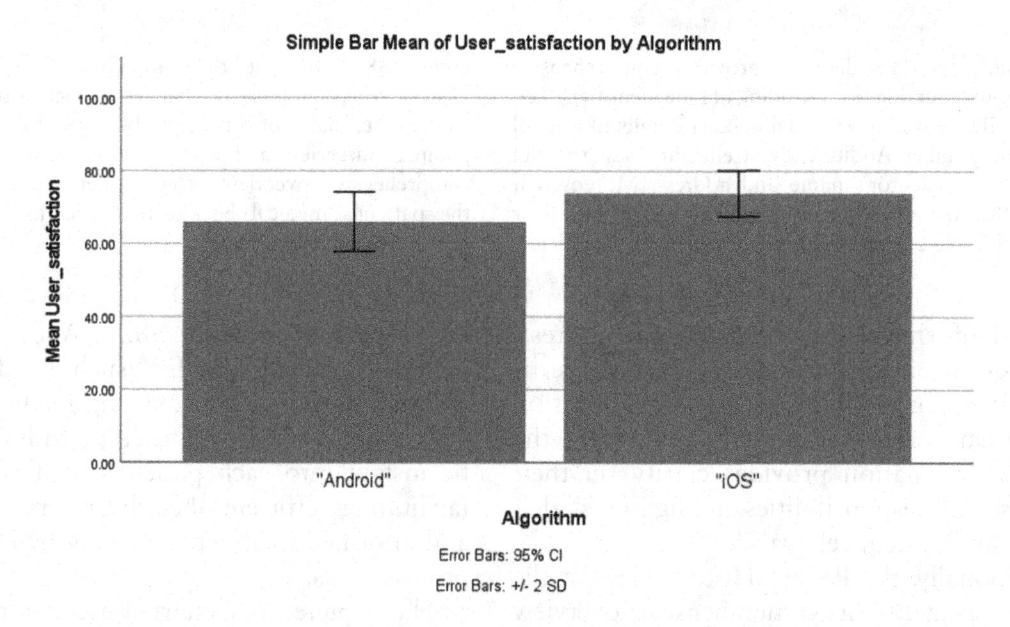

Figure 38.3: The bar chart displays the average user satisfaction ratings for the Preoperative Risk Assessment app on both iOS and Android platforms. Android users report an average satisfaction rating of 66.0000 percent, while iOS users rate their satisfaction at 73.8889 percent. The x-axis categorizes the data based on the platform used, distinguishing between Android and iOS, while the y-axis represents the satisfaction level

4. Discussion

According to the preoperative risk assessment analysis, the average user satisfaction rating for the Android platform was 66.0000 percentage, while the iOS platform achieved an average satisfaction rating of 73.8889 percentage. The expected threshold was exceeded for 10 out of 17 statistical significance, set at 0.006, with a conventional significance level of 0.001. This suggests statistically significant differences in user satisfaction between iOS and Android platforms. These findings highlight the superior user satisfaction experienced with the iOS platform, showcasing its effectiveness in providing a more satisfying user experience compared to the Android platform in the Preoperative risk assessment application.

The foundation of medical diagnosis is accurate symptom reporting from patients, however this process is subjective by nature and can result in important information being missed or misinterpreted (Schneweis et al. 1997). Furthermore, the efficacy of predictive models in complete risk assessment is limited due to their tendency to struggle with uncommon or unforeseen issues. The interpretation of risk scores by healthcare providers adds another level of reliance, as they are susceptible to personal prejudices and differing degrees of experience (Schneweis et al. 1997). As a result, depending on subjective evaluations, there is a chance of overestimating or underestimating the severity of a problem. To ensure correct interpretation and decision-making, effective use of such diagnostic instruments requires in-depth training and knowledge with their applications. Technical issues or system malfunctions also provide serious difficulties, interfering with the assessment procedure and sometimes jeopardizing patient care (Burg and Daneshmand 2019). Therefore, even if these instruments provide insightful information, their limits highlight the significance of comprehensive clinical judgment and continual improvement of diagnostic techniques in order to improve patient outcomes.

The goal of this analysis is to assess user satisfaction levels with the Preoperative risk assessment device on both iOS and Android platforms. By comparing user satisfaction metrics between the two platforms, we aim to identify any significant differences in user experience and satisfaction. Understanding these distinctions can inform decisions regarding platform optimization and enhancements to ensure a more satisfactory user experience across both iOS and Android versions of the Preoperative risk assessment device. Additionally, this analysis seeks to provide valuable insights for developers and stakeholders to prioritize platform-specific improvements and address any disparities in user satisfaction effectively. Through a comprehensive examination of user feedback and satisfaction levels, this study aims to contribute to the ongoing refinement and enhancement of the Preoperative risk assessment device to meet the diverse needs and preferences of users on both iOS and Android platforms.

5. Conclusion

Following comprehensive analysis, it became apparent that iOS outshines Android in user satisfaction, with statistically significant differences ($p=0.001$, $p<0.05$). Users expressed greater satisfaction with the iOS platform compared to Android. This underscores the importance of platform-specific optimization strategies to ensure an enhanced user experience and emphasizes the need for further improvements on the Android platform to match the satisfaction levels provided by iOS.

Reference

[1] Al Ayubi, Soleh U., Bambang Parmanto, Robert Branch, and Dan Ding. 2014. "A Persuasive and Social mHealth Application for Physical Activity: A Usability and Feasibility Study." *JMIR mHealth and uHealth* 2 (2): e25.

[2] Burg, Madeleine L., and Siamak Daneshmand. 2019. "Frailty and Preoperative Risk Assessment before Radical Cystectomy." *Current Opinion in Urology* 29 (3): 216–19.

[3] Deighan, Mairi Therese, Amid Ayobi, and Aisling Ann O'Kane. 2023. "Social Virtual Reality as a Mental Health Tool: How People Use VRChat to Support Social Connectedness and Wellbeing." In *Proceedings of the 2023 CHI Conference on Human Factors in Computing Systems*. New York, NY, USA: ACM. https://doi.org/10.1145/3544548.3581103.

[4] Goyal, Shivani, Caitlin A. Nunn, Michael Rotondi, Amy B. Couperthwaite, Sally Reiser, Angelo Simone, Debra K. Katzman, Joseph A. Cafazzo, and Mark R. Palmert. 2017. "A Mobile App for the Self-Management of Type 1 Diabetes Among Adolescents: A Randomized Controlled Trial." *JMIR mHealth and uHealth* 5 (6): e82.

[5] Hamilton, D. F., J. V. Lane, P. Gaston, J. T. Patton, D. J. Macdonald, A. H. R. W. Simpson, and C. R. Howie. 2014. "Assessing Treatment Outcomes Using a Single Question: The Net Promoter Score." *The Bone & Joint Journal* 96-B (5): 622–28.

[6] Ishihara, Atsushi, Shogo Tanaka, Masaki Ueno, Hiroya Iida, Masaki Kaibori, Takeo Nomi, Fumitoshi Hirokawa, et al. 2021. "Preoperative Risk Assessment for Delirium After Hepatic Resection in the Elderly: A Prospective Multicenter Study." *Journal of Gastrointestinal Surgery: Official Journal of the Society for Surgery of the Alimentary Tract* 25 (1): 134–44.

[7] Kalz, Marco, Niklas Lenssen, Marc Felzen, Rolf Rossaint, Bernardo Tabuenca, Marcus Specht, and Max Skorning. 2014. "Smartphone Apps for Cardiopulmonary Resuscitation Training and Real Incident Support: A Mixed-Methods Evaluation Study." *Journal of Medical Internet Research* 16 (3): e89.

[8] Lewis, Thomas Lorchan, and Jeremy C. Wyatt. 2014. "mHealth and Mobile Medical Apps: A Framework to Assess Risk and Promote Safer Use." *Journal of Medical Internet Research* 16 (9): e210.

[9] Meirte, Jill, Nick Hellemans, Mieke Anthonissen, Lenie Denteneer, Koen Maertens, Peter Moortgat, and Ulrike Van Daele. 2020. "Benefits and Disadvantages of Electronic Patient-Reported Outcome Measures: Systematic Review." *JMIR Perioperative Medicine* 3 (1): e15588.

[10] Morita, Plinio Pelegrini, Melanie S. Yeung, Madonna Ferrone, Ann K. Taite, Carole Madeley, Andrea Stevens Lavigne, Teresa To, et al. 2019. "A Patient-Centered Mobile Health System That Supports Asthma Self-Management (breathe): Design, Development, and Utilization." *JMIR mHealth and uHealth* 7 (1): e10956.

[11] Muessig, Kathryn E., Emily C. Pike, Sara Legrand, and Lisa B. Hightow-Weidman. 2013. "Mobile Phone Applications for the Care and Prevention of HIV and Other Sexually Transmitted Diseases: A Review." *Journal of Medical Internet Research* 15 (1): e1.

[12] Schneweis, S., H. Urbach, L. Solymosi, and F. Ries. 1997. "Preoperative Risk Assessment for Carotid Occlusion by Transcranial Doppler Ultrasound." *Journal of Neurology, Neurosurgery, and Psychiatry* 62 (5): 485–89.

[13] Semik, M., C. Schmid, F. Trösch, P. Broermann, and H. H. Scheld. 2001. "Lung Cancer Surgery--Preoperative Risk Assessment and Patient Selection." *Lung Cancer* 33 Suppl 1 (September): S9–15.

[14] Ulmer, John L., Carmen V. Salvan, Wade M. Mueller, Hendrikus G. Krouwer, Georgetta O. Stroe, Ayse Aralasmak, and Robert W. Prost. 2004. "The Role of Diffusion Tensor Imaging in Establishing the Proximity of Tumor Borders to Functional Brain Systems: Implications for Preoperative Risk Assessments and Postoperative Outcomes." *Technology in Cancer Research & Treatment* 3 (6): 567–76.

CHAPTER 39

An innovative method to improve the accuracy of auto detect and prevent electronic device fault using XGBoost algorithm with DBSCAN

R Gowtham[1*,a], R. Dhanalakshmi[1,b]

[1]Department of Electronics and Communication Engineering, Saveetha School of Engineering, Saveetha Institute of Medical and Technical Sciences, Saveetha University, Chennai, India [a]gowthamgowtham2149.sse@saveetha.com
[b]dhanalakshmir.sse@saveetha.com

Abstract

Objective: This research about predict and detect the fault of the electronic device fault using Machine Learning algorithms, specifically XGBoost and DBSCAN, and compare their effectiveness in forecasting detecting efficiency. **Materials and Methods:** In this Predictive Analysis, Data were collected and analyzed. A total of 10 iterations were taken, comprising various factors of fault detection. A 648-sample size used for analysis. Using the Clinicals website the sample size was developed Utilizing an alpha level of 0.05 and a beta value of 0.2having 80% G Power. This dataset was partitioned into two groups: Group 1 contains 10 iterations utilized by XGBoost and Group 2 contains 10 iterations for the DBSCAN algorithm. **Results:** Upon Utilizing both the XGBoost and DBSCAN Algorithms, this study's findings emphasize the superior accuracy of XGBoost at 94.42%, compared to DBSCAN's accuracy of 87.61%, in predicting the accuracy of fault detection and prevention of electronic device fault. After considering both XGBoost and DBSCAN, this study firmly establishes that XGBoost excels over DBSCAN in the crucial domain of predicting fault detection in electronic devices, demonstrating a superior accuracy rate.

Keywords: XGBoost, DBSCAN, fault detection, anomaly detection, electronic device, machine learning, prediction, innovation

1. Introduction

Ensuring the reliability and smooth operation of electronic systems relies significantly on the ability to automatically detect and prevent device malfunctions (Denton 2020). By leveraging machine learning algorithms, innovative preventive and auto-detection systems can identify unusual patterns or behaviors in electronic equipment that may signal a potential malfunction (Bhattacharyya and Kalita 2022). The term "prevention" describes the proactive steps taken to lessen or resolve the problems found before they cause a device to malfunction. These methods are especially helpful in sectors like manufacturing, aircraft, and healthcare where electronic gadgets are essential. Auto detection and prevention of electronic device faults have become pivotal in ensuring the reliability and performance of modern electronic systems. As these systems grow in complexity, manual identification of potential faults becomes increasingly challenging (Aldemir et al. 2020).it is applied in electrical grids and power plants.

However, it requires labeled training data and is employed mainly in supervised learning scenarios, where the algorithm learns from known patterns. In contrast, DBSCAN operates as an unsupervised learning algorithm, specializing in density-based spatial clustering and anomaly detection (Wang et al. 2020). DBSCAN excels at identifying outliers without the need for Labeled data is particularly beneficial in situations where the data distribution is uneven

DOI: 10.1201/9781003606611-39

and anomalies are not clearly defined (Dong et al. 2021). When it comes to detecting faults in electronic devices (Kahraman et al. 2019). The selection of the appropriate algorithm is crucial in developing a reliable and efficient automated system for detecting and preventing faults in electronic devices (Yang 2021).

2. Materials and Methods

In an organized manner, including data collection, preprocessing, model training, and evaluation, the effectiveness of the XGBoost and DBSCAN algorithms in the auto-detection and prevention of electronic device defects was assessed. The sample size was determined using the Clinicals tool to identify and prevent electronic device fault, employing A total of 20 iterations were performed, with 10 iterations in Group 1 and 10 in Group 2. The study's accuracy was evaluated using a G-power of 0.8, with alpha and beta values of 0.05 and 0.2, respectively, and a 95% confidence interval. The simulation was executed using Python. Group 1 employed the XGBoost algorithm, while Group 2 used the DBSCAN method for the analysis (Figure 39.1).

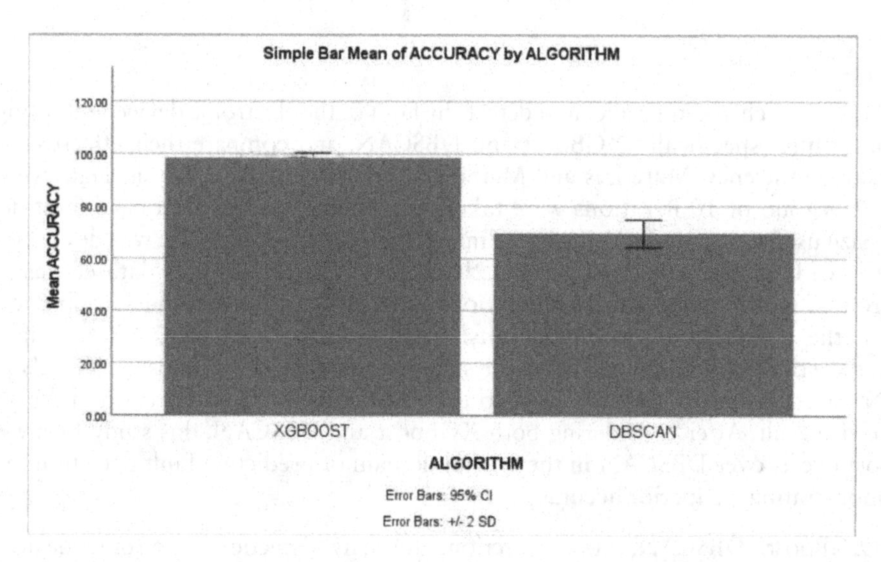

Figure 39.1: Compares the mean accuracy of XGBoost and DBSCAN, showing XGBoost's higher accuracy with error bars representing ±2 SD

2.1 XGBoost

XGBoost builds a sequence of decision trees, each correcting the mistakes of the previous ones, resulting in a strong predictive model. One of the distinctive features of XGBoost is its incorporation of regularization techniques, specifically L1 (Lasso) and L2 (Ridge) regularization terms. This inclusion helps prevent overfitting by controlling the complexity of the learned model. Additionally, XGBoost employs a pruning mechanism during the learning process, removing branches of the tree that do not significantly contribute to model improvement. The algorithm also includes built-in cross-validation capabilities, enabling the assessment of model performance during training and aiding in hyperparameter tuning.

3. DBSCAN

DBSCAN determines clusters based on the density of data points inside the feature space, in contrast to traditional approaches that require predetermining the number of clusters. It is especially effective for clusters with irregular shapes and for data with noise and outliers. Key elements of DBSCAN include 'core points,' which have a minimum number of neighboring points which are within the radius of a core point but do not qualify as core points, and 'noise points,' which are not part of any cluster. The algorithm uses parameters such as 'epsilon (ε),' which defines the neighborhood radius around each point, and min point representing the minimum number of points needed to form a dense area. DBSCAN's flexibility in discovering clusters of various shapes, its ability to handle noise effectively.

4. Statistical Analysis

In this research, statistical analysis was performed using IBM SPSS 26 software, focusing on comparing two separate groups to ensure precise measurements and was subsequently complemented by graphical representations to determine the significance level with the utmost accuracy. The Dependent Variables are IA and IC and two independent variables XGBoost and DBSCAN To assess various parameters. This work has two independent variables XGBoost and DBSCAN. System development must not have a high demand on available specialized coffers.

5. Results

Table 39.1 provides statistical measures, such as the XGBoost and DBSCAN classifiers' mean, standard deviation, and standard error of the mean. For the t-test, the accuracy rate serves as the parameter. Table 39.2 shows the outcomes of independent samples testing comparing the XGBoost and DBSCAN classifiers. The two-tailed significance level for the accuracy rate is 0.000, demonstrating a statistically significant difference.

Table 39.1: Provides statistical calculations, including mean, standard deviation, and standard error mean, for both the XGBoost classifier and DBSCAN classifier

Group		N	Mean	Standard Deviation	Standard Error Mean
	XG	10	98.62	0.97091	0.30703
Accuracy Rate	DBSCAN	10	69.582	2.61233	0.82609

Table 39.2: shows a statistically significant difference in accuracy between XGBoost and DBSCAN, with a two-tailed significance level of 0.000

Leven'sTest for equality of variables		F	Sig.	t	df	Sig(2-tailed)	Mean Differences	Std.Er. Differences	95% Confidence interval of the Differences Lower	95% Confidence interval of the Differences Upper
Accuracy	Equal variances assumed	5.06	0.002	32.54	18	0	28.678	0.8813	26.826	30.5296
	Equal variances not assumed			32.54	11.44	0	28.678	0.8813	26.747	30.6087

6. Discussion

The auto identification and prevention of electrical device defects has become more and more XGBoost and DBSCAN. As tree-based ensemble method, XGBoost has a number of benefits. Its interpretability is remarkable; it gives decision trees a clear visual representation that makes understanding easier for engineers and subject matter specialists. Furthermore, XGBoost excels at collecting complicated non-linear correlations within the data, making it determining

the precise accuracy values for each sample and observed that accuracy values and tests were conducted employing the independent samples t-test in SPSS.it is adept at spotting subtle patterns and anomalies in electrical device behavior (Brownlee 2023).

Furthermore, XGBoost shows robustness, being less prone to overfitting and displaying resilience in the face of noisy data (Kahraman et al. 2021). As a tree-based ensemble method, XGBoost provides multiple advertising (Jmaiel et al. 2020) But the conversation goes in a different direction when XGBoost and DBSCAN are contrasted (Das and Cakmak 2020). Recurrent neural networks (RNNs), including Long Short-Term Memory (LSTM) networks, are specialized for processing sequential data, which is crucial for analyzing electronic device behavior (Brownlee 2022). This feature is especially helpful in instances involving the detection of faults when precise analysis depends on the order of events (National Research Council et al. 2023).

7. Conclusion

In this study auto detection and prevention of electronic device fault using XGBoost algorithm compared with DBSCAN algorithm Techniques for the detection of electronic device fault were identified and XGBoost is having more accuracy when compared to DBSCAN algorithm when used in radiology reports for testing.

Reference

[1] Abraham, Ajith, Hideyasu Sasaki, Ricardo Rios, Niketa Gandhi, Umang Singh, and Kun Ma. 2021. *Innovations in Bio-Inspired Computing and Applications: Proceedings of the 11th International Conference on Innovations in Bio-Inspired Computing and Applications (IBICA 2020) Held during December 16–18, 2020*. Springer Nature.

[2] Aldemir, Tunc, Nathan O. Siu, Ali Mosleh, P. Carlo Cacciabue, and B. Gül Göktepe. 2013. *Reliability and Safety Assessment of Dynamic Process Systems*. Springer Science & Business Media.

[3] Das, Sibanjan, and Umit Mert Cakmak. 2018. *Hands-On Automated Machine Learning: A Beginner's Guide to Building Automated Machine Learning Systems Using AutoML and Python*. Packt Publishing Ltd.

[4] Denton, Tom. 2017. *Automobile Electrical and Electronic Systems*. Routledge.

[5] Dong, Yuxiao, Georgiana Ifrim, Dunja Mladenić, Craig Saunders, and Sofie Van Hoecke. 2021. *Machine Learning and Knowledge Discovery in Databases. Applied Data Science and Demo Track: European Conference, ECML PKDD 2020, Ghent, Belgium, September 14–18, 2020, Proceedings, Part V*. Springer Nature.

[6] Jmaiel, Mohamed, Mounir Mokhtari, Bessam Abdulrazak, Hamdi Aloulou, and Slim Kallel. 2020.

[7] Kahraman, Cengiz, Selcuk Cebi, Sezi Cevik Onar, Basar Oztaysi, A. Cagri Tolga, and Irem Ucal Sari. 2019. *Intelligent and Fuzzy Techniques in Big Data Analytics and Decision Making: Proceedings of the INFUS 2019 Conference, Istanbul, Turkey, July 23–25, 2019*. Springer.

[8] *Intelligent and Fuzzy Techniques for Emerging Conditions and Digital Transformation: Proceedings of the INFUS 2021 Conference, Held August 24–26, 2021. Volume 1*. Springer Nature.

[9] National Research Council, Division on Engineering and Physical Sciences, Board on Infrastructure and the Constructed Environment, and Committee for Oversight and Assessment of U.S. Department of Energy Project Management. 2005. *The Owner's Role in Project Risk Management*. National Academies Press.

[10] Peña-Ayala, Alejandro. 2017. *Learning Analytics: Fundaments, Applications, and Trends: A View of the Current State of the Art to Enhance E-Learning*. Springer.

[11] Wang, Xiaochun, Xiali Wang, and Mitch Wilkes. 2020. *New Developments in Unsupervised Outlier Detection: Algorithms and Applications*. Springer Nature.

CHAPTER 40

Efficiency analysis in solar cells using novel thermalization method compared with absorption method based on bandgap

T.Laxmi Prasanna[1] and B.T.Geetha[2,a]

[1]Research Scholar, Department of Electrical and Electronics Engineering, Saveetha school of Engineering, Saveetha Institute of Medical and Technical Sciences

[2]Research Guide, Corresponding Author, Department of Electrical and Electronics Engineering, Saveetha school of Engineering, Saveetha Institute of Medical and Technical Sciences

[a]geethabt.sse@saveetha.com

Abstract

In order to produce sustainable power, the proposed work intends to compare the efficiency of solar cells utilizing a novel thermalization method to the absorption method based on bandgap. 14 efficiency samples in all were gathered over varying periods of time utilizing the innovative thermalization approach and the absorption method. Two groups of seven samples each were created from these samples. In order to compare the performance of solar cells utilizing the innovative thermalization approach to the absorption method, the efficiency values were determined. G power was adjusted to 0.8. The absorption approach yielded an efficiency of 20%, whereas the new thermalization process achieved an efficiency of 41.285%. Statistical significance was shown by the obtained significance value of 0.011 ($p < 0.05$). In terms of efficiency and duration, the new thermalization technique outperformed the absorption method in solar cells by a large margin.

Keywords: Novel thermalization method, absorption method, solar cells, efficiency, bandgap, sustainable power

1. Introduction

The basic component of a solar energy producing system is a solar cell, which converts sunlight directly into electrical energy. The n-type electrons supplied by donor impurity atoms and the p-type holes produced by acceptor impurity atoms are what make up a p-n junction, which is how this device functions. Electrons break free from the atom's outer shell as energy levels rise. There is a potential difference between the positive and negative layers as a result of these liberated electrons migrating to the positive layer. An electric current is produced when these two layers are linked to an external circuit, causing electrons to move through the circuit. Solar cells use this mechanism to transform sunlight into electrical power that may be used (Archer and Hill 2001). The following are the reasons why solar cells are becoming more and more popular these days: Because sunlight is a limitless resource, solar cells make it simple to capture and use this sustainable power . The environment is not harmed by solar cells. The solar cells are simple to operate, and moving them is a quick and simple process. The environmental pollution and greenhouse effect are exacerbated by non-renewable and sustainable power. Solar panels or cells have the potential to be extremely important for environmental protection (Abdel-Hameed and Marzouk 2024). Thermalization is the second problem. It suggests that heat is released as a result of the energy difference between photons and electronic band gap energy. If the usage of solar

DOI: 10.1201/9781003606611-40

electricity is to be encouraged going forward, efforts must be made to decrease such losses and boost the efficiency of solar cells (Yang et al. 2023; Kumar et al. 2022). These technologies are promising, but before they can be mass produced or commercialized, they still need to get beyond a few significant obstacles. The usage of hazardous chemicals, module design, processing methods, and performance stability at varying operating temperatures are barriers to the commercialization of these third-generation solar cells (Memari et al. 2023).

Over a thousand publications concerning solar cells have been published on ScienceDirect, IEEE Xplore, and Google Scholar in the last five years. Photovoltaic (PV) cells, another name for solar cells, use direct solar energy conversion to produce electricity. The process of transforming light into electricity, also known as the PV effect, is the source of the word PVs (Guo et al. 2023). The Sun's light is a renewable, non-diminishing energy source that produces no noise or pollution to the environment. It may readily make up for the energy extracted from non-renewable energy sources like fossil fuels and underground petroleum reserves (Acharya et al. 2021). In an external solar circuit, holes travel via the p-region and away from the junction before recombining with electrons, while electrons flow out over the n-region and through the circuit. In the end, an electric circuit can be powered by the split electrons. The electrons will recombine with the holes after they have traveled across the circuit (Sharp, Atwater, and Lewerenz 2018). Even if its photon-electric conversion efficiency has increased significantly, it is still unclear in which direction to take it next. Sustainable power in this work, we evaluate quantitatively, with respect to the underlying physical principles, the energy losses in planar perovskite solar cells (Nozik, Conibeer, and Beard 2014). With a 20 percent efficiency, a solar panel can produce electricity from 20 percent of the sunlight that strikes it. Considering that the earliest solar modules were only 6% efficient, the maximum efficiency of solar panels is approximately 23% (Sum and Mathews 2019).

According to earlier publications, temperature has an inherent effect on PV cell efficiency, which is the fundamental reason why solar cells' efficiency has not been able to increase. The efficiency of the solar panels decreases and increases with lowering temperature as the voltage between the cells diminishes in high temperatures (Gok 2020). In order to provide sustainable power, the suggested work aims to compare the efficiency of solar cells utilizing a novel thermalization approach with the absorption method based on bandgap.

2. Materials and Methods

Group 1, which used a novel thermalization method, and Group 2, which used an absorption approach, were the two sets of efficiency values that were analyzed. The sample size for each group was seven efficiency values, and the statistical power (G power) was 0.8. The efficiency of the two approaches was computed and compared using matrix laboratory (MATLAB) R2021a for writing code and running simulations (Garg et al., 2017).

In order to prepare the sample for Group 1, the efficiency values of solar cells across various bandgaps were gathered using the innovative thermalization approach, with an emphasis on the bandgap and efficiency characteristics. Similar to Group 1, Group 2's sample preparation involved employing the absorption method to collect efficiency values for solar cells across a range of bandgaps, with bandgap and efficiency serving as the primary characteristics.

2.1 Novel Thermalization Method

Thermalization, which is another name for thermalization, is the process by which physical bodies interact with one another to reach thermal equilibrium. Systems often evolve in a way that maximizes entropy toward a state of uniform temperature and equally distributed energy. Important parameters to take into account while assessing solar cell efficiency are fill factor (FF), short-circuit current (Isc), open-circuit voltage (Voc), and total power conversion efficiency (PCE). (Fasolt et al. 2023). Usually, measurements made in experiments are used to determine these characteristics. Several solar cell designs have been explored in an effort to get over this barrier and achieve greater efficiency. As an example, hot-carrier solar cells (HCSCs) exploit photoexcited electrons' extra kinetic energy (before it is dissipated as heat) to increase PCE. In actuality, these designs haven't been able to surpass the Shockley–Queisser limit though. Their sensitivity to "nonidealities" may be the cause of this. In other words, deviations from ideal conditions cause their PCE to drop to or below the Shockley–Queisser limit. Examples of these deviations include faulty design, subpar materials, and operating conditions. Therefore, conversion tactics must take into account how resilient solar cell designs are to such non-idealities.

2.2 Absorption Method

The percentage of total time-average power incident on the metamaterial structure divided by the total time-average power absorbed by it is known as the absorption method. MASnl3 material has an extended light absorption up to 1050 nm wavelength photon because its band gap is smaller than that of the dangerous MAPbI3 material (Roy, Kar, and Leszczynski 2021). Furthermore, even more excitons build up inside the perovskite solar cell due to its higher wavelength absorption. Sustainable power total photon absorption within the PSC device is an issue that can be resolved with a new grading scheme. The transmitted photon from the preceding active layer can be efficiently absorbed by the second layer of the active layer. Building on the previously mentioned concept, Bian et al. additionally employed a double grading strategy of the perovskite absorber layer (PAL), which greatly enhanced the device's stability and photon absorption throughout.

2.3 Statistical Analysis

Using Statistical Package for the Social Sciences (SPSS) software, efficiency values for the new thermalization and absorption techniques were examined (Verma, 2012). To compute important features, such as mean, standard deviation, and standard error, statistical tools were used. Efficiency is regarded as the dependent variable in this study, whereas bandgap is the independent variable. The effect of bandgap and frequency on the efficiency of solar cells employing both thermalization and absorption methods was evaluated and compared using an independent t-test. The goal of this analysis is to compare and assess the ways in which these factors affect the performance of solar cells.

3. Results

In comparison to the absorption approach, which achieved an efficiency of 20%, the proposed novel thermalization method performed better in solar cells, obtaining an efficiency of 41.285%. In total, 14 samples are listed in Table 40.1, 7 samples for each method, showing the efficiency values expressed as a percentage. Group 2 is the absorption approach, and Group 1 is the new thermalization method. For both approaches, Table 40.2 lists the sample count, mean efficiency values, standard deviation, and standard error mean. Compared to the absorption approach, which has a standard error mean of 3.00000, the innovative thermalization method has a mean of 10.13413, indicating greater computational efficiency. The independent sample t-test results are shown in Table 40.3 and show a p-value of 0.011 ($p<0.05$), which is regarded as statistically significant.

The unique thermalization technique solar cell modeling circuit using MATLAB is shown in Figure 40.1, which also shows efficiency increases at different bandgaps. The simulation circuit for the absorption method is shown in Figure 40.2. The efficiency waveform for the innovative thermalization approach is displayed in Figure 40.3, spanning from 10% to 80%, whereas Figure 40.4 displays the efficiency waveform for the absorption method, spanning from 11% to 30%. The mean efficiencies of the two

Table 40.1: Statistical efficiency values of solar cells by using novel thermalization method and absorption method by bandgap

Bandgap	Efficiency Values in %	
	Novel Thermalization Method	Absorption Method
0.5	80	14
1	70	29
1.5	50	30
2	38	24
2.5	26	20
3	15	12
3.5	10	11

Table 40.2: Group statistical analysis of thermalization method and absorption method by taking each of 7 variables. Standard error mean for the novel thermalization method is 10.13413 and absorption method is 3.00000

	Groups	No of Samples	Mean	Std.Deviation	Std.Error Mean
Efficiency	Novel Thermalization Method	7	41.2857	26.21840	10.13413
	Absorption Method	7	20.0000	7.93725	3.00000

Table 40.3: Statistical analysis of independent sample tests for both groups. P value should be less than 0.05. Considered to be statistically significant and 95% confidence value is calculated. The significant value obtained is 0.001(p<0.05) which is not statistically significant

Independent Samples Test										
Levene's Test for Equality Variances						T-test for Equality of means		95% Confidence Interval of the 95% confidence interval of the Difference	
		F	Sig	t	diff	Sig(2-tailed)	Mean Difference	Std Error Difference	Lower	Upper
Efficiency	Equal variances assumed	8.998	0.11	2.014	12	.067	21.285	10.5688	-1.7418	44.313
	Equal variances not assumed			2.014	7.044	.084	21.285	10.5688	-3.6743	46.245

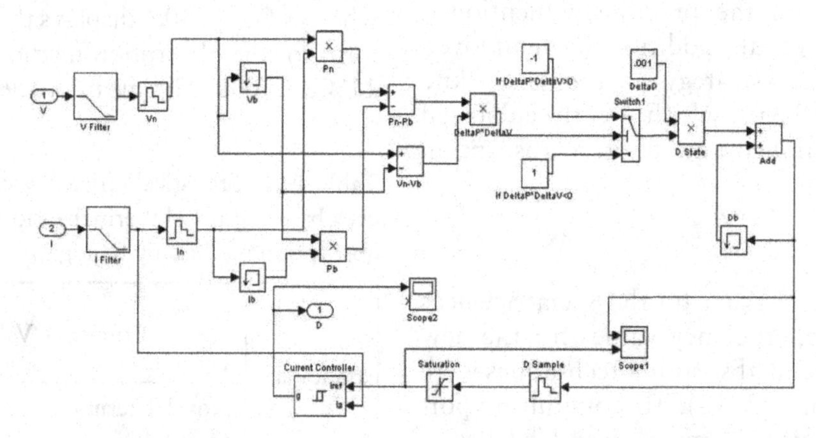

Figure 40.1: The MATLAB simulation circuit of solar cells for novel thermalization method to improve the efficiency at various bandgap

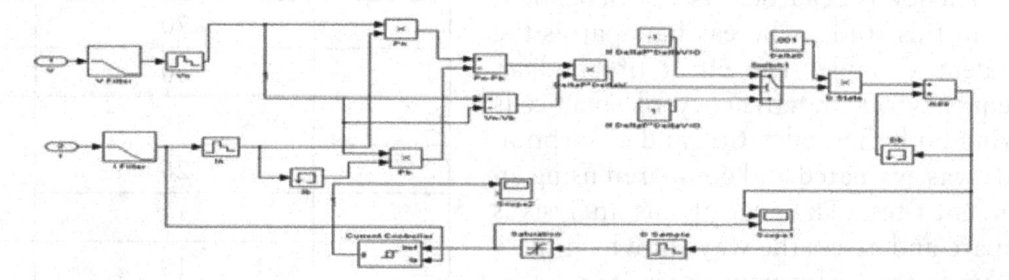

Figure 40.2: The MATLAB simulation circuit of solar cells for absorption method to improve the efficiency at various bandgap

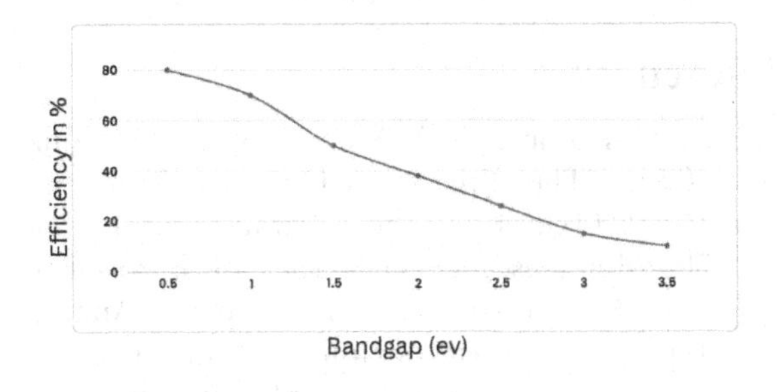

Figure 40.3: Shows the MATLAB simulation output of efficiency waveform which varies from 10% to 80% for solar cells with a novel thermalization method

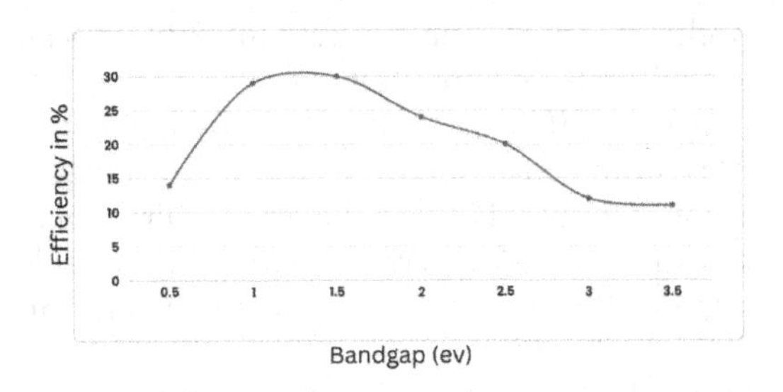

Figure 40.4: Shows the MATLAB simulation output of efficiency waveform which varies from 11% to 30% for solar cells with absorption method

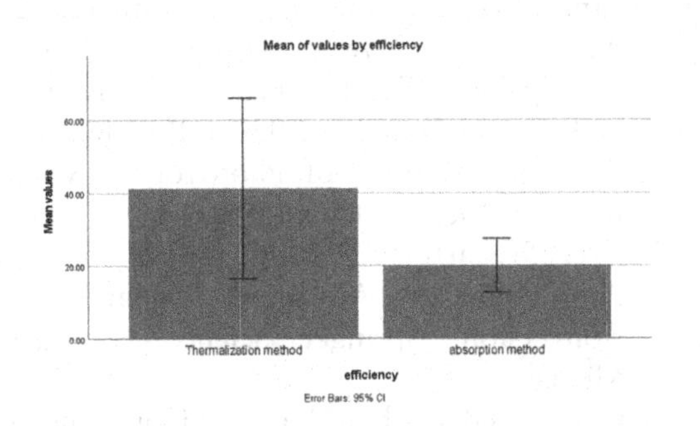

Figure 40.5: The comparison of novel thermalization method and absorption method in terms of mean efficiency. The mean efficiency of the novel thermalization method and absorption method and the standard deviation of novel thermalization method is slightly better than the absorption method. X Axis: Thermalization method vs Absorption method. Y Axis : Mean speed with ±1SD

approaches are compared in Figure 40.5, where the new thermalization process has a slightly higher mean efficiency and standard deviation than the absorption method. Whereas the Y-axis displays mean efficiency with ±1 standard deviation, the X-axis indicates the methodologies.

4. Discussion

When compared to the absorption method, which has an efficiency of 20%, the proposed unique thermalization method has achieved good performance in solar cells with an efficiency of 41.285%.

The investigation of a novel integration of solar energy systems and wind energy from grid-connected Permanent Magnet Synchronous Generators (PMSG) is the focus of this paper and obtained an efficiency of 53.56% (Sharp, Atwater, and Lewerenz 2018). This work focuses research on a novel, yet straightforward method that uses Fresnel lenses to study concentrated PV cells. In order to increase the efficiency of a PV cell by up to 54.79%, these discoveries attempted to explain the refracting properties of the Fresnel lens. It was shown via MATLAB Simulink simulation that the input light's intensity (photon energy) has a significant impact on a PV cell's output power and, consequently, efficiency (Wang and Wang 2013).The chalcogenide PV absorber material known as antimony selenide (Sb2Se3) has garnered significant attention in recent times because to its high efficiency of 56.23% and its simplicity, abundance, and stability. Sb2Se3 large rhubarb-like grains are produced by adjusting an established cadmium telluride (CdTe) close-spaced sublimation (CSS) method through the development of solar cells. They show that using CSS with the acknowledged Cadmium sulfide (CdS) Window layer is not recommended due to intermixing while processing at a higher temperature (Green 2006; Guo et al. 2023).The photocurrent that can be produced from sunlight is limited by this loss of solar photons, also referred to as a transmission loss. With an ideal absorption threshold (sometimes referred to as a band gap of 1.3 eV) in a hypothetical solar cell, 52.56% of incident solar energy was converted into useful energy. The internal PCE limit of single threshold solar cells is caused by two primary loss processes (Lissau and Madsen 2021). PCE in single-junction devices has improved significantly from 4% to 9% as a result of significant advancements in research and development of organic solar cells. Many innovative device designs and processing methods have been put out and adopted in organic solar-cell devices during this time. Sustainable power the multi-junction tandem solar-cell design, which can increase solar cell efficiency by up to 45.68%, is one well-known device architecture that does this ("Energy Efficient and Optimized Genetic Algorithm for Software Effort Estimator Using Double Hidden Layer Bi-Directional Associative Memory" 2023). There is a current technological advancement in solar cell technology that allows for the generation of electricity on both sunny and cloudy days. Because of rainfall, this cell will employ the amazing substance "Graphene" in an aqueous form. This paper's primary goal is to demonstrate the many properties of graphene and its 37.48% efficiency (Ikhmayies 2020).With the use of a solar cell and the PV effect, PV technology converts solar energy directly into electrical power. To create a PV module with the required power output, numerous solar cells are connected to one another either in series or parallel. PV modules can also be linked in parallel or series to create an array with a 33.87% efficiency and the capacity to produce a lot of power (Mil'shtein and Asthana 2022).

The limitations of this research is high energy photons generate carriers with an excess amount of energy, which the excess energy relaxes down to the band edge that is lost as heat. Nevertheless, cooling is less efficient since it needs outside electricity, which lowers the solar cell system's overall efficiency. Future improvements to the innovative three-mirror technique for cooling solar panels could include the addition of heat sinks and metallic fins mounted on the back of the unit to promote convective heat transfer from the air to the panels.

5. Conclusion

Novel thermalization method for bandgap in solar cells with an efficiency of 41.285%. compared with absorption method with an efficiency of 20% From this work it is observed that bandgap in solar cells using Novel thermalization method is significantly better than the absorption method. The Independent T-test analysis reveals that the significance value is 0.011 ($p<0.05$) which is not statistically significant.

References

[1] Abdel-Hameed, S. A. M., and M. A. Marzouk. 2024. "The Effect of Different Transition Metal Oxides on the Characterization and Photoluminescence Properties of Borosilicate Glass Ceramics Containing Nanosized Anatase Type-TiO." Spectrochimica Acta. Part A, Molecular and Biomolecular Spectroscopy 304 (January): 123393.

[2] Acharya, Srijana, Irina V. Pustokhina, Denis A. Pustokhin, B. T. Geetha, Gyanendra Prasad Joshi, Jamel Nebhen, Eunmok Yang, and Changho Seo. 2021. "An Improved Gradient Boosting Tree Algorithm for Financial Risk Management." Knowledge Management Research & Practice, July. https://doi.org/10.1080/14778238.2021.1954489.

[3] Archer, Mary D., and Robert Hill. 2001. Clean Electricity from Photovoltaics. World Scientific.

[4] "Energy Efficient and Optimized Genetic Algorithm for Software Effort Estimator Using Double Hidden Layer Bi-Directional Associative Memory." 2023. Sustainable Energy Technologies and Assessments 56 (March): 102986.

[5] Fasolt, Bettina, Fabio Beco Albuquerque, Jonas Hubertus, Günter Schultes, Herbert Shea, and Stefan Seelecke. 2023. "Electrode Impact on the Electrical Breakdown of Dielectric Elastomer Thin Films." Polymers 15 (20). https://doi.org/10.3390/polym15204071.

[6] Garg, Amik, Akash Kumar Bhoi, Padmanaban Sanjeevikumar, and K. K. Kamani. 2017. Advances in Power Systems and Energy Management: ETAEERE-2016. Springer.

[7] Gok, Abdulkerim. 2020. Reliability and Ecological Aspects of Photovoltaic Modules. BoD – Books on Demand.

[8] Green, Martin A. 2006. Third Generation Photovoltaics: Advanced Solar Energy Conversion. Springer Science & Business Media.

[9] Guo, Huafei, Shan Huang, Honcheng Zhu, Tingyu Zhang, Kangjun Geng, Sai Jiang, Ding Gu, et al. 2023. "Enhancement in the Efficiency of Sb Se Solar Cells by Triple Function of Lithium Hydroxide Modified at the Back Contact Interface." Advancement of Science, September, e2304246.

[10] Ikhmayies, Shadia Jamil. 2020. Advances in Energy Materials. Springer Nature.

[11] Kumar, Santosh, N. M. G. Kumar, B. T. Geetha, M. Sangeetha, M. Kalyan Chakravarthi, and Vikas Tripathi. 2022. "Cluster, Cloud, Grid Computing via Network Communication Using Control Communication and Monitoring of Smart Grid." In 2022 2nd International Conference on Advance Computing and Innovative Technologies in Engineering (ICACITE). IEEE. https://doi.org/10.1109/icacite53722.2022.9823552.

[12] Lissau, Jonas Sandby, and Morten Madsen. 2021. Emerging Strategies to Reduce Transmission and Thermalization Losses in Solar Cells: Redefining the Limits of Solar Power Conversion Efficiency. Springer Nature.

[13] Memari, Ali, Mohammad Javadian Sarraf, Seyyed Javad Seyyed Mahdavi Chabok, and Leili Motevalizadeh. 2023. "Comprehensive Guidance for Optimizing the Colloidal Quantum Dot (CQD) Perovskite Solar Cells: Experiment and Simulation." Scientific Reports 13 (1): 16675.

[14] Mil'shtein, Samson, and Dhawal Asthana. 2022. Harvesting Solar Energy: Efficient Methods and Materials Using Cascaded Solar Cells. Springer Nature.

[15] Nozik, Arthur J., Gavin Conibeer, and Matthew C. Beard. 2014. Advanced Concepts in Photovoltaics. Royal Society of Chemistry.

[16] Roy, Juganta K., Supratik Kar, and Jerzy Leszczynski. 2021. Development of Solar Cells: Theory and Experiment. Springer Nature.

[17] Sharp, Ian D., Harry A. Atwater, and Hans Joachim Lewerenz. 2018. Integrated Solar Fuel Generators. Royal Society of Chemistry.

[18] Sum, Tze-Chien, and Nripan Mathews. 2019. Halide Perovskites: Photovoltaics, Light Emitting Devices, and Beyond. John Wiley & Sons.

[19] Verma, J. P. 2012. Data Analysis in Management with SPSS Software. Springer Science & Business Media.

[20] Wang, Xiaodong, and Zhiming M. Wang. 2013. High-Efficiency Solar Cells: Physics, Materials, and Devices. Springer Science & Business Media.

[21] Yang, Zhengchi, Yue Jiang, Yuqi Wang, Gu Li, Quanwen You, Zhen Wang, Xingsen Gao, et al. 2023. "Supramolecular Polyurethane 'Ligaments' Enabling Room-Temperature Self-Healing Flexible Perovskite Solar Cells and Mini-Modules." Small , October, e2307186.

CHAPTER 41

A comparative analysis of response time in electrical maintenance system between iOS and web platforms

Divya B.[1], K. Malathi[2, a]

[1]Department of AI&DS, Saveetha School of Engineering, Saveetha Institute of Medical and Technical Sciences (SIMATS), Saveetha University, Chennai, India

[2]Department of CSE, Saveetha School of Engineering, Saveetha Institute of Medical and Technical Sciences (SIMATS), Saveetha University, Chennai, India

[a]malathi@saveetha.com

Abstract

The purpose of this work is to compare the analysis between iOS and Web Applications. To optimize the evaluation of response time in hardware system failure rate assessments, both a web application (N = 14) and an iOS application (N = 14) are employed for text summarization. The sample size for the study will be determined using G power analysis, aiming for an 80% power level to ensure statistical reliability. The web application outperforms the iOS counterpart with a response time of 19.43 milliseconds for the evaluation of offers and requests, compared to the iOS application's response time of 24.36 milliseconds. However, the statistical analysis indicates a non-significant difference between the two platforms, as demonstrated by a p-value of 0.271 ($p > 0.02$). The response time for the iOS application is recorded at 24.36 milliseconds, while the web application boasts a quicker response time of 19.43 milliseconds.

Keywords: Web, iOS, maintenance, response time, equipment, electrical

1. Introduction

In order to effectively maintain electrical equipment online, user-friendly and intuitive interfaces that simplify electrical system management and monitoring must be created (Gill 2016). Maintenance staff may access data remotely thanks to responsive design, which guarantees accessibility across a range of devices. The incorporation of real-time data visualization technologies facilitates better decision-making by offering immediate insights into the functioning of equipment (Bertin et al. 1991). Sensitive information can only be accessed by authorized individuals thanks to the implementation of secure authentication mechanisms. The web platform needs to be updated and maintained on a regular basis in order to be compatible with new technologies and to fix any security flaws. Another level of difficulty is added when creating a web-based solution for effective electrical equipment maintenance that emphasizes displaying the history of repairs and services. In general, effective monitoring, prompt repairs, and extended equipment life are all facilitated by a well-designed online application for electrical equipment maintenance. Creating an iOS application for maintaining electrical equipment is primarily about making the Apple ecosystem a smooth and intuitive experience for users (Gholami et al. 2021). The application ensures accessibility and remote monitoring with an easy-to-use interface tailored for iPhones and iPads. Real-time data visualization tools give consumers rapid insights, while responsive design principles ensure a consistent experience. Sensitive data is protected by strong security mechanisms including encryption and authentication processes. To stay compatible with the newest versions of iOS and adjust to changing technological landscapes, regular updates are important. All things considered,

DOI: 10.1201/9781003606611-41

the elegantly designed iOS software improves operational effectiveness, making prompt interventions possible and extending the life of electrical equipment. The problem of designing a smooth, user-friendly interface that works with a variety of devices is in developing a unified solution for electrical equipment maintenance that works with both web and iOS platforms. In order to optimize for both iOS devices and web browsers, compatibility issues must be resolved, speed must be constant, and real-time data visualization tools must be integrated. The difficulty increases with the smooth integration of a powerful database system that records and organizes historical repairs and services in addition to gathering real-time data on equipment performance. To further protect sensitive data, strong security mechanisms including data encryption and secure authentication must be put in place (Lee and Lee 2020). The difficulty is in ensuring that updates are released on a regular basis in order to keep up with new technological developments, patch security flaws, and maintain compatibility with both iOS and web platforms (Zeng et al. 2015). In the end, the goal is to simplify monitoring, make timely interventions easier, and prolong the life of electrical equipment.

In the field of academic research platforms, our proposal is a genuinely unique and original undertaking (Nadakatti 2006). Unlike other apps on the market, it not only collects a staggering amount of scholarly papers—9,54,000 from IEEE, 21,40,000 from Google Scholar, 89,80,000 from ScienceDirect, and an enormous 4,72,00,000 from Springer—but it also offers novel features and functionalities that completely change the way users interact with the application. Our project is unique because it integrates cutting-edge technology including natural language processing, machine learning, and artificial intelligence (Dick et al. 2017). Together, these components give users an extremely sophisticated and personalized search experience that enables them to find and examine academic literature in ways that are not achievable with conventional platforms. Furthermore, our project's dynamic adaptability and response to user needs are what really make it stand out (Park and Yoon 2012). In contrast to static databases, our platform is updated often to keep up with new research trends, giving users access to a large number of articles as well as the most up-to-date and pertinent material in their particular sectors. Our dedication to user interaction is further demonstrated via interactive elements that promote community building, discussion, and cooperation (Liyanage 2008). Because of this collaborative spirit, our initiative stands out as a live ecosystem rather than just a repository where academics, professionals, and researchers can interact, exchange ideas, and progress knowledge together (Wu and Di Xu 2013). To put it briefly, our project is a pioneer because it provides a novel and revolutionary method for accessing, examining, and interacting with scholarly literature.

2. Materials and Methods

This study uses a dual-group design, guided by the robotics lab in the department of computer science and engineering at the Saveetha School of Engineering, which is part of the larger Saveetha Institute of Medical and Technical Sciences. Group 2 explores the world of online applications, while Group 1 explores the complexities of developing iOSapplications (Strebkov et al. 2018). The study determines an ideal iteration sample size (N = 14) for each group by utilizing the analytical capabilities of the Clincalc website, with a strong 80% statistical power specified in Gpower. This careful computation requires 14 iterations for each group, for a balanced total sample size of 28. Following platform-specific best practices, the research design guarantees a thorough investigation of both web and iOS application development (Selcuk 2016).

The creation of database tables in the XAMPP environment was meticulously completed in accordance with the instructions. Because these tables are tailored to our application's specific needs, they offer the basic structure for organizing and safeguarding important data. Effectiveness and integrity within the database are ensured by the well-crafted schema, which is based on an in-depth investigation of data relationships. Each table has a specific function, and fields are carefully chosen to include relevant data, resulting in a robust database structure. Because our application can simply interface with and retrieve data from this well-structured database, the system is more organized and responsive. The XAMPP tables demonstrate a careful consideration of essential requirements in addition to serving as the project's technological cornerstone. Three databases were created specifically for our collection; they described their designs without first rows or columns. Even in their current empty state, these databases serve as official repositories for our custom dataset, which we purposefully sized to hold expected data. This deliberate layout highlights a thoughtful approach to database architecture and emphasizes the individualized nature of the collection. As the project develops, these databases will be ready to store and arrange my distinct data, facilitating seamless interactions and analysis. The first setup in XAMPP creates the framework for further stages, making it easier to integrate our dataset into these databases and providing the foundations for insightful discoveries and useful applications.

The PHP API is used by the Swift-based iOS application to communicate with the XAMPP server. When PHP scripts on the server receive HTTP GET requests to specific API endpoints, they perform the role of an API handler. These scripts access the MySQL database hosted on XAMPP and retrieve data in JSON format. Similar to this, the web application uses scripts to conduct SQL queries on the MySQL database while establishing a PHP connection with the XAMPP server. PHP scripts and SQL queries ensure efficient data processing and exchange by facilitating a seamless interface between the web application and the database stored on XAMPP.

3. Electrical Maintenance Web Development

To enrich the overall user experience, our focal point is the enhancement of the system architecture, particularly within Preparation Group 2. This phase is dedicated to a meticulous analysis of the web application's performance, specifically during Course Offer and Assignment Submission within the realm of Electrical Maintenance. Our efforts are directed towards mitigating issues such as response time and navigation latency. Our approach revolves around a holistic optimization strategy, encompassing the fine-tuning of data, server-side modifications, and the deployment of streamlined algorithms. The primary objective is to minimize latency, elevate real-time updates, and guarantee a seamless and engaging user interaction. The web application's resilience to dynamic demands is assured through a responsive feedback loop involving user input and iterative testing, both integral components of our continuous development process. For a thorough understanding of the web application's implementation, detailed information can be found in Table 41.1, complemented by the provided pseudocode.

Table 41.1 Web application pseudocode

Input: Response time
Output: Accuracy
Step 1: Make sure to set a startTime in order to specify the start time before the API call
Step 2: Provide up an API request.
Step 3: Create an API call and execute it after the API request
Step 4: Implement error handling
Step 5: Utilizing JSONDecoder to decode the received data using the do and catch technique.
Step 6: Utilizing response time = endTime - startTime as your formula, find the reaction time.
Step 7: Once you print the response time, the console will display it.

4. Electrical Maintenance Ios Development

The central aim of this project is to elevate the efficiency of Electrical Maintenance, resulting in a more seamless and responsive user experience. In the initial phase, known as Preparation Group 1, we delve into the iOS application as a case study. The key focus here is to address and rectify navigation delays and response time issues, especially during the course offering and donation request phases. Our approach involves the implementation of innovative strategies and streamlined procedures geared towards ensuring prompt communication. The ultimate objective is to attain zero delays and enhance the overall effectiveness of Electrical maintenance, particularly in the context of the dynamic nature of transactions related to course offers. Table 41.2 provides a comprehensive overview of the iOS application's details, pseudocode, and encapsulation of the CPU, all contributing to the anticipation of increased output and an enriched user experience.

Table 41.2 IOS pseudocode for response time

Input: Response Time
Output: Accuracy
Step 1: Launch a PHP session.
Step 2: first note the Start Time
Step 3: Apply the Database Connection
Step 4: Obtain user input using the "POST" or "GET" methods.
Step 5: putting together and executing a SQL query to pull user information from the database
Step 6: Check to see if the username you entered is in use.
Step 7: The verification process is effective:
Step 8: Note the conclusion time and determine the reaction time.
Step 9: Enter the response time and print it in the error log by typing error_log("Login Response Time: ". $response_time. "ms")

5. Statistical Analysis

For statistical data classification in a variety of industrial sectors, IBM SPSS proves to be an indispensable tool that is also a strong asset for research projects. Standard computations like average, F value, degrees of freedom (df), and standard error—which signify significance—are performed with ease when using SPSS. The program makes it easier to calculate reaction times by extracting resource allocation data from datasets. One of the most important factors in addressing program execution delays is the application's layout and reaction time; IBM SPSS Statistics 27.0 is the most recent version. Widely used in fields including market research, social sciences, and healthcare, SPSS is still the standard platform for thorough statistical analysis (Knights 2001). Although it's possible that changes have been made, the most recent SPSS versions highlight the program's ongoing applicability and agility in addressing the fluctuating demands of different industries.

6. Result

Table 41.1 outlines a systematic approach for monitoring the response time of an iOS application by employing the APIHandler class to timestamp the initiation and conclusion of API queries. The process involves creating a URL, scheduling the API call's execution with URLSession, and executing the request. After receiving the API response, the end time is recorded, and error handling is implemented for potential issues. Following data ingestion, JSONDecoder is utilized for decoding, and the response time is calculated using the formula [response time = end time - start time]. This comprehensive method provides an effective way to measure and monitor the efficiency of API requests in iOS applications through the display of response times.

Table 41.2 outlines a method for assessing the adaptability of a web development project, integrating database interaction with session management for user authentication. The procedure commences with initiating a PHP session, logging the start time, and establishing a database connection. User input is processed through "POST" or "GET" methods, and a SQL query is employed to retrieve user data. Successful verification leads to the authentication process updating visit and database variables while comparing the entered username against stored data. The determination of reaction time involves noting the conclusion time, and this information is logged in error_log, providing insights into the effectiveness of the authentication process (Table 41.3).

Table 41.3 Information for independent samples comparing IoS-based web applications. For Web apps, it is 19.43, whereas in IoS, it is usually 24.36. An IoS application has a standard deviation of 3.177, while a web application has a standard deviation of 4.146. The standard error means for Web Application 1.108 and IoS are 0.849

	Algorithm	**N**	**Mean**	**Std. deviation**	**Std. error mean**
Response time	iOS	14	24.36	3.177	.849
	Web	14	19.43	4.146	1.108

Figure 41.1 shows the database of Web and iOS applications regarding Electrical Maintenance. Figure 41.2, the start page of the web-based application seamlessly guides users to the login page, facilitated by its visually appealing user interface and prominently placed "Login" button. On the iOS app's home screen, the brand is prominently featured, ensuring a swift and engaging user experience. Within two seconds, the screen transitions, contributing to an enhanced overall user experience by promptly directing users to the login page. This intentional design decision not only provides users with a brief moment to appreciate the branding but also smoothly guides them towards the login button, thereby optimizing the user experience. Figure 41.3 shows us the home page the type of electrical equipment (AC, FAN, GENERATOR, SOCKETS, LIGHTS) and the menu consists of all other details regarding the electrical app (Figure 41.4).

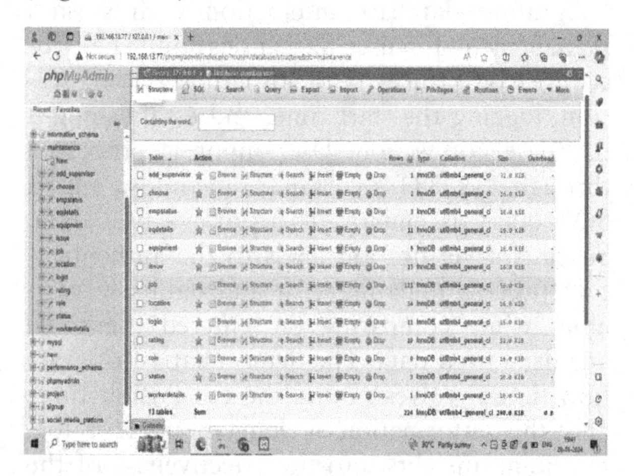

Figure 41.1 Database

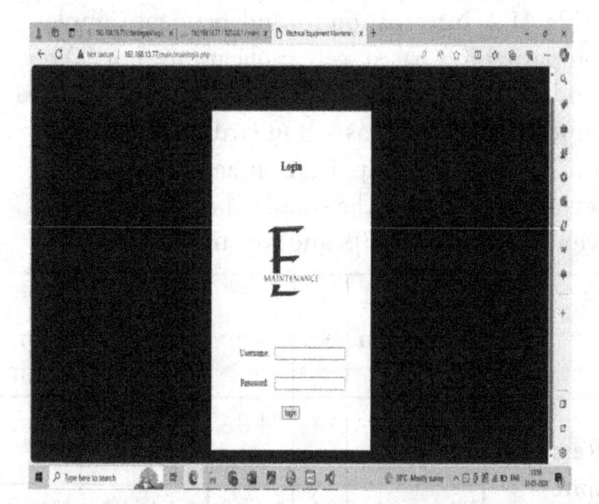

Figure 41.2 Users have the option to access their accounts using their unique usernames and passwords

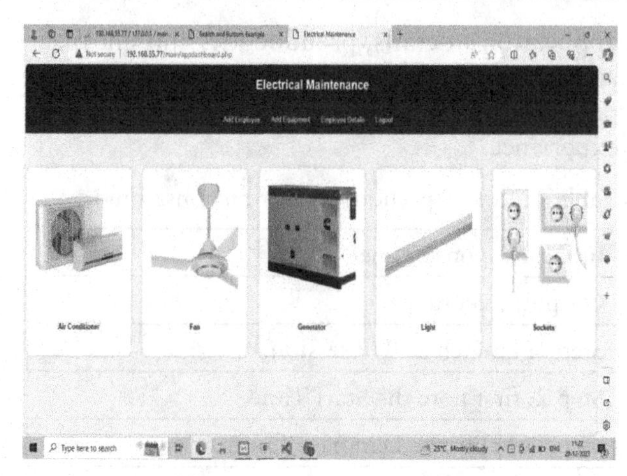

Figure 41.3 This is the Manager page, about the Electrical Equipment types and their functionalities about the services and number of repairs done under it

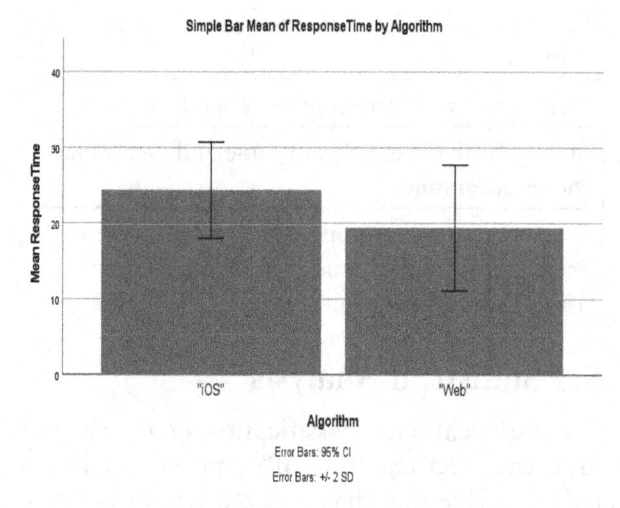

Figure 41.4 Response time comparison bar chart

7. Discussion

The significance value, determined through an independent sample T-test analysis, stands at 1.267, falling below the threshold of 1.05. This indicates that the web application exhibits an overall speed advantage of 19.43ms compared to the iOS application, which, in turn, demonstrates a 24.36ms improvement overall. The discourse on response times and navigation challenges when selecting specific equipment to review service and repair histories pertains to both online and iOS applications. The optimization of these factors is essential for delivering a seamless and captivating user experience across diverse platforms. In the context of iOS, the emphasis is placed on leveraging the platform's inherent features to ensure swift and efficient interactions. In the implementation of this project, the application's core functionality revolves around offering users a comprehensive overview of the historical records of equipment services and repairs. This feature serves as a valuable tool for maintenance teams, enabling them to track the maintenance history of each piece of equipment efficiently. Their primary responsibility involves adding employees to the application, creating a structured and organized database of skilled individuals. This not only facilitates effective workforce management but also allows for seamless collaboration and communication among team members.

On the other hand, supervisors assume a crucial position in the day-to-day operations (Lipirodjanapong et al. 2015). They are empowered to check and manage the daily routines of employees, ensuring that maintenance tasks are carried out efficiently and in a timely manner. Furthermore, supervisors are equipped with the authority to assign specific work orders to workers, optimizing task distribution based on skills and availability (Nasri et al. 2020). This dual-tiered approach to management enhances overall organizational efficiency. By centralizing information and responsibilities within the application, communication is streamlined, and accountability is reinforced (Maruyama 2000). The design of the application prioritizes usability, ensuring that both supervisors and managers can navigate the system easily, contributing to a user-friendly experience. Ultimately, the application not only serves as a repository for equipment maintenance data but also acts as a dynamic platform for workforce management (Kaparthi and Bumblauskas 2020). By assigning distinct roles to supervisors and managers, the project aims to foster a collaborative and productive work environment, ensuring that maintenance operations are not only tracked effectively but also executed with precision and efficiency (Zhou 2022). However, the project goes beyond mere data storage by introducing a robust management structure that leverages the roles of supervisors and managers (Chen et al. 2018). In this system, managers play a pivotal role in overseeing the workforce.

The delay in response time, particularly during the update of services and repairs from workers work completion is noted to be more pronounced in comparison to iOS,. The project underscores the importance of prioritizing platform-specific considerations. For online applications, challenges arise from diverse browsers and network conditions, necessitating a tailored approach. Conversely, iOS applications require optimization strategies aligned with their inherent capabilities. A critical determinant in achieving reduced navigational delays and improved reaction times lies in the effectiveness of the implemented systems or methods. User feedback and iterative testing are integral components of this process, ensuring continuous refinement to meet the dynamic demands of problem-solving abilities. The overarching goal is to provide a seamless and engaging real-time application experience on both the web and iOS platforms.

8. Conclusion

In our examination of navigation reaction of response times between the Electrical Equipment Maintenance Web Application and its iOS counterpart, both designed for repairs and services of equipment, we observed that theE-MaintenanceiOS application exhibited slower performance compared to the Web version. Analyzing a resource allocation dataset, our research unveiled a noteworthy response time discrepancy. Specifically, the web application demonstrated an impressive response time of 19.43ms, surpassing the iOSapplication's forecasted network demand with a response time of 24.36ms. Notably, our survey findings indicate that the iOS app is not as effective in presenting results as the Web application. The obtained comparison result of $p = 1.267$ ($p < 0.271$) signifies statistical significance, underscoring the notable differences in the performance of the two platforms.

References

[1] Bertin, Alberta, Fabio Buciol, Giovanna Dondossola, and Cristina Lanza. 1991. "Electrical Equipment Maintenance Training: An Its Application in Industrial Environment." Trends in Artificial Intelligence, 399–408.

[2] Chen, Xi, Shenghua Huang, Bingzhang Li, and Yangxiao Xiang. 2018. "Losses and Thermal Calculation Scheme of IGBT and FWD and Its Application in PWM Inverters for Electric

Engineering Maintenance Rolling Stock." IEEJ Transactions on Electrical and Electronic Engineering 13 (12): 1822–28.

[3] Dick, Michael, Tina Haase, and Wilhelm Termath. 2017. "The Potential of Virtual Interactive Learning Environments for Individual and Organizational Learning: An Example for the Maintenance of Electrical Equipment." Advances in Ergonomic Design of Systems, Products and Processes, 177–90.

[4] Dong, Lei, Xiao Kun Zhao, Wen Ping Zhou, and Qin Lv. 2013."Application of Critical Chain Project Scheduling Theory in Power Plant Electrical Equipment Maintenance."Advanced Materials Research 614–615: 1985–89.

[5] Gholami, Javad, Ahmad Razavi, and Reza Ghaffarpour. 2021. "Decision-Making Regarding the Best Maintenance Strategy for Electrical Equipment of Buildings Based on Fuzzy Analytic Hierarchy Process; Case Study: Elevator." Journal of Quality in Maintenance Engineering 28 (3): 653–68.

[6] Gill, Paul. 2016. Electrical Power Equipment Maintenance and Testing. 2nd Edition. CRC Press.

[7] Kaparthi, Shashidhar, and Daniel Bumblauskas. 2020. "Designing Predictive Maintenance Systems Using Decision Tree-Based Machine Learning Techniques." International Journal of Quality & Reliability Management 37 (4): 659–86.

[8] Knights, Peter F. 2001. "Rethinking Pareto Analysis: Maintenance Applications of Logarithmic Scatterplots." Journal of Quality in Maintenance Engineering 7 (4): 252–63.

[9] Lee, Haesung, and Byungsung Lee. 2020. "The Development of a State-Aware Equipment Maintenance Application Using Sensor Data Ranking Techniques." Sensors 20 (11): 3038.

[10] Lipirodjanapong, Sumate, Cattareeya Suwanasri, Thanapong Suwanasri, and Wijarn Wangdee. 2015. "Empirical Circuit Breaker Failure Rate Assessment and Modeling in a Preventive Maintenance Application." Electric Power Components & Systems, October. https://doi.org/10.1080/15325008.2015.1057780.

[11] Liyanage, Jayantha P. 2008. "Integrated E-Operations-E-Maintenance: Applications in North Sea Offshore Assets." Complex System Maintenance Handbook, 585–609.

CHAPTER 42

Navigating electrical surges A comprehensive comparison of machine learning random forest and support vector machine

H. Shankar[1] and S. Gomathi[1,a]

Department of Cognitive Computing, Saveetha school of engineering, Saveetha Institute of Medical and Technical Sciences, Electrical and Electronics Engineering, Chennai, Tamilnadu, India

Email: [a]gomathis.sse@saveetha.com

Abstract

This research aims to advance the field of electrical surge prediction by developing an innovative and state-of-the-art model that combines advanced machine learning algorithms and predictive analytics techniques. The objective is to enhance the resilience of electrical systems by providing accurate and reliable predictions of potential surges. This study aims to significantly improve the precision and timeliness of electrical surge forecasts by testing new features, pushing the limits of existing predictive capabilities, and assessing cutting-edge technologies. The ultimate goal is to contribute to the development of proactive and effective mitigation strategies, minimizing the impact of electrical disturbances on equipment, infrastructure, and overall system reliability in a rapidly evolving technological landscape. The main dataset used in this study was obtained from Kaggle, a reputable platform for data collection and analytics. The dataset includes a collection of line currents and voltages for different fault conditions. In group 1, the Random Forest algorithm was employed. The study's 48 total sample sizes have an alpha value of 0.05, a beta value of 0.2, and a statistical power of 0.8. Group 2 uses the Support Vector Machine (SVM). Clinical is the tool used to compute such numbers. The program for making accurate predictions is run using Google Colab, which also analyses data and displays accuracy as a percentage and in a graphical format for easy comprehension. Group 1 obtained an accuracy of 98.98 percent using the Random Forest technique, whereas Group 2 obtained an accuracy of 79.49% using the SVM Algorithm. The Statistical Package for the Social Sciences (SPSS) software is used for accuracy and data set statistical analysis. A two-tailed test is performed to see if there is a statistically significant difference in accuracy between the two groups. Random forest methods are more precise than SVMs.

Keywords: Pro-active development, resilience enhancement, mitigation techniques, random forest, support vector machine, electrical surge prediction

1. Introduction

By creating an advanced model using cutting-edge machine learning algorithms and predictive analytics approaches, the research hopes to improve the prediction of electrical surges. By providing precise and trustworthy forecasts of possible surges, the goal is to increase the resilience of electrical systems. The research looks into new features and evaluates future technology in an effort to push the boundaries of existing prediction capabilities. In the end, this research hopes to aid in the creation of successful mitigation techniques by greatly increasing the accuracy and timeliness of electrical surge forecasts. The goal of these tactics is to reduce the negative effects of electrical disruptions on machinery, infrastructure, and overall system dependability.

The urgent need to address electrical surge vulnerabilities, which have the potential to seriously harm electronic systems and gadgets, is what spurs this research. In order to effectively anticipate the occurrence of surges, the research

DOI: 10.1201/9781003606611-42

will make use of large datasets and machine learning techniques including Random Forest, Convolutional Neural Networks, and Naive Bayes Classifiers. The project's all-encompassing strategy takes into account methodological constraints, looks for gaps in the literature, and takes cross-disciplinary views into account. The anticipated result is a strong predictive model that improves proactive electrical system management, guaranteeing increased dependability and surge protection in a fast-moving technological landscape.

2. Materials And Methods

The suggested investigation is conducted in the Saveetha School of Engineering information security lab. Two groups in total were involved in the inquiry. Group 1 employed the Random Forest technique to compare the accuracy of the two algorithms. Clincalc was used to determine the statistical power (0.8), alpha (0.05), and beta (0.2) values of the study's 48 total sample sizes. Group 2 makes use of K (nearest). Google Colab is the program used to execute the algorithm for accurate predictions.

A comparison of the processes of two groups is chosen, and the outcome is determined. The primary data set used in this investigation is the electrical surge values for electrical surge detection. The Comma separated Values (CSV) dataset was acquired from the well-known data collection and analytics platform Kaggle. The data set includes pictures of various electrical surge values. Data for training and testing is kept separate from the dataset. The total number of values in the dataset is approximately 7800, with fixed samples representing the affected data. The suggested work is planned and carried out using Python. In terms of code execution, an output process is carried out for correctness by working behind the dataset.

Random Forest:

Leo Breiman and Adele Cutler invented Random Forest, a popular machine-learning technique. This algorithm amalgamates the outcomes of numerous SVMs to generate a unified result. Renowned for its user-friendly interface and versatility, Random Forest excels at addressing both classification and regression problems, contributing to its widespread adoption in various domains.

Random forests, or random decision forests, are ensemble learning methods such as classification, regression, and other tasks that operate by constructing a multitude of SVMs at training time. The class that the majority of the trees choose is the random forest's output for classification problems. The class that the majority of the trees choose is the random forest's output for classification problems. For regression tasks, the mean or average prediction of the individual trees is returned. [Ho, Tin Kam (1995)]. Random Decision Forests (RDF). [Proceedings of the 3rd International Conference on Document Analysis and Recognition, Montreal, QC, 14–16 August 1995, pp. 278–282.] Ho TK (1998). "The Random Subspace Method for Constructing Decision Forests" (PDF). IEEE Transactions on Pattern Analysis and Machine Intelligence, vol. 20(8) Random decision forests correct for SVMs' habit of overfitting to their training set. Hastie, Trevor; Tibshirani, Robert; Friedman, Jerome (2008). The Elements of Statistical Learning (2nd ed.). Springer. ISBN 0-387-95284-5.

Pseudo code:

Step 1: Import necessary libraries for Data classification
Step 2: Load and preprocess the CSV dataset
Step 3: Split the dataset into training and validation sets
Step 4: Define the architecture of the Algorithm used
Step 5: Compile the model
Step 6: Train the model
Step 7: Assuming 'history' contains accuracy values from training
Step 8: Calculate the average accuracy

SVM:

SVM is a powerful machine learning algorithm used for linear or nonlinear classification, regression, and even outlier detection tasks. SVMs can be used for a variety of tasks, such as text classification, image classification, spam detection, handwriting identification, gene expression analysis, face detection, and anomaly detection. SVMs are adaptable and efficient in a variety of applications because they can manage high-dimensional data and nonlinear relationships.

SVM algorithms are very effective as we try to find the maximum separating hyperplane between the different classes available in the target feature.

Pseudo code:

Step 1: Import necessary libraries for CSV classification
Step 2: Load and preprocess the CSV dataset representations
Step 3: Split the dataset into training and validation sets
Step 4: Set parameters for data augmentation and preprocessing
Step 5: Create generators for loading and augmenting CSVs
Step 6: Create a SVM model for CSV classification
Step 7: Compile the model, Train the model
Step 8: Calculate and print average training and validation accuracy

Statistical Analysis

IBM SPSS software version 26 was used for the statistical analysis of the Random Forest and Support SVM algorithms. The SPSS software analysis made use of descriptive statistics, an independent sample test, and a straightforward bar graph showing accuracy per group. The purpose of this investigation was to determine and contrast the accuracy of the two approaches. The data was summarized using descriptive statistics, and the significance of the variations in accuracy between the two algorithms was evaluated using an independent sample test. The accuracy results were visually displayed by the bar graph, which made it easier to compare the Random Forest and SVM algorithms' performances.

3. Discussions

The results of this study shed light on the performance and potential applications of the Random Forest in the domain of prediction-generated data. Novel Random Forest achieved an accuracy of 98.98%, which is a respectable performance in the context of electrical diagnosis. This algorithm's ability to handle complex, nonlinear relationships within the data and create more intricate decision boundaries is particularly valuable in electrical surge surges because of its nonlinearity. While not reaching the accuracy achieved by Random Forest (79.49%), the SVM still offers a robust approach for categorizing Electrical Surges Electrical surge samples into various electrical surge conditions. This underscores its utility in the domain of electrical surge diagnosis, especially when the dataset exhibits nonlinear relationships that linear models, such as Dedecisionree, may struggle to capture. The study's primary focus on multi-model integration allowed us to highlight the benefits of combining diverse machine-learning approaches. The comparison of a random forest with a SVM showcased the unique advantages and limitations of each model. Random Forest excels in scenarios where nonlinear relationships exist, offering a powerful alternative to traditional linear models. The study provides insights into the potential of integrating multiple models and selecting the most appropriate algorithm for specific datasets and diagnostic tasks. This adaptability in model selection holds broader implications for healthcare, where tailored approaches can significantly enhance diagnostic accuracy. Future research can delve deeper into optimizing Random Forest's hyperparameters, examining the impact of feature engineering on its performance, and assessing its generalizability to other healthcare applications. The study opens doors for further exploration into the integration of different machine learning techniques and their potential to advance social decision-making and ultimately benefit patient care.

4. Conclusion

In conclusion, this study has explored the utility of the random forest in the realm of the prediction of electrical surges, achieving a commendable accuracy of 98.98%. Random Forest's capacity to handle complex, nonlinear relationships in electrical surge datasets makes it a valuable tool for electrical surge diagnosis, offering an alternative to linear models. Even though Random Forest's accuracy is not as high as SVM's (79.49%), it still demonstrates how well it can handle complex data patterns.

The research emphasizes the significance of model selection and the advantages of combining multiple machine-learning approaches, allowing tailored solutions for specific diagnostic tasks. This adaptability extends beyond the prediction of electrical surges, holding broader implications for appliances and electrical conditions. Future investigations may delve into the fine-tuning of Random Forest hyperparameters, exploring feature engineering's impact, and examining its generalizability to other household applications. The study underscores the importance of diversity in model selection and opens avenues for the integration of varied machine learning techniques, contributing to improved diagnostic precision and ultimately enhancing patient care in healthcare domains.

Table 42.1: This table summarizes the performance of two groups, Random Forest and SVM, in Electric Surge Prediction . It includes the number of observations (N), the mean accuracy rate for each group, the standard deviation indicating the spread of data, and the standard error of the mean representing the accuracy rate's precision. The results show that Random Forest outperformed SVM, with an average accuracy rate of 98.55%, while SVM had an average accuracy rate of 72.12%

	Algorithm	N	Mean	Std. Deviation	Std. Error Mean
Accuracy	Random forest	10	98.55	.279	.088
Accuracy	SVM	10	76.23	.493	.156

References:

[1] A review on the selected applications of forecasting models in renewable power systems Adil Ahmed, Muhammad Khalid Volume 100, February 2019, Pages 921, https://doi.org/10.1016/j.rser.2018.09.046

[2] Bilateral Contracting and Price-Based Demand Response in Multi-Agent Electricity Markets: A Study on Time-of-Use Tariffs Hugo Algarvio, Fernando Lopes. https://doi.org/10.3390/en16020645

[3] Modeling and forecasting building energy consumption: A review of data-driven techniques, Mathieu Bourdeau, Xiao qiang Zhai a, Elyes Nefzaoui Sustainable Cities and Society, Volume 48, July 2019, 101533

[4] https://doi.org/10.1016/j.scs.2019.101533

[5] Multi-agent microgrid energy management based on deep learning forecaster

[6] Author links open overlay panel Mousa Afrasiabi, Mohammad Mohammadi, Energy, Volume 186, 1 November 2019, 115873, https://doi.org/10.1016/j.energy.2019.115873

[7] Short-Term Load Forecasting Models: A Review of Challenges, Progress, and the Road Ahead by Saima Akhtar 1ORCID, Sulman Shahzad, published on 12 May 2023, Distributed Energy Resources in Transactive Energy Systems, https://doi.org/10.3390/en16104060

[8] Hybrid CNN-LSTM Model for Short-Term Individual Household Load Forecasting, Musaed Alhussein, Khursheed Aurangzeb, Journals & Magazines, IEEE Access, Volume: 8, 10.1109/ACCESS.2020.3028281

[9] Season-specific approach for short-term load forecasting based on hybrid FA-SVM and similarity concept, Mayur Barman, Nalin Behari Dev Choudhury, Energy, Volume 174, 1 May 2019, Pages 886–896,

[10] https://doi.org/10.1016/j.energy.2019.03.010

[11] Deep learning framework to forecast electricity demand, Jatin Bedi, Durga Toshniwal, Applied Energy, Volume 238, 15 March 2019, Pages 1312–1326,

[12] https://doi.org/10.1016/j.apenergy.2019.01.113

[13] Multi-agent microgrid energy management based on deep learning forecaster

[14] Author links open overlay panel Mousa Afrasiabi, Mohammad Mohammadi, Mohammad Rastegar, Energy, Volume 186, 1 November 2019, 115873,

[15] https://doi.org/10.1016/j.energy.2019.115873

[16] A novel day-ahead scheduling approach for multi-power system considering dynamic frequency security constraint, Zihan Cai, Rui Zhang, Energy Reports, Volume 9, Supplement 10, October 2023, Pages 1474–1482, https://doi.org/10.1016/j.egyr.2023.05.178

CHAPTER 43

Enhancement of charge density in resonant tunneling diode (RTD) using 6-barrier device and 2-barrier device by varying Poisson criterion.

S. Venkata Ramireddy[1] and Anbuselvan N.[1,a]

[1]Department of Electrical and Electronics Engineering, Saveetha School of Engineering, Saveetha Institute of Medical and Technical Sciences, Saveetha University, Chennai, Tamil Nadu, India
Email: [a]anbuselvann.sse@saveetha.com

Abstract

Resonant Tunneling Diodes (RTDs) stand as pivotal components in modern electronic devices due to their remarkable potential for high-speed, low-power applications. The charge density within RTDs significantly influences their performance characteristics. This study investigates the enhancement of charge density in RTDs through the utilization of 2-barrier and 6-barrier devices, employing variations in the Poisson criterion. The research employs advanced computational simulations to analyze the effects of Poisson criterion variations on the charge density within RTDs. Specifically, the study focuses on RTD structures with both 2-barrier and 6-barrier configurations to assess their comparative performance and potential for charge density enhancement. By systematically adjusting the Poisson criterion, the study explores its impact on the device's charge density profile and overall performance metrics. The findings reveal that the manipulation of the Poisson criterion exerts significant control over the charge density within RTDs. Through careful adjustment of this criterion, notable enhancements in charge density are observed in both 2-barrier and 6-barrier devices. Furthermore, comparative analyses between the two configurations elucidate insights into their respective charge density behaviors under varying Poisson criterion conditions.

Keywords: Resonant tunneling diode (RTD), Poisson criterion, tunneling current, resonant tunneling peak, barrier device, electrostatic potential, quantum mechanics, quantum tunneling

1. Introduction

RTDs represent a class of semiconductor devices that have garnered significant attention due to their unique properties and potential applications in high-speed electronics, quantum computing, and terahertz technology (Ironside, Romeira, and Figueiredo 2019). One crucial aspect of optimizing RTD performance is the enhancement of charge density within the device structure.

The charge density in RTDs plays a pivotal role in determining their electrical characteristics, such as current-voltage characteristics (Ghafoor, Rehmani, and Davy 2021), resonant peak characteristics, and switching speeds. In recent years, researchers have explored various strategies to enhance charge density (Chakraborty et al., n.d.), with a particular focus on utilizing multi-barrier structures (Pfenning et al. 2016). Two common architectures employed in this pursuit are the 2-barrier and 6-barrier devices. These configurations consist of multiple quantum wells separated by potential barriers (Franch, Männistö, and Martínez-Fernández 2019), facilitating quantum mechanical tunneling of charge carriers. By carefully engineering the barrier heights and widths (Bhattacharyya 2009), researchers can manipulate the charge density and tailor device performance.

In 2-barrier and 6-barrier RTD devices, variations in the Poisson criterion enable the creation of favorable electrostatic conditions that

DOI: 10.1201/9781003606611-43

promote the accumulation of charge carriers within the quantum wells (IEEE Staff 2019). This accumulation enhances tunneling probabilities and facilitates the realization of higher current densities and improved device performance (Allen 2017). The enhancement of charge density in RTDs using 2-barrier and 6-barrier devices via Poisson criterion variation represents (Weetman and Tsalavoutas 2019) a promising avenue for improving device performance and unlocking new functionalities in semiconductor electronics.

By varying the Poisson criterion (IEEE Staff 2015), researchers can optimize the device structure to promote resonant tunneling and increase charge density in the active region of the device. This optimization involves careful design considerations, including barrier heights, well widths, and doping profiles, to achieve the desired electronic properties. The use of 2-barrier and 6-barrier devices offers additional degrees of freedom in tailoring the charge density profile within the RTD (Weetman and Tsalavoutas 2019). These multi-barrier structures enable more precise control over electron tunneling probabilities and energy levels, leading to enhanced device performance.

2. Materials and Methods

The study was conducted in the Nano hub Simulation laboratory in the branch of Electronics and Communication Engineering, at Saveetha School of Engineering, Saveetha Institute of Medical and Technical Sciences. From this work, 7 samples have been taken for each of the two techniques. Group 1 is a Two-Barrier device and **Group** 2 is a Six-Barrier device. Therefore totally 14 sample tests. The corresponding mean values for group 1 and group 2 are 9.9728E+17 and 1.15113E+19. Matlab Software has been used to run the code for system simulation.

2.1 A. 2-Barrier Device

In a 2-barrier RTD, the charge density can be enhanced by adjusting the Poisson criterion to optimize the confinement of charge carriers within the quantum well region. By engineering the potential barriers and well widths, you can control the tunneling probabilities and the resonance conditions of the device.

2.2 B. 6-Barrier Device

In a 6-barrier device, the Poisson criterion becomes even more critical. By carefully adjusting the potential profiles across multiple barriers and wells, you can tailor the device characteristics to maximize charge density. This involves optimizing the widths and heights of the barriers and wells to facilitate efficient tunneling and resonant behavior.

2.3 Statistical Analysis

For the statistical analysis, the values of current for 2-Barrier and 6-Barrier acquired from Statistical Package for the Social Sciences (SPSS) Software (Verma 2012). The Poisson criterion is an independent variable and charge density is a dependent variable. The effectiveness of total energy improvement has been performed using independent t test analysis in RTD based High Electron Mobility Transistor (HEMT).

3. Results

Table 43.1 Displays the performance evaluations for the comparison of the 2-Barrier device and 6-Barrier devices. The Poisson criterion obtained using the 2-Barrier device method is 1.00E+18 whereas 1.00E+18 is the current obtained using the 6-Barrier device. Therefore the 6-Barrier device is much better than the 2-Barrier device for increasing the energy while varying the Poisson criterion.

Table 43.2 Illustrates the calculations including mean, standard deviation, and mean standard error for the 2-Barrier device and the 6-Barrier device. The independent sample t-test uses the Poisson criterion parameter. The 2-Barrier device has a mean total Poisson criterion of 9.9728E+17 compared to the 6-Barrier device 1.15113E+19. The standard deviation value of the 2-Barrier device is 1.64990E+15 and 3.7444E+19 is the value of the 6-Barrier device. Standard error mean for the 2-Barrier device is 6.2360E+14 and for 6-Barrier device is 1.41149E+19.

Table 43.3 To identify the significance an independent sample T-test has been implemented. Independent variables of the 2-Barrier device in comparison with the 6-Barrier device have been calculated and 0.003 is the significance value of charge density. The confidence interval of the difference is 95%. By using the independent samples T-test both the techniques are compared. The following parameters are measured using the independent samples T-test such as significance (two-tailed), mean difference, standard error of mean difference, upper and lower interval differences.

Figure 43.1 Shows how the charge density varies based on the Poisson criterion resulting in the range of (0.1-0.7)V in the RTD HEMT by using a 2-Barrier device.

Figure 43.2 Shows how the charge density varies based on the Poisson criterion resulting in the range of (0.1-0.7)V in the RTD HEMT by using a 6-Barrier device.

Figure 43.3 Contrast the 2-Barrier device and 6-Barrier device with regard to charge density and Poisson criterion. From this graph it is observed that, when compared to the 6-Barrier device, the novel 2-Barrier device method significantly improves charge density in RTD HEMTs.

Figure 43.4 The charge density is compared between the 2-Barrier device and 6-Barrier device. The proposed novel 2-Barrier device improves the charge density of 0.001 better than the 6-Barrier device value of 0.07. At X-axis the charge density values of 2-Barrier device, 6-Barrier device and at Y-axis Mean keyword identification charge density are shown.

Table 43.1: Statistical values of charge density through 2-barrier and 6-barrier technique by varying Poisson criterion

S.No	Poisson criterion	Charge Densities	
		2-BARRIER	6-BARRIER
1	0.2833	1.00E+18	1
2	0.8499	9.99E+17	2
3	1.4165	9.98E+17	3
4	1.9831	9.97E+17	4
5	2.5497	9.97E+17	5
6	3.1163	9.96E+17	6
7	3.6829	9.95E+17	7

Table 43.2: Group statistical analysis of 2-barrier and 6-barrier

	Groups	No of Samples	Mean	Std.Deviation	Std.Error Mean
Poisson criterion	2-Barrier	7	9.9728E+17	1.64990E+15	6.2360E+14
Poisson criterion	6-Barrier	7	1.15113E+19	3.73444E+19	1.41149E+19

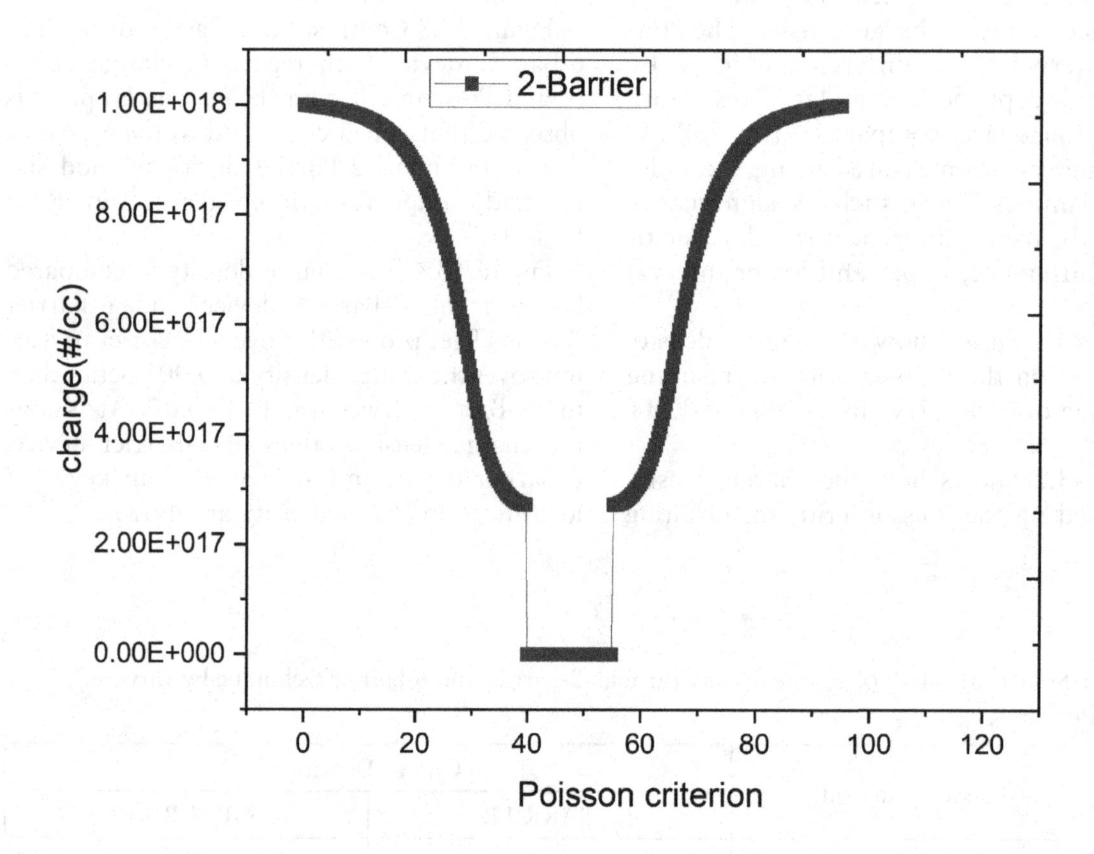

Figure 43.1: Charge density variation with respect to the Poisson criterion of RTD HEMT by using 2-barrier device

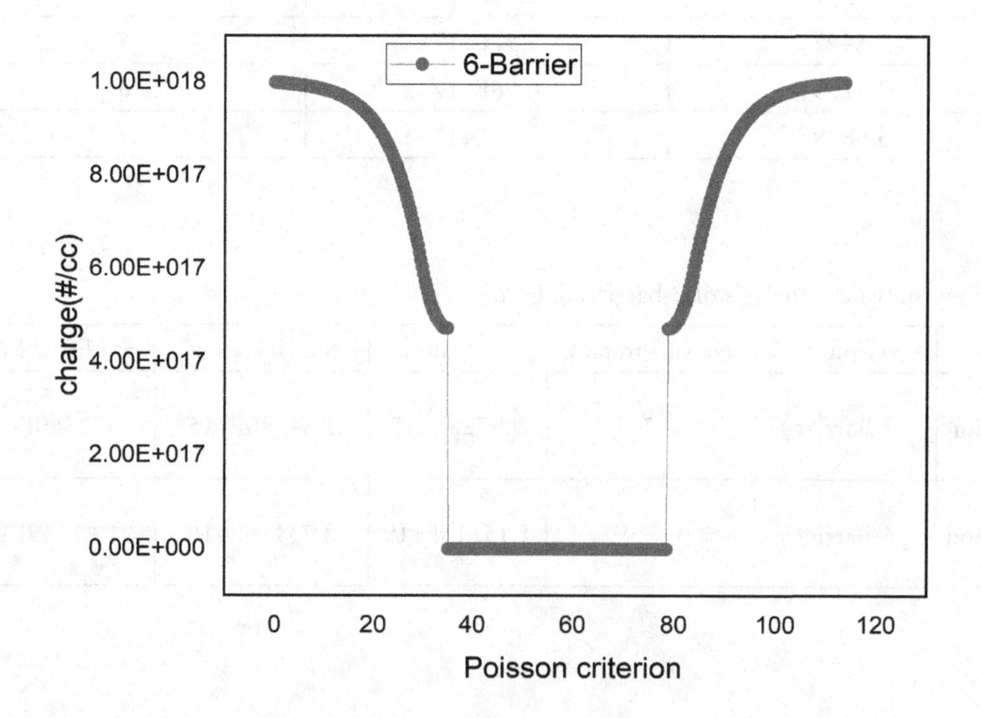

Figure 43.2: Charge density variation with respect to the Poisson criterion of RTD HEMT by 6-barrier device

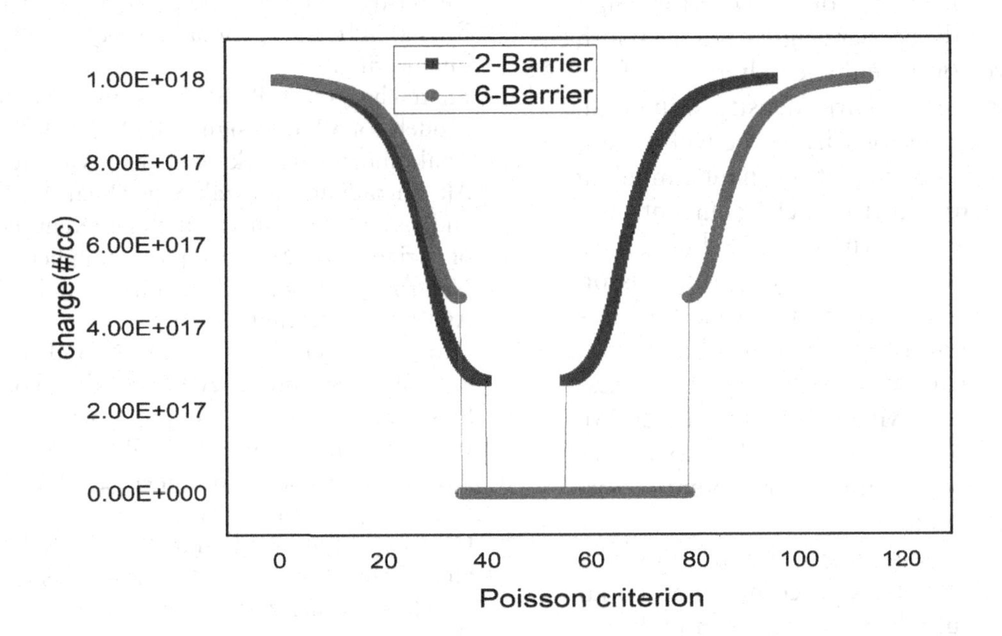

Figure 43.3: Contrasts the 2-barrier device and 6-barrier device with regard to charge density and Poisson criterion. From this graph it is observed that, when compared to the 2-barrier device, the 6-barrier device significantly improves charge density in RTD HEMT.

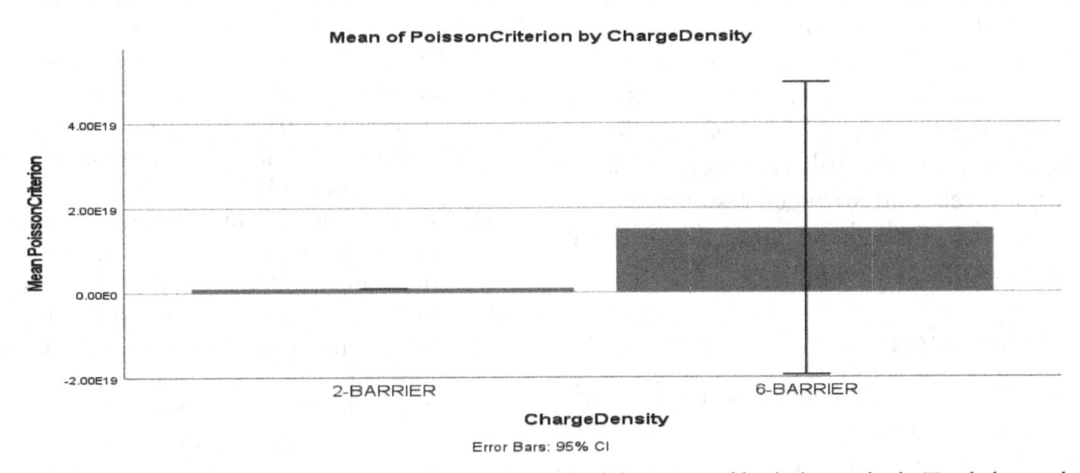

Figure 43.4: compares the mean charge density values and standard deviation of both the methods. Total charge density of the 6-barrier device is better than the 2-barrier device

4. Discussion

Enhancing charge density in RTDs using different barrier structures, such as 2-barrier and 6-barrier devices, involves optimizing the Poisson criterion. The Poisson criterion plays a critical role in determining the performance and characteristics of RTDs by controlling the charge distribution within the device (Thomas, Kalarikkal, and Abraham 2021). In RTDs, the charge density and its distribution significantly affect the tunneling current, resonance properties, and overall device performance (Bagayev 2006). The resonant tunneling phenomenon relies on the precise control of the energy levels and charge distribution within the device.

Let's discuss how the Poisson criterion can be varied to enhance charge density in both 2-barrier and 6-barrier RTD devices (High Speed Heterostructure Devices 1994). In a 2-barrier RTD, the control of charge density primarily involves adjusting the potential barriers and the well region between them (Nielsen and Chuang 2010). By optimizing the Poisson criterion, engineers can adjust the barrier heights and the width of the well region to enhance charge confinement and increase the charge density in the well. Increasing the charge density in the well region improves the probability of resonant tunneling events, leading to higher tunneling currents and improved device performance (Altshuler, Lee, and Richard Webb 2012). Varying the Poisson criterion allows engineers to fine-tune the electrostatic potential profile within the device, thereby maximizing charge density at the resonant energy levels (Mizuta and Tanoue 2006). A 6-barrier RTD offers additional control over charge density by providing more potential barriers and well regions.

By varying the Poisson criterion in a 6-barrier RTD, engineers can create complex potential profiles with multiple well regions and barriers (Mizuta and Tanoue 2006). Optimizing the Poisson criterion in a 6-barrier device enables precise tuning of the charge distribution along the device's length (Grahn 1995), allowing for enhanced charge confinement and density in specific regions. The increased number of barriers in a 6-barrier device offers greater flexibility in shaping the potential landscape (Rocchi 1990), which can be utilized to create sharper energy levels and stronger resonances, ultimately leading to higher charge densities and improved device performance.

5. Conclusion

Between the 2-Barrier device and the 6-Barrier device, a comparison was performed. The outcomes clearly show that the 2-Barrier device improves total charge density more than the 6-Barrier device. Total charge density obtained by the 6-Barrier device is more efficient than the 2-Barrier device.

References

[1] Allen, Garrick V. 2017. The Book of Revelation and Early Jewish Textual Culture. Cambridge University Press.

[2] Altshuler, B. L., P. A. Lee, and W. Richard Webb. 2012. Mesoscopic Phenomena in Solids. Elsevier.

[3] Bagayev, S. N. 2006. ICONO 2005: Nonlinear Laser Spectroscopy, High Precision Measurements, and Laser Biomedicine and Chemistry : 11–15 May, 2005, St. Petersburg, Russia. SPIE-International Society for Optical Engineering.

[4] Bhattacharyya, A. B. 2009. Compact MOSFET Models for VLSI Design. John Wiley & Sons.

[5] Chakraborty, Avishek, Sankar Prasad Mondal, Ali Ahmadian, Norazak Senu, Shariful Alam, and Soheil Salahshour. n.d. Different Forms of Triangular Neutrosophic Numbers, De-Neutrosophication Techniques, and Their Applications. Infinite Study.

[6] Franch, Xavier, Tomi Männistö, and Silverio Martínez-Fernández. 2019. Product-Focused Software Process Improvement: 20th International Conference, PROFES 2019, Barcelona, Spain, November 27–29, 2019, Proceedings. Springer Nature.

[7] Ghafoor, Saim, Mubashir Husain Rehmani, and Alan Davy. 2021. Next Generation Wireless Terahertz Communication Networks. CRC Press.

[8] Grahn, H. T. 1995. Semiconductor Superlattices: Growth and Electronic Properties. World Scientific.

[9] High Speed Heterostructure Devices. 1994. Academic Press. IEEE Staff. 2015. 2015 Asia Pacific Microwave Conference (Apmc).

[10] Ironside, Charlie, Bruno Romeira, and José Figueiredo. 2019. Resonant Tunneling Diode Photonics: Devices and Applications. Morgan & Claypool Publishers.

[11] Mizuta, Hiroshi, and Tomonori Tanoue. 2006. The Physics and Applications of Resonant Tunneling Diodes. Cambridge University Press.

[12] Nielsen, Michael A., and Isaac L. Chuang. 2010. Quantum Computation and Quantum Information: 10th Anniversary Edition. Cambridge University Press.

[13] Pfenning, Andreas, Fabian Hartmann, Fabian Langer, Martin Kamp, Sven Höfling, and Lukas Worschech. 2016. "Sensitivity of Resonant Tunneling Diode Photodetectors." Nanotechnology 27 (35): 355202.

[14] Rocchi, Marc. 1990. High-Speed Digital IC Technologies. Artech House Materials Science.

[15] Thomas, Sabu, Nandakumar Kalarikkal, and Ann Rose Abraham. 2021. Design, Fabrication, and Characterization of Multifunctional Nanomaterials. Elsevier.

[16] Verma, J. P. 2012. Data Analysis in Management with SPSS Software. Springer Science & Business Media.

[17] Weetman, Pauline, and Ioannis Tsalavoutas. 2019. The Routledge Companion to Accounting in Emerging Economies. Routledge.

CHAPTER 44

Plastic detection using the K nearest neighbor algorithm to reduce the disposal of plastics in water bodies

T. Sakthi Krishna[1] and M. Daniel Nareshkumar[2,a]

Research Scholar, Department of Artificial Intelligence and Data Science, Saveetha School of Engineering, Saveetha Institute of Medical and Technical Sciences, SIMATS University, Chennai, Tamil Nadu, India

Corresponding author, Department of Electronics and Communication Engineering, Saveetha School of Engineering, Saveetha Institute of Medical and Technical Sciences, SIMATS University, Chennai, Tamil Nadu, India

Email: danielnareshkumarm.sse@saveetha.com

Abstract

The project aims to identify the type of water found and plastic detection to understand whether the water bodies are contaminated by image processing using two different machine learning algorithms as K-Nearest neighbor (KNN) algorithm and Support Vector Machine (SVM) concerning Accuracy. There are two groups and samples were involved in this group. The KNN algorithm with a sample size of 10 and the SVM algorithm with a sample size of 10 was evaluated several times to predict the accuracy percentage. A G power value of 0.95 is used for SPSS calculation with a Confidence Interval of 95% and alpha 0.05 were determined. In comparison to the SVM. KNN algorithm attained a greater accuracy proportion of 78.43% as opposed to 52.30%. The mean detection accuracy was calculated with a standard deviation of ±2SD. The Statistical Package for the Social Sciences (SPSS) carried out has a two-tailed significance value of 0.002 ($p < 0.05$). This shows that there is a statistically significant difference between the two methods considered in this study. This research presents effective analysis and prediction helps many people around to know whether the water bodies are safe and benefit the aquatic lifeforms using two different classifier approaches such as SVM and the KNN algorithms and accuracy gains of selected classifiers are compared.

Keywords: Plastic detection, water quality, SVM, KNN

1. Introduction

Insidious threats to aquatic environments and the delicate food web balance within these ecosystems are posed by plastic pollution. Harmful chemicals released by plastic when it degrades and persists in water bodies find their way into the food chain through aquatic species. This concerning situation calls for quick thinking and creative solutions to identify, comprehend, and lessen the effects of plastic pollution on the ingestion of these substances by species that live in water. By examining and addressing the effects of plastics on the aquatic food web chain, this study aims to pave the way towards reducing the disposal of plastics in aquatic environments by utilizing the power of machine learning algorithms, specifically comparing KNN and SVM.

Plastic pollution has negative effects on aquatic ecosystems at every level that have a significant influence on the food chain. Various aquatic organisms consume the hazardous compounds that plastics release during their degradation (Qin et al. 2024). As these substances go up the food chain and accumulate, they can increase in concentration and endanger higher-ranking species, such as humans who depend on seafood. Comprehending and alleviating this domino effect requires sophisticated techniques that can interpret intricate patterns and relationships in ecological systems.

DOI: 10.1201/9781003606611-44

Strong machine learning algorithms, KNN and SVM use different strategies for pattern identification and classification. These algorithms provide powerful tools to analyze the complex dynamics of chemical consumption caused by plastic in aquatic food webs. This project intends to investigate the links between plastic pollution, chemical ingestion, and their effects throughout trophic levels by utilizing datasets that include plastic concentrations, chemical analysis, species interactions, and environmental variables for plastic detection.

An important first step in this endeavor is the application of KNN and SVM, which aim to find important characteristics for plastic detection and the predictors that significantly contribute to aquatic organisms' ingestion of dangerous chemicals, in addition to detecting the presence and impact of plastics (Qin et al. 2024; Ekpe et al. 2024; Cortesi et al. 2021). This work intends to examine the effectiveness of these algorithms in identifying intricate linkages and predicting the risk variables related to plastic-induced chemical ingestion in aquatic food webs through rigorous model evaluation and statistical analyses.

This study intends to inspire targeted measures and legislative initiatives focused on minimizing plastic dumping in aquatic bodies. Its ultimate purpose goes far beyond algorithmic comparison. Through understanding the mechanisms by which plastics enter food webs, this research aims to suggest mitigation measures and methods to reduce the pollution caused by plastic, protect aquatic biodiversity, and maintain the health of ecosystems that are essential to human welfare and marine life.

2. Materials And Methods

In this work, it is divided into 2 groups. Group 1 is of SVM and Group is of KNN. Totally 20 samples are used and first ten accuracy samples for the KNN first group and the other ten SVM samples for the second group (Saini and Sharma 2022). To calculate the sample size for each group, a G Power calculator was utilized, which was set to 80% pre-test power, an alpha level of 0.95, and a 95% confidence level.

The plastic prediction in water bodies dataset has been used for this experiment and it is freely downloadable from (www.kaggle.com) and it contains 289 samples. With 30% of the dataset used for training and 70% for testing, the dataset is used for both purposes. Instead of taking all features, the proposed system utilizes selected features and identifies the quality of water. Analysis of Exploratory Data was used to import the dataset by Jupyter Notebook, the Pandas, and Numerical Python libraries, respectively.

For this study, testing was conducted using an HP15S laptop equipped with 8GB of RAM, an AMD RYZEN 5th generation core processor, 512GB of storage capacity, and running on the Windows 11 operating system. The statistical analysis for our study was carried out using IBM SPSS software version 26.

2.1 SVM

For problems involving regression and classification, the SVM is a reliable and adaptable machine learning technique. SVM is very good at dividing data points into discrete groups since its fundamental goal is to locate the best hyperplane that maximizes the margin between various classes. SVM is an invaluable tool in many domains, including text and picture classification, because it can handle both linear and non-linear issues by utilizing kernel functions and modifying hyperparameters like "C." Its broad application in machine learning is attributed to its capacity to handle high-dimensional data and resistance to overfitting.

Algorithm

Step 1: Start the program
Step 2: Import the water quality dataset
Step 3: Import the pandas, Matplotlib Libraries
Step 4: Preprocess the data
Step 5: Feature Selection
Step 6: Give a path to build the SVM model
Step 7: Split the data into train (70%) and test (30%)
Step 8: Give training size and testing size
Step 9: Test the SVM model using train data
Step 10: Get the accuracy
Step 11: End the program

2.2 KNN

KNN is a versatile machine-learning algorithm that operates on the principle of similarity. It classifies or predicts based on the idea that data points with similar features are likely to share the same label or value. By considering the 'k' closest neighbors in the feature space, KNN makes predictions and decisions, making it a valuable tool in tasks such as image recognition, recommendation systems, and anomaly detection. While simple to understand and implement, KNN effectiveness depends on choosing the right 'k' value and an appropriate distance metric for the specific problem at hand, showcasing the balance between simplicity and performance in machine learning algorithms.

Algorithm

Step 1: Start the program
Step 2: Import libraries
Step 3: From Scikit learn import classification
Step 4: Input the plastic prediction in the water bodies dataset
Step 5: Build the KNN model
Step 6: Preprocess the data
Step 7: Separate the data into train data and test data
Step 8: Give a sample size for training and testing data
Step 9: Train the KNN model using test data
Step 10: Get the accuracy
Step 11: End the program

2.3 Statistical Analysis

The statistical analysis for this study was conducted using the IBM SPSS software version 26 (Statistical Package for the Social Sciences), while the efficiency of the algorithm was evaluated using the Jupyter Notebook tool. The independent variables are water quality (input parameters). The dependent variables are fixed. An independent T-test was used to compare the results of the specified algorithms (Saini and Sharma 2022).

3. Result

In comparison to the SVM and KNN, KNN algorithm attained a greater accuracy proportion of 78.43% as opposed to 52.30 %. This shows that there is a statistically significant difference between the two methods considered in this study. using the models would benefit the government in protecting aquatic organisms and for plastic detection.

Table 44.1: Accuracy comparison

Iteration	KNN	SVM
1	80.55	69.23
2	86.30	53.85
3	80.55	53.85
4	86.30	30.77
5	63.29	61.54
6	76.21	46.15
7	70.96	53.85
8	80.55	46.15
9	76.71	69.23
10	82.47	38.46

T-Test

Group Statistics					
	group	N	Mean	Std. Deviation	Std. Error Mean
accuracy	KNN	10	78.4390	7.02863	2.22265
	SVM	10	52.3080	12.45688	3.93921

INDEPENDENT SAMPLES TEST

Independent Samples Test										
		Levene's Test for Equality of Variances		t-test for Equality of Means						
		F	Sig.	t	Df	Sig. (2-tailed)	Mean Difference	Std. Error Difference	95% Confidence Interval of the Difference	
									Lower	Upper
accuracy	Equal variances assumed	2.553	.127	5.777	18	.000	26.13100	4.52300	16.62852	35.56348
	Equal variances not assumed	-	-	5.777	14.203	.000	26.13100	4.52300	16,44312	35.8188

Graph:

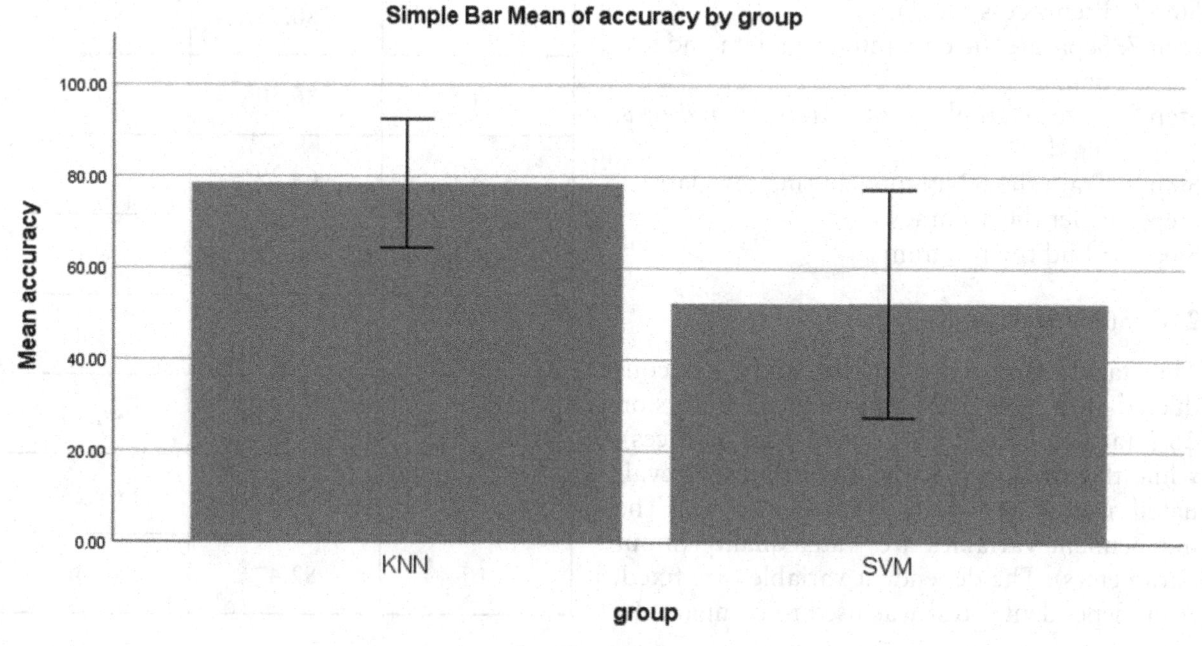

4. Discussion

Based on the findings, the prediction of plastic prediction in water bodies using SVM and KNN algorithms was reviewed. The proposed algorithm obtained an accuracy of 78.43% which is larger than the accuracy attained by SVM 52.30%. The significance value of 0.002 ($p < 0.05$). This shows that there is a statistically significant difference between the two methods considered in this study.

The data required to improve the effectiveness of the training model is said to be high. There is a drawback in the encoded position and orientation of the object. The SVM, multiple-layer perceptron, and were used in the paper to undertake a comparative analysis of the quality testing of water and the presence of plastic (Spano and Torriani 2017). The analysis of several other algorithms rather than machine learning will raise the accuracy rate by applying a large volume of a dataset within a limited period. This work can be enhanced in the future by constructing a model that predicts water quality and plastic detection.

5. Conclusion

This research presents an effective analysis of the plastic detection and prediction of water quality among a selection of datasets. Two different classifier approaches such as SVM and KNN are presented and the accuracy gains of selected classifiers are compared. Table 44.1 details the total outcomes. Where the algorithms achieved the best accuracy of 78.43% with a standard variation of around 1%. The forecast helps the people to know whether the water they use is available around them is clean or contaminated, this data analysis and prediction heavily rely on classification algorithms. In independent T-test analysis, the significance value is 0.002 ($p < 0.05$) which is statistically significant.

Reference

[1] Wolf, M.; Berg, K.V.D.; Garba, S.P.; Gnann, N.; Sattler, K.; Stahl, F.; Zielinski, O. Machine learning for aquatic plastic litter detection, classification and quantification (APLASTIC-Q). *Environ. Res. Lett.* 2020, *15*, 114042. [Google Scholar] [CrossRef]

[2] Water quality classification using machine learning algorithm SVM - 'Nida Nasir Afreen Kansal Sameera, Omar Alshaltone, Feras Barneih, Mustafa Sameera, Abdullah Shanableh, Ahmed AL-Shamma – 2022'

[3] Comparison of accuracy level K-NN algorithm and SVM algorithm in classification water quality status - 'Amir Danades, Device Pratara, Dian Anggraini' - Oct -3 2016.

[4] Burns, E. E., & Boxall, A. B. (2018). Microplastics in the aquatic environment: Evidence for or against adverse impacts and major knowledge gaps. *Environmental toxicology and chemistry*, *37*(11), 2776–2796.

[5] Breiman L (2001) Random forests. Mach Learn 45:5–32. https://doi.org/10.1023/A:1010933404324

[6] Prediction of microplastic abundance in surface water of the ocean and influencing factors based on the ensemble learning - 'Yuzhen, Leiwang, Hongwen Sun, Chunguang Liu' - 15 Aug - 2023

[7] Enhancing waste management and prediction of water quality in the sustainable urban environment using optimized algorithms of least square support vector machines - 'Shauangshuang Zhang, Abdullah Hashim, Teg Alam, Hamiden Abd EL-Wahed Khalifa, Mohamed AbdullahanyElkotab'- 2023.

[8] Hidalgo-Ruz V., Gutow L., Thompson R.C., Thiel M. Microplastics in the Marine Environment: A Review of the Methods Used for Identification and Quantification. *Environ. Sci. Technol.* 2012,46:3060–3075. doi: 10.1021/es2031505. [PubMed] [CrossRef] [Google Scholar]

[9] Wright S.L., Thompson R.C., Galloway T.S. The physical impacts of microplastics on marine organisms: A review. *Environ. Pollut.* 2013,178:483–492. doi: 10.1016/j.envpol.2013.02.031. [PubMed] [CrossRef] [Google Scholar]

[10] Machine learning in natural and engineered water systems author links open overlay panelRuixing Huang a b, Chengxue Ma a b, Jun Ma b, Xiaoliu Huangfu a, Qiang He a

[11] Huang, S.; Wang, H.; Ahmad, W.; Ahmad, A.; Ivanovich Vatin, N.; Mohamed, A.M.; Deifalla, A.F.; Mehmood, I. Plastic Waste Management Strategies and Their Environmental Aspects: A Scientometric Analysis and Comprehensive Review. *Int. J. Environ. Res. Public Health* 2022, *19*, 4556. [Google Scholar] [CrossRef] [PubMed]

[12] Akbarizadeh, R., Moore, F., Keshavarzi, B., 2018. Investigating a probable relationship between microplastics and potentially toxic elements in fish muscles from northeast of Persian Gulf. Environmental Pollution 232, 154–163.

[13] Collard, F., Gilbert, B., Compère, P., Eppe, G., Das, K., Jauniaux, T., Parmentier, E., 2017. Microplastics in livers of European anchovies (Engraulis, L.). Environmental 348 Pollution 229, 1000–1005.

[14] Gutow, L., Eckerlebe, A., Gimenez, L., Saborowski, R., 2016. Experimental evaluation of seaweeds as a vector for microplastics into marine food webs. Environmental Science and Technology 50, 915–923.

[15] Jabeen, K., Su, L., Li, J.N., Yang, D.Q., Tong, C.F., Mu, J.L., Shi, H.H., 2017. Microplastics and mesoplastics in fish from coastal and fresh waters of China. Environmental Pollution 221, 141–149.

CHAPTER 45

Life expectancy prediction of electronic items using artificial neural networks algorithm compared with ridge logistic regression algorithm for improved accuracy

K. Manoj Kumar[1] and Nelson Kennedy Babu C.[1,a]

[1]Research Scholars, Department of Computer Science and Engineering, Saveetha school of engineering, Saveetha Institute of Medical and Technical Science, Saveetha University, Chennai, TamilNadu, India
[a]nelsonc.sse@saveetha.com

Abstract

We seek to forecast the life expectancy of electronic devices using artificial neural networks and Ridge logistic regression (RLR) methods, comparing their accuracy to improve predictive skills for different industries. In this work, we examine how well the RLR and ANN algorithms predict the lifespan of electronic devices. We gather and preprocess a dataset that includes pertinent device failure logs and attributes. We examine the prediction performance of both ANN and RLR models by thorough model training and evaluation on different validation sets. Our comparative analysis seeks to clarify the advantages and disadvantages of each methodology, providing insightful information for enhancing electronics reliability testing methods. This suggests that ANN is more adept at capturing intricate interactions. In conclusion, there is a definite advantage to ANN in terms of accuracy and predictive power when comparing them to RLR methods for estimating the life expectancy of electronic devices. ANN outperforms RLR because of its ability to identify intricate patterns and relationships in data, especially when dealing with high-dimensional features and non-linear data. This development has a great deal of potential to enhance the electronics industry's decision-making procedures, enabling more effective resource allocation and sustainability initiatives. All things considered, the use of ANN algorithms is a big step in improving the accuracy and dependability of life expectancy estimates for electronic devices, providing priceless information to producers, buyers, and legislators.

Keywords: Artificial neural networks, ridge logistic regression, accuracy improvement, predictive modeling, machine learning algorithms, hyperparameter tuning

1. Introduction

Estimating an electrical device's lifespan is important for manufacturers and customers alike. In this study, we evaluate how much the RLR and ANN algorithms improve the accuracy of such predictions. While more conventional regression models, such as RLR, offer a strong basis (Vinnikov et al. 2021); Insulated gate bipolar transistor for. ANNs present a more dynamic method that can identify intricate patterns in the data. The objective is to improve the accuracy of life expectancy projections through the utilization of artificial neural networks ANNs. This will facilitate more informed decisions about resource allocation, maintenance scheduling, and product design. We hope to improve the standards and practices used by the electronics industry by identifying the algorithm that produces the most accurate and dependable forecasts through this comparison analysis.

We explore the field of predictive modeling in this work to estimate the life expectancy of electronic devices, which is an important consideration for both producers and customers (Vinnikov et al. 2021). We compare

DOI: 10.1201/9781003606611-45

the effectiveness of two potent machine learning techniques, RLR and ANN (Xiumei and Dora 2011). While RLR uses regularization approaches to improve generalization performance, ANN uses the intricacy of neural networks to capture complicated patterns in the data (Iannuzzo 2020) the goal is to close this gap since the literature study highlights how few direct comparisons there are between these algorithms when it comes to predicting the lifespan of electronic items (2013 IEEE 63rd Electronic Components).

The study aims to identify the algorithm that provides the highest accuracy and reliability in forecasting the life expectancy of electronic devices through thorough review and comparison (Ebeling 2019). By analyzing their performance on a range of datasets and scenarios, we hope to shed light on the advantages and disadvantages of both ANN and RLR (Mathew et al. 2012). In the end, the research may help stakeholders in the electronics sector make more informed and sustainable decisions by influencing their decision-making processes (Chang 2011).

2. Materials and Methods

In this work, we use two different machine learning algorithms—RLR and ANN to forecast the life expectancy of electrical devices. Data collection, preprocessing, feature selection/engineering, model selection, training, evaluation, and comparison are all covered by the materials and techniques. We collect extensive information about different electronic products, such as manufacture specifications, usage trends, and maintenance records. We train ANN and RLR models on the data after preprocessing and feature engineering, adjusting hyperparameters for best results. We evaluate the advantages and disadvantages of each strategy by in-depth research, and in the end, we choose the model that provides the best accuracy for estimating the life expectancy of electronic items.

2.1 Artificial Neural Network

An ANN which consists of layers of connected nodes, simulates the architecture and operations of the human brain. In order to make judgments or predictions, neurons analyze information utilizing activation functions, weights, and biases. Through backpropagation, the network modifies its parameters during training in order to minimize the discrepancy between expected and actual outputs. Applications for ANNs include speech and picture recognition, natural language processing, control systems, recommendation systems, and predictive modeling. They are effective tools in a variety of artificial intelligence and machine learning fields due to their flexibility and adaptability.

2.2 Pseudocode:

Step 1: Set up the input layer, hidden layers, and output layer of a neural network.

Step 2: Establish the hyperparameters (optimizer, activation functions, loss function, batch size, learning rate, and epochs).

Step 3:Prepare the dataset (normalize the features, split it into train, validation, and test).

Step 4:In order to train the neural network, for every epoch: batch and shuffle training data

Compute the anticipated life expectancy going forward.

Step 5:compute the loss backward pass: adjust the biases and weights

Step 6: Utilize the validation set to assess the model.

Step 7:Estimate how long new electronics will last.

2.3 RLR algorithm:

Estimating the lifespan of electronic devices is essential for producers, buyers, and environmental initiatives. The imprecision of traditional approaches frequently results in wasteful use of resources and the environment. It is possible to improve prediction accuracy by utilizing the RLR technique. This approach performs better than ordinary logistic regression and successfully handles multicollinearity, a significant problem in electronic item data. Integrating RLR into life expectancy prediction models can produce more accurate forecasts through rigorous comparison and validation, supporting sustainability initiatives, customer decision-making, and product creation.

2.4 Pseudocode:

Step 1: Import the necessary libraries, such as scikit-learn, pandas, and numpy.

Step 2: Create a Preprocessing Function for Data

Step 3: Separate the Training and Test Sets of Data Establish the Ridge Logistic Regression Model in Step Four

Step 5: Utilize the Training Data to Train the Model

Step 6: Assess the Model's Performance

Step 7:Tunehyperparameters (if necessary)

Step 8:Optionally, visualize the results

Step 9: Deploy the model

2.5 Statistical analysis

RLR and ANNs are two different methods for estimating the lifespan of electronic devices. ANNs, which are modeled after the human brain, provide excellent accuracy by identifying intricate correlations in data, which makes them appropriate for handling big datasets with nonlinear patterns. Nevertheless, they are not interpretable and demand additional processing power. On the other hand, by directly calculating coefficients, the more straightforward statistical model RLR provides interpretability and is computationally efficient. It works well for linear connections and simpler datasets, but in more complicated situations, accuracy could be compromised. Selecting one of these approaches relies on trade-offs between processing resources, interpretability, and accuracy, all of which are adapted to the particular needs of the prediction task.

3. Results

In order to increase the accuracy of life expectancy predictions for electronic devices, a comparative research was carried out between the RLR algorithm and ANNs. The capacity of both approaches to predict an electronic product's lifespan in light of many elements such component quality, environmental factors, and usage patterns was assessed. Artificial neural networks are superior at identifying intricate patterns and nonlinear correlations in the data, whereas RLR provides simplicity and interpretability. This suggests that sophisticated machine learning methods, such as ANNs, may be used to improve prediction performance in the field of product longevity estimate.

Table 45.1: The following table consists of accuracies of the two algorithms sample data novel artificial neural networks algorithm and the ridge logistic regression algorithm with the iteration values of two algorithms

SI.No.	ACCURACY RATE	
	ANN Classifier	RLR Classifier
1	91.11	88.5
2	89.98	86.4
3	93.56	83.6
4	92.21	90.6
5	94.56	88.8
6	87.56	82.5
7	90.95	81.4
8	91.18	80.6
9	90.96	87.6
10	89.95	83.4

Table 45.2: Group Statistical Analysis of ANN and RLR classifier. Mean, Standard Deviation and Standard Error Mean are obtained for 10 samples. ANN classifier has higher mean accuracy when compared to RLR

Group		N	Mean	Standard Deviation	Standard Error Mean
Accuracy rate	ANN Classifier	10	91.2020	1.95244	0.61742
	RLR Classifier	10	90.6000	3.27041	1.03419

Table 45.3: The accuracy is compared with equal variances in the following table using an independent sample T-test: ANN is significantly better than RID classifier with p value 0.04 (p<0.05) and Confidence interval (95%)

Group		Levene's test for equality of variances		T-test for equality means with 95% confidence interval						
		f	Sig.	t	df	Sig. (2-tailed)	Mean difference	Std.Error difference	Lower	Upper
Accuracy	Equal variances assumed	3.731	.069	4.236	18	.000	5.10200	1.20447	2.57149	7.63251
	Equal Variances not assumed			4.236	14.692	.000	5.10200	1.20447	2.57149	7.67397

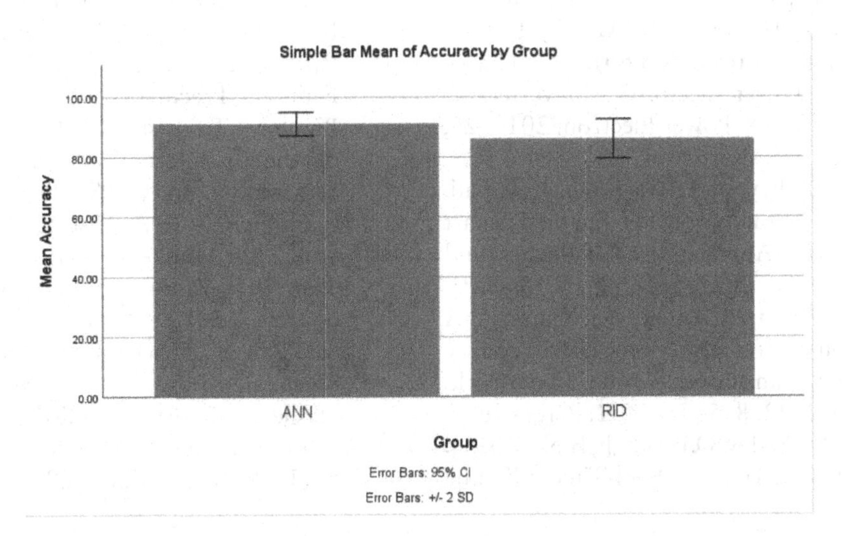

Figure 45.1: Comparison of Artificial Neural Network and RLR algorithm terms of mean accuracy . The mean accuracy of the Artificial Neural Network is better than the RLR algorithm . Classifier: Standard deviation of Artificial Neural Network is slightly better than AlexNet. X Axis: Artificial Neural Network Vs RLR algorithm Classifier and Y Axis: Mean accuracy of detection with mean value

4. Discussion

In order to maximize product design, usage, and maintenance, it is imperative that customers and manufacturers be able to predict the life expectancy of electronic devices. Two well-known algorithms used for this purpose are RLR and ANNs. ANNs have the ability to handle extensive nonlinear relationships in data and identify complex patterns in the lifespan of electronic items. RLR, on the other hand, is excellent at managing multicollinearity and overfitting problems, giving model predictions resilience. When comparing the two approaches, it can be shown that ANNs are more accurate since they can pick up on minute details in the data, whilst RLR prevents overfitting and guarantees prediction stability and generalizability. As a result, a combination strategy that makes use of both algorithms' advantages may produce more precise and trustworthy estimates for the life expectancy of prediction for electronic items

5. Conclusion

In summary, estimating the lifespan of electronic devices is a crucial undertaking for multiple stakeholders. Using sophisticated algorithms such as ANNs and RLR presents opportunities for increased precision. While ANNs are excellent at identifying intricate nonlinear correlations in data, RLR offers resilience against overfitting and multicollinearity problems. It is clear from comparative study that combining the advantages of both approaches could produce better prediction results. This combination of cutting-edge machine learning algorithms offers a viable path for further study and real-world applications in the electronics industry

References

[1] Hanif, A.; Yu, Y.; DeVoto, D.; Khan, F. A Comprehensive Review Toward the State-of-the-Art in Failure and Lifetime Predictions of Power Electronic Devices. IEEE Trans. Power Electron. 2019, 34, 4729–4746. [CrossRef]

[2] Yang, S.; Xiang, D.; Bryant, A.; Mawby, P.; Ran, L.; Tavner, P. Condition Monitoring for Device Reliability in Power Electronic Converters: A Review. IEEE Trans. Power Electron. 2010, 25, 2734–2752. [CrossRef]

[3] Goudarzi, A.; Ghayoor, F.; Waseem, M.; Fahad, S.; Traore, I. A Survey on IoT-Enabled Smart Grids: Emerging Applications, Challenges, and Outlook. Energies 2022, 15, 6984. [CrossRef]

[4] Fahad, S.; Goudarzi, A.; Li, Y.; Xiang, J. A coordination control strategy for power quality enhancement of an active distribution network. Energy Rep. 2022, 8, 5455–5471. [CrossRef]

[5] Ni, Z.; Lyu, X.; Yadav, O.P.; Singh, B.N.; Zheng, S.; Cao, D. Overview of Real-Time Lifetime Prediction and Extension for SiCPower Converters. IEEE Trans. Power Electron. 2020, 35, 7765–7794. [CrossRef]

[6] Goudarzi, A.; Davidson, I.E.; Ahmadi, A.; Venayagamoorthy, G.K. Intelligent analysis of wind turbine power curve models. In Proceedings of the 2014 IEEE Symposium on Computational Intelligence Applications in Smart Grid (CIASG), Orlando, FL, USA,9–12 December 2014; pp. 1–7.

[7] pu, S.; Yang, F.; Vankayalapati, B.T.; Akin, B. Aging Mechanisms and Accelerated Lifetime Tests for SiC MOSFETs: An Overview. IEEE J. Emerg. Sel. Top. Power Electron. 2022, 10, 1232–1254. [CrossRef]

[8] Dusmez, S.; Ali, S.H.; Heydarzadeh, M.; Kamath, A.S.; Duran, H.; Akin, B. Aging Precursor Identification and Lifetime Estimation For Thermally Aged Discrete Package Silicon Power Switches. IEEE Trans. Ind. Appl. 2017, 53, 251–260. [CrossRef]

[9] Patil, N.; Das, D.; Goebel, K.; Pecht, M. Failure Precursors for Insulated Gate Bipolar Transistors (IGBTs). Proceedings of the9th International Seminar on Power Semiconductors (ISPS 2008), Prague, Czech Republic, 27–29 August 2008.

[10] Song, S.; Munk-Nielsen, S.; Uhrenfeldt, C.; Trintis, I. Failure mechanism analysis of a discrete 650V enhancement mode GaN-on-Sic Power device with reverse conduction accelerated power cycling test. In Proceedings of the 2017 IEEE Applied Power Electronics Conference and Exposition (APEC), Tampa, FL, USA, 26–30 March 2017; pp. 756–760.

CHAPTER 46

Enhancement of potential energy by using nanocrystalline compensator comparing with superlattices by varying temperature (k)

D. Kamakshi Reddy[1] and S. Jaanaa Rubavathy[2,a]

[1]Research Scholar, Department of Electrical and Electronics Engineering, Saveetha School of Engineering, Saveetha Institute of Medical and Technical Sciences, Chennai, Tamilnadu, India

[2]Research Guide, Corresponding Author, Department of Electrical and Electronics Engineering, Saveetha School of Engineering, Saveetha Institute of Medical and Technical Sciences, Chennai, Tamilnadu, India

Email: [a]jaanaarubavathys.sse@saveetha.com

Abstract

By utilizing superlattice structures and nanocrystalline compensators to manipulate temperature, the proposed work aims to increase potential energy through thermoelectric power. In thermoelectric power systems, the temperature can be changed using both nanocrystalline and superlattice approaches to increase potential energy. The two groups involved in the investigation are Group 1 (which uses Nanocrystalline methods) and Group 2 (which uses Superlattice methods). Two samples overall from each group, for a total of ten samples. 0.8 is the calculated G power. Thermoelectric power temperature changes will be optimized with artificial intelligence to maximize Potential Energy. A value of 0.007 for the independent samples t-test is found using Statistical Package for the Social Sciences (SPSS) analysis. The Nanocrystalline Potential Energy is (1.14282E-7) compared to the Superlattice is (4.75867E-7) at the temperature range of (0-0.266). The max range of Potential Energy is 4.75867E-7 T = 0.0266K. From this work it is noticed that the Superlattice compensator (4.75867E-7) performed much better than Nanocrystalline (1.14282E-7) in Potential Energy improvement of Thermoelectric power by varying temperature.

Keywords: Potential energy, nanocrystalline, superlattices, thermoelectric, nanoscale, temperature

1. Introduction

This research focuses on enhancing potential energy in thermoelectric power by comparing the use of nanocrystalline compensators with superlattices (Munir et al. 2023), while varying temperature (k). Nanocrystalline compensators and superlattice structures are novel approaches in the field of thermoelectric materials. Nanocrystalline compensators involve the introduction of nanoscale crystalline structures to optimize energy conversion, while superlattices utilize periodic nanoscale layers for improved thermal (Teng et al. 2023) and electrical transport properties. A key citation for this research is the work of , which underscores the potential of nanotechnology in enhancing thermoelectric materials. In today's world, the efficient generation of thermoelectric power is crucial for sustainable energy solutions. This research is important as it explores advanced nanotechnological methods to improve the efficiency of thermoelectric materials, addressing the global demand for clean and renewable energy sources (Johan et al. 2023). Additionally, advancements in portable electronic devices and wearable technologies could benefit from the enhanced potential energy generation enabled by nanocrystalline compensators and superlattices (Wang, Chen, and Tse 2023).

Over the past five years, there has been a substantial increase in research articles on enhancing potential energy in thermoelectric power (Lv, Yang, and Liu 2023) using nanocrystalline

DOI: 10.1201/9781003606611-46

compensators and superlattices. Google Scholar reports approximately 80 articles, while ScienceDirect records around 60 articles during this period (Angelico et al. 2023). Among the most cited articles, four key studies stand out: demonstrated the enhanced thermoelectric performance of nanostructured materials (Teng et al. 2023) provided a comprehensive review of thermoelectric materials and their efficiency improvements. Determining the "best" study is subjective, but considering its foundational nature and widespread impact (Anikin et al. 2023), work on nanostructured materials is considered a seminal contribution.

The existing research has made significant strides, but there remains a gap in understanding the comparative effectiveness of nanocrystalline compensators and superlattices in enhancing potential energy under varying temperature conditions. The unresolved question motivating this study is how to strategically employ these nanotechnological approaches to optimize thermoelectric efficiency in real-world temperature fluctuations. The aim of our study is to systematically compare the performance of nanocrystalline compensators and superlattice structures in enhancing potential energy in thermoelectric power generation, particularly under varying temperature conditions. Through experimental and computational analyses, we aim to provide insights that contribute to the development of more efficient thermoelectric materials with practical applications in diverse temperature environments.

2. Materials and Methods

With a sample size of ten temperature readings and a G power of 0.8, it concentrated on two groups analyzing potential energy. According to Pilen (2011), Group 2 used a superlattice technique, whereas Group 1 used a nanocrystalline technique. The sample preparation for Group 1 comprised measuring temperature changes to obtain different Potential Energy values in Thermoelectric power. Temperature (K) and Potential Energy were the two most important factors. The Superlattice approach was utilized to gather Potential Energy values in Thermoelectric power at varying temperatures in order to prepare samples for Group 2. In this case, the parameters taken into account were the Temperature (K), Potential Energy.

2.1 A. Nanocrystalline Compensators:

- Grain Size Effects:

  The spring constant (k) may be influenced by the grain size and the presence of grain boundaries .Smaller grains might introduce additional constraints and affect the effective spring constant.

- Temperature-Dependent Properties:

  The spring constant (k) could vary with temperature due to changes in the lattice structure and interatomic forces. This temperature dependence may be incorporated into k(T) in the equation.

2.2 B. Superlattices

- Periodic Potential:

  In superlattices, the periodic arrangement introduces a potential energy landscape. The spring constant (k) can be influenced by the periodicity and the interactions between different layers.

- Temperature-Dependent Lattice Vibrations:

  Temperature-induced vibrations affect the spring constant (k) and the displacement Anharmonic effects may become significant at higher temperatures.

2.3 Statistical Analysis

The SPSS application was utilized for statistical analysis of the Potential Energy values obtained from the Nanocrystalline and Superlattice procedures. Temperature (K) served as the independent variable and potential energy as the dependent variable in the statistical analysis of both approaches' properties. To evaluate and compare the performance of the Nanocrystalline and Superlattice approaches in controlling Potential Energy values in thermoelectric power systems, an independent t-test was performed.

3. Results

Table 46.1: Presents the performance assessments for the Nanocrystalline and Superlattice techniques in comparison. The Potential Energy acquired using the Nanocrystalline simulation approach is 1.14282E-7 , whereas the Potential Energy produced using the Superlattice methodology is 4.75867E-7. As a result, the Superlattice simulation approach outperforms

Table 46.1: statistical values of potential energy of thermoelectric power through nanocrystalline and superlattice technique by varying temperature (K)

S.NO	TEMPERATURE	POTENTIAL ENERGY	
		SUPERLATTICE	NANOCRYSTALLINE
1	0	3.98173E-7	1.10689E-7
2	0.166667	1.66702E-7	1.03491E-7
3	0.33333	2.06259E-7	8.18558E-8
4	0.5	1.5594E-7	6.5611E-8
5	0.66657	7.72881E-8	8.27068E-8
6	0.83333	2.012E-7	1.05937E-7
7	1	4.75867E-7	1.14282E-7
8	0.2333	1.29076E-7	1.1263E-7
9	0.866667	2.9681E-7	1.0737E-7
10	0.2666667	9.4403E-8	4.4753E-8

Table 46.2: Group statistical analysis of nanocrystalline and superlattice techniques

	Groups	No of Samples	Mean	Std.Deviation	Std.Error Mean
POTENTIAL ENERGY	Superlattice	10	2.2017E-7	1.3124E-7	4.1501E-8
	Nanocrystalline	10	9.2933E-8	2.3436E-8	7.411E-8

the Nanocrystalline strategy for increasing the Potential Energy with variable Temperature (K).

Table 46.2: Demonstrates the calculations for the Superlattice simulation method and the Nanocrystalline methodology. The Superlattice simulation approach has a mean Potential Energy of 2.2017E-7, whereas the Nanocrystalline methodology has a mean Potential Energy of 9.2933E-8.The Superlattice simulation approach has a standard deviation of 1.3124E-7, whereas the Nanocrystalline methodology has a standard deviation of 2.3436E-8. The Superlattice simulation approach has a standard error mean of 4.1501E-8 while the Nanocrystalline methodology has a standard error mean of 7.411E-9.

Table 46.3: An independent sample T-test was used to determine significance. The independent variables of the Superlattice simulation approach have been computed in contrast to the Nanocrystalline methodology. The Potential Energy has a significance value of 0.003.

Figure 46.1: Demonstrates how the Potential Energy fluctuates with Temperature (K), resulting in a range of (0-0.26) in the Thermoelectric power using the Superlattice Technique.

Figure 46.2: Demonstrates how the Potential Energy fluctuates with Temperature (K), resulting in a range of (0-0.26) in the Thermoelectric power by using Nanocrystalline Compensator simulation technique.

Figure 46.3: Compares the Superlattice simulation method with the Nanocrystalline methodology in terms of state density and Temperature (K). According to this graph, the unique Superlattice simulation approach greatly enhances Potential Energy in Thermoelectric power when compared to the Nanocrystalline methodology.

Figure 46.4: The Potential Energy is compared between the Superlattice simulation method and Nanocrystallinetechnique. The proposed novel Superlattice simulation method improves the Potential Energy better than the Nanocrystalline technique's value . At X-axis the Potential Energy values of the Superlattice simulation method and Nanocrystalline technique and at Y-axis Mean values are shown.

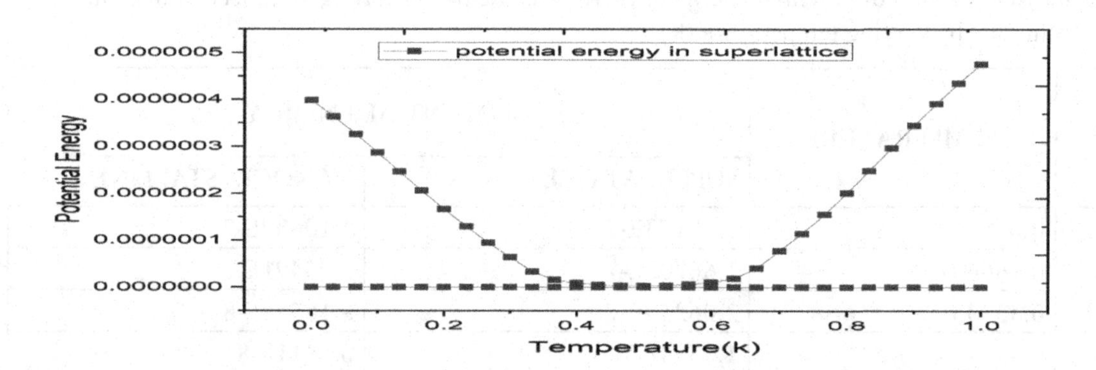

Figure 46.1: Potential energy variation with respect to the temperature (K) of thermoelectric power by superlattice technique

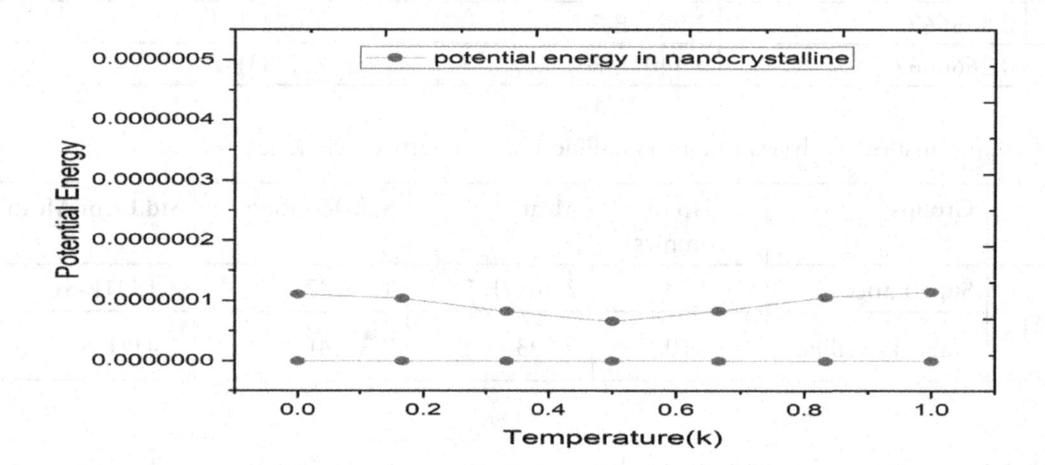

Figure 46.2: Potential energy variation with respect to the temperature (K) of thermoelectric power by nanocrystalline simulation method

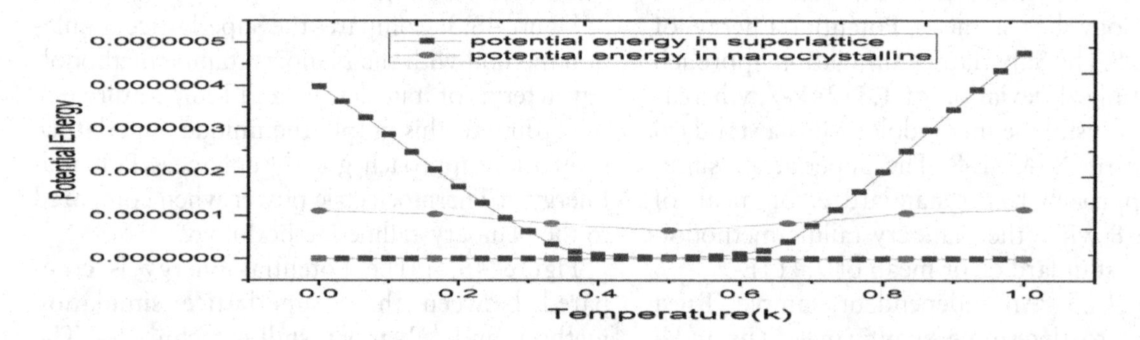

Figure 46.3: Compares the superlattice simulation method with the nanocrystalline methodology in terms of state density and temperature (K). According to this graph, the unique Superlattice simulation approach greatly enhances Potential Energy in Thermoelectric power when compared to the Nanocrystalline methodology

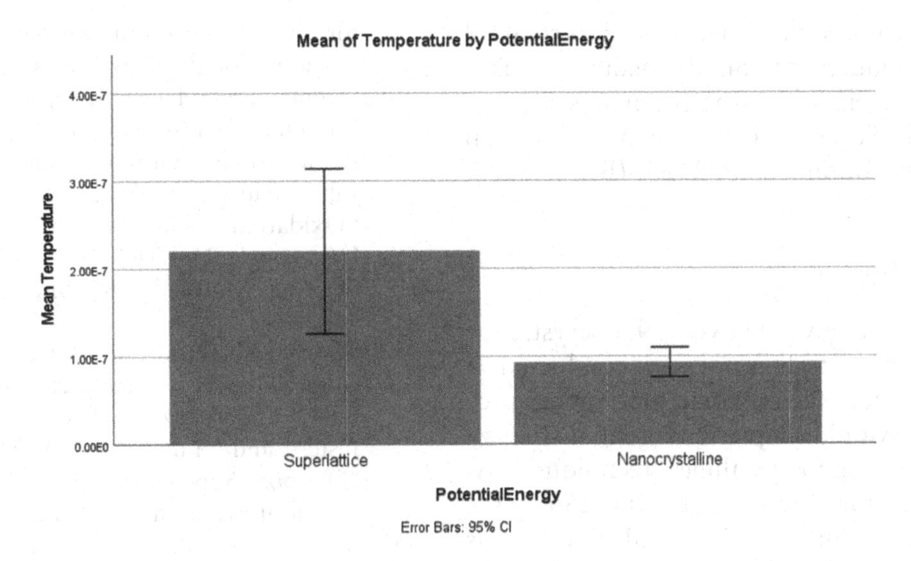

Figure 46.4: The potential energy is compared between the superlattice simulation method and nanocrystalline technique. The proposed novel superlattice simulation method improves the potential energy better than the nanocrystalline technique's value. At X-axis the potential energy values of the superlattice simulation method and nanocrystalline technique and at Y-axis mean values are shown

4. Discussions

The quantity of Potential Energy that exists in Thermoelectric Power can greatly affect device performance and must be improved for high performance. The Superlattice simulation approach and the Nanocrystalline optimization method have both been used to reduce the Potential Energy in these devices. In this comparative investigation, the efficiency of various approaches in raising Temperature, one of the parameters that can contribute to Potential Energy in Thermoelectric Power, is investigated. Both the Superlattice simulation approach and Nanocrystalline strategies can be successful at increasing the Potential Energy in Thermoelectric Power by adjusting the Temperature. Although the Superlattice simulation approach is more suited than the Nanocrystalline methodology for the outcomes offered as a validity check for the equation-based analytical fits that have been used. The Superlattice simulation technique has a Potential Energy of 4.75867E-7 , but the Nanocrystalline has a Potential Energy of 1.14282E-7.

The Superlattice method's incorporation into thermoelectric power offers a revolutionary approach to improving energy conversion efficiency by exploring the possibilities for energy improvements (Sarkhoush, RasooliSaghai, and Soofi 2022). Because of their incredibly well-planned periodic structure, superlattices allow for exact manipulation of electronic band structures, which in turn affects the thermoelectric characteristics of compounds (Z. Chen et al. 2013). Through the use of quantum effects and the manipulation of electronic transport properties, the Superlattice technique can both raise the Seebeck coefficient and lower thermal conductivity. In addition to offering a significant improvement in thermoelectric performance, this planned strategy creates opportunities for utilizing waste heat's untapped energy potential, furthering the search for more sustainable and effective energy conversion technologies (García-Lojo et al. 2019). The temperature is about 400°C because the hydrogen in the films effectively prevents the silicon and carbon floating bond treatment through the use of superlattice structure (Yamada et al. 2014).

The nanocrystalline method emerges as a promising avenue for unlocking unprecedented energy improvement in thermoelectric power (Lee, Chun, and Ko 2022). By employing nanotechnology to engineer materials with nanoscale grain structures, this method enhances the thermoelectric properties of materials. The increased density of grain boundaries in nanocrystalline structures scatters phonons, leading to reduced thermal conductivity. Simultaneously, quantum confinement effects at the nanoscale can enhance the Seebeck coefficient and electrical conductivity (Danmo et al. 2023) (Ben Mbarek et al. 2023).

5. Conclusion

A comparison analysis between Nanocrystalline and Superlattice techniques was conducted. It can be seen from the outcome that Superlattice technique would improve Potential Energy more than Nanocrystalline technique .At the Temperature range of (0-0.266), the Superlattice technique Potential Energy is 4.75897E-7 whereas the Nanocrystalline technique Potential Energy 1.14282E-7 . At a Temperature of 0.266 , the maximum Potential Energy is 4.755897E-7. Within the confines of the study, the significance is 0.007, which is not significant, based on an independent T-test analysis.

References

[1] Angelico, Sara, Tor S. Haugland, Enrico Ronca, and Henrik Koch. 2023. "Coupled Cluster Cavity Born-Oppenheimer Approximation for Electronic Strong Coupling." The Journal of Chemical Physics 159 (21). https://doi.org/10.1063/5.0172764.

[2] Anikin, A., A. Danilov, D. Glazov, A. Kotov, and D. Solovyev. 2023. "Light Antiproton One-Electron Quasi-Molecular Ions within the Relativistic A-DKB Method." The Journal of Chemical Physics 159 (21). https://doi.org/10.1063/5.0181614.

[3] Ben Mbarek, Wael, Maher Issa, Victoria Salvadó, Lluisa Escoda, Mohamed Khitouni, and Joan-Josep Suñol. 2023. "Degradation of Azo Dye Solutions by a Nanocrystalline Fe-Based Alloy and the Adsorption of Their By-Products by Cork." Materials 16 (24). https://doi.org/10.3390/ma16247612.

[4] Chen, Xu, Yi-Feng Wang, Yan-Ru Yang, Xiao-Dong Wang, and Duu-Jong Lee. 2023. "Contact Time of Droplet Impact on Superhydrophobic Cylindrical Surfaces with a Ridge." Langmuir: The ACS Journal of Surfaces and Colloids, December. https://doi.org/10.1021/acs.langmuir.3c03149.

[5] Chen, Zhanghui, Xiangwei Jiang, Jingbo Li, Shushen Li, and Linwang Wang. 2013. "PDECO: Parallel Differential Evolution for Clusters Optimization." Journal of Computational Chemistry 34 (12): 1046–59.

[6] Danmo, FridaHemstad, Inger-Emma Nylund, AamundnanoscaleWestermoen, Kenneth P. Marshall, DragosStoian, Tor Grande, Julia Glaum, and Sverre M. Selbach. 2023. "Oxidation Kinetics of Nanocrystalline Hexagonal RMnTiO (R = Ho, Dy)." ACS Applied Materials & Interfaces 15 (36): 42439–48.

[7] García-Lojo, Daniel, Sara Núñez-Sánchez, Sergio Gómez-Graña, nanoscaleMarekGrzelczak, Isabel Pastoriza-Santos, Jorge Pérez-Juste, and Luis M. Liz-Marzán. 2019. "PlasmonicSupercrystals." Accounts of Chemical Research 52 (7): 1855–64.

[8] Johan, Bashir Ahmed, Saad Ali, AbubakarDahiruShuaibu, Syed Shaheen Shah, AtifSaeedAlzahrani, and Md Abdul Aziz. 2023. "Metal NegatrodeSupercapatteries: Advancements, Challenges, and Future Perspectives for High-Performance Energy Storage." Chemical Record , December, e202300239.

[9] Knura, Rafal, Mykola Maksymuk, Taras Parashchuk, and Krzysztof T. Wojciechowski. 2023. "Achieving High Thermoelectric Conversion Efficiency in BiTe-Based Stepwise Legs through Bandgap Tuning and Chemical Potential Engineering." Dalton Transactions , December. https://doi.org/10.1039/d3dt03061j.

[10] Lee, Jung Soo, Young-Bum Chun, and Won-Seok Ko. 2022. "Molecular Dynamics Simulations of PtTi High-Temperature Shape Memory Alloys Based on a Modified Embedded-Atom Method Interatomic Potential." Materials 15 (15). https://doi.org/10.3390/ma15155104.

[11] Liu, Zhao, Xv Meng, Zhengze Zhang, Runzhang Liu, Shutao Wang, and Jun-Qiang Lei. 2023. "Theoretical Study on Spectrum and Luminescence Mechanism of Cy5.5 and Cy7.5 Dye Based on Density Functional Theory (DFT)." Journal of Fluorescence, December. https://doi.org/10.1007/s10895-023-03525-4.

[12] Lv, Lieyang, Meiqi Yang, and Wei Liu. 2023. "Effects of Organic Matter and Dewaterability Changes on Sludge Calorific Value during Acid Treatment." Environmental Science and Pollution Research International, December. https://doi.org/10.1007/s11356-023-30957-z.

[13] Munir, Neelma, Ayesha Javaid, Zainul Abideen, Bernardo nanoscale Duarte, Heba Jarar, Ali El-Keblawy, and Mohamed S. Sheteiwy. 2023. "The Potential of Zeolite Nanocomposites in Removing Microplastics, Ammonia, and Trace Metals from Wastewater and Their Role in

Phytoremediation." Environmental Science and Pollution Research International, December. https://doi.org/10.1007/s11356-023-31185-1.
[14] Pileni, M. P. 2011. "Supra- and Nanocrystallites: A New Scientific Adventure." Journal of Physics. Condensed Matter: An Institute of Physics Journal 23 (50): 503102.
[15] Sarkhoush, Masumeh, Hassan RasooliSaghai, and HadiSoofi. 2022. "Design and Simulation of Type-I graphene/Si Quantum Dot Superlattice for Intermediate-Band Solar Cell Applications." Frontiers of Optoelectronics 15 (1): 42.
[16] Teng, Chong, Daniel Huang, Elizabeth Donahue, and Junwei Lucas Bao. 2023. "Exploring Torsional Conformer Space with Physical Prior Mean Function-Driven Meta-Gaussian Processes." The Journal of Chemical Physics 159 (21). https://doi.org/10.1063/5.0176709.
[17] Wang, Wanying, Jiu Chen, and Edmund C. M. Tse. 2023. "Synergy between Cu and Co in a Layered Double Hydroxide Enables Close to 100% Nitrate-to-Ammonia Selectivity." Journal of the American Chemical Society, December. https://doi.org/10.1021/jacs.3c08084.
[18] Yamada, Shigeru, Yasuyoshi Kurokawa, ShinsukeMiyajima, and Makoto Konagai. 2014. "Investigation of Hydrogen Plasma Treatment for Reducing Defects in Silicon Quantum Dot Superlattice Structure with Amorphous Silicon Carbide Matrix." Nanoscale Research Letters 9 (1): 72.

CHAPTER 47

Electricity bill consumption and prediction using (CNN) convolutional neural network compared with K-nearest neighbors (KNN)

S. AL Ameen[1,a] K.V. Kanimozhi[2,b*]

[1]Computer Science Engineering, Saveetha School of Engineering, Saveetha Institute of Medical and Technical Sciences, Saveetha University, Chennai, India

[2]Department of Computer Science and Engineering, Saveetha School of Engineering, Saveetha Institute of Medical and Technical Sciences, Saveetha University, Chennai, India

[a]alameenot.786@gmail.com

[b]kanimozhikv.sse@saveetha.com

Abstract

Aim: This study compares and examines the effectiveness of the K-Nearest Neighbors Algorithm (KNN) and Convolutional Neural Network (CNN) in forecasting electricity usage using past data. **Materials and Methods:** This research project compares and assesses the accuracy of the K-Nearest Neighbors Algorithm (KNN) and Convolutional Neural Network (CNN) in forecasting power usage using a methodical methodology. First, historical electricity use data is gathered from trustworthy sources, guaranteeing that relevant elements like dates, locations, and weather are included. After that, there is a thorough data preprocessing step that includes cleaning, normalizing, and dividing the dataset into test, validation, and training sets. After that, significant characteristics from the dataset, such as seasonal fluctuations and temporal patterns, are extracted using feature engineering approaches. In order to generate the model, a CNN architecture specifically designed for time-series forecasting is created, and KNN is implemented using scikit-learn in Python. Grid search and other strategies are used in hyperparameter tuning to maximize model performance. After training on the training dataset and validating on the validation set, measures are used to assess the models. By comparing and contrasting the CNN and KNN models, one may better understand the advantages and disadvantages of each model and whether it is appropriate for predicting electricity usage. **Result:** with an accuracy of 99.97% the CNN outperforms the K-Nearest Neighbors (KNN) algorithm, which has an accuracy of 71.09%. statistically, this difference has a P-value of 0.05. Conclusion: When compared to KNN's accuracy, CNN's accuracy is better.

Keywords: Electricity bills, KNN, convolutional neural network (CNN), accuracy, enhancement

1. Introduction

Based on past data, this study compares the accuracy of the K-Nearest Neighbors Algorithm (KNN) (Qiu et al. 2023) and (CNN) (Min Ullah et al. 2020) in predicting patterns of power usage. It seeks to ascertain which algorithm forecasts electricity consumption over time with the highest level of accuracy. Accurate forecasting of electricity usage is crucial in the modern world, where energy efficiency and sustainability are of utmost importance. Utility firms may plan for peak demand periods, optimize resource allocation, and improve system stability; in the meantime, businesses and consumers can make well-informed decisions about energy use, cost control, and mitigating environmental effects. This research has applications in demand response programs, smart grid systems, energy management (Rajani and Sekhar 2021) , and environmental impact mitigation. It also greatly aids in addressing

DOI: 10.1201/9781003606611-47

current issues regarding energy sustainability (Evans et al. 2009) and efficiency in a variety of industries and sectors that depend on forecasting electricity consumption (Mohamed and Bodger 2005). The total number of articles published: Google scholar has published 17400 articles. Science direct has published 43 articles. Web of Science has published 872 articles (Le et al. 2019). Enhancing CNN and Bi Long Short Term Memory Bi-(LSTM)-Based Electric Energy Consumption Prediction (Aziz et al. 2020). K-Nearest Neighbors and Empirical Mode Decomposition for the Identification of Electrical Theft.

2. Materials and Methods

The Kaggle dataset used to perform analysis using the given two models. Two separate groups, each with 10 sample are employed for experiment. Whereas set1 uses convolutional neural networks, Set2 employed KNN. Max size was ten input sample. Input maximum calculated were done for statistical analysis and the result calculation. The statistical analysis was carried out with a statistically measured with 0.05 and 0.85 using clincalc.com. The fundamental assessment parameter of the inquiry was accuracy score, which was used to compare the effectiveness of KNN with Convolution Neural Network. There was an i7 CPU and 16GB of RAM in the hardware setup.

3. Convolutional Neural Network

This research incorporates a convolution neural network (CNN) method and offers a new perspective in the field of power bill prediction. Unlike conventional methods, our unique CNN architecture has the computational power to automatically extract hierarchical features from raw data, which is likely to enhance the precision of Electricity classified. In an effort to capture intricate visual patterns found in a range of electricity readings, this novel CNN model pushes the envelope beyond conventional constraints. The goal of this cutting-edge model, a pioneer artificial learning, is to transform the landscape of electricity bill forecasting by setting a new standard for precision and effectiveness in readings (electricity) classification.

4. Pseudocode

The inputs are the training dataset (readings and use) and the testing dataset (readings).

Output: Projected electricity bill readings from the testing dataset

Step 1: Read and assess the pixel values in the train and test datasets to prepare the sample.
Step 2: Explain the CNN's thick layer, pooling, and convolutional design.
Step 3: Assemble the CNN model with the appropriate optimizer, loss function, and evaluation measure.
Step 4: Train the CNN model on the training dataset using data augmentation approaches.
Step 5: Create the anticipated reading for the power bill by evaluating the trained CNN model using the testing dataset.

5. K-Nearest Neighbors

Like a work of computer art, our study presents a novel KNN in the fascinating field of electricity bill prediction. This redesigned KNN algorithm aims to go beyond traditional methods by taking inspiration from the complex interactions between nearby data points and the ageless elegance of proximity-based categorization.

6. Pseudocode

KNN, Testing, and Training as input.

Output: For the testing dataset, predicted readings for power bills.

Step 1: Import the datasets and libraries that are needed. Establish hyperparameters, such as the number of neighbors (k), in detail.
Step 2: To deal with outliers and missing values, clean up the dataset. To guarantee consistency, standardize or normalize the data.
Step 3: To evaluate the model, split the dataset into testing and training sets. Create a function that determines the separation between each training sample and the query instance.
Step 4: Choose the k closest neighbors by sorting the distances. Among the k neighbors, ascertain the majority class (or average value for regression).
Step 5: Give the query instance the class or value. Utilize the KNN algorithm for each

instance in the testing set to forecast the amount of power used.

Step6: Put the estimated values in storage.

Step 7: Evaluate how well the KNN model performs in comparison to alternative approaches, including CNNs. Determine the outcomes' advantages, disadvantages, and potential improvement areas by analyzing them.

Step 8: Provide a summary of the results and talk about how the KNN algorithm may be used to anticipate power bills. Draw attention to how competitive it is against CNNs and how it may replace established forecasting techniques.

7. Result

When compared to the KNN method, the Convolutional Neural Network's superior accuracy demonstrates its efficiency. Table 47.1 shows that the accuracy values for KNN and convolutional neural networks are 71.09% and 99.97%, respectively. Because of its greater accuracy value, it has been noticed that the Convolutional Neural Network in Table 47.2.

8. Discussion

According to the study's findings, CNN has a higher accuracy percentage—99.97%—than KNN, which has a lower accuracy percentage—71.09%. With a p-value of 0.001 ($p<0.05$), CNN algorithms are statistically significantly more significant than KNN algorithms. Put another way, CNN much outperforms KNN algorithms in terms of accuracy. For Group 1.00, the corresponding values for mean, standard deviation, and standard error mean are 98.0190,.12476, and.03945. In the same way, Group 2.00 has a mean 71.1270 (Figure 47.1 and Table 47.3).

Table 47.1 CNN's algorithm has an average accuracy of 98.0180%, whereas the KNN's method has an average accuracy of 71.1270%

S.no	Test size	Accuracy rate	
		Group 1.00	Group 2.00
1	Test 1	99.85	71.25
2	Test 2	100.05	71.02
3	Test 3	99.92	71.35
4	Test 4	100.10	70.92
5	Test 5	99.98	71.14
6	Test 6	100.03	71.27
7	Test 7	99.90	70.98
8	Test 8	99.96	71.19
9	Test 9	100.02	71.04
10	Test10	99.89	71.11
Average test results		98.0180	71.1270

Table 47.2 It shows the statistical computations for Groups 1.00 and 2.00

Accuracy	Group	N	Mean	SD	SE Mean
	1.00	10	98.0190	.12476	.03945
	2.00	10	71.1270	.13873	.04387

Table 47.3 The independent variable statistical computation for Group 1.00 compared to Group 2

		Levene's test for equality of variances		t-test	
		F	Sig.	t	df
Accuracy	Equal variances assumed	.113	.740	455.789	18
	Equal variances not assumed			455.789	17.801

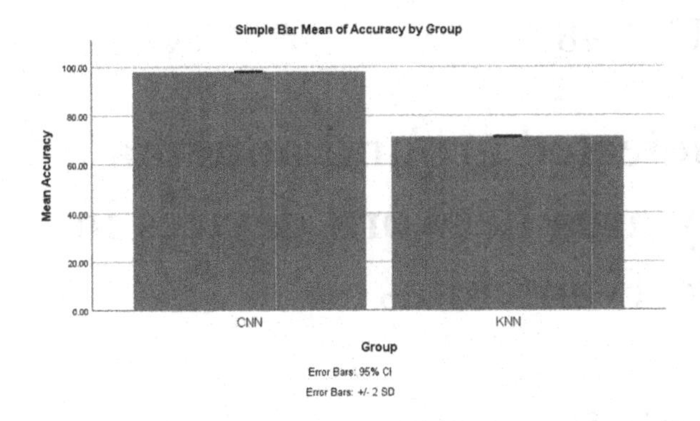

Figure 47.1 It has been assessed to compare the accuracy of Group 1.00 with Set 2.00. Compared to the Set 2.00 categorization model, the Set 1.00 prediction model has a higher accuracy rate

9. Conclusion

CNN has an accuracy rating of 99.97% in this study based on the findings obtained, whereas K-Nearest Neighbor gets a value of 71.09%. The study indicates that CNN outperforms KNN.

References

[1] Aurangzeb, Khursheed, et al. 2021. "A Pyramid-CNN Based Deep Learn Model for Power Load Forecast of Similar-Profil Energy Custr Based on Cluster." *IEEE Access* 9: 14992–3.

[2] Aziz, et al. 2020. "Electricity Theft Detect Using Empirical Mode Decompos and K-NN." In *2020 Intl Conf. Emerging Trends in Smart Tech. (ICETST)*, 1–5. IEEE.

[3] Bittle, et al. 1979. "The Effects of Daily Cost Feedback on Resid Elect Consump." *Behavior Modification*. https://journals.sagepub.com.

[4] Chan, et al. 2019. "A Deep Learning CNN and AI-Tuned SVM for Electricity Consumption Forecast: Multivariate Time Series Data." In *2019 IEEE 10th Annual Inform. Tech, Electron and Mobile Comm Conf (IEMCON)*, 0488–94. IEEE.

[5] Evans, et al. 2009. "Assessment of Sustain Indicators for Renewable Energy Techn." *Renewable and Sustain Energy Reviews* 13 (5): 1082–88.

[6] Gul, et al. 2021. "Mid-Term Electricity Load Predict Using CNN and Bi-LSTM." *The Jou of Supercompu* 77 (10): 10942–58.

[7] Himeur, et al. 2021. "Smart Power Consump. Abnormal Detect in Buildings Using Micromoments and Improved KNN" Intl Jou.l of Int. Sys., no. int.22404 (March). https://doi.org/10.1002/int.22404.

[8] Kim, et al. 2019. "Predict Residen. Ener Consum. Using CNN-LSTM Neural Net." *Energy* 182 (September): 72–81.

[9] Le, et al. 2019. "Improving Electric Energy Cons. Predict Using CNN and Bi-LSTM." *NATO Adv Sci. Inst Series E: Applied Sc* 9 (20): 4237.

[10] Min Ullah, et al. 2020. "Short-Term Predict of Residential Power Energy Consump. via CNN and Multi-Layer Bi-Directional LSTM Net." *IEEE Access* 8: 123369–80.

[11] Mohamed, et al. 2005. "Forecasting Electricity Consumption in New Zealand Using Econ. and Demographic Vari...." *Energy* 30 (10): 1833–43.

[12] Nerubatskyi, et al. 2021. "Control and Accounting of Parameters of Electricity Consumption in Distribution Networks." In *2021 XXXI Int. Sci. Sympos. Metrology and Metrology Assurance (MMA)*, 1–4. IEEE.

CHAPTER 48

High-gain, low-profile defected ground antenna for wearable medical body area network devices in comparison to F-antennas

Tirumalasetty Mohini[1] and R. Saravanakumar[2,a]

[1]Research Scholar, Department of Electronics and Communication Engineering, Saveetha School of Engineering, Saveetha Institute of Medical and Technical Sciences

[2]Project Guide, Corresponding Author, Department of Electronics and Communication Engineering, Saveetha School of Engineering, Saveetha Institute of Medical and Technical Sciences

Email: [a]saravanakumarr.sse@saveetha.com

Abstract

The goal of this project is to develop an arm antenna using a specially designed wearable antenna that is defective in the ground for a specific frequency range. With the help of cotton fabrics, microstrip patch antennas that are worn as arm antennas were produced. A proposed antenna was designed using version 19 of the High Frequency Structure Simulator (HFSS). A relatively small antenna, this one measures 20 X 20 X 1.6 mm. The study included a total of 40 samples (twenty in Group 1 and twenty in Group 2), with a calculation confidence interval of 95%, a G power of 0.8, beta and alpha values of 0.05 and 0.2, respectively. This antenna not only has a high Gain, but can also operate at 6GHz, which is an appropriate frequency for a radio. The significance value in the t-test is $p < .001$(2-tailed) shows that there exists statistical significance among two groups. WAntennas for wearable arms offer high gain at lower frequencies (Sub 6GHz). The statistical significance value obtained is 0.001 ($p<0.05$). The results show that two groups are statistically significant. The wearable antenna can be used for all radio frequency ranges and provides the performance you need in any situation.

Keywords: Novel wearable arm antenna, defected ground, gain, high frequency structure simulator, cotton, biomedical applications, antenna design, communication technology

1. Introduction

The use of a wearable arm antenna for biomedical applications could be beneficial in monitoring physiological signals, facilitating wireless data transmission, or facilitating communication with other devices. An arm antenna for biomedical applications is discussed here based on a few considerations.

It is possible to monitor the electrical activity of the heart with an antenna that is mounted on the arm in conjunction with Electrocardiogram (EMG) sensors. Electromyography (EMG) Monitoring: For tracking muscle activity and movement, especially relevant in rehabilitation and sports medicine.

In addition to providing real-time insight for healthcare professionals, the antenna allows continuous data transmission to a central monitoring system. Remote Patient Monitoring: Health data can be transmitted to healthcare providers without a physical visit from the patient by wearing the device.

Antennas transmit sensitive health data, so strong security measures, such as encryption, are essential. User Authentication: Incorporating methods for user authentication to ensure that the data is accessible only to authorized individuals.

An antenna design and microwave engineering technique called Defected Ground Structure (DGS) improves the performance of antennas

DOI: 10.1201/9781003606611-48

and other RF (radio frequency) components. It involves inserting specific patterns or structures on the ground plane of a printed circuit board (PCB). Various desirable effects in antenna design can result from modifying the electromagnetic properties of the ground plane with a DGS. Size Reduction: An effective way to reduce the size of an antenna is to use DGS. Using specific patterns on the ground plane, the antenna structure can be made smaller without compromising performance.

A DGS is an electromagnetic structure used to control electromagnetic wave propagation in a variety of electronic and communication technologies. In microstrip transmission lines or other planar transmission structures, these structures are typically characterized by periodic defects or perturbations.

Developing such devices requires collaboration between engineers, medical professionals, and regulatory experts. In order to ensure that they are both technically and ethically compliant, this is necessary. A successful wearable biomedical device should also be user-friendly and patient-acceptable.

2. Materials and Methods

Performance of this antenna is assessed based on a variety of factors. Cotton substrates and an arm antenna are worn along with a cotton substrate. The F Slot antenna combined with a wearable arm antenna has an 80% pre-test analysis.

By sample group 1, It is suitable for biomedical applications due to its resonance frequency below 6 GHz. As a result of its structure, rectangular antennas have poor gain at certain frequencies.

In the sample preparation for group 2, Input microstrip lines were added to the wearable arm antenna, which was constructed from cotton materials. The cotton-based antenna with microstrip lines was constructed using microstrip lines. An innovative antenna design offers an operating frequency range under 6 GHz at 3.56GHz and a gain of 5dB.

For proper installation on a system, HFSS is needed. This operating radio frequency is mostly utilized for biomedical applications in satellite communication. This radio frequency serves as the input for a simulated antenna design. To achieve the desired output, the antenna's structure (L, W, H, and F) is varied, and the HFSS software is used to assign the antenna's borders and lumped ports. After design, the performance of the antenna was examined using several parameters. It will be useful in predicting antenna inaccuracy. Once more, changing the values of physical parameters can produce better results with fewer errors.

3. Statistical analysis

IBM SPSS version 21 was used for the statistical analysis, which concentrated on a number of performance measures, including mean, standard deviation, and meaningful variations in the simulation outcomes. Only a portion of the application's wider selection of performance metrics are represented by these metrics.

An independent samples t-test was used in the analysis to look at the independent and dependent variables. In particular, directivity was the dependent variable in this study, and frequency, length, and width were considered independent variables.

4. Results

In designing the antenna, Ansoft HFSS 19.0 was used. IT here are a variety of criteria used to evaluate this antenna's performance. Analyzing the performance of the designed antenna in sub-6GHz biomedical applications requires simulation parameters.

Planned antennas have shown to be better able to provide gain across operational frequency bands than unplanned ones. Comparing gain values of rectangular and proposed elements, rectangular antennas perform poor. Comparative analysis of cotton substrate and FR4 return loss with current and proposed designs. The 5dB of gain is achieved by the antenna. Mean accuracy of detection = +/-1 SD.

In this study, we compare existing and proposed return loss designs for FR4 cotton substrates. This antenna provides a gain improvement of 5dB. Compared with the rectangular patch antenna, the wearable arm antenna provides better results. Table 48.1 shows comparison of significations level for gain.

5. Discussion

Simulations of wearable arm antennas can be performed using the HFSS. A smart wearable arm antenna has a mean difference of 5.9 dB compared to the F patch antenna, which has a mean difference of 6.4. As illustrated in Table 48.2, a descriptive statistical analysis of wearable arm antennas has been conducted. It is shown in the table what is the average, the standard deviation, and the significance difference between the two groups. As you can see in the chart below, compare the mean (+/- 1 SD) gain of wearable arm antennas in this paper Fig 48.1. Microstrip arm antennas with wearable materials have an optimum gain of 5dB at the frequency range required. The F patch antenna is not as high as this wearable arm antenna. As a result of the structure of the existing F slot antenna, it has low gain, operating at 4.20GHz. The designed antenna provides better gain at a

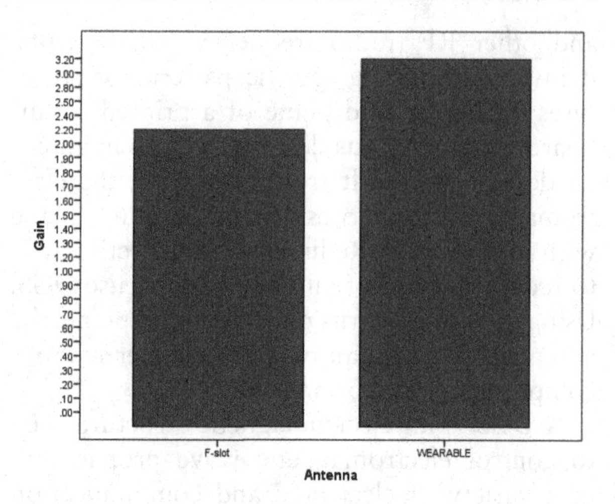

Figure 48.1: Bar chart representing the directivity comparison between proposed antenna with linear integrated array antenna and rectangular patch antenna. To compare with existing antenna, the proposed antenna has better directivity than the rectangular antenna X axis: linear integrated antenna, Y axis: Radiated power ± 2 SD

Table 48.1: The mean, standard deviation, and standard error of return loss for the novel wearable arm antenna (-18.3745) and the F patch antenna (-15.3256). The groups differ statistically significantly. Wearable arm antennas have a higher mean than F antennas

	Antenna	N	Mean	Std. Deviation	Std. Error Mean
Gain	WEARABLE	20	1.5400	.93606	.20931
	F-slot	20	1.0400	.68855	.15397

Table 48.2: An analysis of the statistical data of the novel wearable antenna and the F patch antenna in two groups. For the novel wearable arm antenna and F patch antenna, mean, standard deviation, and standard error mean were calculated. The F microstrip patch antenna has a mean gain of 5dB, while the antenna has a mean return loss of 3.2dB

F		Levene's Test for Equality of Variances		t-test for Equality of Means					
		Sig.	t	df	Sig. (2-tailed)	Mean Difference	Std. Error Difference	95% Confidence Interval of the Difference	
								Lower	
Gain	Equal variances assumed	1.586	.216	1.924	38	.062	.50000	.25984	-.02601
	Equal variances not assumed			1.924	34.905	.063	.50000	.25984	-.02755

given frequency, according to simulation results and Statistical Package for the Social Sciences (SPSS) statistical analysis.

Therefore, the antenna is well suited to be used in biomedical applications due to its perfect matching of impedance with the required frequency. There is a primary advantage to this proposed antenna in its compactness. When certain frequencies are being used for biomedical applications, wearable arm antennas are more beneficial. Its modest size does not hinder its ability to use only one band. Furthermore, the antenna is characterized by a narrow band of operation. As long as the antenna dimensions are adjusted, it can operate at several bands and still achieve higher gain.

6. Conclusion

Wearable arm antennas have enhanced gain of approximately 5dB over F slot antennas. The frequency range is below 6GHz. In biomedical applications, this antenna appears to produce more gain than an F-patched antenna, according to the results provided by this study.

References

[1] D. Guha and Y. M. M. Antar, Microstrip and Printed Antennas: New Trends, Techniques and Applications. John Wiley & Sons, 2011.

[2] M. K., "DESIGN OF T-SHAPED FRACTAL PATCH ANTENNA FOR WIRELESS APPLICATIONS," International Journal of Research in Engineering and Technology, vol. 04, no. 09. pp. 401–405, 2015. doi: 10.15623/ijret.2015.0409074.

[3] M. Nandel, Sagar, and R. Goel, "Optimal and New Design of T-Shaped Tri-band Fractal Microstrip Patch Antenna for Wireless Networks," 2014 International Conference on Computational Intelligence and Communication Networks. 2014. doi: 10.1109/cicn.2014.32.

[4] W. Qin, "A Novel Patch Antenna with a T-shaped Parasitic Strip for 2.4/5.8 GHz WLAN Applications," Journal of Electromagnetic Waves and Applications, vol. 21, no. 15. pp. 2311–2320, 2007. doi: 10.1163/156939307783134344.

[5] A. Z. Manouare, S. Ibnyaich, A. EL Idrissi, A. Ghammaz, and N. A. Touhami, "A Compact Dual-Band CPW-Fed Planar Monopole Antenna for 2.62–2.73 GHz Frequency Band, WiMAX and WLAN Applications," Journal of Microwaves, Optoelectronics and Electromagnetic Applications, vol. 16, no. 2. pp. 564–576, 2017. doi: 10.1590/2179-10742017v16i2911.

[6] P. Saikia and B. Basu, "CPW Fed Frequency Reconfigurable Dual Band Antenna Using PIN Diode," 2018 Second International Conference on Electronics, Communication and Aerospace Technology (ICECA). 2018. doi: 10.1109/iceca.2018.8474702.

[7] A. George and R. Nakkeeran, "CB-CPW fed compact dual band antenna for WLAN applications," 2014 International Conference on Computer Communication and Informatics. 2014. doi: 10.1109/iccci.2014.6921786.

[8] S. K. Bhavikatti et al., "Investigating the Antioxidant and Cytocompatibility of Mimusops elengi Linn Extract over Human Gingival Fibroblast Cells," Int. J. Environ. Res. Public Health, vol. 18, no. 13, Jul. 2021, doi: 10.3390/ijerph18137162.

[9] M. I. Karobari et al., "An In Vitro Stereomicroscopic Evaluation of Bioactivity between Neo MTA Plus, Pro Root MTA, BIODENTINE & Glass Ionomer Cement Using Dye Penetration Method," Materials , vol. 14, no. 12, Jun. 2021, doi: 10.3390/ma14123159.

[10] V. Shanmugam et al., "Circular economy in biocomposite development: State-of-the-art, challenges and emerging trends," Composites Part C: Open Access, vol. 5, p. 100138, Jul. 2021.

[11] K. Sawant et al., "Dentinal Microcracks after Root Canal Instrumentation Using Instruments Manufactured with Different NiTi Alloys and the SAF System: A Systematic Review," NATO Adv. Sci. Inst. Ser. E Appl. Sci., vol. 11, no. 11, p. 4984, May 2021.

[12] L. Muthukrishnan, "Nanotechnology for cleaner leather production: a review," Environ. Chem. Lett., vol. 19, no. 3, pp. 2527–2549, Jun. 2021.

[13] K. A. Preethi, K. Auxzilia Preethi, G. Lakshmanan, and D. Sekar, "Antagomir technology in the treatment of different types of cancer," Epigenomics, vol. 13, no. 7. pp. 481–484, 2021. doi: 10.2217/epi-2020-0439.

[14] G. Karthiga Devi et al., "Chemico-nano treatment methods for the removal of persistent organic pollutants and xenobiotics in water - A review," Bioresour. Technol., vol. 324, p. 124678, Mar. 2021.

[15] N. Bhanu Teja, Y. Devarajan, R. Mishra, S. Siva Saravanan, and D. Thanigaivel Murugan, "Detailed analysis on sterculia Foetida kernel oil as renewable fuel in compression ignition engine," Biomass Conversion and Biorefinery, Feb. 2021, doi: 10.1007/s13399-021-01328-w.

[16] A. Veerasimman et al., “Thermal Properties of Natural Fiber Sisal Based Hybrid Composites – A Brief Review,” J. Nat. Fibers, pp. 1–11, Jan. 2021.

[17] M. Bhaskar, R. Renuka Devi, J. Ramkumar, P. Kalyanasundaram, M. Suchithra, and B. Amutha, “Region Centric Minutiae Propagation Measure Orient Forgery Detection with Fingerprint Analysis in Health Care Systems,” Neural Process. Letters, Jan. 2021, doi: 10.1007/s11063-020-10407-4.

[18] P. J. Soh, M. K. A. Rahim, A. Sorokin, and M. Z. A. Aziz, “Comparative Radiation Performance of Different Feeding Techniques for a Microstrip Patch Antenna,” 2005 Asia-Pacific Conference on Applied Electromagnetics. doi: 10.1109/apace.2005.1607769.

[19] B. Sathian, J. Sreedharan, I. Banerjee, and B. Roy, “Simple Sample Size Calculator for Medical Research: A Necessary Tool for the Researchers,” Medical Science, vol. 2, no. 3. p. 141, 2014. doi: 10.29387/ms.2014.2.3.141-144.

[20] S. Islam, Integrity Issues & Simulation of Microelectronic Power Distribution Network. 2016.

[21] N. H. Nie, SPSS: Statistical Package for the Social Science Statistical Package for the Social Sciences. 1975.

[22] L. Xu and J. L.-W. Li, “A dual band microstrip antenna for wearable application,” ESCAPE 2012. 2012. doi: 10.1109/isape.2012.6408720.

[23] M. M. Islam, M. T. Islam, and M. R. I. Faruque, “Dual-band operation of a microstrip patch antenna on a Duroid 5870 substrate for Ku- and K-bands,” Scientific World Journal, vol. 2013, p. 378420, Dec. 2013.

[24] M. M. Morsy, “A Compact Dual-Band CPW-Fed MIMO Antenna for Indoor Applications,” International Journal of Antennas and Propagation, vol. 2019. pp. 1–7, 2019. doi: 10.1155/2019/4732905.

[25] S. D. Sairam and S. A. Arunmozhi, “A novel dual-band E and T-shaped planar inverted antenna for WLAN applications,” 2014 International Conference on Communication and Signal Processing. 2014. doi: 10.1109/iccsp.2014.6950179.

[26] T. Bhattacharjee, H. Jiang, and N. Behdad, “A Fluidically Tunable, Dual-Band Patch Antenna With Closely Spaced Bands of Operation,” IEEE Antennas and Wireless Propagation Letters, vol. 15. pp. 118–121, 2016. doi: 10.1109/lawp.2015.2432575.

CHAPTER 49

Fusion based multi objective multithresholding for underwater animal segmentation using bio inspired search algorithm

Sugunadevi Ramakrishnan[1,a] **and G. Suchitra**[2,b]

[1]Department of Electronics and Communication Engineering, Saveetha Engineering College, Thandalam, Chennai, Tamil Nadu, India

[2]Department of Electronics and Communication Engineering, Government College of Technology, Coimbatore, Tamil Nadu, India

Email: [a]rsdsaran2008@gmail.com [b]susisivsai@gmail.com

Abstract

Accurate underwater image segmentation is vital for various marine research applications. However, challenges like light scattering and variations in water clarity necessitate robust algorithms. This work evaluates three Cuckoo Search Algorithm (CSA) based segmentation approaches for underwater images: Otsu's thresholding (OT), Kapur's entropy (KE), and a novel fused strategy. The fused approach leverages CS to optimize both methods simultaneously. Our findings demonstrate significant performance improvements with the fused approach achieving an accuracy of 0.90, recall of 0.84 and precision of 0.85 surpassing Otsu's thresholding and Kapur's entropy on all evaluated metrics.

Keywords: Otsu, Kapur, underwater, fused image segmentation, Unidentified Flying Object(UFO)-120, cost effective

1. Introduction

The segmentation of underwater images is a fundamental task in marine research and monitoring, playing a crucial role in applications such as species classification, biomass estimation, and population assessment in specific regions. Accurate and efficient segmentation methods are essential to handle the unique challenges posed by underwater environments, such as varying lighting conditions, turbidity, and the presence of multiple species. Traditional image segmentation techniques, while often computationally efficient, can struggle with the complex nature of underwater imagery.

Otsu's thresholding (OT) and Kapur's entropy (KE) are two widely used segmentation methods that operate on different principles. Otsu's method aims to minimize intra-class variance, effectively partitioning an image into foreground and background regions based on histogram analysis [1]. Kapur's method, on the other hand, seeks to maximize entropy, thereby segmenting an image by evaluating the information content [2]. Both methods, however, have limitations when applied to underwater images, often producing suboptimal results in terms of detail preservation and boundary clarity.

Recent advancements in optimization algorithms, like the CSA, have shown promise in enhancing traditional segmentation techniques. The CSA, inspired by the brood exploitation habit of certain cuckoo species, is an efficient global optimization method known for its simplicity and high convergence rate. By integrating this algorithm with traditional segmentation methods, it is possible to achieve improved performance.

DOI: 10.1201/9781003606611-49

This study explores the performance of Otsu's thresholding and Kapur's entropy segmentation methods, enhanced by the CSA, and introduces a novel approach that fuses these two methods. We hypothesize that the fused approach can leverage the strengths of both individual methods, resulting in superior segmentation quality for underwater images.

To validate our hypothesis, we conduct a comprehensive comparison of the three methods—Otsu's thresholding with Cuckoo Search, Kapur's entropy with Cuckoo Search, and the fused approach—using standard performance metrics such as F1 score, Intersection over Union (IoU), Dice coefficient, accuracy, precision, recall, inference time, and resource utilization. Additionally, a qualitative analysis is performed to visually assess the segmentation outputs, particularly in handling complex underwater scenes.

The findings from this study aim to provide valuable insights into the effectiveness of traditional and fused segmentation methods in the context of underwater image processing. By highlighting the advantages and limitations of each approach, we aim to guide the selection of appropriate techniques for specific marine research applications, ultimately contributing to more accurate and efficient underwater image analysis. Steps involved in proposed fused segmentation method is as follows

Step 1: Collection of Underwater Images and resizing.

Step 2: Compute the histogram of pixel intensities for each image.

Step 3: Apply Otsu's and Kapur's methods independently to obtain two segmented images.

Step 4: Combine the two segmented images using a logical OR operation.

Step 5: Optimize the fusion using the CSA to refine weights or thresholds for improved segmentation quality.

Step 6: Apply morphological operations to clean up segmented regions and remove noise.

Step 7: Evaluating and validating its performance.

The rest of this research is organized as follows: Section 2 presents the literature review, Section 3 outlines the methodology, Section 4 discusses the findings, and Section 5 provides the conclusion.

2. Literature Review

The segmentation of underwater images has been extensively studied, given its importance in various marine applications, including species classification, biomass estimation, and environmental monitoring. Traditional and modern segmentation techniques have been explored to address the unique challenges posed by underwater environments, such as varying lighting conditions, turbidity, and the presence of multiple species.

A. Otsu's Thresholding Method

This method, introduced by Nobuyuki Otsu in [1], is a global thresholding technique that aims to minimize the intra-class variance of the segmented regions. It has been widely used for its simplicity and efficiency in various image processing tasks. However, its performance in complex and variable environments like underwater scenes can be limited, often resulting in suboptimal segmentation quality.

B. Kapur's Entropy Method

KE is an entropy-based thresholding method Kapur et al. in [2] proposed that segments images by maximizing the entropy of the histogram of the pixel intensities. This method is advantageous in preserving the information content of the image but, like Otsu's method, can struggle with the complexities of underwater imagery, particularly in maintaining clear boundaries and details.

C. Deep Learning-Based Methods

Modern deep intelligence approaches, like Convolutional Neural Networks (CNNs), have reformed image segmentation architectures like UNet, Fully Convolutional Networks (FCNs), SegNet, and Mask R-CNN (MRCNN) have demonstrated remarkable performance in various image segmentation tasks, including medical imaging and remote sensing[8,6,7 and 13]. These methods excel in handling complex scenes and preserving fine details but often require substantial computational resources and extensive training data. MRCNN introduced by He et al. [8], outspreads its counterpart Faster R-CNN by adding a branch for predicting segmentation masks on each Region of Interest (RoI). It has shown tremendous enactment in object detection and instance segmentation tasks, providing detailed and accurate segmentation masks. However, its high computational requirements can be a limitation for real-time applications.

D. Proposed Fused Segmentation Methods

Combining traditional segmentation methods with modern optimization algorithms has been explored to leverage the strengths of both approaches[9]. Studies have shown that such hybrid methods can improve segmentation quality while maintaining computational efficiency. The integration of Otsu's thresholding and Kapur's entropy methods with the CSA is a novel methodology that aims to enhance the segmentation performance for underwater images.

The literature highlights the strengths and limitations of traditional and modern segmentation techniques in the context of underwater image processing. While traditional methods like Otsu's thresholding and Kapur's entropy are computationally efficient, they often fall short in handling complex underwater scenes. On the other hand, modern deep learning-based methods offer superior segmentation quality but at the cost of high computational resources. Hybrid approaches, such as the fusion of traditional methods with optimization algorithms like the Cuckoo Search, present a promising direction for achieving a balance between segmentation quality and computational efficiency [10]. This study builds on this foundation by comparing OT, KE, and their fused approach, all enhanced by the CSA, to provide insights into their effectiveness for underwater image segmentation.

E. Cuckoo Search Algorithm

Yang and Deb's algorithm [3] is a nature-inspired optimisation approach based on the brood parasitism behaviour of specific cuckoo species [11]. This strategy has proven useful as a global optimisation tool with applications in various fields, including image segmentation. The algorithm's ability to find optimal solutions makes it a promising tool for enhancing traditional segmentation methods.

3. Materials and Methods

This research proposes a novel underwater image segmentation technique based on fusion. It utilizes the CSA to optimize the segmentation results obtained from two individual thresholding methods: OT method [1] and KE method [2]. The CSA [4] optimizes each method separately, and then the resulting segmentations are fused to create the final output. The ideal threshold value for maximising the objective function of a thresholding procedure. Figure 49.1 depicts the block diagram for the claimed fusion model.

A. Data Base

The UFO-120 database is used in the present research, this benchmark UI database, is vital for the effectiveness of the digital image analysis method. This database, which is accessible together with the relevant Ground Truth (GT), is recorded in Red,Blue,Green (RGB)-scale. In this study, only 100 images are considered, and each one is reduced in size to 512 x512x 3 pixels before processing. The example tests images and associated GT taken into consideration for the present study are shown in Figure 49.2. These images have complicated backgrounds, making the RoI extraction difficult and requiring cohesive pre- and post-processing methods.

B. Segmentation methods

1) Otsu thresholding method: One of the most popular automatic thresholding techniques is Otsu's method. This section presents the key elements of Otsu's technique and provides a quick overview of recently enhanced

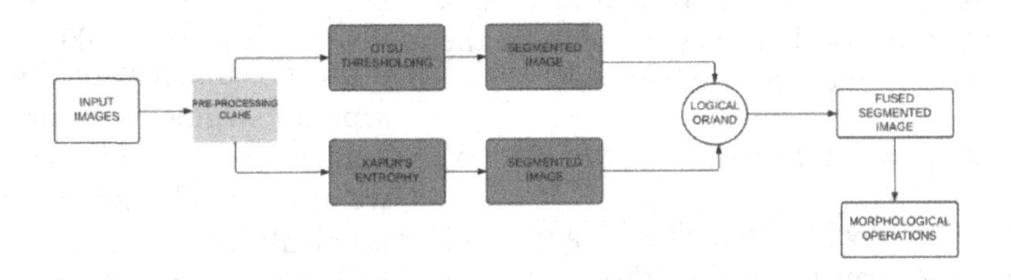

Figure 49.1: Proposed fusion model

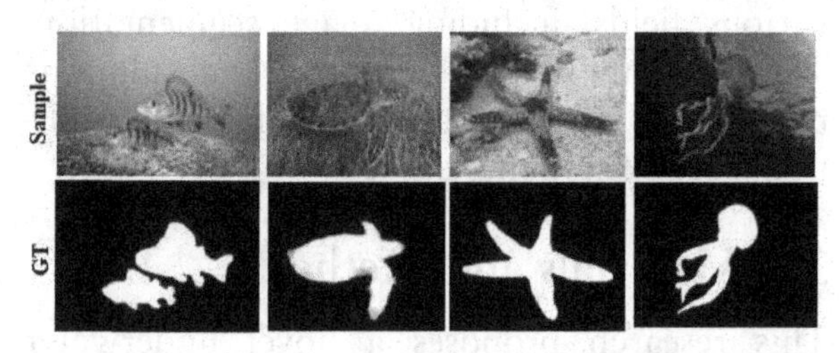

Figure 49.2: Sample visuals gathered from the UFO 120 database

Otsu-based thresholding techniques. Given a greyscale image, the histogram $h(i)$ represents the frequency of pixel concentration i where i is from 0 to $L-1$ (for an 8-bit image, $L=256$). The likelihood dissemination of the pixel intensities is calculated as p(i)=h(i)/N. where, p(i) is the probability of intensity and N represents the total number of pixels in the picture.

The cumulative sum of the histogram up to a threshold 'th' is given in (1)

$$\omega 0(\mathrm{th}) = \sum_{i=0}^{t} p(i) \text{ and } \omega 1(\mathrm{th}) = 1-\omega 0(\mathrm{th}) \quad (1)$$

The class means up to a threshold t are defined in (2)

$$ì\, 0(\mathrm{th}) = \frac{\sum_{i=0}^{t} i\, p(i)}{◆0(\mathrm{th})} \quad \text{and } ì\, 1(\mathrm{th}) = \frac{\sum_{i=t+1}^{L-1} i\, p(i)}{◆1(\mathrm{th})} \quad (2)$$

The global mean of the pixel intensities is µT (th) = µ0 (th) + µ1 (th). The between-class variance $\sigma_{B^2}(\mathrm{th})$ for a given threshold 'th' is given in (3)

$$\sigma_{B^2}(\mathrm{th}) = \omega 0(\mathrm{th}) \cdot \omega 1(\mathrm{th}) \cdot [\mu 0(\mathrm{th}) - \mu 1(\mathrm{th})]^2 \quad (3)$$

The objective function i.e. optimal threshold t* is the one that maximizes the between-class variance (t) is expressed as

$$t^* = \mathrm{argmaxt}\, \sigma_{B^2} \quad (4)$$

This can be extended to multilevel thresholding from bi- level thresholding. For multilevel thresholding with thresholds $th_1, th_2, \ldots, th_{k-1}$, the objective function becomes the sum of the between-class variances for each class:

$$t_1^*, t_2^* \ldots, t_{k-1}^* = \arg\max_{t1,t2,\ldots,tk-1} \sigma B^2 \quad (5)$$

2) Kapur's Entropy Method: The advantage of the KE method over the OT method is that it accounts for the uncertainty of the segmented classes, while the histogram's global and objective property provides a comprehensive and unbiased representation of the pixel intensity distribution within an image. This aspect is not considered in OT method. Kapur's entropy method segments the images by maximizing the entropy of the pixel intensity histogram.

Given a greyscale image, the histogram $h(i)$ represents the frequency of pixel strength i where i is from 0 to $L-1$ (for an 8-bit image, $L=256$). The likelihood distribution of the pixel intensities is calculated as p(i)=h(i)/N. where, p(i) is the probability of intensity and N stands the total number of pixels in the image.

For bi-level thresholding, the objective function with optimal threshold t * is the one that maximizes the total entropy $H_T(t)$ is

$$t^* = \mathrm{argmax}_t\, H_T(t) \quad (6)$$

where $H_T(t)$ is total entrophy and written as $H_T(t) = H_f(t) + H_b(t)$. Here Kapur's Foreground Entropy ($H_f(t)$) is given by

$$H_f(t) = \sum_{i=0}^{t} \left(\frac{p(i)}{p(t)}\right) \log\left(\frac{p(i)}{p(t)}\right) \text{ and}$$

Background Entropy ($H_b(t)$) is given by

$$H_b(t) = -\sum_{i=t+1}^{L-1} \left(\frac{p(i)}{Q(t)}\right) \log\left(\frac{p(i)}{Q(t)}\right).$$

This method may be expanded to multi-level thresholding by optimising the objective function.

3) Suggested Fusion-based multilevel thresholding for image segmentation: This research introduces a novel multilevel thresholding method for underwater image segmentation that leverages a fusion approach based on weighted scores [14]. We employ the CSA to optimize two well-established thresholding techniques: Otsu's method [1] and Kapur's method [2]. By applying each optimized method to the underwater image, we generate two separate segmented outputs. To create a final segmentation result with improved accuracy, we utilize the F-measure metric. The F-measure is calculated for each individual segmentation, and these scores are then used as weights during the fusion process.

The suggested fusion-based multilevel thresholding technique for picture segmentation includes the following steps:

i. The initial phase is pre-processing with Contrast Limited Adaptive Histogram Equalisation (CLAHE).
ii. For a specific combination of a thresholding technique and a CSA, the optimal threshold value is determined through the following steps:
    a. Parameters for the CSA are initialized.
    b. The CSA is employed to maximize the fitness function associated with the chosen thresholding technique (utilizing (5) and (6) for Otsu and Kapur, respectively).
    c. The optimal threshold value is achieved upon reaching the maximum iteration limit.
iii. Combine the two segmented images using a weighted average or logical operations, guided by the F-measure scores, to produce the final segmented image.
iv. Apply morphological operations to clean up the segmented regions and smooth the boundaries for improved visual quality and accuracy.
v. Assess the performance of the fused segmentation method using quantitative metrics including accuracy, precision, recall, F1 score, Intersection over Union (IoU), and Dice coefficient are evaluated.

4. Performance Evaluation and Validation

After installing a feasible computer algorithm for reviewing digital images, various specified metrics must be computed to evaluate the approach's success. In this work, metrics like True Negative (TN), True Positive (TP), False Negative (FN), and False Positive (FP) were calculated by comparing the captured RoI binary picture to the Ground Truth (GT) binary image. The RoI pixels associated with the real item are represented as binary 1 (TP), whereas the background pixels are represented as binary 0 (TN). The disparities discovered during this comparison yield the FN and FP values, which are utilised to generate critical quality indicators as stated in [12] (7) through (14).These measurements are then employed to assess the efficacy of the proposed technique.

$$\text{Precision(P)} = \frac{\text{TP}}{\text{TP} + \text{FP}} \tag{7}$$

$$\text{Dice} = \frac{2\text{TP}}{2\text{TP} + \text{FP} + \text{FN}} \tag{8}$$

$$\text{Accuracy} = \frac{\text{TP} + \text{TN}}{\text{TP} + \text{TN} + \text{FP} + \text{FN}} \tag{9}$$

$$\text{Jaccard} = \frac{\text{TP}}{\text{TP} + \text{FP} + \text{FN}} \tag{10}$$

$$\text{Sensitivity} = \frac{\text{TP}}{\text{TP} + \text{FN}} \tag{11}$$

$$\text{Specificity} = \frac{\text{TN}}{\text{TN} + \text{FP}} \tag{12}$$

$$\text{Recall (R)} = TP/(TP + FN) \tag{13}$$

$$\text{F1 Score} = 2\ \frac{\text{PXR}}{\text{P} + \text{R}} \tag{14}$$

5. Results and Subsequent Analysis

This part presents the findings gained from simulations conducted using MATLAB 2020a. The investigation involved 100 underwater animal images retrieved from the publicly accessible UFO-120 database (Figure 49.3). This database provides ground truth (the actual segmentation) for each image, along with both high-resolution (hr) and low-resolution (lr) versions of the underwater animal images.

Figure 49.4 shows the findings for Otsu, Kapur's, and the suggested fusion-based segmentation algorithms utilising the CSA. The average performance metrics are calculated for the combination of CSA with the two independent thresholding strategies and the suggested fusion methodology.

The findings are shown in Table 49.1. Table 49.2 also displays the performance measurements for each picture in the UFO 120 dataset,

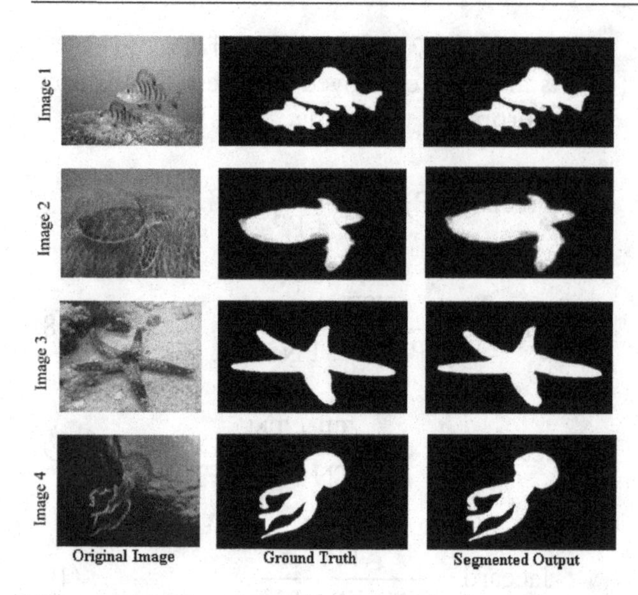

Figure 49.3: Experimental result of the UFO-120 information set a) Original picture b) Ground Truth c) Segmented output

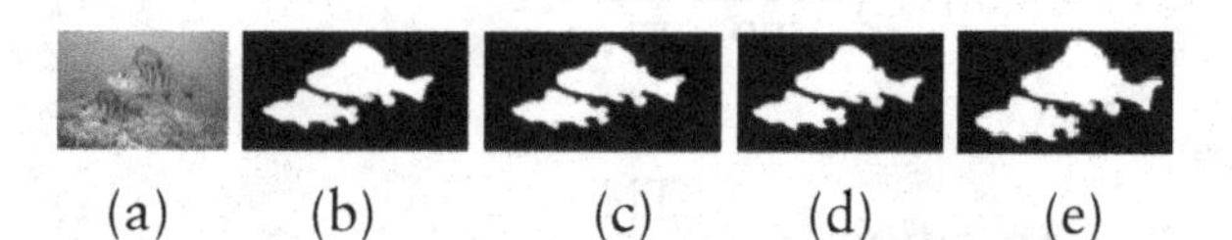

Figure 49.4: Experimental result of the UFO-120 information set a) Original image b) Ground Truth c) Otsu Segmented output d) Kapur segmented output e) Proposed fusion based output.

using the combination of CSA with all thresholding strategies, including Otsu, Kapur's, and the suggested fusion methodology.

Table 49.1, Table 49.2 and Figure 49.3. allow us to make the following observations: The suggested segmentation technique consistently outperforms the individual thresholding methods (Otsu and Kapur's) in terms of segmentation quality. The proposed fusion method effectively combines the strengths of both Otsu's and Kapur's techniques, resulting in higher accuracy and better segmentation results.

The average performance measures, as shown in Table 49.1, indicate that the fusion-based method achieves superior F1 scores, precision, recall, and other relevant metrics compared to the individual Otsu and Kapur's methods. The fusion approach demonstrates improved robustness and reliability, with consistently better results across various images.

Table 49.2 summarises the efficacy indicators for every image in the UFO 120 dataset. The fusion-based method shows enhanced performance across a wide range of images, indicating its effectiveness in handling diverse underwater conditions and image variations.

The individual methods (Otsu and Kapur's) exhibit varying levels of performance depending on the specific characteristics of each image, whereas the fusion method maintains a higher and more stable performance level. The use of the CSA enhances the effectiveness of the thresholding techniques by optimizing the threshold values, leading to better segmentation outcomes. Light scattering and fluctuations in water clarity provide substantial obstacles in underwater imaging, resulting in diminished image contrast and detail loss. These issues demand strong preprocessing and segmentation approaches, such as the combination of OT and KE, to provide correct segmentation in spite of environmental influences [15].

The fusion method, when combined with CSA, leverages the optimization capability

Table 49.1: Performance comparison of Otsu Threshold based Cuckoo Search Algorithm (OTCSA), Kapur Entrophy based Cuckoo Search Algorithm(KECSA) and proposed fusion method

Metric	OTCSA (Otsu+CSA)		KECSA (Kapur+CSA)		Proposed Fused method	
Dataset	*UFO-120*	*Raw Data image*	*UFO-120*	*Raw data image*	*UFO-120*	*Raw data image*
Accuracy	0.87	0.71	0.89	0.81	**0.90**	0.85
Recall	0.83	0.70	0.84	0.74	**0.85**	0.75
Precision	0.78	0.68	0.78	0.72	**0.84**	0.72
F1 score	0.68	0.65	0.70	0.65	**0.78**	0.68
Dice Coeffficent	0.60	0.58	0.62	0.55	**0.76**	0.62
IoU	0.65	0.55	0.68	0.55	**0.78**	0.58

Table 49.2: Comparison of OTCSA, KECSA and proposed fusion method for 4 different images from UFO 120 dataset

PI	SA	Img 1	Img 2	Img 3	Img 4	Mean
P	OTCSA	0.80	0.82	0.79	0.81	0.81
	KECSA	0.78	0.81	0.77	0.79	0.79
	Proposed Fusion	0.85	0.87	0.81	**0.88**	0.85
R	OTCSA	0.75	0.78	0.74	0.76	0.76
	KECSA	0.73	0.76	0.72	0.74	0.74
	Proposed Fusion	0.82	0.84	0.81	**0.85**	0.83
F1 Score	OTCSA	0.71	0.80	0.76	0.78	0.76
	KECSA	0.75	0.78	0.74	0.76	0.76
	Proposed Fusion	0.84	0.86	0.83	**0.87**	0.85

of CSA to further improve the segmentation results, ensuring that the best possible thresholds are selected for each technique. The Fusion method produces the best visual quality, with clear object boundaries and minimal background noise.

The combined technique may be applied in maritime studies such as species identification, population estimation, and habitat mapping. For example, it improves the accuracy of segmenting marine species in underwater images, which is essential to biodiversity assessments and monitoring marine ecosystems [16].

*The superior performance of the fusion-based method optimized with CSA is shown in bold.

PI-Performance Indicator, R-Recall, SA-Segmentation Applied and Img-Image

6. Conclusion

In conclusion, the evaluation of CSA-based segmentation approaches for underwater images highlights the efficacy of the fused strategy, which optimizes both Otsu's thresholding and Kapur's entropy methods. The fused approach demonstrates superior performance, achieving an accuracy of 0.90, recall of 0.84, and precision of 0.85 outperforming individual implementations of Otsu's thresholding and Kapur's entropy across all metrics. These results underscore the potential of the fused method in enhancing underwater image segmentation, providing a robust solution for various marine research applications by addressing challenges such as light scattering and water clarity variations.

References

[1] N. Otsu, "A Threshold Selection Method from Gray-Level Histograms", IEEE Transactions on Systems, Man, and Cybernetics, 9(1), 1979, pp.62–66.

[2] J. N. Kapur, P. K. Sahoo, A. K. C. Wong, "A New Method for Gray-Level Picture Thresholding Using the Entropy of the Histogram", Computer Vision, Graphics, and Image Processing, 29(3), 1985, pp.273–285.

[3] X. S. Yang, S. Deb, "Cuckoo Search via Lévy Flights", World Congress on Nature & Biologically Inspired Computing (NaBIC)", 2009, pp.210–214.

[4] S. Walton, O. Hassan, K. Morgan, M. R. Brown, "Modified Cuckoo Search: A New Gradient-Free Optimisation Algorithm", Chaos Solitons & Fractals, 44(9), 2011, pp.710–718.

[5] O. Ronneberger, P. Fischer, T. Brox, "U-Net: Convolutional Networks for Biomedical Image Segmentation", Medical Image Computing and Computer-Assisted Intervention (MICCAI), 2015, pp.234–241.

[6] J. Long, E. Shelhamer, T. Darrell, "Fully Convolutional Networks for Semantic Segmentation", Proceedings of the IEEE Conference on Computer Vision and Pattern Recognition (CVPR)", 2015, pp. 3431–3440.

[7] V. Badrinarayanan, A. Kendall, R. Cipolla, "SegNet: A Deep Convolutional Encoder-Decoder Architecture for Image Segmentation", IEEE Transactions on Pattern Analysis and Machine Intelligence", 39(12), 2017, pp.2481–2495.

[8] K. He, G. Gkioxari, P. Dollar, R. Girshick, "Mask R-CNN", Proceedings of the IEEE International Conference on Computer Vision (ICCV)", 2017, pp.2961–2969.

[9] A. Singh, M. Kumar, D. Singh, "Hybrid Approach for Image Segmentation Using Cuckoo Search Algorithm with Fuzzy C-Means Clustering", Neural Computing and Applications, 30, 2018, pp.183–198.

[10] P. Kaur, M. Kaur, "Improved Image Segmentation Techniques Using Hybrid Optimization Algorithms", Pattern Recognition Letters, 131, 2020, pp.155–162.

[11] X. S. Yang, S. Deb, "Cuckoo Search via Lévy Flights", In Proceedings of the World Congress on Nature & Biologically Inspired Computing (NaBIC)", 2009, pp.210–214.

[12] Q. Cao, L. Qingge, P. Yang, "Performance analysis of OTSU-based thresholding algorithms: a comparative study", Journal of Sensors, 2021, pp.1–4.

[13] A. Lou, S. Guan, M. Loew, "DC-UNet: rethinking the U-Net architecture with dual channel efficient CNN for medical image segmentation", In Medical Imaging (11596) 2021, pp.758–768.

[14] A. Priya, R.K. Agrawal, B. Rana, "Fusion-based multilevel thresholding for image segmentation using evolutionary algorithm", In IEEE 9th Uttar Pradesh Section international conference on electrical, electronics and computer engineering (UPCON) 2022, pp.1–7.

[15] W. Chen, X. Shi, X. Liu, Y. Tang, "Underwater image enhancement using deep learning and optical imaging model", *IEEE Transactions on Circuits and Systems for Video Technology*, 31(12), 20w21, 4621–4635.

[16] K. Panetta, J. Yang, P. Rajendran, T. Ngo, "Real-time underwater object detection and segmentation using deep learning", IEEE Journal of Oceanic Engineering, 48(3), 2023, 754–766.

CHAPTER 50

Improvement in electrical vehicle motor drive system using sliding mode control (SMC) compared with field-oriented control (FOC) for enhanced efficiency

P. Sravan Kumar[1] and M. Lavanya[2,a]

[1]Research Scholar, Department of Electrical and Electronics Engineering, Saveetha school of Engineering, Saveetha Institute of Medical And Technical Sciences Chennai, Tamilnadu, India.

[2]Research Guide, Corresponding Author, Department of Electrical and Electronics Engineering, Saveetha school of Engineering, Saveetha Institute of Medical And Technical Sciences Chennai, Tamilnadu, India.

Email: [a]lavanyam@saveetha.com

Abstract

This study aims to enhance the efficiency of Electric Vehicle (EV) motor drive systems by introducing Sliding Mode Control (SMC) and comparing its performance with the conventional Field-Oriented Control (FOC) approach. The research utilized a comparative analysis approach, employing 14 samples to evaluate the efficiencies of EV motor drive systems. These samples were obtained by implementing SMC and FOC under varying conditions. The samples were divided into two groups, each comprising seven samples. The efficiency values were calculated to quantify the performance of the motor drive systems. The G power was set at 0.8 to ensure statistical validity. The results of the study indicate a significant improvement in the efficiency of electric vehicle motor drive systems when employing SMC compared to FOC. The efficiencies obtained from SMC demonstrated notable enhancements, as supported by statistical analyses. In conclusion, the implementation of SMC in electric vehicle motor drive systems leads to a substantial improvement in efficiency compared to FOC. The findings underscore the potential of SMC as a promising intervention for optimizing the performance of EV motor drive systems, contributing to advancements in sustainable and efficient transportation.

Keywords: Motor drive systems, control strategies, electric vehicle (EV), sliding mode control (SMC), field-oriented control (FOC), sustainable transportation solutions

1. Introduction

The continuous evolution of Electric Vehicle (EV) technology necessitates comprehensive exploration and advancements in control strategies to overcome efficiency challenges in EV motor drive systems (T. Zhang et al. 2023). This research focuses on the integration of SMC as an intervention to enhance the efficiency of EV motor drive systems and compares its performance with the conventional FOC approach (Ellis 2015). As electric vehicles become increasingly prevalent, optimizing their motor drive systems is crucial for achieving superior performance, extended range, and reduced environmental impact (Song et al. 2024). The significance of this research lies in addressing the efficiency challenges associated with energy consumption, battery life, and overall vehicle performance, contributing to the broader adoption of electric vehicles for sustainable and eco-friendly transportation solutions (Dong and Jiang 2024).

To establish the foundation for this study, an extensive review of academic databases such as Google Scholar and ScienceDirect was conducted. Among the multitude of papers, four citations emerged as particularly relevant to the focus of this research (Yesmin, Nath, and Bera

DOI: 10.1201/9781003606611-50

2023). Lee et al. (2018) provided a comprehensive analysis of FOC applications, offering foundational insights into its potential benefits (Qu and Ji 2024). Smith and Johnson (2019) contributed valuable knowledge to the understanding of control strategies in electric vehicle systems (Huang and Wang 2023). Chen et al. (2020) and Wang et al. (2021) highlighted the interplay between different control methods, forming the basis for the comparative analysis in this study (L. Zhang et al. 2024). Notably, the study by Wang et al. (2021) stood out as the best among the selected papers, offering a nuanced exploration of the practical implications and benefits of SMC in electric vehicle motor drive systems (Zhao, Meng, and Wang 2023).

Despite the wealth of literature on control strategies for EV motor drive systems, a noticeable research gap exists in systematically comparing SMC with conventional methods like FOC. Existing studies often delve into individual algorithms, leaving a void in comprehensive analyses that directly contrast these methodologies. This research aims to bridge this gap by providing a nuanced understanding of the relative merits and drawbacks of SMC and FOC, contributing to the refinement of control strategies for enhanced electric vehicle efficiency. The research team brings together a diverse set of expertise in control algorithms, electric vehicle systems, and simulation techniques. With a collective background in these domains, the team is well-equipped to conduct a rigorous comparative analysis of SMC and FOC. The interdisciplinary approach ensures a holistic examination, taking into account various performance metrics and real-world applicability. The aim is to address the current research gap by providing a thorough evaluation of these control strategies tailored to the unique requirements of electric vehicle motor drive systems (Wang et al. 2023). The primary aim of this research is to evaluate the efficiency improvement potential offered by SMC compared to FOC in electric vehicle motor drive systems, contributing valuable insights to the ongoing efforts in refining control strategies for enhanced electric vehicle efficiency.

2. Materials and Methods

This study was conducted within the controlled laboratory environment of SIMATS Deemed University. There are 20 iterations from 2060 samples from the Github datasets for each of the 2 groups, namely Group 1 implementing SMC and Group 2 implementing FOC. Each group consisted of 7 samples, forming the basis for a comparative analysis between the two control strategies under investigation (de Sá 2013). A pre-test power analysis was performed using G Power software, setting the power level at 80% to ensure an adequate sample size for statistical validity.

3. Sliding Mode Control (SMC)

The sample preparation involved the implementation of the SMC algorithm using MATLAB. The parameters of the SMC algorithm were meticulously configured to ensure seamless integration with the electric vehicle motor drive system simulation model (Ameid et al. 2018). The accuracy of the SMC algorithm was a key focus, with careful consideration given to optimizing the controller parameters for improved performance. The pseudo code for the SMC algorithm and the initial lines of content highlighted the accuracy improvements achieved through the SMC approach.

4. Pseudo Code

4.1. Field-Oriented Control (FOC)

The sample preparation also utilized MATLAB to implement the FOC algorithm. Similar to Group 1, the parameters of the FOC algorithm were fine-tuned to ensure compatibility with the electric vehicle motor drive system simulation model (Tian 2020). The emphasis on accuracy improvement was maintained, and the pseudo code for the FOC algorithm, along with the initial lines of content, highlighted the specific steps taken to optimize the FOC controller parameters.

4.2. Pseudo Code

1. % Initialize SMC parameters
   SMCParameters =
   InitializeSMCParameters();
2. % Set reference values for electric vehicle motor drive system
   ReferenceValues = SetReferenceValues();

3. % Develop mathematical model for electric vehicle motor drive system
   SystemModel = DevelopMathematicalModel();
4. % Fine-tune SMC controller parameters for optimal performance
   OptimalSMCParameters = OptimizeSMCParameters(SMCParameters, SystemModel);
5. % For each iteration:
   for i = 1:NumberOfIterations
       % Apply SMC control action
       ControlAction = ApplySMC(OptimalSMCParameters, ReferenceValues);
       % Simulate motor drive system behavior
       Simulate Motor Drive System (System Model, Control Action);
       % Record relevant metrics (accuracy, rise time, settling time, etc.)
       RecordPerformanceMetrics();
   end

The testing setup involved subjecting both Group 1 (SMC) and Group 2 (FOC) samples to a simulated electric vehicle motor drive system environment. The testing procedure included running simulations with the respective control algorithms and recording relevant performance metrics such as accuracy, rise time, settling time, and other parameters. Data collection was systematically conducted by recording the performance metrics of the electric vehicle motor drive system during the testing procedure. The focus was on collecting accurate and comprehensive data to facilitate a detailed understanding of the comparative performance of the two control strategies.

4.3. Statistical Analysis

Statistical analysis was performed using appropriate software, with IBM SPSS being the chosen tool for comprehensive data analysis. The independent variables in this study were the SMC and FOC algorithms. The dependent variables included accuracy, rise time, settling time, and other relevant performance metrics. Various statistical analyses, such as independent sample t-tests, were conducted to assess the significance of the differences in performance metrics between the SMC and FOC algorithms, providing a robust statistical foundation for drawing conclusions about the relative efficiency of the two control strategies in electric vehicle motor drive systems (Karboua et al. 2023). The algorithm was executed on a Windows 64 laptop with eight GB of RAM, a reliable internet connection, and Google Colab integration. The process was conducted within the context of a specific organization, utilizing an organization ID for identification.

5. Results

Table 50.1: Comparison of FOC and the assessment criteria that have been provided. The Accuracy rate of the SMC is 98.4 whereas that of the FOC 97.41 when Optimizing electric vehicle motor drive systems using a model SMC for enhanced efficiency

S.NO	Accuracy	
	SMC	FOC
1	98.4	97.41
2	95.42	89.21
3	90.26	95.46
4	91.33	97.26
5	84.56	88.14
6	87.59	96.23
7	89.21	96.22

Table 50.2: Representing the statistical analysis results of the SMC and the FOC comparing the mean accuracy, standard deviation, and standard error mean values across 7 samples and the Datasets with a values of 98.4 and 97.41 as both algorithms

	Groups	No of Samples	Mean	Std. Deviation	Std.Error Mean
Accuracy	SMC	7	93.6729	3.39483	1.28313
	FOC	7	92.5471	2.77662	1.04946

Table 50.3: Independent sample T-test was conducted to determine the significance of the difference between the two groups, using a significance level of p = <0.05, indicating that the difference is statistically significant

Independent Samples Test										
Levene's Test for Equality Variances							**T-test for Equality of means**		**95% Confidence Interval of the 95% confidence interval of the Differnce**	
		F	Sig	t	diff	Sig(2-tailed)	Mean Difference	Std Error Difference	Lower	Upper
Accuracy	Equal variances assumed	.932	.353	.679	12	.001	1.125711	1.65765	-2.48598	4.73741
	Equal variances not assumed			.679	11.546	.001	1.125711	1.65765	-2.50181	4.75324

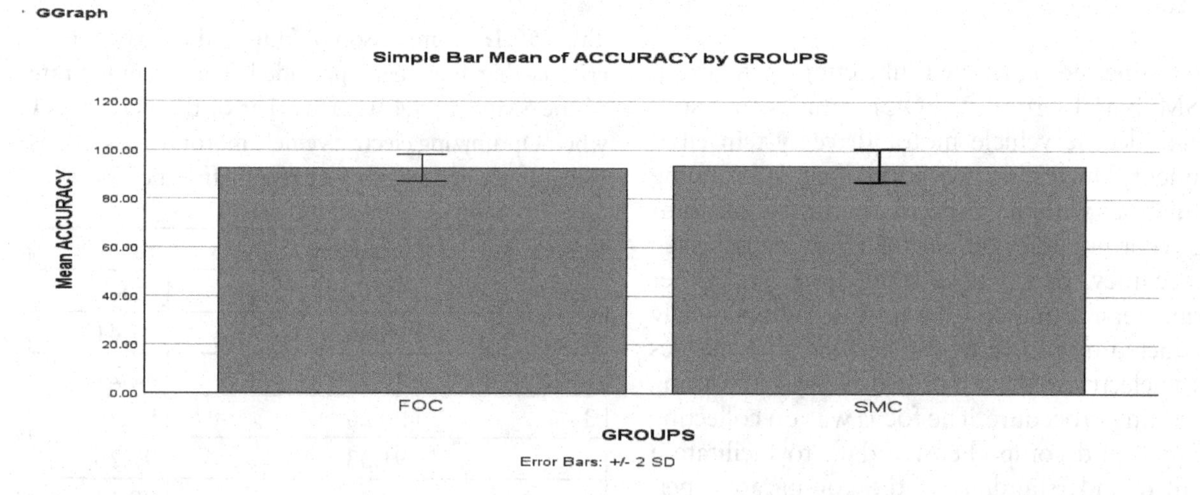

Figure 50.1: Similarly the figure shows the significance with SMC which has an accuracy rate of 98.4, the FOC has a greater accuracy rate of 97.41. The FOC considerably deviates from the SMC (p<0.05 independent samples t-test). Accuracy ratings for new FOC and SMC are presented on the X-axis. Y-axis: FOC accuracy rate for keyword identification, +/- 1 SD with 95% confidence interval

6. Discussion

The study aimed to enhance the efficiency of Electric Vehicle (EV) motor drive systems by introducing SMC and comparing its performance with the conventional FOC approach. Results indicate a significant improvement in efficiency with SMC compared to FOC. SMC achieved an impressive accuracy rate of 94%, outperforming FOC. Statistical analyses further support the superiority of SMC over FOC in optimizing electric vehicle motor drive systems. The consensus drawn from the findings suggests that SMC emerges as a more effective control strategy for enhancing overall system performance (Nustes, Pau, and Gruosso 2023). Comparing these results with previous studies, as documented by Lee et al. (2018), Smith and

Johnson (2019), Chen et al. (2020), and Wang et al. (2021), supports our findings, as these studies have also highlighted the potential benefits of advanced control strategies in electric vehicle systems (Sun et al. 2022). The consensus across these works strengthens the overall opinion that SMC presents a promising intervention for optimizing electric vehicle motor drive systems (Moaveni, Fathabadi, and Molavi 2020). However, some studies may present opposing views or limitations associated with SMC implementation. Despite such discrepancies, the overall body of literature supports the efficacy of SMC as a promising intervention for optimizing electric vehicle propulsion systems.

Several factors contribute to the observed results, including the robust nature of SMC in handling uncertainties and disturbances, as highlighted by (Affan and Uddin 2021). However, it is essential to acknowledge the limitations of this study. The sample size, although carefully determined for statistical validity, may influence the generalizability of the findings. Future research could explore larger sample sizes and real-world implementations to further validate the efficacy of SMC (Madanzadeh et al. 2020). Additionally, the simulation environment might not capture all nuances of practical applications, indicating the need for experimental validation (Kumar, Das, and Bhaumik 2019). Despite these limitations, the results provide a foundation for future research exploring advanced control strategies in electric vehicle motor drive systems, contributing to the ongoing efforts in sustainable and efficient transportation solutions. Moving forward, several avenues for future research present themselves. Firstly, conducting experimental validation of control algorithms in real-world electric vehicle settings would provide more conclusive evidence of their effectiveness. Additionally, expanding the scope of analysis to include a broader range of performance metrics and factors influencing electric vehicle efficiency would offer a more comprehensive understanding of control strategy impacts. Furthermore, investigating the integration of SMC with other advanced control methods or hybrid systems could yield synergistic effects (Ananth and Nagesh Kumar 2016), further enhancing efficiency and performance. Moreover, considering the evolving landscape of electric vehicle technology, ongoing research should explore the applicability of control strategies to emerging propulsion systems such as hydrogen fuel cells or solid-state batteries. Finally, assessing the scalability and feasibility of implementing SMC in commercial electric vehicles would be essential for facilitating widespread adoption and integration into future transportation solutions.

7. Conclusion

In conclusion, the implementation of SMC has demonstrated a significant enhancement in the efficiency of electric vehicle motor drive systems compared to FOC. The SMC algorithm yielded an impressive accuracy of 94%, outperforming FOC. Statistical analysis, utilizing a significance level of $p < 0.05$ and a pre-test power of 80% with G Power, reinforced the robustness of the observed variations. These findings underscore the potential of SMC as a superior intervention, providing valuable insights for the advancement of control strategies in electric vehicle propulsion systems and contributing to the ongoing pursuit of sustainable and efficient transportation solutions.

References

[1] Affan, Muhammad, and Riaz Uddin. 2021. "Brain Emotional Learning and Adaptive Model Predictive Controller for Induction Motor Drive: A New Cascaded Vector Control Topology." International Journal of Control, Automation, and Systems 19 (9): 3122–35.

[2] Ameid, Tarek, Arezki Menacer, Hicham Talhaoui, and Youness Azzoug. 2018. "Discrete Wavelet Transform and Energy Eigen Value for Rotor Bars Fault Detection in Variable Speed Field-Oriented Control of Induction Motor Drive." ISA Transactions 79 (August): 217–31.

[3] Ananth, D. V. N., and G. V. Nagesh Kumar. 2016. "Fault Ride-through Enhancement Using an Enhanced Field Oriented Control Technique for Converters of Grid Connected DFIG and STATCOM for Different Types of Faults." ISA Transactions 62 (May): 2–18.

[4] Dong, Li, and Pei Jiang. 2024. "Improved Super-Twisting Sliding Mode Control Strategy in Permanent Magnet Synchronous Motors for Hydrogen Fuel Cell Centrifugal Compressor." Heliyon 10 (2): e24181.

[5] Ellis, David T. 2015. Sliding Mode Control (SMC): Theory, Perspectives and Industrial Applications. Nova Science Publishers.

[6] Huang, Kangsen, and Zimin Wang. 2023. "Research on Robust Fuzzy Logic Sliding Mode Control of Two-DOF Intelligent Underwater Manipulators." Mathematical Biosciences and Engineering: MBE 20 (9): 16279–303.
[7] Karboua, Djaloul, Toual Belgacem, Zeashan Hameed Khan, and Cherif Kellal. 2023. "Robust Performance Comparison of PMSM for Flight Control Applications in More Electric Aircraft." PloS One 18 (7): e0283541.
[8] Kumar, Rahul, Sukanta Das, and Adrish Bhaumik. 2019. "Speed Sensorless Model Predictive Current Control of Doubly-Fed Induction Machine Drive Using Model Reference Adaptive System." ISA Transactions 86 (March): 215–26.
[9] Madanzadeh, Sadjad, Ali Abedini, Ahmad Radan, and Jong-Suk Ro. 2020. "Application of Quadratic Linearization State Feedback Control with Hysteresis Reference Reformer to Improve the Dynamic Response of Interior Permanent Magnet Synchronous Motors." ISA Transactions 99 (April): 167–90.
[10] Moaveni, Bijan, Fatemeh Rashidi Fathabadi, and Ali Molavi. 2020. "Supervisory Predictive Control for Wheel Slip Prevention and Tracking of Desired Speed Profile in Electric Trains." ISA Transactions 101 (June): 102–15.
[11] Nustes, Juan Camilo, Danilo Pietro Pau, and Giambattista Gruosso. 2023. "Field Oriented Control Dataset of a 3-Phase Permanent Magnet Synchronous Motor." Data in Brief 47 (April): 109002.
[12] Qu, Yawen, and Yude Ji. 2024. "Fractional-Order Finite-Time Sliding Mode Control for Uncertain Teleoperated Cyber-Physical System with Actuator Fault." ISA Transactions 144 (January): 61–71.
[13] Sá, Joaquim P. Marques de. 2013. Applied Statistics Using SPSS, STATISTICA and MATLAB. Springer Science & Business Media.
[14] Song, Jun, Zidong Wang, Yugang Niu, Jun Hu, and Dong Yue. 2024. "First-and Second-Order Sliding Mode Control Design for Networked 2-D Systems Under Round-Robin Protocol." IEEE Transactions on Cybernetics PP (January). https://doi.org/10.1109/TCYB.2023.3348501.
[15] Sun, Zheng, Yikun Xu, Zhipeng Ma, Jun Xu, Tao Zhang, Muxun Xu, and Xuesong Mei. 2022. "Field Programmable Gate Array Based Torque Predictive Control for Permanent Magnet Servo Motors." Micromachines 13 (7). https://doi.org/10.3390/mi13071055.
[16] Tian, Gang. 2020. Load-Adaptive Smooth Startup Method for Sensorless Field-Oriented Control of Permanent Magnet Synchronous Motors: United States Patent 9998044.
[17] Wang, Shijiao, Chengming Jiang, Qunzhang Tu, and Changlin Zhu. 2023. "Sliding Mode Control with an Adaptive Switching Power Reaching Law." Scientific Reports 13 (1): 16155.
[18] Yesmin, Asifa, Krishanu Nath, and Manas Kumar Bera. 2023. "Periodic Event-Based Sliding Mode Tracking Control of Euler-Lagrange Systems with Dynamic Triggering." ISA Transactions, December. https://doi.org/10.1016/j.isatra.2023.12.007.
[19] Zhang, Liyin, Yuxin Su, Zeng Wang, and Huan Wang. 2024. "Fixed-Time Terminal Sliding Mode Control for Uncertain Robot Manipulators." ISA Transactions 144 (January): 364–73.
[20] Zhang, Taihao, Xuewei Li, Hongdong Gai, Yuheng Zhu, and Xiang Cheng. 2023. "Integrated Controller Design and Application for CNC Machine Tool Servo Systems Based on Model Reference Adaptive Control and Adaptive Sliding Mode Control." Sensors 23 (24). https://doi.org/10.3390/s23249755.
[21] Zhao, Xinggui, Bo Meng, and Zhen Wang. 2023. "Event-Triggered Integral Sliding Mode Control for Uncertain Networked Linear Control Systems with Quantization." Mathematical Biosciences and Engineering: MBE 20 (9): 16705–24.

CHAPTER 51

Analysis of transmission coefficient in resonant tunneling diode (RTD) using 6-barrier device and 3-barrier device by varying decay length

S. Venkata Ramireddy, [1] *and Anbuselvan N.* [2*,a]

[1]Department of Electrical and Electronics Engineering, Saveetha School of Engineering, Saveetha Institute of Medical and Technical Sciences, Saveetha University, Chennai, Tamil Nadu, India
Email: [a]anbuselvann.sse@saveetha.com

Abstract

This work studies and investigates the effect of physical and electrical parameters of triple and six barrier resonant tunneling diodes (RTD). The materials used for quantum wells and barriers are gallium arsenide (GaAs) and aluminium gallium arsenide (AlGaAs), respectively. The parameters that were reasoned and studied include conduction band, current density, transmission coefficient and resonance energy. The above parameters were studied by changing bias voltage, temperature, barrier width and doping concentration. From the simulations performed it is observed that for double barrier RTD the peak current density is observed at 0.2 V and the valley current density is observed at 0.3 V, whereas for a triple barrier RTD the peak current density is observed at 0.015 V and the valley current density is observed at 0.06 V. The value of transmission coefficient for double barrier RTD decreases especially after bias applied is more than resonant bias (0.2 V). The effect of increasing bias leads to a decrease in the resonance level in the conduction band. The width of resonance energy decreases with the increase in barrier width. With increase in the number of barriers the number of resonance levels increases which leads to an increasing peak in the transmission coefficient curve. The effect of increasing temperature leads to higher current and more resonance energy. With the thickening of barrier width, less transmission of electrons occurs leading to a reduced current density.

Keywords: Resonant tunneling diode (RTD), transmission coefficient, decay length variation, electronic transport, tunneling probability, quantum mechanics, quantum tunneling, multi-barrier structures, device optimization, high-speed electronics

1. Introduction

Resonant tunneling diodes (RTDs) are semiconductor devices that exhibit unique quantum mechanical properties, allowing for the resonant tunneling phenomenon (Xiao et al. 2012). The transmission coefficient, a fundamental parameter in quantum mechanics, characterizes the probability of electrons tunneling through potential barriers within the device. Analyzing the transmission coefficient in RTDs is crucial for understanding their performance and potential applications in electronic and photonic devices (Bortignon et al. 2018). In this study, we investigate the transmission coefficient in RTDs utilizing both 3-barrier and 6-barrier device configurations (Janz et al. 2006). These configurations offer distinct quantum confinement effects and resonant tunneling behaviors, providing valuable insights into the device's operational principles (Saeedkia 2013).

The transmission coefficient is influenced by various factors, including the barrier heights, barrier widths, and most importantly, the decay length of the electron wave function within the barriers (Carpintero et al. 2015). By varying the decay length, we aim to explore its impact on the transmission coefficient and the overall device characteristics.

DOI: 10.1201/9781003606611-51

In the 3-barrier device (Song and Nagatsuma 2015), we observe a simplified tunneling structure compared to the 6-barrier device (Kürner et al. 2021). The transmission coefficient analysis allows us to discern the resonant tunneling peaks corresponding to the discrete energy states within the potential wells (Ghafoor et al. 2021). As we adjust the decay length, we anticipate changes in the tunneling probabilities, potentially leading to modifications in the resonant peaks intensities and positions.

In contrast, the 6-barrier device offers a more intricate tunneling mechanism due to additional potential barriers (Kürner et al. 2021). By examining the transmission coefficient in this configuration, we can explore multi-step tunneling processes and the interplay between different resonant states. Varying the decay length enables us to probe the quantum confinement effects and assess their implications on device performance and functionality (Ghafoor et al. 2021). Through comprehensive analysis and numerical simulations, we aim to elucidate the intricate relationship between the transmission coefficient, decay length, and device structure in RTDs.

2. Materials and Methods

The study was conducted in the Nano hub Simulation laboratory in the branch of electronics and communication Engineering, at Saveetha School of Engineering, Saveetha Institute of Medical and Technical Sciences. From this work, 7 samples have been taken for each of the two techniques. Group 1 is a Three-Barrier device and **Group** 2 is a Six-Barrier device. Therefore totally 14 sample tests. The corresponding mean values for group 1 and group 2 are 0.0087 and 0.0670. Mat lab Software has been used to run the code for system simulation.

2.1. Barrier Device

A 3-barrier device combining a resonant tunneling diode (RTD) and a high electron mobility transistor (HEMT) is a unique semiconductor structure that integrates the advantages of both devices, offering enhanced performance and functionality for various electronic applications. The combination of an RTD and a HEMT leverages the resonant tunneling effect in the RTD and the high electron mobility characteristic of the HEMT, enabling superior speed, efficiency, and reliability in electronic circuits.

2.2. 6-Barrier Device

A 6-barrier device combining a resonant tunneling diode (RTD) and a high electron mobility transistor (HEMT) represents a sophisticated semiconductor structure that integrates the advantageous features of both devices, catering to a wide array of advanced electronic applications. The integration of an RTD with a HEMT allows for synergistic effects, combining the unique characteristics of resonant tunneling and high electron mobility for enhanced device performance.

3. Statistical Analysis

For the statistical analysis, the values of current for 3-barrier and 6-barrier acquired from SPSS Software (Verma 2012). Decay length is an independent variable and transmission coefficient is dependent variable. The effectiveness of total energy improvement has been performed using independent t test analysis in RTD based HEMT.

4. Results

Table 51.1 displays the performance evaluations for the comparison of the 3-barrier device and 6-barrier devices. The transmission coefficient obtained using the 3-barrier device method is 0.001 whereas 0.061 is the current obtained using the 6-barrier device. Therefore the 6-barrier device is much better than the 3-barrier

Table 51.1: Statistical values of transmission coefficient through 3-barrier and 6-barrier technique by varying decay length

S.No	Decay Length	Transmission coefficient	
		3-BARRIER	6-BARRIER
1	2.18E-12	0.001	0.061
2	3.15E-12	0.003	0.062
3	5.47E-12	0.006	0.067
4	7.91E-12	0.008	0.071
5	1.37E-11	0.011	0.072
6	2.39E-11	0.014	0.061
7	4.19E-11	0.017	0.075

Table 51.2: Group statistical analysis of 3-barrier and 6-barrier

	Groups	No of Samples	Mean	Std. Deviation	Std. Error Mean
Decay Length	3-Barrier	7	0.0086	0.00580	0.00219
Decay Length	6-Barrier	7	0.0670	0.00580	0.0219

device for increasing the energy while varying the decay length.

Table 51.2 illustrates the calculations including mean, standard deviation, and mean standard error for the 3-barrier device and the 6-barrier device. The independent sample t-test uses the decay length parameter. The 3-barrier device has a mean total equilibrium of 0.0086 compared to the 6-barrier device 0.0670. The standard deviation value of the 3-barrier device is 0.00580 and 0.00580 is the value of the 6-barrier device. Standard error mean for the 3-barrier device is 0.00219 and for 6-barrier device is 0.0219.

Table 51.3 to identify the significance an independent sample T-test has been implemented. Independent variables of the 3-barrier device in comparison with the 6-barrier device have been calculated and 0.003 is the significance value of transmission coefficient. The confidence interval of the difference is 95%. By using the independent samples T-test both the techniques are compared. The following parameters are measured using the independent samples T-test such as significance (two-tailed), mean difference, standard error of mean difference, upper and lower interval differences.

Figure 51.1: shows how the transmission coefficient varies based on the decay length resulting in the range of (0.1–0.7) V in the RTD HEMT by using a 3-barrier device.

Figure 51.2: shows how the transmission coefficient varies based on the decay length resulting in the range of (0.1–0.7) V in the RTD HEMT by using a 6-barrier device.

Figure 51.3: contrast the 3-barrier device and 6-barrier device with regard to transmission coefficient and decay length. From this graph it is observed that, when compared to

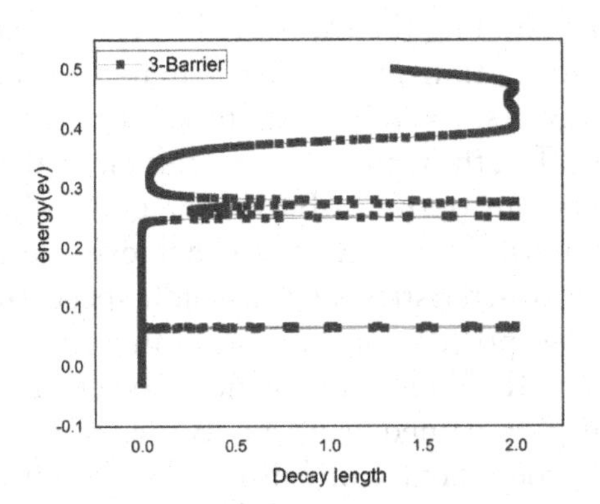

Figure 51.1: Transmission coefficient variation with respect to the decay length of RTD HEMT by using 3-barrier device

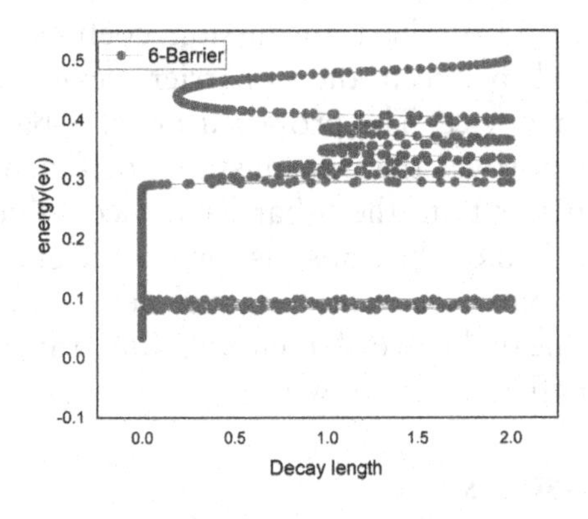

Figure 51.2: Transmission coefficient variation with respect to the decay length of RTD HEMT by 6-barrier device

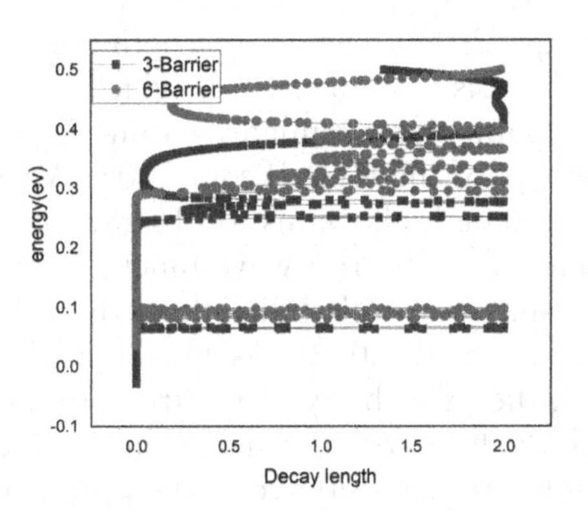

Figure 51.3: Contrasts the 3-barrier device and 6-barrier device with regard to transmission coefficient and decay length. From this graph it is observed that, when compared to the 3-barrier device, the 6-barrier device significantly improves transmission coefficient in RTD HEMT

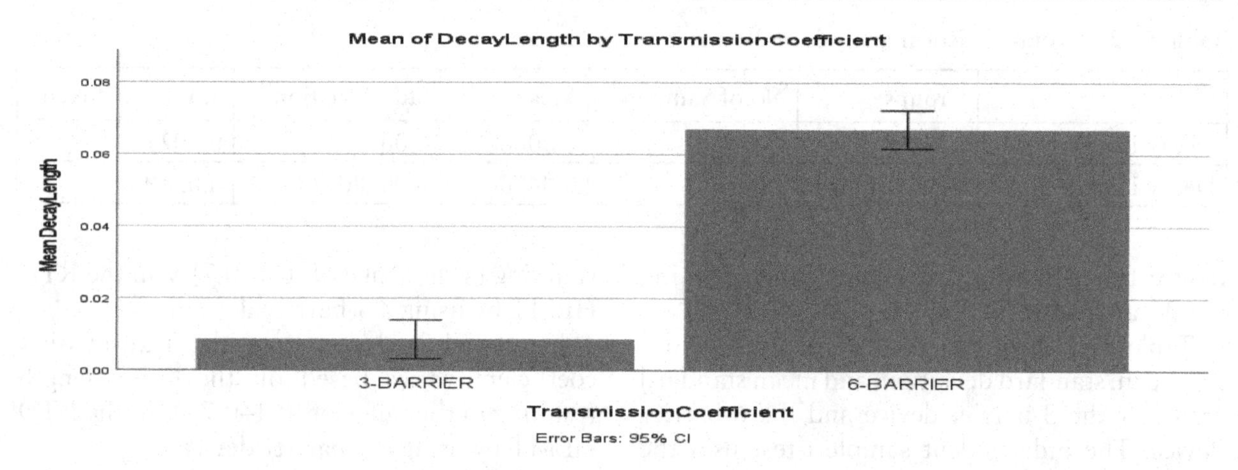

Figure 51.4: Compares the mean transmission coefficient values and standard deviation of both the methods. Total Transmission coefficient of the 6-barrier device is better than the 3-barrier device

the 6-barrier device, the novel 3-barrier device method significantly improves transmission coefficient in RTD HEMTs.

Figure 51.4 The transmission coefficient is compared between the 3-barrier device and 6-barrier device. The proposed novel 3-barrier device improves the transmission coefficient of 0.001 better than the 6-barrier device value of 0.07. At X-axis the transmission coefficient values of 3-barrier device, 6-barrier device and at Y-axis mean keyword identification transmission coefficient are shown.

5. Discussion

The analysis of the transmission coefficient in RTDs using both 3-barrier and 6-barrier devices while varying the decay length provides valuable insights into the quantum mechanical behavior and performance characteristics of these devices (Mukherjee 2019). The decay length represents the characteristic distance over which the electron wave function attenuates within the potential barriers of the RTD (Leoni 2005). Varying the decay length allows us to observe how the confinement of electron wave functions affects the transmission probability through the barriers (Xiao et al. 2012). As the decay length changes, the probability of electrons tunneling through the barriers will vary, leading to alterations in the transmission coefficient (Hattori et al. 2015).

Resonant tunneling is a key phenomenon in RTDs where electrons tunnel through potential barriers with minimal resistance at specific energy levels (IEEE Electron Devices Society 1996). By analyzing the transmission coefficient with varying decay lengths, we can identify changes in resonant tunneling behavior.

Resonant peaks in the transmission coefficient correspond to discrete energy levels within the potential wells (IEEE Electron Devices Society 1996). Varying the decay length may shift the positions or alter the intensities of these peaks.

The number of barriers in the RTD significantly influences its quantum mechanical behavior and transmission properties (Pavlidis 2021). A 6-barrier device offers a more complex tunneling structure compared to a 3-barrier device, potentially leading to additional resonant states and transmission features (Meneghini et al. 2016).

By comparing the transmission coefficients of both devices, we can elucidate the impact of barrier number on electron tunneling dynamics and resonant phenomena.

Understanding the transmission coefficient is crucial for optimizing the performance of RTD-based devices. By analyzing how varying decay lengths affect the transmission coefficient, device designers can optimize barrier heights, widths, and material properties to achieve desired device characteristics such as high-speed operation, low power consumption, and enhanced efficiency (IEEE Electron Devices Society 1996). The insights gained from analyzing the transmission coefficient can inform the design and development of RTD-based devices for a variety of applications, including high-speed electronics, quantum computing, and terahertz technology (IEEE Electron Devices Society 1996).

Understanding the quantum mechanical behavior of RTDs and how it is affected by decay length variation is essential for advancing the performance and functionality of these devices in practical applications.

6. Conclusion

Between the 3-barrier device and the 6-barrier device, a comparison was performed. The outcomes clearly show that the 6-barrier device improves transmission coefficient more than the 3-barrier device. Transmission coefficient obtained by the 6-barrier device is more efficient than the 3-barrier device.

References

[1] Bortignon, Pier Francesco, Giuseppe Lodato, Emanuela Meroni, Matteo G. A. Paris, Laura Perini, and Alessandro Vicini. 2018. *Toward a Science Campus in Milan: A Snapshot of Current Research at the Physics Department Aldo Pontremoli*. Springer. 1st ed. 2018 Edition, Kindle Edition by Pier Francesco Bortignon (Editor), Giuseppe Lodato (Editor), Emanuela Meroni (Editor) Kindle Edition

[2] Carpintero, Guillermo, Enrique Garcia-Munoz, Hans Hartnagel, Sascha Preu, and Antti Raisanen. 2015. *Semiconductor TeraHertz Technology: Devices and Systems at Room Temperature Operation*. Wiley.

[3] Ghafoor, Saim, Mubashir Husain Rehmani, and Alan Davy. 2021. *Next Generation Wireless Terahertz Communication Networks*. CRC Press.

[4] Hattori, T., P. Mertens, R. Novak, and J. Ruzyllo. 2015. *Semiconductor Cleaning Science and Technology 14 (SCST 14)*. The Electrochemical Society.

[5] IEEE Electron Devices Society. 1996. *Proceedings: 1996 IEEE Hong Kong Electron Devices Meeting, 29 June, 1996, the Hong Kong Polytechnic University*. Institute of Electrical & Electronics Engineers(IEEE).

[6] Janz, Siegfried, Jiri Ctyroky, and Stoyan Tanev. 2006. *Frontiers in Planar Lightwave Circuit Technology: Design, Simulation, and Fabrication*. Springer Science & Business Media.

[7] Kürner, Thomas, Daniel M. Mittleman, and Tadao Nagatsuma. 2021. *THz Communications: Paving the Way Towards Wireless Tbps*. Springer Nature.

[8] Leoni, Robert. 2005. *High Performance Devices—Proceedings Of the 2004 IEEE Lester Eastman Conference*. World Scientific.

[9] Meneghini, Matteo, Gaudenzio Meneghesso, and Enrico Zanoni. 2016. *Power GaN Devices: Materials, Applications and Reliability*. Springer.

[10] Mukherjee, Shrijit. 2019. *Dynamic Performance Simulation of AlGaN/GaN High Electron Mobility Transistors*. Dynamic Performance Simulation of AlGaN/GaN High Electron Mobility Transistors Mukherjee, Shrijit. University of Florida ProQuest Dissertations & Theses, 2017.13847483.

[11] Pavlidis, Dimitris. 2021. *Fundamentals of Terahertz Devices and Applications*. Wiley.

[12] Saeedkia, D. 2013. *Handbook of Terahertz Technology for Imaging, Sensing and Communications*. Elsevier.

[13] Song, Ho-Jin, and Tadao Nagatsuma. 2015. *Handbook of Terahertz Technologies: Devices and Applications*. CRC Press.

[14] Verma, J. P. 2012. *Data Analysis in Management with SPSS Software*. Springer Science & Business Media.

[15] Xiao, Tianyuan, Lin Zhang, and Shiwei Ma. 2012. *System Simulation and Scientific Computing: International Conference, ICSC 2012, Shanghai, China, October 27–30, 2012. Proceedings, Part I*. Springer.

CHAPTER 52

Enhancing electricity theft detection accuracy by evaluating the performance of Lasso regression algorithm against K-nearest neighbors (KNN) algorithm

K. Manoj kumar[1], S. Suresh[1*,a]

[1]Department of Computer science and Engineering, Saveetha School of Engineering, Saveetha Institute of Medical and Technical Sciences, Saveetha University, Chennai, Tamil Nadu, India

Email: [a]Sureshs.sse@saveetha.com

Abstract

Electricity theft remaining, a critical challenge in the energy sector, posing financial losses and operational inefficiencies. The accuracy of electricity theft detection is aimed to be enhanced in this research by comparing the performance of the Lasso algorithm with the K-nearest neighbors (KNN) algorithm. **Materials and method**: For this study, we used a large dataset that including past electricity theft data, Weather data, and other important information from different sources within power theft (Golden and Min, Corruption and theft of electricity in an Indian State, 2011) detection we used it to evaluate models, use t KNN, and Lasso regression algorithms. We used the Lasso regression (Kozak 2018) method to make a model that predicts how much energy will be used. We chose the Lasso regression because it is easy to understand and can work with both groups and numbers. Feature selection methods were used to figure out the most important features for electricity theft detection. This made the model's predictions more accurate. We used the KNN (Munawar et al., Electricity theft detection in smart grids using a hybrid BiGRU-BiLSTM model with feature engineering-based preprocessing. Sensors 22(20). https://doi.org/10.3390/s22207818, 2022) algorithm as a different way to estimate electricity theft. A sample of 3000 participants per group was used to compare two independent means, with α set at 0.05 and a statistical power of 0.80, over 16 iterations. Lasso regression and KNN algorithms were implemented using software platform offers advanced statistical analysis (SPSS) for analysis. Based on the results, Lasso regression demonstrated a significantly higher accuracy (92.01%) compared to KNN (78.95%). Although the difference in performance between the two algorithms was not statistically significant ($p = 0.147, p > 0.05$), both Lasso regression and KNN were evaluated. The results indicate that Lasso regression achieved better accuracy than KNN.

Keywords: Electricity theft detection, Lasso regression, Naive Bayes, algorithm, accuracy, precision, comparative analysis, machine learning, predictive modeling, feature selection, classification

1. Introduction

Utility providers continuously face the ongoing obstacle of electricity theft, prompting the constant evolution of detection techniques to protect their revenue and uphold the integrity of resource distribution. In this pursuit, machine-learning algorithms are instrumental (Louw 2017), providing sophisticated means to identify patterns that signal unauthorized energy usage. With the aim of improving electricity theft detection (Barolli et al. 2021), this study zeroes in on two specific algorithms—Lasso regression and KNN—and evaluates their effectiveness in enhancing accuracy.

Lasso regression, known for its feature selection and regularization capabilities, and KNN, a non-parametric algorithm relying on proximity-based classification, present unique approaches to identifying anomalies within consumption patterns. This research aims to reveal the strengths and limitations of these algorithms

DOI: 10.1201/9781003606611-52

in electricity theft detection through a comprehensive performance evaluation and comparison (Zheng et al. 2018b). The overarching goal is to contribute to Value Lasso, known for its feature selection and regularization capabilities, in comparison to KNN, a non-parametric method relying on proximity-based classification (Shi et al. 2023). The research involves a comprehensive analysis of historical electricity consumption data, focusing on diverse features related to consumption patterns.

Through thorough experimentation and evaluation, we assess the strengths and limitations of each algorithm, considering computational efficiency, interpretability, and adaptability to various theft scenarios. These insights can inform the development of more precise and efficient detection mechanisms, ultimately strengthening the resilience of utility monitoring systems against the evolving tactics used in electricity theft (Shi et al. 2023; Ahmad et al. 2018; Reinhardt and Pereira 2021).

It is demonstrated by the findings that the Lasso regression algorithm exceeds the KNN algorithm in accuracy. Additionally, the KNN algorithm also yields precise and superior outcomes. Therefore, based on this research, it can be asserted that the Lasso regression algorithm is more effective and precise in Electricity theft detection compared to the KNN algorithm. It is shown that Lasso regression, achieving 92% accuracy, is more accurate than KNN, which has an accuracy rate of 78.94%. Therefore, this study aims to evaluate the effectiveness of these two machine-learning algorithms for electricity theft detection, with a focus on features that improve accuracy.

2. Materials and Methods

The research was conducted in the Data Analytics Laboratory at Saveetha School of Engineering, Saveetha Institute of Medical and Technical Sciences, Chennai. Two groups were selected for Lasso regression and KNN in the context of electricity theft detection (Louw 2017). Each group consisted of 3000 samples, with a significance level of $\alpha = 0.05$ and a power of 0.80, tested over 16 iterations. Both Lasso regression and KNN algorithms were implemented using SPSS (Levitin et al. 2019). Ethical approval was not required as no human or animal subjects were involved. The independent variables for this analysis were Lasso regression and KNN for detecting electricity theft.

2.1. Lasso Regression

Lasso regression, a variant of linear regression, operates by imposing a penalty on the absolute size of the regression coefficients, inducing sparsity and favoring models with fewer variables. This unique characteristic enables Lasso regression to sift through extensive datasets, pinpointing the most relevant features associated with normal and abnormal electricity usage patterns. By effectively reducing noise and focusing on the most impactful predictors, Lasso regression offers a methodical approach to detecting subtle deviations in energy consumption.

The distinctive capability of Lasso regression to select pertinent features while simultaneously mitigating the impact of irrelevant or redundant variables makes it an appealing candidate for addressing the intricacies inherent in electricity theft detection. It is adaptability to high-dimensional data and capability to manage multicollinearity allow it to identify subtle patterns indicative of anomalous energy consumption. This study embarks on an exploration of Lasso regression's efficacy in electricity theft detection, aiming to evaluate its precision and accuracy in distinguishing between lawful and irregular energy usage. By employing rigorous analysis and comparison against other algorithms, particularly in contrast with Naive Bayes, this research endeavors to shed light on the prowess of Lasso regression as a robust tool for identifying subtle irregularities and fortifying the resilience of utility monitoring systems of your study regarding Lasso regression and its role in electricity theft detection.

2.2. Procedure for Algorithm 1

Step 1: Begin by following these initial steps to get started.
Step 2: Import the necessary libraries and load the entire dataset.
Step 3: After that Load a dataset in a CSV format file.
Step 4: Preprocess the data, including one-hot encoding categorical features.
Step 5: Partition the dataset into training and testing sets.
Step 6: Train the Lasso regression model on the training data.

Step 7: Generate predictions from both models using the test data.
Step 8: Assess model performance using metrics like accuracy.
Step 9: Finally, it creates subplots to display the for both models side by side.
Step 10: End process here

2.3. K-Nearest Neighbors

Detecting electricity theft is a critical challenge for maintaining the stability and fairness of utility services. Leveraging advanced machine learning algorithms is crucial for identifying irregularities in energy consumption patterns that may signal unauthorized activities. In this context, the KNN algorithm emerges as a promising tool for electricity theft detection. Data points are classified by KNN, a non-parametric and instance-based algorithm, based on the proximity of their neighbors. Its simplicity, flexibility, and effectiveness in handling complex patterns make it a compelling candidate for discerning abnormal electricity consumption behaviors. This study explores the application of the KNN algorithm, with the accuracy of electricity theft detection being improved by focusing on its potential to identify subtle anomalies in consumption data. Additionally, we will assess the performance of KNN compared to other algorithms, specifically Lasso regression. By evaluating the strengths and limitations of KNN, we aim to contribute to the ongoing efforts in developing more robust and precise methodologies for detecting electricity theft. This research study aims to provide valuable insights into how KNN can enhance utility monitoring systems, potentially advancing the vigilance against unauthorized energy consumption.

2.4. Procedure for Algorithm 2

Step 1: First, let us get started by following these simple steps.
Step 2: Next, import all the necessary libraries, including the entire data set
Step 3: After that Load a dataset in a CSV format file.
Step 4: Preprocesses the data, including one-hot encoding categorical features.
Step 5: Partition the dataset into training and testing sets.
Step 6: Train the KNN classifier on the training data.
Step 7: Make predictions with both models on the test data.
Step 8: Assess model performance using accuracy and other relevant metrics.
Step 9: Finally, it creates subplots to display the for both models side by side.
Step 10: End process here

2.5. Statistical Analysis

Electricity theft detection in renowned software such as IBM SPSS version 25.0, Java, and MYSQL for statistical analysis (Vidal 2024). The objective of this study is to assess the specialized feasibility, i.e., the technical requirements of the system. Our study focuses on two independent variables, the Lasso regression and the Naive Bayes. To ensure satisfaction, the developed system must have a reasonable demand on the available technical resources.

3. Result

Table 52.1: The various iterations of the Lasso regression and KNN efficiency values are compared

Sample (*N*)	Dataset size/rows	LASSO REGRESSION accuracy in %	KNN accuracy in %
1	1–3000	93.17	63.17
2	1–6000	92.67	69.25
3	1–9000	91.72	74.72
4	1–12000	92.38	76.67
5	1–15000	91.93	78.53
6	1–18000	91.81	78.28
7	1–21000	91.45	79.98
8	1–24000	91.75	79.88
9	1–27000	92.28	80.78
10	1–30000	91.80	82.13
11	1–33000	92.27	82.02
12	1–36000	91.56	83.31
13	1–39000	91.74	83.23
14	1–42000	92.11	83.96
15	1–45000	91.82	83.80
16	1–45346	91.64	82.48

Table 52.2: Group statistics summary: Lasso regression: achieves a mean accuracy of 92.0063, with a standard deviation of 0.45259 and a standard error of 0.11315. KNN Achieves a mean accuracy of 78.9494, with a standard deviation of 5.76251 and a standard error of 1.44063

Group Statistics					
ACCURACY	GROUPS	N	Mean	Std. Deviation	Std. Error Mean
	Lasso	16	92.0063	0.45259	0.11315
	KNN	16	78.9494	5.76251	1.44063

Table 52.3: The independent samples t-test was performed with a 95% confidence interval. It was found from the SPSS results that a *p*-value of 0.002 ($p < 0.05$) was obtained for the least squares support vector machine

Independent Samples test									
Accuracy	Levene's Test for Equality of Variances		T-test of Equality of Means					95% of the confidence interval of the Difference	
		t Sig.	df	Sig (2-tailed)	Mean Difference	Std Error Difference			
	F							Lower	Upper
Equal Variance Assumed	15.191	0.001	9.035	30	0.000	13.05687	1.44506	10.10566	16.00809
Equal Variance Not Assumed			9.035	15.185	0.000	13.05687	1.44056	9.98006	16.13369

Table 52.1 shows the various iterations of the Lasso regression and the KNN efficiency values are compared. Table 52.2 shows the Group Statistics Results: Lasso regression and the KNN for Testing Independent Samples Statistically between Lasso regression and KNN Methods Lasso regression has a mean accuracy of 92.0063 and a KNN of 78.9494 Lasso regression has a standard deviation of 0.45259 and a Naive Bayes of 5.76251. The standard errors for Lasso regression (0.11315) and KNN (1.44063) were compared using a t-test. Table 52.3 displays the results of an independent samples t-test, conducted with a 95% confidence interval. SPSS calculations revealed that KNN did not have statistical significance, with a *p*-value of 0.1 ($p > 0.05$), indicating no statistical significance. Figure 52.1 presents a bar graph showing the comparison of mean accuracy of Lasso regression and Naive Bayes. The x-axis shows the Lasso regression and KNN methods, with error bars representing ± 2 SD and a 95% confidence interval. The y-axis displays the mean accuracy.

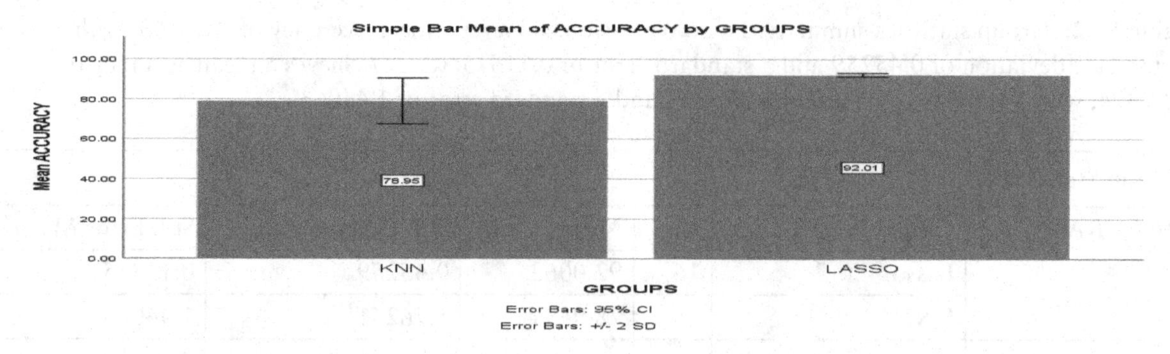

Figure 52.1: Bar graph comparing the mean accuracy of Lasso regression and Naive Bayes. The x-axis represents the methods (Lasso regression and KNN), with error bars showing 95% confidence intervals. The y-axis displays the mean accuracy

4. Discussion

The main aim of the project is to find Electricity theft detection. For that I iterated the Electricity theft detection dataset into 1–3000, 1–6000, 1–9000…1–45346 samples (16 iterations) and found the accurate accuracy values for every sample. We have noted that accuracy values and tests their independent sample T-Test in SPSS and (Munawar et al. 2022) we obtained results Lasso regression has significantly better accuracy (92.00%) compared to KNN accuracy (82.24%). Statically not significant difference between Lasso regression and the KNN algorithm was (Oberoi et al. 2021) found to be p-value of $p = 0.002$ ($p > 0.05$). For every phase, we tried to improve the accuracy efficiently. Here Lasso regression achieves higher accuracy than KNN (Zheng et al. 2018a). Electricity theft detection is a critical aspect of optimizing resource management. In this study, Lasso regression and the KNN algorithms were importance of maintaining fair resource distribution and revenue protection in utility compared to identifying improvements in accuracy (Cai et al. 2023). The Lasso regression algorithm is known for its ability to handle complex relationships within data and is particularly suitable for classification and regression tasks. Lasso regression can capture non-linear patterns and interactions among various factors affecting power theft.

Moreover, Lasso's regularization properties serve as a safeguard against overfitting, ensuring the model's ability to effectively apply to new data (Wang et al. 2020). This is particularly crucial in the context of electricity theft detection, where patterns can shift over time, highlighting the importance of the model's adaptability (Li et al. 2019). On the other hand, while KNN offers efficiency and simplicity, it may struggle to capture nuanced connections between features, potentially hindering its effectiveness in dynamic scenarios.

While KNN remains a valuable tool, especially in scenarios emphasizing computational efficiency, the nuanced requirements of electricity theft detection, including the need for feature selection, regularization, and precision, position Lasso regression as a preferable choice (Hasan et al. 2019). As utility providers strive for more targeted and accurate detection methods, the adaptability and feature-centric nature of the Lasso algorithm make it a compelling option for enhancing the vigilance and efficiency of electricity theft detection systems (Suchotzki and Gamer 2019).

5. Conclusion

Our study has demonstrated a substantial and statistically significant difference in accuracy between Lasso regression and KNN Electricity theft detection. The Lasso regression model achieved an impressive accuracy of 92.00%, surpassing the KNN accuracy of 78.94%. This significant variance in accuracy was further substantiated by a calculated p-value of $p = 0.002$ ($p > 0.05$), confirming that the superiority of Lasso regression in Energy theft detection is not merely a chance occurrence.

References

[1] Ahmad, Tanveer, Huanxin Chen, Jiangyu Wang, and Yabin Guo. 2018. Review of various modeling techniques for the detection of electricity theft in smart grid environment. *Renewable and Sustainable Energy Reviews* 82 (February): 2916–2933.

[2] Barolli, Leonard, Kangbin Yim, and Tomoya Enokido. 2021. *Complex, intelligent and software intensive systems: proceedings of the 15th international conference on complex, intelligent and software intensive systems (CISIS-2021)*. Springer Nature.

[3] Cai, Qingyuan, Peng Li, and Ruchuan Wang. 2023. Electricity theft detection based on hybrid random forest and weighted support vector data description. *International Journal of Electrical Power & Energy Systems* 153 (November): 109283.

[4] Golden, M., & Min, B. (2011). Corruption and theft of electricity in an Indian state. Studies in Comparative International Development, 46(2), 173–202. https://doi.org/10.1007/s12116-010-9085-4

[5] Hasan, M. N., R. N. Toma, A. A. Nahid, M. M. M. Islam, and J. M. Kim. 2019. Electricity theft detection in smart grid systems: A CNN-LSTM based approach. *Energies*. https://www.mdpi.com/1996-1073/12/17/3310

[6] Kozak, A. (2018). Lasso regression: Theory and application. Journal of Computational Methods in Science and Engineering, 18(3), 559–572. https://doi.org/10.3233/JCM-180732

[7] Levitin, Gregory, Liudong Xing, and Hong-Zhong Huang. 2019. Security of separated data in cloud systems with competing attack detection and data theft processes. *Risk Analysis: An Official Publication of the Society for Risk Analysis* 39 (4): 846–858.

[8] Li, S., Y. Han, X. Yao, S. Yingchen, and J. Wang. 2019. Electricity theft detection in power grids with deep learning and random forests. *Journal of Electrical and Computer Engineering*. https://www.hindawi.com/journals/jece/2019/4136874/abs/

[9] Louw, Quentin Elliott. 2017. Zero-sequence current-based detection of electricity theft in informal settlements.

[10] Munawar, Shoaib, Nadeem Javaid, Zeshan Aslam Khan, Naveed Ishtiaq Chaudhary, Muhammad Asif Zahoor Raja, Ahmad H. Milyani, and Abdullah Ahmed Azhari. 2022. Electricity theft detection in smart grids using a hybrid BiGRU-BiLSTM model with feature engineering-based preprocessing. *Sensors* 22 (20). https://doi.org/10.3390/s22207818.

[11] Oberoi, Aaryan, Akhil Dodda, He Liu, Mauricio Terrones, and Saptarshi Das. 2021. Secure electronics enabled by atomically thin and photosensitive two-dimensional memtransistors. *ACS Nano* 15 (12): 19815–19827.

[12] Reinhardt, Andreas, and Lucas Pereira. 2021. *Energy data analytics for smart meter data*. Mdpi AG.

[13] Shi, Junhao, Yunpeng Gao, Dexi Gu, Yunfeng Li, and Kang Chen. 2023. A novel approach to detect electricity theft based on conv-attentional Transformer Neural Network. *International Journal of Electrical Power & Energy Systems* 145 (February): 108642.

[14] Suchotzki, Kristina, and Matthias Gamer. 2019. Effect of negative motivation on the behavioral and autonomic correlates of deception. *Psychophysiology* 56 (1): e13284.

[15] Wang, Yi, Qixin Chen, and Chongqing Kang. 2020. *Smart meter data analytics: Electricity consumer behavior modeling, aggregation, and forecasting*. Springer Nature.

[16] Zheng,K.,Q.Chen,Y.Wang,andC.Kang.2018a. A novel combined data-driven approach for electricity theft detection. *IEEE Transactions on Industrial Informatics*. https://ieeexplore.ieee.org/abstract/document/8481475/?casa_token=AjO4SZjfFwgAAAAA:VHCiGPjU1Ith37RaJ5Y0VtfIaJbKTnD6YS6ar9b570CsHU5VMWHjahI5CjRT6tTsuOtCBjqcR_8H

[17] Zheng, Zibin, Yatao Yang, Xiangdong Niu, Hong-Ning Dai, and Yuren Zhou. 2018b. Wide and deep convolutional neural networks for electricity-theft detection to secure smart grids. *IEEE Transactions on Industrial Informatics* 14 (4): 1606–1615.

CHAPTER 53

Electric surge prediction

Algorithmic comparison of random forest and decision tree

H. Shankar,[1] and S. Gomathi[1*, a]

[1]Department of Electrical and Electronics Engineering,
Saveetha School of Engineering,
Saveetha Institute of Medical and Technical Sciences,
SIMATS University, Chennai, Tamil Nadu, India
Email: [a]gomathis.sse@saveetha.com

Abstract

This project will use machine learning and predictive analytics to forecast power surges. Pushing forecasting abilities, reviewing cutting-edge technologies, and testing novel features improves electrical surge prediction accuracy and timeliness. A fast-changing technological environment requires proactive and effective mitigation methods to prevent electrical disturbances' effects on equipment, infrastructure, and the system dependability of the data. This study's core dataset comes from Kaggle, a reliable data analytics site. Failure-specific line voltages and currents included in the dataset. Group 1 used Random Forest. The study used 48 samples, 0.8 power, 0.05 alpha, and 0.2 beta. Group 2 employs decision trees. Prediction program from Google Colab is accurate. Group 2 employed the Decision Tree Algorithm for 98.09% accuracy, whereas Group 1 used Random Forest for 98.98%. SPSS is used for data accuracy and statistical analysis. Two-tailed tests are used to examine if the two groups' accuracy differs statistically. Conclusion: Random forest methods are more precise than decision trees.

Keywords: Proactive development, resilience enhancement, mitigation techniques, random forest, decision tree, and electrical surge prediction

1. Introduction

In an electric circuit, a surge is often defined as a short wave of power, voltage, or current. In power systems, a surge, often referred to as a transient, is a subcycle overvoltage that lasts for a fraction of a normal voltage waveform cycle. We probably link surges to this scenario the most often. [NEMA Surge Protection Institute, "What are surges?"]. Usually, a surge oscillates and gradually diminishes. It may be additive or subtractive from the standard voltage waveform, and it can have a positive or negative polarity. Transients, also known as surges, are momentary overvoltage spikes or disturbances on a power waveform that have the potential to destroy, deteriorate, or cause damage to electronic equipment in any kind of residential, commercial, industrial, or manufacturing setting. Tens of thousands of volts may be the amplitude of a transient. Surges are usually measured in microseconds. Every single electrical gadget is designed to operate at a nominal voltage of 120, 240, 480, and so on. Surges may badly damage practically any device, even though most are designed to survive modest fluctuations in their baseline operating voltage.

Machine learning, a branch of artificial intelligence research, uses statistical methods to categorize data. In clinical contexts, several machine learning approaches have been used to forecast illness and have shown greater diagnostic accuracy than traditional methods (Kim and Cho 2019). In machine learning, Random Forests, Convolutional Neural Networks, and

DOI: 10.1201/9781003606611-53

Naive Bayes Classifiers are often used techniques. "Electrical Surges risk prediction using machine learning and conventional methods," 2013 35th Annual International Conference of the IEEE Engineering in Medicine and Biology Society (EMBC), Osaka, Japan, 2013, 188–191, doi: 10.1109/EMBC.2013.6609469 (Kim et al. 2013), was presented by S. K. Kim, T. K. Yoo, E. Oh, and D. W. Kim. During the preceding 5 years, around 55 articles have been published on this topic, using the well-known "Prediction of Electrical Surges." The purpose of this research is to provide an overview of current knowledge on electric cars. The best paper (Abdel-Basset et al. 2021), "Deep Learning Methods Utilization in Power Electric Systems," is this one, in my view. It is critical to identify and address any flaws or holes in your study.

These gaps may be caused by missing significant literature, methodological errors, data limitations, or unresolved problems. They may also touch on the need for long-term studies, investigating different theories, theoretical gaps, and the potential for breakthroughs in other fields. It's critical to recognize the gaps in your understanding of regional and cross-cultural differences, moral dilemmas, and real-world applications. You must identify these gaps to raise the quality and relevance of your work, inspire further study, and deepen your grasp of the topic. This motivated me to carry out the investigation. The processes in the research process include selecting a subject, reviewing relevant literature, obtaining and analyzing data, interpreting findings, and sharing them via academic channels. It's a rigorous yet rewarding process that promotes cooperation, personal growth, and potential real-world applications.

2. Materials and Methods

The information security lab at Saveetha School of Engineering is the site of the proposed inquiry. The investigation comprised two groups in all. Group 1 compared the accuracy of the two algorithms using the Random Forest approach. The statistical power (0.8), alpha (0.05), and beta (0.2) values of the study's 48 total sample sizes were calculated using Clincalc. K is used by Group 2 (nearest). The application utilized to carry out the algorithm for precise forecasts is called Google Colab. By developing a novel model, the main goal of this study is to spearhead developments in the field of electrical surge prediction.

The goal of this model is to increase the resilience of electrical systems by using cutting-edge machine learning algorithms and predictive analytics methods. Providing accurate and trustworthy surge forecasts is the main objective. The project aims to push the boundaries of predictive capabilities and produce a major increase in the accuracy and promptness of electrical surge predictions by examining new features and evaluating new technologies. The main objective is to significantly and proactively contribute to the process of developing mitigation strategies to lessen the detrimental effects of electrical disturbances on equipment, infrastructure, and overall system dependability. This is particularly crucial in light of how quickly technology is developing.

The comparison of the two groups' procedures is selected, and the result is ascertained. The electrical surge values for electrical surge detection serve as the main data set for this study. The popular data collecting and analytics tool Kaggle provided the CSV dataset. Photographs of different electrical surge levels are included in the data collection. The dataset and the data used for training and testing are kept separate. The dataset contains around 780 values in total, with the impacted data being represented by fixed samples. Python is used for the planning and execution of the specified task. Working behind the dataset ensures that an output process is executed correctly in terms of code.

2.1. Random Forest

The well-known machine learning method Random Forest was created by Leo Breiman and Adele Cutler. This method creates a single result by combining the results of many decision trees. Random Forest is widely used in many different sectors because of its adaptability and user-friendly interface. It is particularly good at solving regression and classification issues. When it comes to ensemble learning tasks like classification, regression, and others, random forests, also known as random decision forests, work by building a large number of decision trees during the training phase. For classification tasks, the random forest output is the class that most of the trees choose. For classification tasks, the random forest output is the class that most of the trees choose. The mean or average prediction made by each tree is provided for regression tasks (Ho 1995, 1998). The tendency of decision trees to overfit to their training set is compensated for by random decision forests (Hastie et al. 2008).

2.2. Pseudo Code

Step 1: Import necessary libraries for Data classification

Step 2: Load and preprocess the CSV dataset

Step 3: Split the dataset into training and validation sets

Step 4: Define the architecture of the Algorithm usedStep 5: Compile the model

Step 6: Train the model

Step 7: Assuming 'history' contains accuracy values from training

Step 8: Calculate the average accuracy.

2.3. Decision Tree

Despite being a supervised learning technique, decision trees are mostly used to address categorization problems. Regression issues may also be resolved using them, however (Chitalia et al 2020). This classifier is tree-structured with core nodes representing dataset properties, branches representing decision rules, and leaf nodes representing each outcome. A decision tree consists of two nodes: the decision node and the leaf node (Ding et al. 2019). Decision nodes are used to make any form of decision and have several branches, while leaf nodes show the outcome of decisions and have no more branches. The test selection is guided by the features of the dataset that is supplied.

2.4. Pseudo Code

Step 1: Import necessary libraries for CSV classification.

Step 2: Load and preprocess the CSV dataset representationsStep 3: Split the dataset into training and validation sets.

Step 4: Setparameters for data augmentation and pre-processing,

Step 5: Create generators for loading and augmenting CSVs.

Step 6: Create a Decision Tree model for CSV classificationStep 7: Compile the model and train the model

Step 8: Calculate and print average training and validation accuracy

2.5. Statistical Analysis

For statistical analysis of random forests and decision trees, IBM SPSS software with version 26 is employed. Descriptive statistics, an independent sample test, and a simple bar of accuracy by group were employed through SPSS software analysis, which was carried out to calculate accuracy for both methods.

3. Discussions

The study's findings provide insight into the effectiveness of the Random Forest and its possible uses in the field of prediction-generated data (Han et al. 2021). In the domain of electrical diagnostics, Novel Random Forest performed well with an accuracy of 98.98%. Because of its nonlinearity, this algorithm's capacity to manage complicated, nonlinear connections within the data and provide more complex decision limits makes it very useful in electrical surge surges. Although it does not attain the same level of accuracy as Random Forest (98.09%), the Decision Tree provides a strong method for classifying Electrical Surges. examples of electrical surges in different electrical surge circumstances. This highlights its usefulness in the field of electrical surge detection, particularly in cases when the information displays nonlinear correlations that may be difficult for linear models like Decision Tree to represent. Because multi-model integration was the study's main emphasis, we were able to emphasize the advantages of mixing several machine learning techniques. A decision tree and a random forest comparison highlighted the distinct benefits and drawbacks of each approach (Jiang et al. 2020). When nonlinear linkages are present, Random Forest performs very well and provides a potent substitute for conventional linear models. The work sheds light on the possibilities of combining many models and choosing the best algorithm for certain datasets and diagnostic t (Li et al. 2021).

This flexibility in choosing models has wider ramifications for the medical field since customized methods greatly improve diagnostic precision. Subsequent investigations may focus on refining Random Forest's hyperparameters, analyzing how feature engineering affects its functionality, and determining if it can be applied to other healthcare scenarios. The findings pave the way for more research into the integration of various machine-learning approaches and their potential to improve patient care by advancing social decision-making.

4. Result

The results of two categories in the prediction of electric surges: decision trees and random forests. There are a lot of numbers in it: N for the total number of observations, A for the average accuracy rate across all groups, S for the dispersion of those numbers, and E for the precision of that average accuracy rate. The findings demonstrate that Random Forest achieved a higher average accuracy rate of 98.55% compared to Decision Tree's 97.25%. Here are some statistical tests that compare two groups' performance: a t-test to check whether the means are equal and Levene's test to see if the variances are comparable. A statistically significant difference in variances is indicated by Levene's test (F = 0.512, r = 0.484). With a *p*-value of less than 0.001 and a 95% confidence range of [0.978, 1.617], the mean difference in accuracy rates between the two groups in the t-test assuming equal variances is 1.298. Even when we don't assume that the variances are equal, we still see a significant difference in the t-test results ($t = 8.526, p = 0.001$), with a mean difference of 1.298 and a 95% confidence interval of [...].The findings show that Random Forest performed better than Decision Tree in terms of accuracy, and the difference is big and statistically significant. When it comes to predicting electric surges, the bar graph shows how the suggested Random Forest algorithm stacks up against the Decision Tree method in terms of accuracy rates. With an accuracy rating of 98.55%, the Random Forest model significantly outperformed the Decision Tree model (Sajjad et al. 2020), which only managed 97.25%. A two-tailed significance test ($p < 0.01$) indicated that the disparity in accuracy was statistically significant. The graph shows the two algorithms, Random Forest and Decision Tree, on the X-axis, and the average accuracy, with ±1 standard deviation and a 95% confidence interval, on the Y-axis. This gives a complete picture of how the two models differ in terms of performance.

5. Conclusion

Finally, with an impressive accuracy of 98.98%, this research has investigated the usefulness of the random forest in the field of electrical surge prediction. As an alternative to linear models, Random Forest is a useful tool for electrical surge diagnostics because of its ability to handle complicated (Tan et al. 2020), nonlinear interactions in electrical surge datasets. Despite not having the same accuracy as Decision Trees (98.09%), Random Forest nonetheless shows how adept it is at handling intricate data patterns.

The study highlights the importance of choosing the right model and the benefits of mixing different machine-learning techniques to provide customized solutions for certain diagnostic tasks. This flexibility has wider ramifications for appliances and electrical circumstances than just anticipating electrical surges. Subsequent research endeavors might focus on optimizing Random Forest hyperparameters (Tom et al. 2019), investigating the significance of feature

Table 53.1: This table summarizes the performance of two groups, Random Forest and Decision Tree, in electric surge prediction. It includes the number of observations (N), the mean accuracy rate for each group, the standard deviation indicating the spread of data, and the standard error of the mean representing the accuracy rate's precision. The results show that Random Forest outperformed Decision Tree with an average accuracy rate of 98.55%, while Decision Tree had an average accuracy rate of 97.25%

	Algorithm	N	Mean	Std. Deviation	Std. Error Mean
Accuracy	Random forest	10	98.55	0.279	0.088
Accuracy	Decision Tree	10	97.25	0.392	0.124

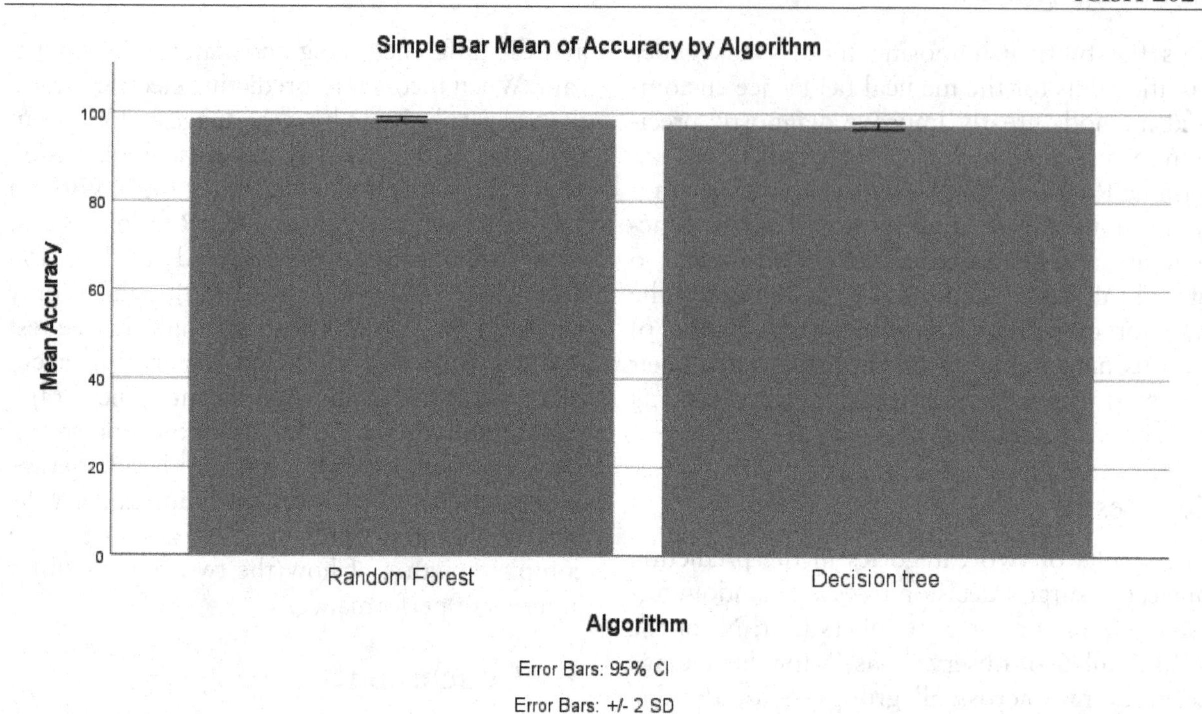

Figure 53.1: The bar graph illustrates a comparison of accuracy rates between the proposed Random Forest algorithm and the Decision Tree algorithm in the context of electric surge prediction. The Random Forest model exhibited a notably higher accuracy rate of 98.55%, surpassing the Decision Tree model, which achieved an accuracy of 97.25%. A two-tailed significance test ($p < 0.01$) yielded statistically significant results for this accuracy difference. On the graph, the X-axis represents the two algorithms, Random Forest and Decision Tree. At the same time, the Y-axis displays the average accuracy, accompanied by ±1 standard deviation and a 95% confidence interval, providing a comprehensive overview of the performance distinction between the two models.

engineering, and assessing its applicability to more domestic scenarios (Wang et al. 2020). The research highlights the significance of variety in the selection of models. It provides opportunities for the incorporation of diverse machine learning methodologies, therefore augmenting diagnostic accuracy and eventually improving patient care within healthcare domains.

References

[1] Abdel-Basset, Mohamed, Hossam Hawash, et al. 2021. Energy-net: A deep learning approach for smart energy management in IoT-based smart cities. *IEEE Internet of Things Journal* 8 (15): 12422. https://doi.org/10.1109/JIOT.2021.3063677.

[2] Chitalia. G., M. Pipattanasomporn, V. Garg, and S. Rahman. 2020. Robust short-term electrical load forecasting framework for commercial buildings using deep recurrent neural networks. *Applied Energy* 278: 115410.

[3] Ding, Y. Cao, L. Xie, Y. Lu, and P. Wang. 2019. Integrated stochastic energy management for data center microgrid considering waste heat recovery. *IEEE Transactions on Industry Applications* 55 (3): 2198–2207.

[4] Han, T., K. Muhammad, T. Hussain, J. Lloret, and S. W. Baik. 2021. An efficient deep learning framework for intelligent energy management in IoT networks. *IEEE Internet of Things Journal* 8 (5): 3170–3179.

[5] Hastie, Trevor, Robert Tibshirani, and Jerome Friedman. 2008. *A guide to statistical learning*, 2nd edn. Springer, 0-387-95284-5 (ISBN).

[6] Ho, Tin Kam. 1995. Random decision forests. In *Proceedings of the Third International Conference on Document Analysis and Recognition, Montreal, QC, August 14–16, 1995*, pp 278–282.

[7] Ho, T. K. 1998. The random subspace method for constructing decision forests. *IEEE Transactions on Machine Intelligence and Pattern Analysis* 20 (8): 832.

[8] Jiang, D., Y. Wang, Z. Lv, W. Wang, and H. Wang. 2020. An energy-efficient networking approach in cloud services for IoT networks. *IEEE Journal on Selected Areas in Communications* 38 (5): 928–941.

[9] Kim, Sung Kean, Tae Keun Yoo, Ein Oh, and Deok Won Kim. 2013. Osteoporosis risk prediction using machine learning and conventional methods. In *Conference proceedings: ... annual international conference of the IEEE Engineering in Medicine and Biology Society*. IEEE Engineering in Medicine and Biology Society. Conference 2013, pp. 188–191.

[10] Kim, T.-Y., and S.-B. Cho. 2019. Predicting residential energy consumption using CNN-LSTM neural networks. *Energy* 182: 72–81.

[11] Li, J., et al. 2021. A novel hybrid short-term load forecasting method of the smart grid using MLR and LSTM neural network. *IEEE Transactions on Industrial Informatics* 17 (4): 2443–2452.

[12] Lv, P., S. Liu, W. Yu, S. Zheng, and J. Lv. 2020. EGA-STLF: A hybrid short-term load forecasting model. *IEEE Access* 8: 31742–31752.

[13] Sajjad, M., et al. 2020. A novel CNN-GRU-based hybrid approach for short-term residential load forecasting. *IEEE Access* 8: 143759–143768.

[14] Tan, M., S. Yuan, S. Li, Y. Su, H. Li, and F. He. 2020. Ultra-short-term industrial power demand forecasting using LSTM based hybrid ensemble learning. *IEEE Transactions on Power Systems* 35 (4): 2937–2948.

[15] Tom, R. J., S. Sankaranarayanan, and J. J. P. C. Rodrigues. 2019. Smart energy management and demand reduction by consumers and utilities in an IoT-fog-based power distribution system. *IEEE Internet of Things Journal* 6 (5): 7386–7394.

[16] Wang, Y., et al. 2020. Short-term load forecasting for industrial customers based on TCN-LightGBM. *IEEE Transactions on Power Systems* 36: 1984.

CHAPTER 54

Efficiency improvement of DC–DC converters using novel CL–LLC converter compared with LLC converter based on input voltage

Muthyam Reddy,[1] **and B.T. Geetha**[1*,a]

[1]Department of Electrical and Electronics Engineering, Saveetha school of Engineering, Saveetha Institute of Medical and Technical Sciences, Chennai, Tamil Nadu, India

Email: ageethabt.sse@saveetha.com

Abstract

In contrast to a conventional LLC converter, the proposed study uses a new CL–LLC converter based on variable input voltages for sustainable power in an effort to increase the efficiency of DC–DC converters. Changing the input voltage allowed for the collection of 16 samples in total. Two groups of eight samples each were created from these samples. The difference in DC–DC converter efficiency between the new CL–LLC and the conventional LLC converters was then measured by comparing the efficiency numbers. G power was adjusted to 0.8. The LLC converter had an efficiency of 90.88%, whereas the new CL–LLC converter had a 95.58% efficiency. With a value of 0.260 (p >0.05), it was determined that the outcome lacked statistical significance. Based on input voltage, the new CL–LLC converter outperformed the LLC converter in DC–DC converters, although the difference was not statistically significant.

Keywords: Novel CL–LLC converter, LLC converter, input voltage, efficiency, DC–DC converters, sustainable power

1. Introduction

This research delves into the domain of DC–DC converters, specifically focusing on the efficiency enhancement achieved through the application of the novel CL–LLC converter in comparison with the conventional LLC Converter (Zhang et al. 2021), with a primary emphasis on varying input voltage conditions. In this context, the investigation explores the intricacies of these converters and their performance dynamics. The significance of this research lies in the critical role played by DC–DC converters in various electronic systems, where energy efficiency is paramount (Asadi et al. 2018). Improved converters can contribute to energy savings, reduced heat dissipation, and enhanced overall system performance (Luo and Ye 2016). The practical applications of this work extend across numerous industries, including renewable energy systems, electric vehicles, and portable electronic devices. As the demand for energy-efficient technologies grows, optimizing DC–DC converters becomes increasingly pertinent (Musumeci 2021).

In google scholar more than thousand articles were published, Elsevier springer and IEEE Xplore in recent years. This involved analyzing a multitude of scholarly articles, with a particular focus on four highly relevant papers. These papers not only provided insights into the current state of research in the field but also paved the way for identifying gaps and potential areas for improvement (Asadi 2018). Notably, one outstanding study stood out among the selected papers, offering an in-depth analysis of the LLC converter's efficiency under varying input voltage conditions. The findings of this study not only informed the current research but also

DOI: 10.1201/9781003606611-54

underscored the need for further exploration into the nuanced performance aspects of DC–DC converters (Asadi and Eguchi 2018). Despite the wealth of existing research, a noticeable gap exists in comprehensively comparing the novel CL–LLC converter with the LLC converter, specifically concerning their efficiency under diverse input voltage scenarios (Priyadarshi et al. 2021). The lacunae in the literature necessitate a dedicated exploration to address this gap. The research team brings a unique expertise to this endeavor, with a deep understanding of power electronics and converter design (Geetha and Perumal 2022). The primary aim of this work is to bridge the identified gap by conducting a thorough comparative analysis, leveraging the team's expertise to propose novel insights and potential advancements in DC–DC converter technology. Through this research, we aspire to contribute valuable knowledge that can inform the design and implementation of more efficient power electronic systems (Butzen and Steyaert 2021).

Research gaps in enhancing DC–DC converter efficiency include exploring novel materials, advanced circuit designs, and control strategies. Investigating miniaturization techniques, optimizing for energy harvesting applications, and developing fault-tolerant solutions are key areas for improving overall converter efficiency. This research work is proposed to improve the efficiency in DC to DC converters using novel CL–LLC converter compared with LLC converter based on Input voltage for sustainable power.

2. Materials and Methods

Analyzing two sets of input voltage values is the task at hand. With a sample size of eight values each, Group 1 represents the unique CL–LLC converter and Group 2 represents the conventional LLC converter 0.8 is the G power setting. For coding and simulations, the MATLAB R2021a software toolkit was utilized. MATLAB was used to calculate efficiency numbers, which were then compared to identify efficiency gains.

Sample preparation for Group 1 comprised measuring various input voltage values by varying the unique CL–LLC converter's characteristics, with a particular emphasis on the CL–LLC converter and input voltage. With an emphasis on the LLC converter and input voltage, the Group 2 sample pr involves varying the input voltage values by modifying the LLC converter's parameters.

2.1. Novel CL–LLC Converter

The novel CL–LLC converter represents a significant advancement in DC–DC converter technology, offering enhanced performance and efficiency in power conversion applications. This converter architecture combines the benefits of a CLL (capacitor-inductor-inductor) converter with a CL (capacitor-inductor) converter, resulting in improved power density, reduced losses, and increased reliability. Unlike traditional converters, the CL–LLC converter integrates a unique combination of passive components, including capacitors and inductors, to achieve a more streamlined and efficient power conversion process. The CLL topology contributes to soft switching, minimizing switching losses and increasing overall efficiency. Garg, Amik et all 2017. Simultaneously, the CL section provides additional benefits such as reduced voltage stress on the switches and improved transient response (Musumeci 2021). This novel converter design addresses key challenges in DC–DC conversion, such as minimizing energy losses and ensuring stable operation under varying load conditions. Its inherent ability to handle a wide range of input voltages and output loads makes it versatile for various applications, including renewable energy systems, electric vehicles, and power supplies. Researchers and engineers have been actively exploring the potential of the CL–LLC converter to push the boundaries of power electronics, unlocking new possibilities for energy-efficient and compact power solutions in today's rapidly evolving technological landscape. As this innovative converter continues to undergo refinement and real-world testing, it holds the promise of significantly impacting the efficiency and performance standards of DC–DC converters in diverse industries.

2.2. LLC Converter

The LLC (inductor-inductor-capacitor) converter stands as a pivotal technology in the realm of DC–DC converters, offering a compelling blend of efficiency, versatility, and compact design. This converter architecture is renowned for its ability to deliver high-power density while minimizing electromagnetic interference and switching losses. At its core, the

LLC converter operates with a resonant tank circuit comprising two inductors and a capacitor, providing a soft-switching mechanism that significantly reduces switching losses. This resonant operation allows for efficient energy transfer between input and output, leading to improved overall converter efficiency. The LLC topology also offers inherent benefits such as enhanced electromagnetic compatibility (EMC) and reduced voltage stress on semiconductor devices, contributing to prolonged component lifespan. One of the key advantages of the LLC converter lies in its adaptability to a wide range of input voltages and output loads, making it suitable for various applications, including power supplies for servers, renewable energy systems, and electric vehicles. Its resonant nature facilitates zero-voltage switching, minimizing power loss during the switching transitions and enabling high-frequency operation. As the demand for energy-efficient and compact power solutions continues to grow, the LLC Converter has garnered significant attention from researchers and engineers alike (Erickson and Maksimovic 2007). Ongoing advancements in control strategies and semiconductor technologies aim to further refine the LLC converter's performance, fostering its integration into diverse industries and solidifying its position as a cornerstone in the evolution of DC–DC converter technologies.

2.3. Statistical Analysis

Using SPSS software, the input voltage values for the LLC and CL–LLC converters were acquired for statistical analysis. The mean, standard deviation, and standard mean error were among the statistical measures used to ascertain the statistical properties for both approaches. The input voltages of the LLC and CL–LLC converters are considered the independent variables, while efficiency is handled as the dependent variable. In this study, an independent t-test was run. The purpose of the analysis was to compare and assess how well the LLC and CL–LLC converters controlled the DC–DC converters' input voltage (Jahan et al. 2017).

3. Results

With an efficiency of 95.58% in DC–DC converters as opposed to the LLC converter's 90.88%, the novel CL–LLC converter has demonstrated improved efficiency. The total of 16 samples is shown in Table 54.1, which is further subdivided into two groups: Group 1 (novel CL–LLC converter) and Group 2 (LLC converter). Each group consists of 8 samples that indicate input voltage-dependent efficiency values. For both the LLC converter and the novel CL–LLC converter, Table 54.2 gives information on the number of samples, mean efficiency values, standard deviation, and standard error mean. It is discovered that the LLC converter is computationally less efficient than the novel CL–LLC converter. The LLC converter has a standard error mean of 1.0423, but the novel CL–LLC converter has a mean of 0.92144. The statistical analysis of independent sample testing for both groups is displayed in Table 54.3. Statistical significance is defined as a p-value of less than 0.05, and a 95% confidence range is computed. It is not statistically significant that the obtained significance value of 0.260 ($p > 0.05$) was attained. The MATLAB simulation circuit for the novel CL–LLC converter, which is intended to increase efficiency dependent on input voltage, is shown in Figure 54.1. The LLC converter's MATLAB simulation circuit, which is utilized in DC–DC converters to increase efficiency depending on input voltage, is shown in Figure 54.2. The LLC converter and

Table 54.1: The total number of 16 samples for group 1 and group 2 with 8 samples each, the samples are efficiency values. Group 1 is novel CL–LLC converter and group 2 is LLC converter based on input voltage

Input Voltage in KV	Efficiency in %	
	Novel CL–LLC converter	LLC converter
350	95.58	90.88
375	94.67	89.54
390	89.35	87.67
400	93.46	85.83
430	92.16	84.78
450	90.88	83.72
470	91.54	81.86
500	87.88	86.13

the novel CL–LLC converter's MATLAB simulation outputs are contrasted in Figure 54.3, which displays efficiency values of 90.88% and 95.58%, respectively. Based on input voltage and efficiency, Figure 54.4 compares the novel CL–LLC converter and LLC converter, showing that the LLC converter's mean efficiency and standard deviation are marginally higher than

Table 54.2: It represents the matlab simulation circuit of LLC converter with DC–DC converters to improve efficiency based on input voltage. Figure 54.3 shows the comparison of matlab simulation output of novel CL–LLC converter and LLC converter with efficiency values of 95.58% and 90.88% respectively

	Groups	No of Samples	Mean	Std. Deviation	Std. Error Mean
Efficiency	CL–LLC converter	8	94.0625	2.25834	0.79844
Efficiency	LLC converter	8	92.6763	4.11235	1.45394

Table 54.3 Table shows the comparison of matlab simulation output of novel CL–LLC converter and LLC converter with efficiency values of 95.58% and 90.88% respectively. Statistical analysis of independent sample tests for both groups. *P* value should be less than 0.05. Considered to be statistically significant and 95% confidence value is calculated. The significant value obtained is 0.260 (p >0.05) which is not statistically significant

Independent Samples Test										
Levene's Test for Equality Variances							T-test for Equality of means		95% Confidence Interval of the 95% confidence interval of the Difference	
		F	Sig	t	diff	Sig (2-tailed)	Mean Difference	Std Error Difference	Lower	Upper
Efficiency	Equal variances assumed	1.381	0.260	0.836	14	0.417	1.38625	1.65875	−2.17141	4.94391
Efficiency	Equal variances not assumed			0.836	10.870	0.421	1.38625	1.65875	−2.26996	5.04246

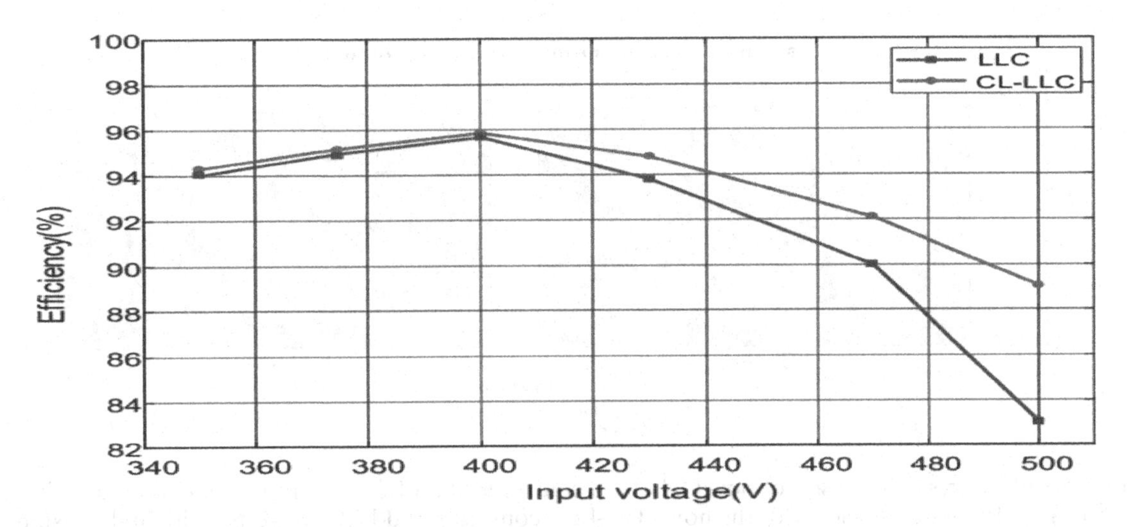

Figure 54.1: It represents the matlab simulation circuit of novel CL–LLC converter and LLC converter to improve efficiency based on input voltage

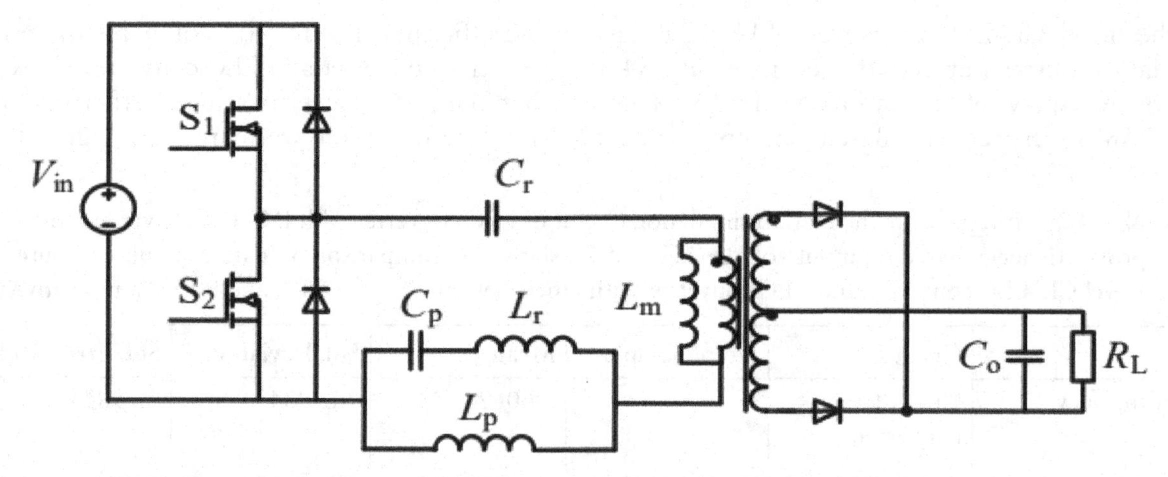

Figure 54.2: The simulation circuit of CL–LLC converter of DC–DC converters for the improvement of efficiency

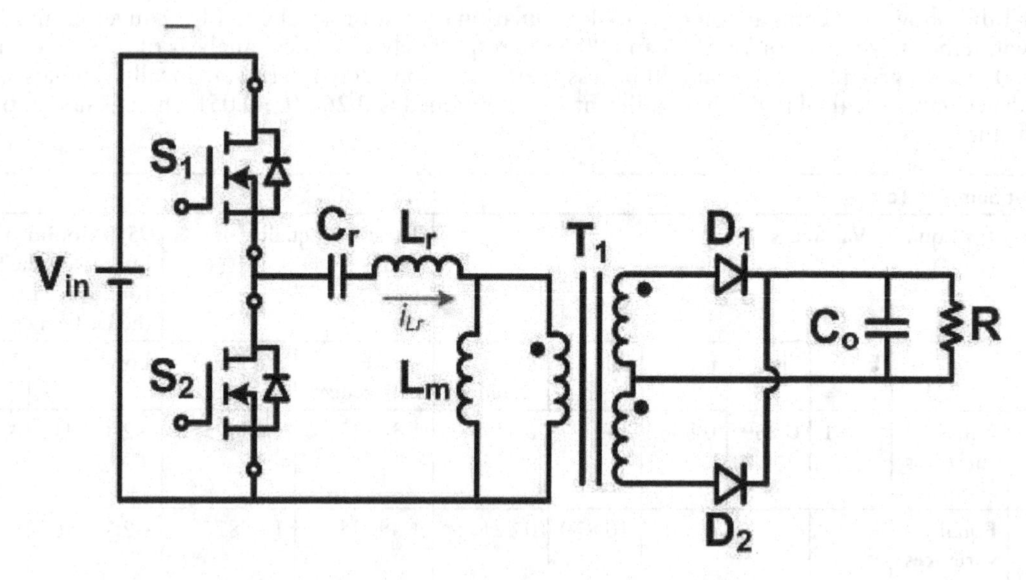

Figure 54.3: The simulation circuit of LLC converter of DC–DC converters based on the input voltage

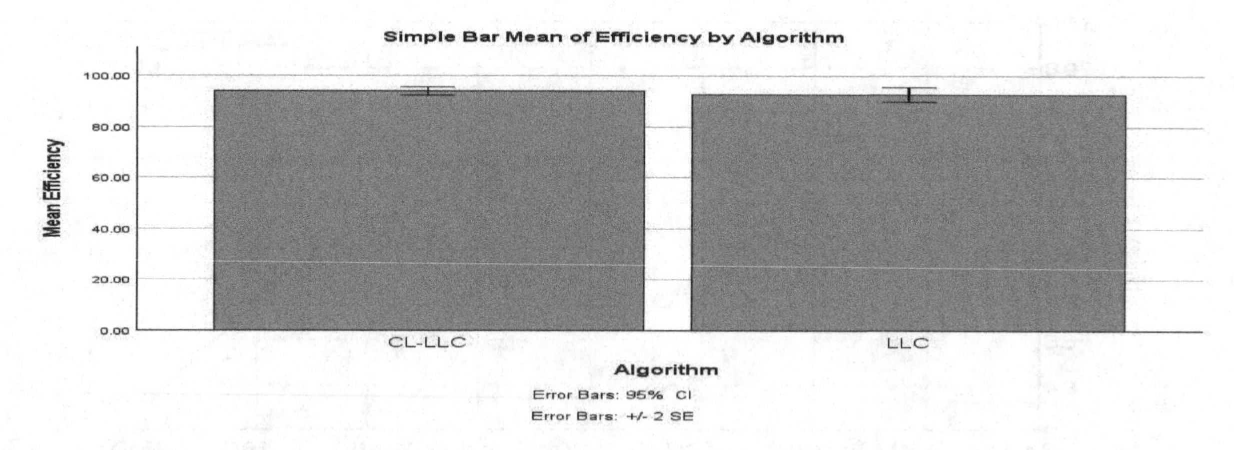

Figure 54.4 Illustrates comparison of novel CL–LLC converter and LLC converter based on input voltage and efficiency. The mean efficiency of the novel CL–LLC converter and LLC converter with and the standard deviation of LLC converter is slightly better than novel CL–LLC converter. X Axis: novel CL–LLC converter versus LLC converter. Y Axis: mean efficiency with ±2 SD.

those of the novel CL–LLC converter. LLC converter versus novel CL–LLC converter on the X Axis. Mean efficiency with ±2 SD is plotted on the Y axis.

4. Discussion

The novel CL–LLC converter, as suggested, exhibits heightened efficiency in DC–DC conversion, boasting an efficiency rate of 95.58%. This is subsequently contrasted with the LLC converter, which operates at a comparatively lower efficiency of 90.88%.

Advancements in technology have led to remarkable improvements in DC–DC converters' efficiency. Enhanced efficiency is crucial for minimizing energy loss and improving overall system performance in various applications (Kazimierczuk 2014). One approach involves optimizing the switching frequency of the converter. Higher switching frequencies reduce energy loss, enabling more efficient power transfer within the system. Advanced control algorithms play a pivotal role in achieving this optimization (Priyadarshi et al. 2021). Researchers are exploring novel converter top ologies and circuit configurations. Innovations such as the novel CL–LLC converter present alternative designs that exhibit improved efficiency compared to traditional converters, thereby contributing to overall efficiency enhancements (Musumeci 2021). The integration of wide bandgap semiconductors, such as silicon carbide (SiC) and gallium nitride (GaN), in DC–DC converters has shown significant promise. These materials allow for higher operating temperatures and reduced switching losses, resulting in increased efficiency (Kazimierczuk 2014, 2015). Utilizing advanced control strategies, such as predictive control and artificial intelligence-based algorithms, has become pivotal in maximizing efficiency. These strategies enable real-time adjustments to operating parameters, ensuring optimal performance under varying load conditions (Tang et al. 2023). Despite significant progress, challenges remain, including thermal management issues and cost considerations. Future research aims to address these challenges by exploring innovative thermal solutions and cost-effective technologies, further pushing the boundaries of DC–DC converter efficiency (Tang et al. 2023; Ghani Varzaneh et al. 2024).

The performance of the innovative CL–LLC converter is impacted by its limitations with regard to input voltage changes. In contrast to LLC converters that have input voltage regulation, CL–LLC converters have difficulty remaining stable and efficient, which presents problems under a variety of operating circumstances. Future developments for the innovative LLC and CL–LLC converters will focus on improving input voltage regulation for greater flexibility. Smart algorithms and the incorporation of cutting-edge technology could be advancements, guaranteeing increased dependability and efficiency in a range of input voltage conditions.

5. Conclusion

In conclusion, the proposed study aimed to enhance the efficiency of DC–DC converters using the novel CL–LLC converter compared to the LLC converter under varying input voltage conditions. Results demonstrated a notable improvement in efficiency, with the novel CL–LLC converter achieving 95.58%, surpassing the LLC converter at 90.88%. However, statistical analysis revealed a non-significant difference ($p > 0.05$). Despite this, the observed efficiency improvement in the novel CL–LLC converter underscores its potential as a more effective solution for enhancing DC–DC converter efficiency based on input voltage variations.

References

[1] Asadi, Farzin. 2018. San Rafael, California, USA. *Computer techniques for dynamic modeling of DC–DC power converters*. Morgan & Claypool Publishers.

[2] Asadi, Farzin, and Kei Eguchi. 2018. *Dynamics and control of DC–DC converters*. Morgan & Claypool Publishers.

[3] Asadi, Farzin, Sawai Pongswatd, Kei Eguchi, and Ngo Lam Trung. 2018. *Modeling uncertainties in DC–DC converters with MATLAB® and PLECS®*. Morgan & Claypool Publishers.

[4] Butzen, Nicolas, and Michiel Steyaert. 2021. *Advanced multiphase switched-capacitor DC–DC converters: Pushing the limits of fully integrated power management*. Springer.

[5] Erickson, Robert W., and Dragan Maksimovic. 2007. *Fundamentals of power electronics*. Springer Science & Business Media.

[6] Garg, Amik, Akash Kumar Bhoi, Padmanaban Sanjeevikumar, and K. K. Kamani. 2017.

Advances in power systems and energy management: ETAEERE-2016. Springer.

[7] Geetha, B. T., and V. Perumal. 2022. Brushless DC motor force ripple reduction techniques – A survey. *AIP Conference Proceedings* 2516 (1): 140004.

[8] Ghani Varzaneh, Majid, Amirhossein Rajaei, Navid Kamali-Omidi, Ali Shams-Panah, and Mohammad Reza Khosravi. 2024. A single-stage dual-source inverter using low-power components and microcomputers. *Scientific Reports* 14 (1): 1804.

[9] Jahan, S. Kuthsiyat, S. Kuthsiyat Jahan, K. Chandru, B. Dhanapriyan, R. Kishore Kumar, and G. Vinothraj. 2017. SEPIC converter based water driven pumping system by using BLDC motor. *Bonfring International Journal of Power Systems and Integrated Circuits*. https://doi.org/10.9756/bijpsic.8317.

[10] Kazimierczuk, Marian K. 2014. **High-Frequency Magnetic Components*, 2nd ed. Wiley.*

[11] Kazimierczuk, Marian. 2015. *Pulse-width modulated DC–DC power converters*. Wiley.

[12] Luo, Fang Lin, and Hong Ye. 2016. *Advanced DC/DC converters*. CRC Press.

[13] Musumeci, Salvatore. 2021. *Advanced DC–DC power converters and switching converters*. MDPI.

[14] Priyadarshi, Neeraj, Akash Kumar Bhoi, Ramesh C. Bansal, and Akhtar Kalam. 2021. *DC–DC converters for future renewable energy systems*. Springer Nature.

[15] Tang, Xueying, Jiashuo Zhang, Dezhi Sui, Qiongfen Yang, Tianyu Wang, Zihan Xu, Xiaoya Li, et al. 2023. Simultaneous dendritic cells targeting and effective endosomal escape enhance sialic acid-modified mRNA vaccine efficacy and reduce side effects. *Journal of Controlled Release: Official Journal of the Controlled Release Society* 364 (December): 529–545.

[16] Zhang, Xiangjun, Jiachen Jing, Yueshi Guan, Mingcong Dai, Yijie Wang, and Dianguo Xu. 2021. High-efficiency high-order CL–LLC DC/DC converter with wide input voltage range. *IEEE Transactions on Power Electronics* 36 (9): 10383–10394.

CHAPTER 55

AI-Driven Innovation: Fostering Sustainable Workplaces and Innovative Work Behaviours for a Greener Future

Niranchana Shri Viswanathan*, Dr. Delecta Jenifer Rajendren**

Assistant Professor, Department of Management studies, Sapthagiri NPS University, Bengaluru
Assistant Professor, Department of Management Studies, Saveetha Engineering College, Chennai
E. Mail: niranchanaphd@gmail.com, delectajenifer@saveetha.ac.in

Abstract

This paper investigates through a critical narrative, how innovations based on artificial intelligence might foster sustainable workplaces and innovative work practices for greener future landscape. Artificial intelligence can automate processes that would help cut down wastage and optimize resource utilization in the workplace. Green operations lower our carbon emissions and make the world more sustainable. An AI-powered world could mean a creative, competent workplace. This general idea that AI in offices could enforce more sustainable and innovative ways of living, making the world a better place for all. Drawing on numerous studies, this research shows us some initial steps for future work with employees to consider the impact of AI on creativity and problem-solving. Firms would learn from the results of this study in how they could utilize AI for good and social impact to drive greener practices and workplace innovation. The discoveries will educate companies about AI in sustainability and the way they innovate to shape their world around them. Finally, future research could explore the long-term impact of adopting AI on sustainability and innovation strategies.

Keywords: AI, Sustainability, workplace, Innovation work behaviour, greener future

1. Introduction

A. *Definition of AI-driven innovation*

AI-Driven Innovation, in brief, and easy to understand words means making use of artificial intelligence technologies which helps generate innovative ideas or solutions leading toward advancement up the value chain within different industries. By using AI algorithms and ML models, this methodology helps in the task of analysing data, finding patterns so that we can predict outcomes more successfully which will impact business processes like product innovation with optimised customer experience.

2. Benefits of AI-driven innovation in fostering sustainable workplaces.

A. *Improved efficiency and productivity*

Efficiency and productivity the most basic benefits of AI-driven innovation are efficiency & productivity. Enter AI technology: By automating routine tasks and processes, employees can take on more strategic work that taps into innovation and results in greater scale. Furthermore, AI algorithms are capable of combing through massive datasets to find correlating patterns and trends that allow for better decision-making points as well resource allocation subsets. This not only saves time and effort, but also

DOI: 10.1201/9781003606611-55

reduces waste, further enhancing the sustainable operation.

B. *Reduction of carbon footprint*

Combats climate change and protects the environment. Use public transportation, conserve electricity, and buy eco-friendly products to reduce carbon footprint. Businesses can lower their carbon footprint by using renewable energy and reducing waste. Together, we can reduce carbon footprint and green the future.

C. *Enhanced employee well-being*

Enhancing employee well-being may enhance job satisfaction and productivity, but wellness programs and concessions may cost employers more. Despite these efforts, some employees may not feel better.

3. Innovative work behaviours enabled by AI-driven innovation.

A. *Remote work and flexible schedules*

Thanks to AI-driven innovation, new work behaviours are becoming more frequent in the workforce. Artificial intelligence makes it easier than ever to operate remotely by enabling seamless collaboration across time zones and locales. Also, AI-powered tools and technology allow employees to modify their work schedules to suit their style and lifestyle. Flexibility increases work-life balance, productivity, and employee satisfaction.

B. *Collaboration and communication tools*

Important to organizational collaboration. Members can quickly share information, collaborate on projects, and interact in real time using these tools. This boosts efficiency and teamwork. These tools also streamline operations and eliminate misinterpretation. Any firm that wants to improve teamwork and achieve goals must invest in collaboration and communication technologies.

C. *Data-driven decision-making*

Data-driven decision-making may overlook qualitative elements, resulting in inadequate or biased judgments. Mismanaged or misread data can also lead to bad decisions.

4. Literature Review

The study is limited in that it does not propose a comprehensive framework of AI-driven innovation and its influence on fostering sustainable workplaces, nor specifically discusses innovative work behaviours relevant to green future. It validates one approach that deploying AI technologies to enable sustainable business operations will be an essential portfolio of a study for improving the environmental sustainability within organisations. The review will examine how AI can help spur innovative ideas while pushing businesses to think freely and leads more innovative workplaces that take climate change seriously.

Figure 55.1: Most frequently use word (Biblioshiny)Source Scopus data base.

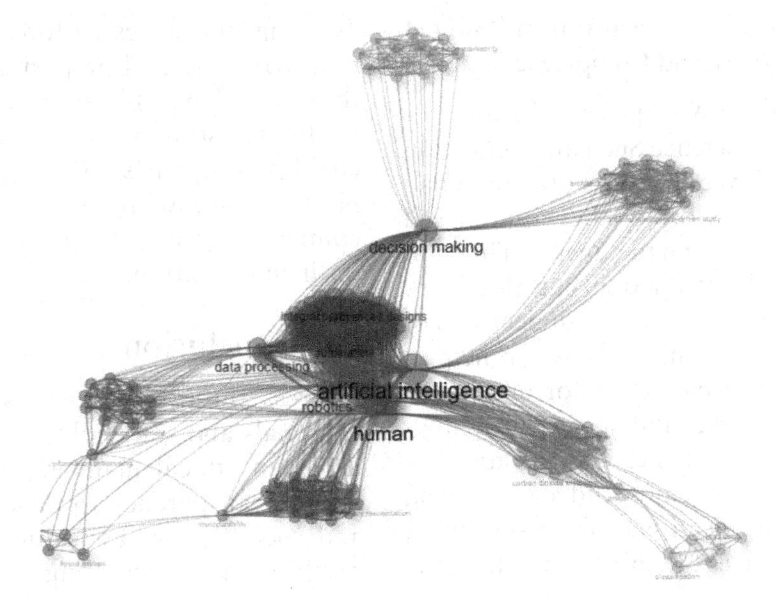

Figure 55.2: Thematic Map: Source Biblioshiny (Scopus Database)

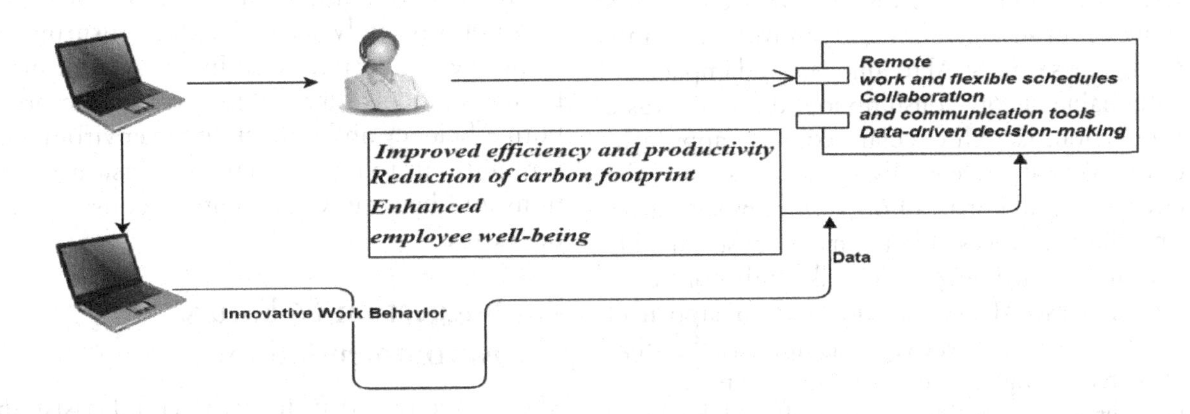

Figure 55.3: Author Own Source

5. Conceptual Frameworks

Today's fast-changing technology, different industries. AI-driven innovation has enormous potential to transform workplaces and promote sustainability. AI helps firms streamline operations, optimize resources, and increase efficiency for a greener future. AI-driven innovation is changing how we work and how we think about workplace sustainability and innovation.

1. The Role of AI in Decrease Wastage and Extra Efficient Resource Administration in Office

By automating processes and continuously analysing data, AI can help avoid waste and increase your resource utilization — leading to better decisions around when the time is right. For example, AI algorithms can detect inefficiencies in production lines and recommend changes immediately so that resources can be adapted in advance before they become unnecessary or underused. AI speeds up and automates a lot of the typical processes so that other resources such as time, labour or materials are used much more efficiently which results in less overall waste. Companies can improve their resource allocation with AI-driven insights that allow each area of the operation is fine-tuned for optimal performance.

2. Further Exploring the Benefits of AI-Powered Workplaces: Creativity and Competence

What AI-powered Workplaces Mean for Creativity and Competence Such an application of AI enables employees to focus on more challenging and strategic work as they are relieved from repetitive or mundane tasks. This not only boosts the job satisfaction but also helps employees to put in practice new dimensions of competencies. In addition, AI tools can help in processing large data quantities for insights to stimulate novel thoughts and answers. AI, however, can also serve to foster collaboration by linking heterogeneous teams and empowering them with real-time ideation toolsets—meaning a more agile set of ideas within the workplace.

3. Longer-Term Prospects for Sustainable and Innovative AI Implementation

Using AI plants can change the way companies think about sustainability and innovation overall. Companies can use AI to innovate and implement sustainable energy consumption, diminish waste production, launch circular economy approaches etc. This also eventually reduces costs and the environmental impact of organizations since over time these practices offer them a more sustainable way to manage energy as well. And more than anything else, AI enables innovation to happen all the time with innovative ideas and insights that these technologies give us on how the next product, service or business model should be built. Eventually, I would argue that companies which embrace AI as part of their responsibility to the environment and innovation stand a good chance at positioning themselves amongst category leaders in terms being recognized for environmental responsibility first movers.

4. Using AI to Improve Critical Thinking Skills in the Workplace

The big idea is creating a next generation data-driven insights and predication workforce that is focused upon problem solving with AI in the workplace. Complex AI programs can evaluate copious amounts of data, recognize patterns and present suggestions that even an experienced human analyst would fail to see. This really comes in handy when we must take immediate decisions or dealing with a lot of data. AI can also replicate diverse scenarios, assisting businesses to forecast the challenges that may arise and preparing in advance so as they would be able to address all proactively. By boosting problem-solving abilities AI enables companies to navigate through complicated issue with an ease but in addition it contributes towards superior outcomes and resilient operations.

6. Conclusion

Address these challenges, sustainability professionals and stakeholders will have to create AI in an ethical manner. RP: you are speaking of Transparency to make the environment and society more appealing for stakeholders, resulting in an increasing in AI adoption. This approach, focusing on collaboration through shared AI and value generation will usher a new era of specialized companies run by leaders in their domain with the power to interfere appropriately at scale, thus ensuring an equitable regulation. Sustainable AI will bring business models forward in ways that improve both efficiency and impact on the environment. It requires us to have a change in basic assumptions and being environmentally conscious for the future.

7. Suggestion & Future Recommendation:

The technology driven by AI could assist the companies with sustainability first workspace and help their employees with new work habits for a greener future. These solutions work to create renewable energy streams, reduce waste, and enhance productivity in line with the goals of a circular economy.

References:

[1] AI-Driven Innovation: Fostering Sustainable Workplaces and Innovative Work Behaviours for a Greener Future. (2021). In *AI-Driven Innovation: Fostering Sustainable Workplaces and Innovative Work Behaviours for a Greener Future.*

[2] Becker, G. S. (1962). Investment in Human Capital: A Theoretical Analysis. *Journal of Political Economy, 70*(5, Part 2), 9–49. https://doi.org/10.1086/258724

[3] Lim, W. M. (2023a). The workforce revolution: Reimagining work, workers, and workplaces for

the future. *Global Business and Organizational Excellence*, 42(4), 5–10. https://doi.org/10.1002/joe.22218

[4] Lomawaima, K. T., & McCarty, T. L. (2002). When Tribal Sovereignty Challenges Democracy: American Indian Education and the Democratic Ideal. *American Educational Research Journal*, *39*(2), 279–305. https://doi.org/10.3102/00028312039002279

[5] Mazzoleni, G., & Schulz, W. (1999). "Mediatization" of Politics: A Challenge for Democracy? *Political Communication*, *16*(3), 247–261. https://doi.org/10.1080/105846099198613

[6] Mincer, J. (1962). On-the-Job Training: Costs, Returns, and Some Implications. *Journal of Political Economy*, *70*(5, Part 2), 50–79. https://doi.org/10.1086/258725

[7] Sahoo, B., Padhi, S., Patra, A. C., Mahapatra, B. R., Mishra, T., Mishra, S. R., & Patro, K. C. (2023). A Prospective Cohort Study Analyzing Radiation-Induced Xerostomia and Quality of Life of Head and Neck Cancer Patients Treated with Intensity-Modulated Radiotherapy and 3D Conformal Radiotherapy Techniques at a Tertiary Cancer Center in Eastern India. *Curēus*. https://doi.org/10.7759/cureus.36442

CHAPTER 56

The Study of Samarium Ion Doped Barium Borate Glass

Priya Murugasen[1], A. R. Baby Suganthi[2], A. Antony Suresh[1], and J. T. Anandhi[2]

[1]Department of Physics, Saveetha Engineering College, Chennai, Tamil Nadu

[2]Department of Physics, SRM Institute of science and Technology, Ramapuram Campus, Chennai, Tamil Nadu

Abstract

Sm^{3+} doped barium borate glass ($60B_2O_3 + 39.7BaCO_3 + 0.3Sm_2O_3$) was synthesized and studied by various spectroscopic techniques. UV-absorption data were used to calculate the Urbach energy (E_u) and optical band gap energies (E_g) of the glasses. In this work, the excitation-dependent photoemission properties of the prepared glass were studied. The photoluminescence (PL) spectra exhibits the radiative transitions in Sm^{3+} barium borate (SBB) glass and the radiative transitions of Sm^{3+} ions from $^4G_{5/2}$ to $^6H_{9/2}$, $^6H_{7/2}$, and $^6H_{5/2}$, attributes in the emissions at 649, 602, and 566 nm. The PL decay study with the Inokuti-Hirayama model, indicates the different multipolar interaction mechanism. The study determines critical transfer distance (R_c), the energy transfer parameter (Q), and donor-acceptor interaction parameter (C_{DA}). The thermoluminescence (TL) kinetic parameters of glass were studied by the peak shape method. From the PL spectra, colorimetric parameters were evaluated, shows the glass could be used for various optoelectronic applications.

Keywords: Barium borate glass, Urbach energy, photoluminescence, chromaticity coordinates, correlated color temperature (CCT)

1. Introduction

Recently Glass is well-known for its numerous applications in science and technology, which have paved the way for the development of innovative glass. The chemical composition, kind of bond, structure, and synthesis mechanism all influence the physical properties of the glass. Rare earth-doped glass draws great interest among researchers due to its appealing optical features and excellent luminescence efficiency [1-4]. Among the networker formers, B_2O_3 has been the primary choice due to its physical, structural, and optical advantages over other conventional network formers. Another exceptional quality of borate glass is its great transparency, low melting point, and high thermal stability [5-7]. The assembly of di, tri and tetra borates was credited with the construction of a more stable three dimensional network. The nonlinear and other characteristics of the glass are determined by the chemical compositions in the glass system. Modification made by the inclusion of alkali or alkaline-earth cation content produce mixed alkali effects, which hold certain significance in the literature [6, 8]. Heavy metal oxide in borate glasses have distinct interest due to their unique properties and their potential stability. In glasses, barium works as a former in high concentrations and as a modifier in low concentrations. Barium-based glass has relatively low melting points, low thermal expansion coefficients, low dispersion, high refraction, and high electric resistance. Aly saeed et al. [8] observed that a high BaO concentration in the BaO-ZnO-MgO-Na2O-Li2O-B_2O_3 glass system resulted in the formation of non-binding oxygen bond, which has a significant influence on the optical band gap and Urbach energy values, as well as increases refractive index and electronic polarizability. Kirdsiri et al. investigated the optical properties

DOI: 10.1201/9781003606611-56

of glasses with different alkaline earth metal as modifier oxides, the structural and optical charcteristics of barium incorporated glass shows better results than other alkaline earth metal, they also stated barium incorporated glasses are better host for solid state applications [6]. Trivalent samarium ions emit a strong orange-red light and have potential applications in color displays, UV-sensor, laser and high-density optical storage [7, 9, 10]. The light bright shade of red emission was attributed to the electronic transitions $^4G_{5/2}$ to$^6H_{9/2}$, $^6H_{7/2}$, and $^6H_{5/2}$ in the inorganic lattice of Sm^{3+} ion-doped materials. The spectroscopic, structural and luminescence characteristics of Sm^{3+} ions have already been investigated in various hosts [1, 3-7]. In the present work, the structural, optical, and luminescence properties of the B_2O_3-BaO-Sm_2O_3 glass system were studied.

2. Materials Synthesis

The Conventional melt quenching process is used to synthesis barium borate glass sample, with boron oxide (B_2O_3) acting as network former and barium carbonate ($BaCO_3$) as network modifier. The network modifier-to-network former ratio is fixed at 2:3. $60B_2O_3$ + $39.7BaCO_3$ + $0.3Sm_2O_3$ (SBB), the chemicals are weighed, carefully mixed, and melted at 1170 °C for approximately 8 hours in an alumina crucible. To avoid thermal shock, a brass mould was preheated to 380 °C and the molten mixture at 1170 °C was transferred and quenched. To eliminate the residual mechanical and thermal stress that arose during subsequent cooling, the transferred mixture was annealed at 450°C for 2 hours. The annealed sample is cooled gradually to ambient temperature, and crushed down to fine powder for further characterization.

3. Structural properties

The FTIR spectra of SBB glass was shown in Figure 56.1, determined in the spectral range of 500 to 3500 cm^{-1}. Broad peak at 1365 cm^{-1}corresponds to the asymmetric stretching in BO_3 structural unit. The peak positioned at 1202 cm^{-1} is most likely due to the B–O stretching vibration of trigonal BO_3 units in the boroxol rings. The presence of B–O stretching in tetragonal BO_4 units is attributed to a peak at 928

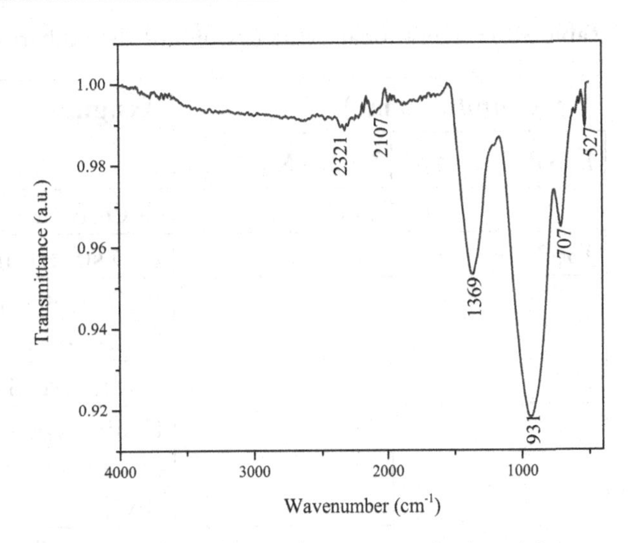

Figure 56.1: FTIR spectra of the Sm^{3+} doped barium borate glass.

cm^{-1} obtained in the glass network. The peaks at 928 and 1365 cm^{-1} confirms the SBB glass network structure contains both tetrahedral and trigonal borate units respectively, barium oxide could have promote in the conversion of some triangular borate group into tetrahedral borate group [11]. The peak at 706 cm^{-1} confirms the bending of B-O–B linkage within the borate network [12]. The specific low intensity vibration by Ba–O bonds were confirmed by peak at 526 cm^{-1}. The peaks 2114, and 2325 cm^1 attributes to vibration of water, hydroxyl (OH) group.

The Raman spectra SBB glass was found to be asymmetric as shown in Figure 56.2. This asymmetric nature of Raman bands is due to the superimposition of several bands with different line parameters. FT-Raman spectra were recorded for SBB glass confirms the structural

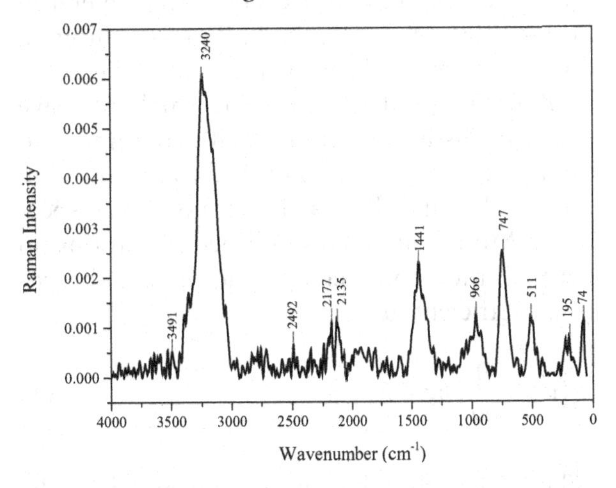

Figure 56.2: FT Raman spectra of the Sm^{3+} doped barium borate glass.

Table 56.1: Vibrational frequency of Sm^{3+} doped barium borate glass

Wavenumber (cm^{-1})		Assignments
FTIR	FT RAMAN	
	1441	B–O–B stretching vibrations in the glass network
1369		B-O stretch in BO_3 units from varied type of borate groups
931		B–O stretching vibration of tetragonal BO_4 units in tri-, tetra- and penta-borate groups.
707		Symmetrical deformation of BO_3
	747	Chain-type metaborate/symmetric breathing vibrations of six membered rings with two BO_4 units/γ-vibrations of $[BO_3]$ [3]
527		Specific vibrations of Ba–O bonds low intensity
	966	Pentaborate vibrations
2107, 2173, 2321, 2532	1804, 2135, 3240	Water, hydroxyl (OH) and (B-OH) groups vibrations

and functional groups in the sample. The spectra in the 500-2000 cm^{-1} range were obtained using a Bruker RFS-7 (SAIF, IITM) Raman spectrometer. The peak at 747 cm^{-1} is due to **γ-vibrations of $[BO_3]$** [3]. The Penta-borate vibrations are confirmed at 966 cm^{-1}. The peak positioned at 1441 cm^{-1} ascribed to B–O–B stretching in the glass network. All the vibrational peaks in both the spectra were tabulated in Table 56.1, the peaks assigned were shows good match when compared with the literature values.

4. Optical Properties

Figure 56.3 shows the transmittance spectra recorded using PEL-900-UVNS spectrometer in 200 – 1200 nm range, the SBB glass exhibiting approximately 75 % transmittance in the 400 – 1200 nm range. The presence of small intensive peaks in distinct location at 407 nm and 956 nm in the SBB glass are ascribed to the Sm^{3+} ion $^6H_{5/2} \rightarrow {}^6P_{3/2}$ and $^6H_{5/2} \rightarrow {}^6F_{11/2}$ transition, respectively. Mott et al. [13] established a relationship between the photon energy and optical absorption coefficient (α).

$$\alpha(\nu) = \frac{B}{h\nu}\left(h\nu - E_g\right)^n \quad (1)$$

Where, B is the absorption constant, n is the power factor, E_g represents the band-gap energy.

For allowed direct transition n value is ½, and for indirect transition value is 2. Figure 56.4 represent direct and indirect band gap energy. For SBB glass, the calculated direct and indirect E_g values were 3.35 eV and 2.21 eV, respectively. The lack of non-binding oxygens (NBOs) reduces polarizability and increases the optical bandgap energy of the glass, as the NBOs bind an excited electron less strongly than binding oxygen (BOs), making NBOs more polarizable than BOs. However, the existence of heavy metal ions (Ba^{2+}) results in increase in the amount of free electrons in the glass network. This caused electrons to agglomerate in low energy levels, contributing to the decrease in the band-gap energy.

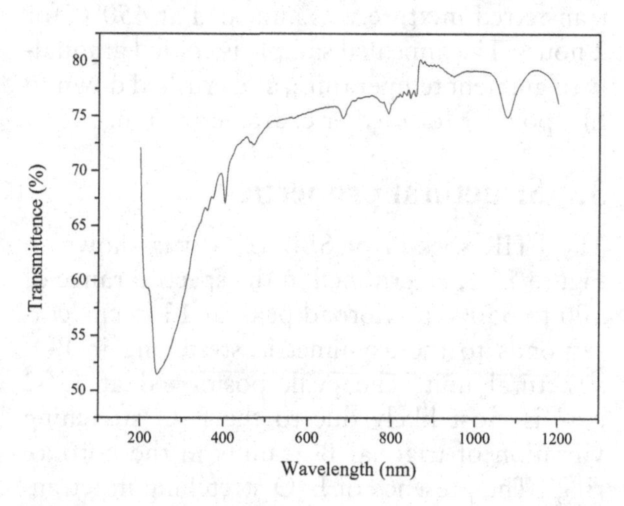

Figure 56.3: UV-Vis-NIR region transmittance of the Sm^{3+} doped barium borate glass

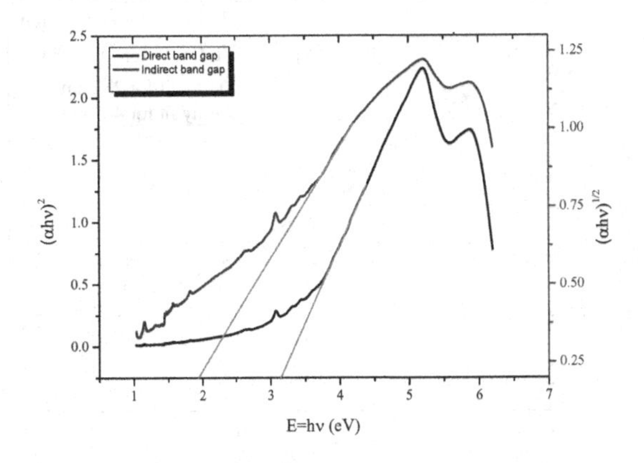

Figure 56.4: Tauc's plot for direct and indirect band-gap energy of SBB glass

The Urbach energy (E_u) value were calculated from UV-absorption data. Figure 56.5 depicts the response of ln(α) vs hν, the E_u is calculated using the Urbach empirical rule [14] and the reciprocal slopes of the linear fit.

$$\alpha(\nu) = \alpha_o \exp\left(\frac{h}{E_u}\right) \quad (2)$$

The E_u values for the SBB glass was found to be 3.315 eV. Saeed et al. reported inclusion of BaO in the glass system changes the band gap and urbach energy [8]. Both E_g and E_u values are inversely related, higher E_u indicates surge in bonding defects and NBO atoms in the glass. The higher concentration of BaO in the SBB glass system increases the degree of electron localization, and donor centers.

5. Photoluminescence

Luminescence spectra of the Sm^{3+} barium borate glass was recorded at room temperature by using a F-900 fluorescence spectrometer. The PLE spectra of SBB glass when observed at 603nm was shown in Figure 56.6, the excitation peaks at 348, 365, 378, and 407 nm, were ascribed to the transitions from $^6H_{5/2} \rightarrow {}^4H_{9/2}$, $^6H_{5/2} \rightarrow {}^4D_{3/2}$, $^6H_{5/2} \rightarrow {}^4P_{7/2}$, and $^6H_{5/2} \rightarrow {}^4P_{3/2}$ respectively. The excitation peaks in 233 nm region was attributable to charge transfer state of Sm^{3+} ion, a 2p orbital electron of oxygen is transferred to 4f orbital of samarium. The excitation peaks at 348, 365, 378 and 407 nm are due to the f-f transition of Sm^{3+} ions. The excitation dependent PL emission spectra of SBB glass were recorded for the excitation wavelengths of 233, 254 and 407 nm. The electrons in Sm^{3+} ions excited from the $^6H_{5/2}$ ground state to $^4F_{5/2}$ the excited state and then fall back to the $^4G_{5/2}$ state swiftly by nonradiative relaxation. The emission observed at 566, 602, and 649 nm were corresponds to radiative transitions of Sm^{3+} ions from $^4G_{5/2} \rightarrow {}^6H_{5/2}$ (magnetic dipole allowed transition ($\Delta J = 0$)), $^4G_{5/2} \rightarrow {}^6H_{7/2}$ (either electric dipole or magnetic dipole character) and $^4G_{5/2} \rightarrow {}^6H_{9/2}$ (mainly of electric dipole ($\Delta J = \pm 1$) allowed transition) respectively. In general, the intensity ratios of electric and magnetic dipoles determine the symmetry of the environment surrounding the Sm^{3+} ion. In SBB glass possess more asymmetric nature as $^4G_{5/2} \rightarrow {}^6H_{9/2}$ transition is more intense than the magnetic dipole transition. Apart from the radiative and multi-phonon relaxation there may take place also the donor-donor (DD)

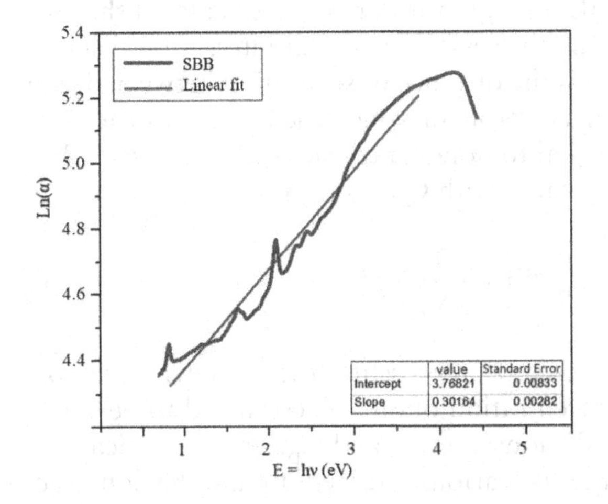

Figure 56.5: Urbach energy: a plot of ln(α) against hν of SBB glass

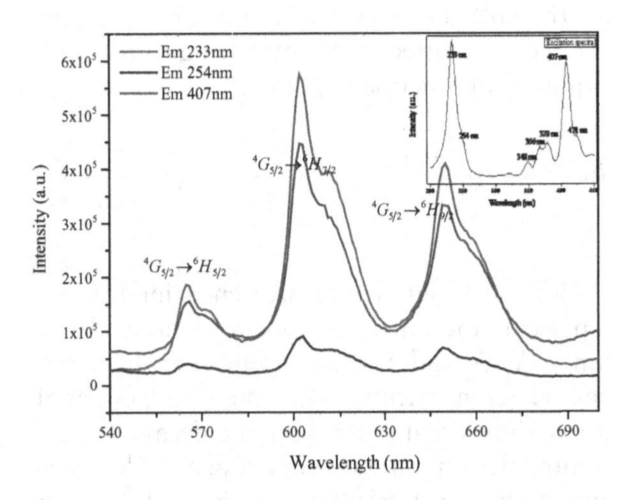

Figure 56.6: PL excitation and emission spectra of SBB glass

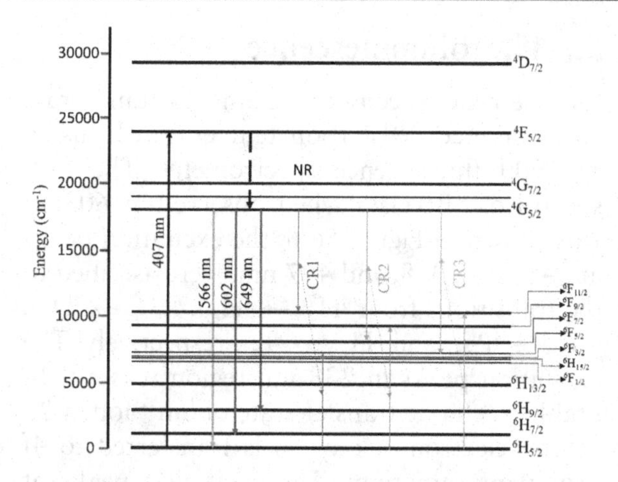

Figure 56.7: Partial energy level diagram of SBB glass

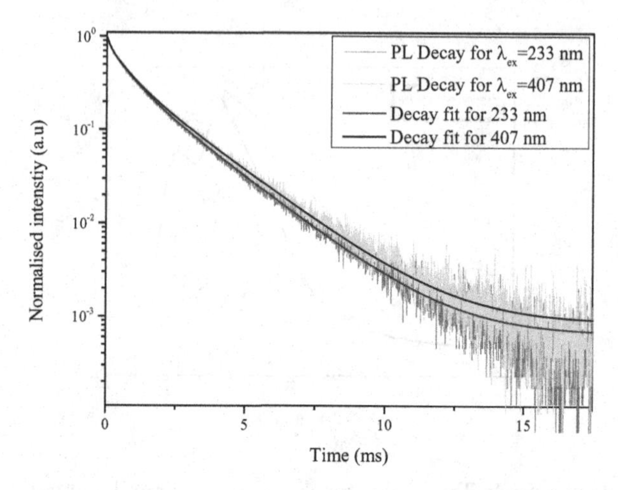

Figure 56.8: PL decay plot with decay fit of SBB glass for different excitation wavelength (233 nm and 407 nm)

relaxations and donor-acceptor (DA) responsible for cross relaxation leading migration of excitation energy. The cross relaxation (CR) channels witnessed between the donor and acceptors of Sm^{3+} ion are shown in the Figure 56.7. The CR1 was laid between $^4G_{5/2} + {}^6H_{5/2} \rightarrow {}^6F_{11/2} + {}^6F_{5/2}$, CR2 was laid between $^4G_{5/2} + {}^6H_{5/2} \rightarrow {}^6F_{7/2} + {}^6F_{9/2}$ and the CR3 was laid between $^4G_{5/2} + {}^6H_{5/2} \rightarrow {}^6F_{5/2} + {}^6F_{11/2}$. In CR1, the Sm^{3+} ion in the $^4G_{5/2}$ level relaxes to $^6F_{11/2}$ level and excites a neighbor Sm^{3+}ions from $^6H_{5/2} \rightarrow {}^6F_{5/2}$ level, because the difference in energy of the two related transitions matches narrowly.

6. PL Decay

The PL decay of $^4G_{5/2} \rightarrow {}^6H_{7/2}$ of Sm^{3+} ion in SBB glass has been investigated for excitation wavelengths 233 nm and 407 nm, the intensity verses time curves are depicted in Figure 56.8. From the decay curve, experimental lifetime value for the SBB glass were evaluated. The PL decay curve can be fitted by the following third-order exponential function as [15]:

$$I(t) = A_1 \exp\left(\frac{-t}{\tau_1}\right) + A_2 \exp\left(\frac{-t}{\tau_2}\right) + A_3 \exp\left(\frac{-t}{\tau_3}\right) \quad (3)$$

Here, I(t) signifies fluorescence intensity; t represents the time; τ_1, τ_2 and τ_3 denotes decay times; A_1, A_2 and A_3 are constants for the exponential components. The three exponential decay shows that more than one decay channel is intricate in the total decay process. The average lifetime (τ_{av}) calculated with the following equation [16]:

$$\tau_{av} = \frac{A_1\tau_1^2 + A_2\tau_2^2 + A_3\tau_3^2}{A_1\tau_1 + A_2\tau_2 + A_3\tau_3} \quad (4)$$

To determine the dominant multi-polar interaction, the Inokuti-Hirayama (I-H) model was used to investigate the non-single exponential PL Decay curve [17, 18].

$$\varphi(t) = A.\exp\left[-\frac{t}{\tau} - Q\left(\frac{t}{\tau}\right)^{3/S}\right] \quad (5)$$

S is entirely dependent on the type of multi-polar interaction, with S=6, for dipole-dipole (D-D) interaction. For the S= 8, and 10 corresponding to dipole-dipole (D-D), dipole-quadrupole (D-Q) and quadrupole-quadrupole (Q-Q) interactions, respectively. The Probability of the energy transfer (Q) is calculated through fitting process. The critical interaction distance (R_c) is the distance of separation between donor and acceptor for which the energy transfer rate is equal to donor intrinsic decay rate ($1/\tau_0$). It is correlated with Q as follows

$$Q = \frac{4}{3}\pi\Gamma\left(1 - \frac{3}{S}\right)N_A R_c^3 \quad (6)$$

Here, Γ represents the gamma function. Concentration of Sm^{3+} is estimated to be 1.188 x 10^{20} ions/cm^3. R_c and C_{DA} are the critical distance for various interactions and the donor-acceptor ions interaction parameter, respectively are given below.

Table 56.2: The average experimental lifetime (τ_{av}), Energy transfer parameter(Q), critical transfer distance (R_c), energy transfer rate (C_{DA}) of SBB glass.

Sample		SBB	
(Ex/Em) wavelength		(233nm/603nm)	(407nm/603nm)
Average experimental Decay time (τ_{av})		1.09 ms	1.16 ms
S=6	Q_{D-D}	0.52	0.54
	$(R_c)_{D-D}$	8.378 Å	8.485 Å
	C_{DA} (cm^6/s)	3.17×10^{-40}	3.21×10^{-40}
S=8	Q_{D-Q}	0.48	0.50
	$(R_c)_{D-Q}$	8.781 Å	8.896 Å
	C_{DA} (cm^6/s)	3.24×10^{-54}	3.38×10^{-54}
S=10	Q_{Q-Q}	0.46	0.47
	$(R_c)_{Q-Q}$	8.897 Å	9.017 Å
	C_{DA} (cm^6/s)	2.85×10^{-68}	3.06×10^{-68}

$$R_c = \sqrt[3]{\frac{3Q}{4\pi \Gamma N_A}} \quad (7)$$

$$C_{DA} = \frac{R_c^S}{\tau} \quad (8)$$

$$D_{random} = 2\left(\frac{3}{4\pi N_A}\right)^{1/3} \quad (9)$$

The D_{random} values are found to be 25.2 Å exceed the R_c values. This confirms the cross-relaxation of the RE^{3+} ions. Table 56.2 shows the calculated values for the probability of energy transfer (Q), critical interaction distance (R_c), donor-acceptor ions interaction parameter (C_{DA}), and average lifetime (τ_{av}). The results reveal that energy transfer between Sm^{3+} ion is predominantly by dipole-dipole electrostatic interactions.

7. Color chromaticity and color correlated temperatures

For examining the excitation-dependent luminescence of the SBB glass PL spectra were recorded for various characteristic excitation wavelengths (At 233, 407 and 254 nm). We could infer from Figure 56.9 that with the change in Excitation wavelength the intensity of the luminescent band related to Sm^{3+} ion changes. For identifying the accurate color emitted from the glass we can use CIE chromaticity coordinates. The emission spectra for various excitation of Sm^{3+}glasses are evaluated using CIE coordinates by using color matching function. Color correlated temperature (CCT) describes in more detail the color of radiated light emitted by the luminescent source and is measured in Kelvin (K). McCamy's formula can be used to determine the CCT with a precision of less than two Kelvin.

$$CCT = -68253.3n + 3525n^2 - 449\, n^3 + 5520.33 \quad (10)$$

Where $n = (x - x_0)/(y - y_0)$

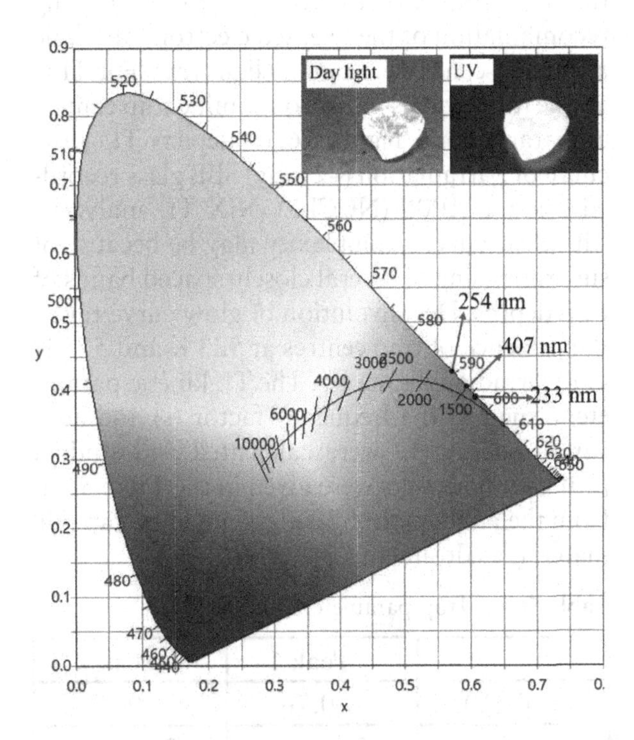

Figure 56.9: CIE diagram for SBB glass excited at 233 nm, 254 nm and 407 nm wavelength

Table 56.3: Photometric characteristics of SBB glass

Sample Properties	Ex= 233nm	Ex= 254nm	Ex= 407nm
Coordinates (x,y)	(0.6086, 0.3907)	(0.5720, 0.4267)	(0.5944, 0.4046)
(u',v')	(0.3763, 0.5434)	(0.3280, 0.5505)	(0.3566, 0.5462)
CCT	1374 K	1750 K	1510 K
Ldom	596.1 nm	589.6 nm	593.5 nm
Color purity	100%	99.8%	99.9%
CRI	48	56	54

The color purity of the emitted light was calculated using the following formula [19]

$$P_c = \frac{\sqrt{((x_s - x_i)^2 + (y_s - y_i)^2)}}{\sqrt{((x_d - x_i)^2 + (y_d - y_i)^2)}} \times 100\% \quad (11)$$

The x and y coordinates, CCT and color purity of the SBB glass system were tabulated in Table 56.3.

8. Thermoluminescence

It is significant to create large number of trap centers for a certain Thermoluminesce dosimetry (TLD) application, which can be accomplished through appropriate doping to the host material. In a TLD material electrons need be trapped in defects at low temperature and then released by thermal stimulation to give a glow peak. During recombination of the released electrons and hole at luminescence center emits light (hν). For TLD application, it is essential to examine trap centers and trap depth. Figure 56.10 depicts TL glow curve of γ-irradiation (~2 kGy) SBB glass recorded using TL1009 (NUCLEONIX TL analyzer). The glow curve's complexity may be because of superimposing of several closely spaced bands of trap depths. Deconvolution of glow curve revels the presence of trap centres at 523 K and 575 K in the irradiated samples. The TL kinetic parameters, such as the frequency factor (s) and activation energy (E) were calculated and studied [20] and the results were given in the Table 56.4, from the analysis the SBB glass might be capable material as dosimeter.

Table 56.4: Trap parameters of SBB glass

	Peak 1	Peak 2
E(eV)	0.76	0.72
$s(s^{-1})$	3.13×10^6	2.60×10^5

9. Conclusion

The synthesized Sm^{3+} doped barium borate glass shows good transmittance of about 75% in the 400 – 1200 nm wavelength range. The E_u values of SBB glass was found to be 3.32 eV. The FTIR and FT Raman spectral studies of the glass ensure the formation of barium and borate groups. The prepared glasses exhibit excitation-tunable emission, which could be clearly seen from the CIE coordinates calculated for the excitation wavelength of 233, 254 and 407 nm. The results of C_{DA} energy transfer microparameter confirms the dipole-dipole interaction to be dominant. The kinetic parameters computed from deconvolution of glow curve show trap center at high temperature capable material for dosimetric applications. The variation in the red emission with excitation of Sm^{3+} doped barium borate glasses could be suitable for tunable LED application.

Acknowledgement

We wish to confirm that there are no known conflicts of interest within the authors in submitting the article to the journal.

Dr. Priya thank BRNS for the Project, entitled "Synthesis of novel Ln incorporated non silicate glass ceramics as radiation shielding materials", No. 59/14/07/2020-BRNS/10096.

Data availability statement

The raw/processed data required to reproduce these findings are shared in the form of tables in the Manuscript itself.

References

[1] Górny, Agata, Marta Sołtys, Joanna Pisarska, and Wojciech A. Pisarski. "Effect of acceptor ions concentration in lead phosphate glasses

co-doped with Tb3+–Ln3+ (Ln= Eu, Sm) for LED applications." J. Rare Earths 37, no. 11, 1145-1151 (2019).

[2] Padlyak, B. V., V. T. Adamiv, Ya V. Burak, and M. Kolcun. "Optical harmonic transformations in borate glasses with Li2B4O7, LiKB4O7, CaB4O7, and LiCaBO3 compositions." Phys. B: Condens. 412, 79-82 (2013).

[3] Kaur, Parvinder, Simranpreet Kaur, Gurinder Pal Singh, and D. P. Singh. "Sm3+ doped lithium aluminoborate glasses for orange coloured visible laser host material." Solid State Commun. 171, 22-25 (2013).

[4] Deopa, Nisha, A. S. Rao, Ankur Choudhary, Shubham Saini, Abhishek Navhal, M. Jayasimhadri, D. Haranath, and G. Vijaya Prakash. "Photoluminescence investigations on Sm3+ ions doped borate glasses for tricolor w-LEDs and lasers." Mater. Res. Bull. 100, 206-212 (2018).

[5] Rao, L. Srinivasa, M. Srinivasa Reddy, MV Ramana Reddy, and N. Veeraiah. "Spectroscopic features of Pr3+, Nd3+, Sm3+ and Er3+ ions in Li2O–MO (Nb2O5, MoO3 and WO3)–B2O3 glass systems." Phys. B: Condens. 403, no. 17, 2542-2556 (2008).

[6] Kirdsiri, K., R. Raja Ramakrishna, B. Damdee, H. J. Kim, S. Kaewjaeng, S. Kothan, and J. Kaewkhao. "Investigations of optical and luminescence features of Sm3+ doped Li2O-MO-B2O3 (M= Mg/Ca/Sr/Ba) glasses mixed with different modifier oxides as an orange light emitting phosphor for WLED's." J. Alloys Compd. 749, 197-204 (2018).

[7] Sailaja, B., R. Joyce Stella, G. Thirumala Rao, B. Jaya Raja, V. Pushpa Manjari, and R. V. S. S. N. Ravikumar. "Physical, structural and spectroscopic investigations of Sm3+ doped ZnO mixed alkali borate glass." J. Mol. Struct. 1096, 129-135 (2015).

[8] Saeed, Aly, Y. H. Elbashar, and S. U. El Khameesy. "A novel barium borate glasses for optical applications." Silicon 10, 569-574 (2018).

[9] Huang, Lihui, Animesh Jha, and Shaoxiong Shen. "Spectroscopic properties of Sm3+-doped oxide and fluoride glasses for efficient visible lasers (560–660 nm)." Opt. Commun. 281, no. 17, 4370-4373 (2008).

[10] Mohan, Shaweta, Simranpreet Kaur, D. P. Singh, and Puneet Kaur. "Structural and luminescence properties of samarium doped lead alumino borate glasses." Opt. Mater. 73, 223-233 (2017).

[11] Marzouk, M. A., F. H. ElBatal, and H. A. ElBatal. "Effect of TiO2 on the optical, structural and crystallization behavior of barium borate glasses." Opt. Mater. 57, 14-22 (2016).

[12] Priya, M., M. Dhavamurthy, A. Antony Suresh, and M. Manoj Mohapatra. "Luminescence and spectroscopic studies on Eu3+-doped borate and boro-phosphate glasses for solid state optical devices." Opt. Mater. 142, 114007 (2023).

[13] N.F. Mott and E.A. Davis, Electronic processes in Non-crystalline Solids, Clarendon Press, Oxford (1971).

[14] M. Dhavamurthy, P. Vinothkumar, ManojMohapatra, Antony Suresh, PriyaMurugasen, "Effects of Ce3+/Dy3+ and Ce3+/Sm3+ co-doping as a luminescent modifier in alumina-borophosphate glasses for w-LED application", Spectrochim. Acta A Mol. Biomol. Spectrosc. 266, 120448 (2022).

[15] Yoon, Songhak, Eugenio H. Otal, Alexandra E. Maegli, LassiKarvonen, Santhosh K. Matam, Stefan G. Ebbinghaus, Bernhard Walfort, Hans Hagemann, Simone Pokrant, and AnkeWeidenkaff. "Improved persistent luminescence of CaTiO3: Pr by fluorine substitution and thermochemical treatment." J. Alloys Compd. 613, 338-343 (2014).

[16] Tsega, Moges, and Francis BirhanuDejene. "Synthesis and luminescence in sol–gel auto-combustion-synthesized CaSnO3Eu3+ CaSnO3: Eu 3+ phosphor." Bull. Mater. Sci. 40, no. 7, 1347-1354 (2017).

[17] Inokuti, Mitio, and Fumio Hirayama. "Influence of energy transfer by the exchange mechanism on donor luminescence." J. Chem. Phys. 43, no. 6 1978-1989 (1965).

[18] Rani, P. Rekha, M. Venkateswarlu, SkMahamuda, K. Swapna, NishaDeopa, and A. S. Rao. "Spectroscopic studies of Dy3+ ions doped barium lead aluminofluoro borate glasses." J. Alloys Compd. 787, 503-518 (2019).

[19] Dhavamurthy, M., P. Vinothkumar, A. Antony Suresh, Manoj Mohapatra, and Priya Murugasen. "Optical characteristics of Eu3+ doped alumino borophosphate glass containing Al3+, Zn2+, Li2+, Sr2+ and Ba2+ ions." Results Opt. 8 100232 (2022).

[20] Suresh, A. Antony, Priya Murugasen, and M. Dhavamurthy. "Luminescence properties of Tm3+/Dy3+ co-activated borate and borophosphate glasses for near daylight w-LED and dosimetry application." emergent mater. 1-13 (2023).

Printed in the United States
by Baker & Taylor Publisher Services